Artificial Intelligence and Image Processing in Medical Imaging

Developments in Biomedical Engineering and Bioelectronics

Artificial Intelligence and Image Processing in Medical Imaging

Edited by

Walid A. Zgallai

Faculty of Engineering Technology and Science, Higher Colleges of Technology, Dubai, United Arab Emirates

Dilber Uzun Ozsahin

Department of Medical Diagnostic Imaging, College of Health Science, University of Sharjah, Sharjah, United Arab Emirates; Research Institute for Medical and Health Sciences, University of Sharjah, Sharjah, United Arab Emirates; Operational Research Centre in Healthcare, Near East University, Nicosia/TRNC, Mersin-10, Turkey

Series Editor

Dennis Fitzpatrick

School of Engineering, University of East Anglia, Norwich, United Kingdom

Academic Press is an imprint of Elsevier
125 London Wall, London EC2Y 5AS, United Kingdom
525 B Street, Suite 1650, San Diego, CA 92101, United States
50 Hampshire Street, 5th Floor, Cambridge, MA 02139, United States

ISBN: 978-0-323-95462-4

For Information on all Academic Press publications
visit our website at https://www.elsevier.com/books-and-journals

Publisher: Mara E. Conner
Acquisitions Editor: Carrie L. Bogler
Editorial Project Manager: Isabella C. Silva
Production Project Manager: Anitha Sivaraj
Cover Designer: Christian J. Bilbow

Typeset by MPS Limited, Chennai, India

Contents

List of contributors

Hussein Abdul-Rahman
Faculty of Engineering Technology and Science, Higher Colleges of Technology, Dubai, United Arab Emirates

Khaldoon Alhusari
Department of Computer Science and Engineering, American University of Sharjah, Sharjah, United Arab Emirates

Mohanad Alkhodari
Healthcare Engineering Innovation Center (HEIC), Department of Biomedical Engineering, Khalifa University, Abu Dhabi, United Arab Emirates; Cardiovascular Clinical Research Facility (CCRF), Radcliffe Department of Medicine, University of Oxford, Oxford, United Kingdom

Wafaa Abdulhameed Al-Olofi
Department of Biomedical Engineering and Systems, Faculty of Engineering, Cairo University, Giza, Egypt

Sekeroglu Boran
Department of Software Engineering, Near East University, Nicosia, Northern Cyprus, Mersin-10, Turkey

Salam Dhou
Department of Computer Science and Engineering, American University of Sharjah, Sharjah, United Arab Emirates

Kamil Dimililer
Department of Electrical and Electronic Engineering, Near East University, Nicosia, Northern Cyprus, Mersin-10, Turkey; Applied Artificial Intelligence Research Centre, Near East University, Nicosia, Northern Cyprus, Mersin-10, Turkey; Center for Science Technology and Engineering, Near East University, Nicosia, Northern Cyprus, Mersin-10, Turkey

Binnur Demir Erdem
Department of Automotive Engineering, Near East University, Nicosia, Northern Cyprus, Mersin-10, Turkey

Emmanouil Evangelopoulos
Department of Orthodontics, College of Dental Medicine, University of Sharjah, Sharjah, United Arab Emirates

Devrim Kayali
Department of Electrical and Electronic Engineering, Near East University, Nicosia, Northern Cyprus, Mersin-10, Turkey

Heba Mohamed
Faculty of Health Sciences, Higher Colleges of Technology, Dubai, United Arab Emirates

Mostafa Moussa
Healthcare Engineering Innovation Center (HEIC), Department of Biomedical Engineering, Khalifa University, Abu Dhabi, United Arab Emirates

Sayan Murat
Faculty of Medicine, Clinical Laboratory, PCR Unit, Kocaeli University, Kocaeli, Turkey; DESAM Research Institute, Near East University, Nicosia, Northern Cyprus, Mersin-10, Turkey

Mubarak Taiwo Mustapha
Operational Research Centre in Healthcare, Near East University, Nicosia/TRNC, Mersin-10, Turkey; Department of Biomedical Engineering, Near East University, Nicosia/TRNC, Mersin-10, Turkey

Dilber Uzun Ozsahin
Department of Medical Diagnostic Imaging, College of Health Science, University of Sharjah, Sharjah, United Arab Emirates; Research Institute for Medical and Health Sciences, University of Sharjah, Sharjah, United Arab Emirates; Operational Research Centre in Healthcare, Near East University, Nicosia/TRNC, Mersin-10, Turkey

Ilker Ozsahin
Operational Research Centre in Healthcare, Near East University, Nicosia/TRNC, Mersin-10, Turkey; Brain Health Imaging Institute, Department of Radiology, Weill Cornell Medicine, New York, NY, United States

Snigdha Pattanaik
Department of Preventive and Restorative Dentistry, College of Dental Medicine, University of Sharjah, Sharjah, United Arab Emirates

Oluwaseun Priscilla Olawale
Department of Software Engineering, Near East University, Nicosia, Northern Cyprus, Mersin-10, Turkey

Hani Qusa
Faculty of Computer Information Science, Higher Colleges of Technology, Dubai, United Arab Emirates

Hossein Rabbani
Medical Image and Signal Processing Research Center, School of Advanced Technologies in Medicine, Isfahan University of Medical Sciences, Isfahan, Iran

Bashar Rajoub
Faculty of Engineering Technology and Science, Higher Colleges of Technology, Dubai, United Arab Emirates

Muhammad Ali Rushdi
Department of Biomedical Engineering and Systems, Faculty of Engineering, Cairo University, Giza, Egypt; School of Information Technology, New Giza University, Giza, Egypt

Debarchita Sarangi
Department of Prosthetic Dentistry, Institute of Dental Sciences, Siksha O Anusandhan (Deemed to be University), Bhubaneswar, Odisha, India

Shruti Singh
Department of Dentistry, All India Institute of Medical Sciences, Raebareli, Uttar Pradesh, India

Natacha Usanase
Operational Research Centre in Healthcare, Near East University, Nicosia/TRNC, Mersin-10, Turkey; Department of Biomedical Engineering, Near East University, Nicosia/TRNC, Mersin-10, Turkey

Berna Uzun
Operational Research Centre in Healthcare, Near East University, Nicosia/TRNC, Mersin-10, Turkey; Department of Mathematics, Near East University, Nicosia/TRNC, Mersin-10, Turkey

Mohammad Hossein Vafaie
Medical Image and Signal Processing Research Center, School of Advanced Technologies in Medicine, Isfahan University of Medical Sciences, Isfahan, Iran

Melize Yuvali
Operational Research Centre in Healthcare, Near East University, Nicosia/TRNC, Mersin-10, Turkey; Department of Biostatistics, Near East University, Nicosia/TRNC, Mersin-10, Turkey

CHAPTER

Introduction to machine learning and artificial intelligence

1

Mubarak Taiwo Mustapha[1,2], Ilker Ozsahin[1,3] and Dilber Uzun Ozsahin[1,4,5]

[1]*Operational Research Centre in Healthcare, Near East University, Nicosia/TRNC, Mersin-10, Turkey*

[2]*Department of Biomedical Engineering, Near East University, Nicosia/TRNC, Mersin-10, Turkey*

[3]*Brain Health Imaging Institute, Department of Radiology, Weill Cornell Medicine, New York, NY, United States*

[4]*Department of Medical Diagnostic Imaging, College of Health Science, University of Sharjah, Sharjah, United Arab Emirates*

[5]*Research Institute for Medical and Health Sciences, University of Sharjah, Sharjah, United Arab Emirates*

1.1 Comprehensive introduction to machine learning and artificial intelligence

Since the invention of programmable computers, artificial intelligence (AI) has been a hot topic of discussion (French, 2000). Academics and philosophers have cast doubt on the distinctions between man and machine. Could we program the human brain and implement all its complexities into a computer? Will a computer be able to think at that point? We have not yet found an answer to these fascinating, mind-numbing questions, but we have made strides toward making computers more intelligent. However, some may claim that even the most advanced computers lack the intelligence of a cockroach. Even the most intelligent computers are incapable of performing multiple tasks concurrently. Instead, they excel at one task they are programmed for (Boucher, 2020).

Most often, AI and machine learning are often used interchangeably. However, AI is not machine learning. AI is a technique for building systems capable of performing tasks that usually require human intelligence. These tasks encompass a wide range, such as perceiving visual information, comprehending speech, making decisions, translating languages, and many others (Burns et al., 2022). The principle involves mimicking human behavior. AI focuses on knowledge of engineering to replicate human intelligence. This knowledge is a critical component of AI research. Machines equipped with AI are expected to be able to solve problems in the same way that humans do. This means that they will be able to learn from experience, adapt to new situations, and use their knowledge to come up with creative solutions to problems. The machine must thoroughly understand the real world to accomplish this. In other words, the machines must comprehend concepts like the relationship between objects and circumstances, the characteristics

Artificial Intelligence and Image Processing in Medical Imaging. DOI: https://doi.org/10.1016/B978-0-323-95462-4.00001-7

of an event, and cause and effect. This data is then analyzed and fed into a computer program that analyzes it and makes decisions as humans do. The objective is to transfer human expertise to a computer program capable of processing the same data and arriving at the same conclusions as humans. The modeling process is sometimes referred to as providing data to a software program and having it make human-like decisions. The model is continuously modified until its conclusions are comparable to those reached by humans.

As shown in Fig. 1.1, machine learning and deep learning are subsets of AI. Today's job market favors those with expertise in machine learning, one of technology's fastest-growing fields (Terra, 2023). An AI system can be created using machine learning or deep learning algorithms. These algorithms include support vector machine, logistic regression, linear regression, K-means, K-nearest neighbor (KNN), random forest, decision tree, and neural network. The algorithms can suggest user actions and recommendations, diagnose infectious diseases and cancer, and conduct market segmentation (Grau et al., 2009; Kumar, 2022; Sarker, 2021a). Machine learning is frequently used when explicit programming is too stiff or impractical.

Machine learning begins with preexisting data. It process the data using algorithms to identify behavioural patterns and outcome trends. It then interprets those patterns to forecast future events. These predictions are utilized to determine the machine learning algorithm's next phase. This decision generates a result, which is evaluated and added to the data pool. The additional data will affect future predictions and decisions. Machine learning can make predictions from massive datasets, optimize utility functions, and extract hidden data structures through data classification. Ultimately, a computer program can learn and make future predictions.

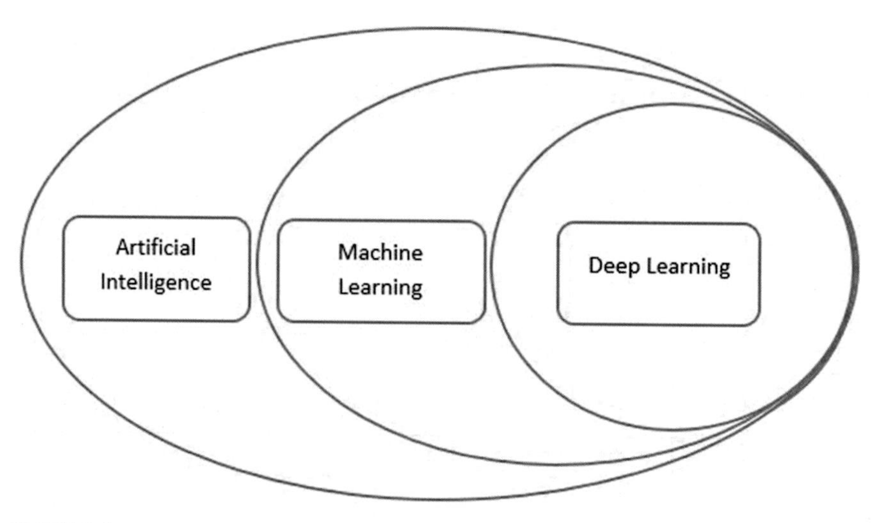

FIGURE 1.1

Artificial intelligence, machine learning, and deep learning.

1.1.1 Types of machine learning

Given that machine learning is a crucial component of AI, it is classified as supervised, unsupervised, or reinforcement learning according to the nature of the learning, as shown in Fig. 1.2

1. In supervised learning, an algorithm is trained using labeled input. The targets are then utilized to predict the correct response when new instances are presented. This is identical to human learning when a teacher is present. The teacher provides appropriate examples for the student to memorize, and the student uses these specific examples to deduce general rules.
2. Unsupervised learning occurs when an algorithm learns from basic instances that do not have an accompanying response, allowing the program to infer data patterns independently. This technique restructures the data into new features representing a class or a unique sequence of uncorrelated values. They benefit humans by providing insights into data meaning and novel inputs to supervised machine learning algorithms. As a form of learning, it is comparable to how people determine if two things or events belong to the same class, such as by comparing their degree of resemblance. Specific recommendation systems used in marketing automation on the web are based on this form of learning (Isinkaye et al., 2015).
3. By developing a reward system, reinforcement learning teaches machines, through trial and error, to take the best action. Through reinforcement learning, models can be trained to play games or control autonomous vehicles by providing feedback to the machine when it makes a correct decision, allowing it to learn and understand the appropriate actions to take.

Machine learning is further classified into classification, regression, and clustering based on the desired output of a machine-learned system.

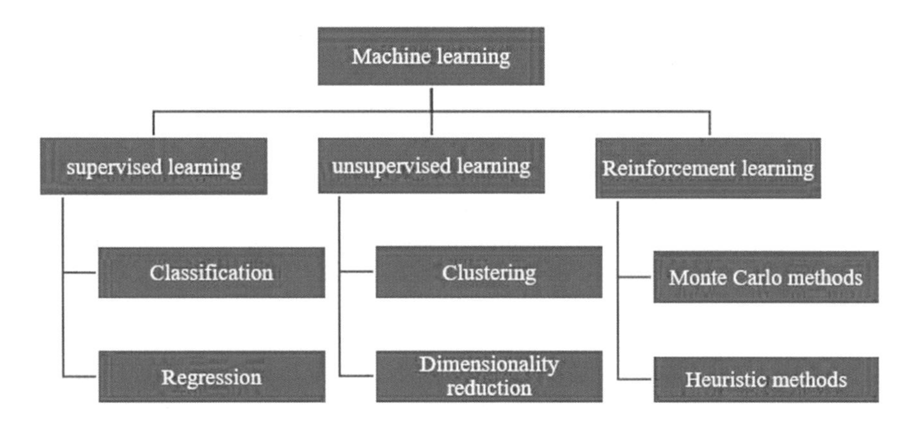

FIGURE 1.2

Machine learning types.

1. Classification, a type of supervised learning, involves sorting data into different categories. It can be applied to both structured and unstructured data (Waseem, 2022). The process starts with identifying the classes for the given data items, commonly referred to as targets, labels, or categories. Some widely used classification algorithms are random forest, decision tree, logistic regression, and support vector machine (Gong, 2022).
2. Regression is a statistical technique that enables continuous outcome prediction based on the values of one or more predictor variables. Due to its ease of use in predicting and forecasting, linear regression is the most often used type of regression analysis (Kurama, 2019).
3. Clustering is an unsupervised machine learning technique that allows identifying data patterns without prior knowledge. Unlike predictive modeling, which is supervised learning, clustering algorithms examine the input data and discover natural groups or clusters in the feature space.

1.1.2 Machine learning algorithm

1.1.2.1 Support vector machine

The support vector machine (SVM) is a popular supervised learning algorithm commonly used for classification problems (Yan, 2016). However, it can also be applied to solve regression problems (Ray, 2021). The SVM identifies the optimal border (or "hyperplane") that separates distinct data classes. The optimal border is the one that maximizes the margin, which is the distance between the boundary and the nearest data points (support vectors) from each class, as shown in Fig. 1.3. These support vectors are the key elements of the algorithm and are used to define the boundary. Once the boundary is established, new data can be classified by determining which side of the boundary it falls into. SVM can also be extended to handle nonlinearly separable data by using the kernel trick, which projects the data into a higher-dimensional space where a linear boundary can be found. The SVM can be used in real-life applications, including text categorization, fraud detection, and image classification (Sarker, 2021a).

1.1.2.2 Logistic regression

The logistic regression algorithm is widely used, easy to comprehend, and is one of the most basic machine learning algorithms, particularly in binary classification tasks. Despite its name, logistic regression is a classification algorithm and not a regression algorithm. It predicts the output of a categorical dependent variable by using a set of independent variables. Logistic regression works by finding the best model representing the relationship between the input features and the output class label. It is a linear model used to predict a probability of an event occurring; it's called logistic because it uses the logistic function (also known as the sigmoid function) to model this probability, as shown in Fig. 1.4. Logistic regression models the probability that an input belongs to a particular class by estimating the parameters of the model (weights and

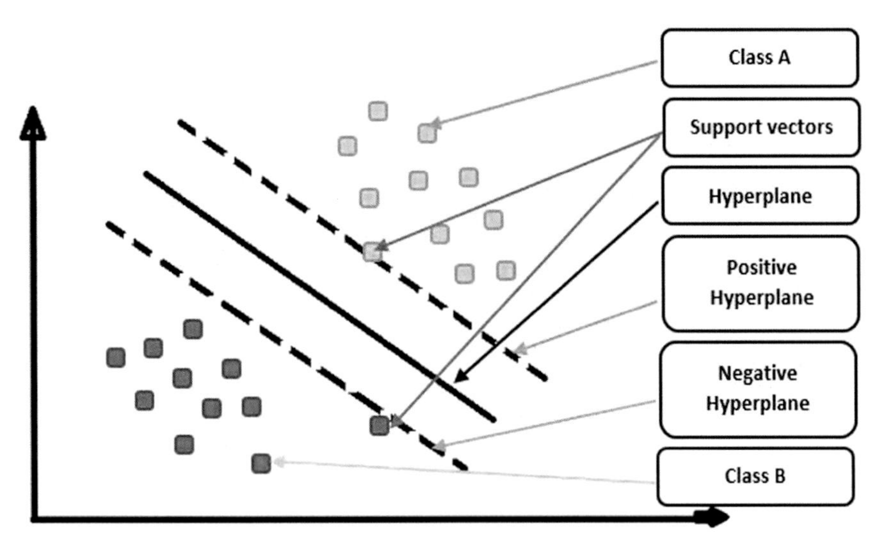

FIGURE 1.3

The support vector machine.

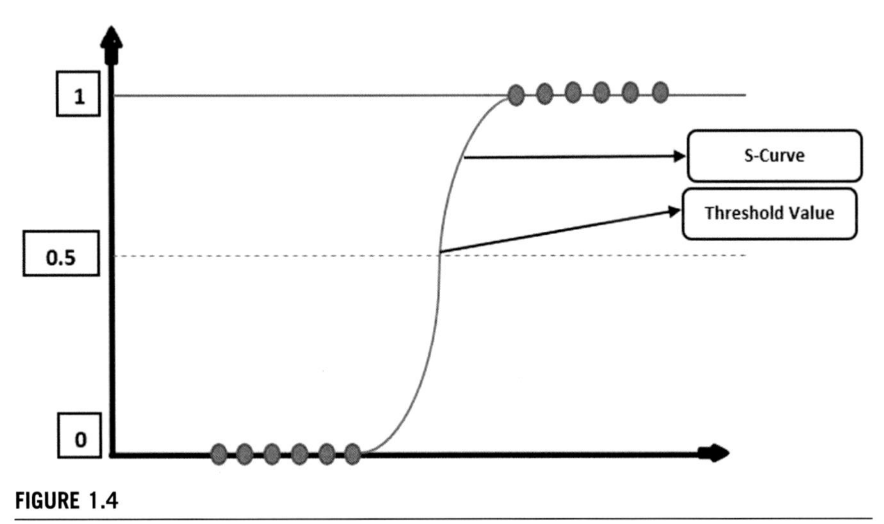

FIGURE 1.4

Logistic regression.

bias) that maximize the likelihood of the observed data. Once the model is trained, it can be used to predict the probability of an input belonging to a particular class, and a threshold is used to decide the final class label of the input (Logistic regression in machine learning, 2022).

1.1.2.3 Linear regression

Linear regression is a statistical method used to predict a continuous outcome variable based on one or more predictor variables (Maulud & Abdulazeez, 2020). It is a linear model that tries to find the best linear relationship between the predictor variables and the outcome variable. The best linear relationship is defined as the one that minimizes the residuals (the difference between the predicted values and the actual values). Linear regression assumes that the relationship between the predictor and the outcome variables is linear, which means that the change in the outcome variable is proportional to the change in the predictor variable.

The principle of linear regression operation is to find the best-fitting line representing the relationship between the predictor and outcome variables, as shown in Fig. 1.5. This is done by minimizing the sum of the squared residuals. Once the model is trained, it can be used to predict the outcome variable for new data using the equation of the best-fitting line. Linear regression can be extended to handle multiple predictor variables, known as multiple linear regression. It can also model nonlinear relationships by transforming the predictor variables.

1.1.2.4 K-means clustering

K-means clustering is a popular unsupervised machine learning algorithm for clustering or grouping similar data points. The algorithm first initializes k centroids, where k is the number of clusters the user wants to create. These centroids are chosen randomly from the data set. The algorithm then iteratively assigns each data point to the cluster whose centroid is closest, as shown in Fig. 1.6, and then recalculates the centroid for each cluster based on the data points assigned.

The algorithm repeats these two steps until the clusters no longer change or a maximum number of iterations is reached. The final result is k clusters, each with

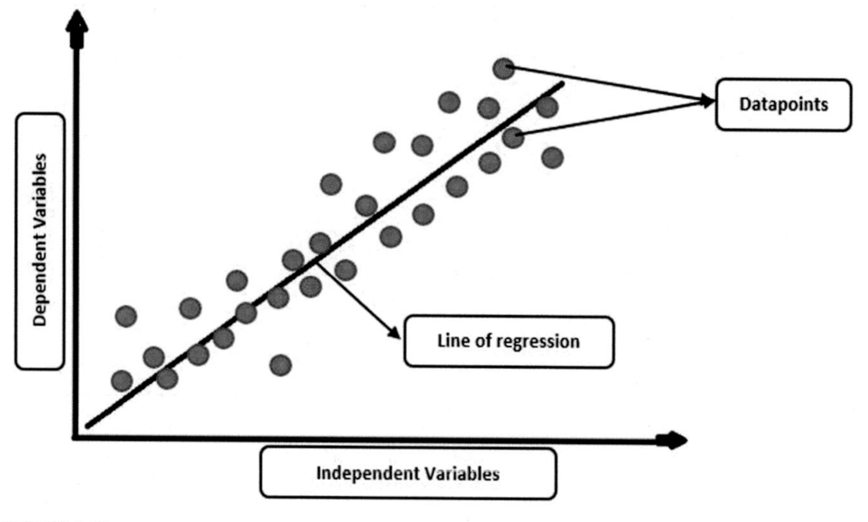

FIGURE 1.5

Linear regression.

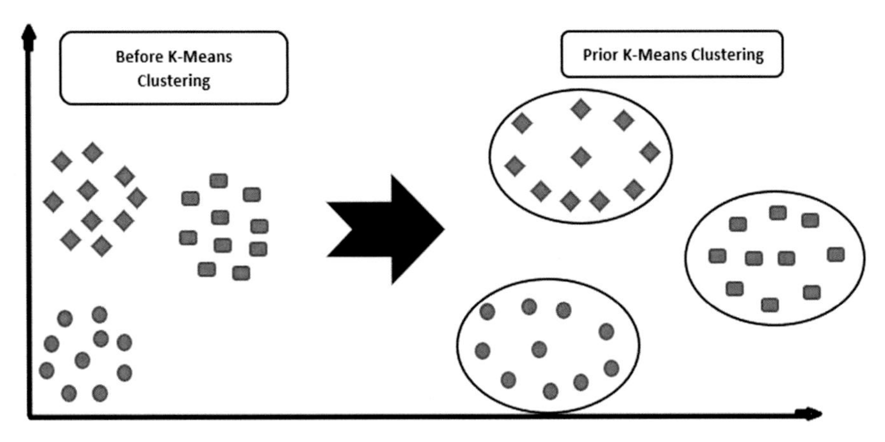

FIGURE 1.6

K-means clustering.

a unique centroid and a set of data points assigned to it. The algorithm is sensitive to the initialization of the centroids, so it is common to run the algorithm multiple times with different initial centroids to ensure the best results. The k-means algorithm is commonly used in image compression, image segmentation, market research, and other applications. The time complexity of the k-means algorithm is $O(nkI^*d)$ where n is the number of data points, k is the number of clusters, I is the number of iterations, and d is the number of attributes.

1.1.2.5 K-nearest neighbor

The KNN algorithm is a basic supervised learning method mainly used for classification tasks. Nevertheless, it is only sometimes applied in regression problems (Sarker, 2021a). The KNN algorithm is considered nonparametric as it does not make any assumptions about the underlying data. Additionally, the KNN is known as a lazy learner because it does not immediately utilize the training data. Instead, it saves it and uses it for classification later on (Zhang, 2022). The KNN algorithm classifies new data by comparing it to previously classified cases and determining the most similar ones. New data points are then classified based on their similarities to the previously stored data. When new data is acquired, the KNN algorithm can quickly classify it into the appropriate category.

1.1.2.6 Decision tree

The decision tree algorithm, a supervised learning method, is suitable for classification and regression problems but is more commonly used for classification scenarios (Sarker, 2021a). Like a tree, a decision tree algorithm consists of a root node that spreads to form a tree-like structure. In a decision tree, the internal node reflects the dataset's feature, while branches and leaf nodes indicate decision

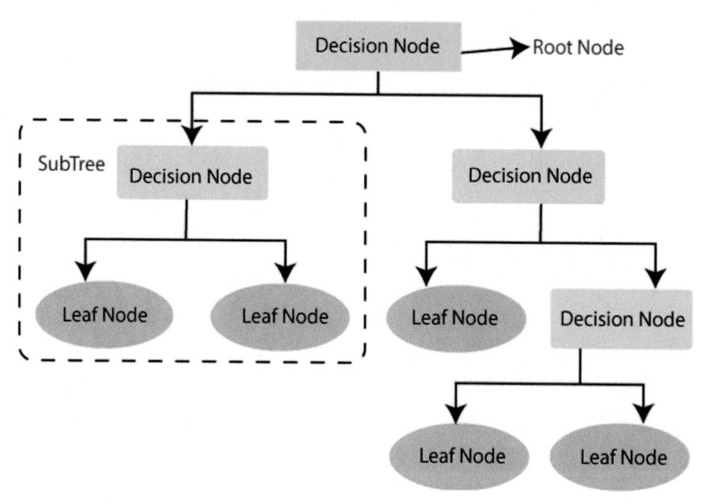

FIGURE 1.7

The decision tree.

From Machine learning decision tree classification algorithm - javatpoint. (n.d.). Retrieved November 5, 2022, from https://www.javatpoint.com/machine-learning-decision-tree-classification-algorithm.

rules and outcomes, respectively. The decision tree consists of the leaf node and the decision node, as shown in Fig. 1.7. The decision node contains several branches and is responsible for making decisions. However, the leaf node has no branches and is the output of the decision node's decision. The decision tree algorithm got the name because of the resemblance of its operation to that of a tree.

1.1.2.7 Random forest

The random forest algorithm, a supervised learning method, is used for classification and regression problems and is based on ensemble learning (Iacovazzi & Raza, 2022). This technique uses multiple classifiers. The random forest algorithm combines a set of decision trees created from different subsets of the input data and takes the average of these trees to enhance the prediction accuracy of the dataset. Instead of just one decision tree, the random forest aggregates the predictions from each tree and, based on most predictions, determines the outcome, as illustrated in Fig. 1.8. In random forests, more trees indicate better accuracy and less chance of overfitting (Zhu, 2020). The random forest algorithm is used in banking, finance, medicine, and logistics (Hussin Adam Khatir & Bee, 2022).

1.1.3 Deep learning

Deep learning is a branch of machine learning that solves tasks using multilayered neural networks (Sarker, 2021b). Deep learning is a step beyond machine learning. Rather than telling the computer what features to look for, deep learning lets the machine

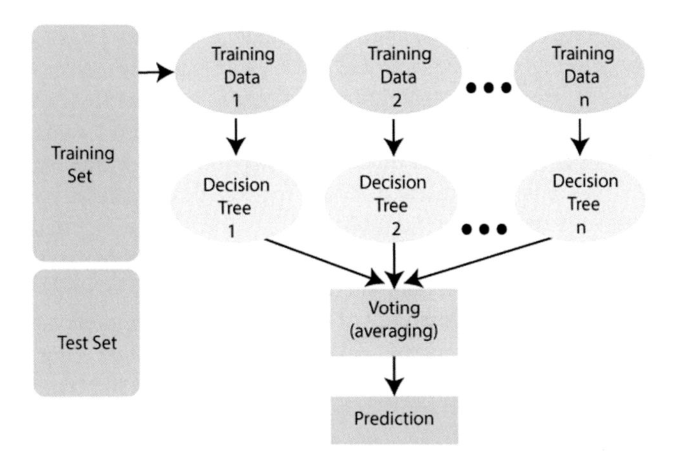

FIGURE 1.8

Random forest.

From Machine learning random forest algorithm. (n.d.). Retrieved November 5, 2022, from https://www.javatpoint.com/machine-learning-random-forest-algorithm.

define them using the data provided. Our brain processes information through the use of neurons. Deep learning operates similarly but with artificial processing structures known as artificial neural networks (Silvestrini & Lavagna, 2022). It constructs this structure from the data it analyzes and then infers features about the subject matter. It then weights these features based on their similarities and hierarchy and organizes them into layers of order and relationships with one another. The previous layer's output is input for the subsequent layer. Deep learning algorithms can be supervised and assist in data classification or unsupervised and do pattern analysis (Silvestrini & Lavagna, 2022).

In AI, more data equates to improved accuracy (Seng, 2022). Nonetheless, that is only sometimes true (Mustapha, 2022; Ozsahin, 2022; 2020; Seyer Cagatan, 2022; Uzun Ozsahin, 2022a; 2022b). The most widely used and developed machine learning algorithm, deep learning consumes the most data and has outperformed humans in specific cognitive tasks (Li et al., 2021). Deep learning has emerged as the approach with the most potential in the AI area due to these characteristics. Deep learning algorithms have enabled substantial advancements in computer vision and speech recognition (Moshayedi, 2022).

1.1.4 Terminologies in machine learning

1.1.4.1 Bias and variance

An error in machine learning measures how accurately an algorithm can make predictions given a previously unknown dataset (Sarker, 2021b). The machine learning model that performs the best on the given dataset is selected based on these errors. Typically, during the training process, a machine learning model identifies patterns in the data and makes predictions. During the model evaluation phase, there will be a discrepancy between the predicted and actual values, known as an error due to bias (Singh, 2022). Bias is the inability of the model to correctly identify the true relationship among data points (Barcelos, 2022). A model with low bias will make fewer assumptions about the target function's shape than one with a high bias, leading to a model that fails to represent essential features of the dataset accurately.

A variance is a prediction variation from various training data (Barcelos, 2022). It describes the amount by which a random variable deviate from its predicted value. Typically, a model should remain unchanged from one training dataset to the next. This necessitates the algorithm to comprehend the hidden mapping between input and output variables. A low variance suggests that the prediction of the target function is insensitive to changes in the training data set. In contrast, a high variance indicates that the prediction of the target function varies significantly as the training dataset changes (Brownlee, 2019). A model with high variance learns extensively and performs well on the training dataset but does not generalize well to new data (Catrambone, 2022). Therefore it has a high accuracy on the training dataset but a high error rate on the test dataset. With a high variance, the model learns too much from the dataset, resulting in overfitting and increased model complexity (Singh, 2022). Linear regression, logistic regression, and linear discriminant analysis are common machine learning models with low variance, while those with high variance have a decision tree, SVM, and KNN (Varghese, 2018). To reduce high variance, measures including increasing training data, regularizing terms, and reducing input features must be implemented (Varghese, 2018).

1.1.4.2 Overfitting and underfitting

The ultimate goal of every machine learning model is to obtain optimal performance by generalizing well. Generalization implies producing an accurate prediction when provided with an unseen dataset. Overfitting and underfitting are the two major concerns for data scientists, researchers, and machine learning engineers as they degrade model performance, thereby causing the model to be unable to generalize well (Brownlee, 2019; Li et al., 2021). Hence, underfitting and overfitting are the terms that need to be checked to know whether the model is generalizing well. Overfitting occurs when a model attempts to account for every data point in a given dataset. As a result, the model accumulates noise and inaccurate values in the dataset, reducing the model's efficiency and precision. An overfitting model has low bias and high variance (Barcelos, 2022). To avoid overfitting, it is encouraged to train with more datasets, employ cross-validation, adopt early stopping and callbacks, apply regularization, and use ensemble models (Barcelos, 2022).

In contrast, underfitting occurs when a model cannot capture the trends in the dataset. This implies that the model does not get enough information from the training data, leading to inaccurate predictions. To prevent overfitting, the training process of a machine learning model can be halted prematurely, which means the model may require additional data to learn from. This may necessitate assistance in identifying the optimal fit for the dominant trend in the data. A model with low variance and high bias is considered to be underfitting. To avoid underfitting a model, it is encouraged to increase the number of data and employ feature selection to reduce features (Barcelos, 2022).

1.1.4.3 Principal component analysis

Principal component analysis (PCA) is a widely used unsupervised learning approach that effectively decreases data dimensions and improves interpretability while minimizing information loss (Washizawa, 2012). Its main goal is to identify principal components that can describe the data points. Although the principal components are vectors, their selection process is not random. The first principal component is calculated to capture as much of the original features' variation as possible to increase its efficiency. The second principal component, orthogonal to the first one, captures most of the remaining variance. Additionally, PCA reduces noise by reducing a large number of features to just a few principal components (Katayama et al., 2022). PCA has various applications, such as computer vision and image compression. Also, it can be useful in identifying hidden patterns in data with high dimensions (Guyon et al., 2012).

1.1.4.4 Cross-validation

Cross-validation is a method for evaluating the performance of a model by training it on a subset of the input data and testing it on a different, previously unseen subset of the input data (Kucheryavskiy, 2020). The ability of a statistical model to generalize to a new dataset is tested via cross-validation. Regularly checking your models' reliability is essential when working with machine learning. To do this, it is necessary to put the model through its paces using data that wasn't used during training. Cross-validation is similar to but different from the general train-test split. The train-test split technique divides the dataset into two parts: the training and test set, with the majority ratio being the training set while the minority ratio has the test set, for example, 70:30, 80:20, etc. The most significant disadvantage of the train-test split technique is that it provides a high variance. The cross-validation technique is separated into validation set approach, leave-P-out cross-validation, leaves-one-out cross-validation, K-fold cross-validation, and stratified k-fold cross-validation.

1.1.4.5 Gradient descent

In training a machine learning or deep learning model, gradient descent is often used as an optimization algorithm. Its primary objective is to use a convex function as a starting point and iteratively modify its parameter to reduce a given function to its local minimum. To do this, a set of initial parameter values is defined, and calculus is implemented by the gradient descent algorithm to iteratively modify the value iteratively, thereby minimizing the given cost function. The three most commonly used gradient descents include:

1. Mini-batch gradient descent: Mini-batch gradient descent, a combination of stochastic gradient descent (SGD) and batch gradient descent, is the most widely used gradient descent algorithm. The training dataset is broken down

into smaller groups and updated incrementally in mini-batch gradient descent. This creates a balance between SGD's robustness and batch gradient descent's efficiency.

2. Batch gradient descent: Batch gradient descent evaluates the error for each group within a training dataset but only updates the model once all the training groups have been assessed. Batch gradient descent has certain benefits, including its processing efficiency and stability to both the error gradient and the convergence. One drawback is that the stable error gradient may cause the model to converge to a state that could be more optimal. It also necessitates that the complete training dataset is loaded into the algorithm's working memory.
3. SGD: Unlike batch gradient descent, SGD updates the parameters for each training example one by one. Depending on the task, this can make SGD a faster alternative than batch gradient descent. Regular reports are helpful since they provide a precise measure of progress. However, the computational cost of the frequent updates is more than that of the batch gradient descent method.

1.1.4.6 Cost function

The effectiveness of a machine learning model on a certain dataset is heavily influenced by a parameter called the cost function (Adams, 2017). The cost function determines the discrepancy between the predicted and expected values and expresses the result as a single real number (Cost function, 2017). Even though various accuracy measures can indicate how well or how poorly a model performs, they often cannot guide how to enhance it. A cost function can determine when the model is most accurate by finding the between the overtrained and undertrained. There are three main types of cost functions: binary, multiclass, and regression cost function. A binary cost function predicts categorical variables such as cat or dog, 0 or 1, and yes or no. The multiclass cost function is used when there are two or more classes, while the regression cost function applies to weather, a house's price, and loan predictions.

1.1.4.7 Parameter and hyperparameter

Arguably, parameter and hyperparameter are among the two most confusing terms among beginners and experts alike in machine learning. While parameters are fitted configurations and variables peculiar to a model, hyperparameters are adjustable configurations liable to tuning for the optimal performance of the model. Parameters are the values a machine learning algorithm independently alters as it learns. These include biases and weight in a neural network, cluster centroid in clustering, coefficients linear regression, and support vectors in SVM. Hyperparameters are specific configurations that influence the training process, and they include the learning rate in a neural network, K in the KNN algorithm. Table 1.1 summarizes the differences between parameters and hyperparameters.

Table 1.1 Differences between parameter and hyperparameter in machine learning.

	Parameters	Hyperparameters
Definition	Are configurations internal to a model.	Are configurations explicitly specified to regulate the training process.
Purpose	For making predictions.	For optimizing a model.
Usage	They are specified during training.	They are specified before training.
Location	They are internal to the model.	They are external to the model.
Implementation	They are independent and set by the model itself.	They are dependent and are set by the data scientist, researcher, or machine learning engineer.
Dependency on dataset	Dependent on the training dataset.	Independent on dataset.
Examples	Coefficients in logistic and linear regression, support vectors in SVM, and weights in neural networks.	The kernel in SVM, K in KNN, and learning rate in neural networks.

KNN, *K-nearest neighbor;* **SVM**, *support vector machine.*

1.1.4.8 Transfer learning

Transfer learning is a method that involves employing a previously trained model, often called a pretrained model, as the basis for an entirely new task. As a result, the pretrained model optimizes the second task. Transfer learning can outperform newly developed models trained with a smaller number of data when applied to a new problem. Because of this, it is rare to see a machine-learning model trained from scratch for image classification-related tasks. Instead, data scientists and researchers like starting with a pretrained model that has been taught how to classify objects and has picked up on common features, including shapes, edges, and contours. Common models that have the basis of transfer learning include AlexNet, ResNet, VGG16 & 19, and InceptionV3.

1.1.4.9 Performance evaluation matrix

Evaluating the performance of a machine learning model is crucial in developing an effective model. Various measures are used to evaluate the quality or performance of a model. These performance measures help understand a model's performance on a specific dataset. By adjusting these hyperparameters, the performance of the model can be improved. The ultimate goal of every machine learning model is to perform well on unseen data, and performance evaluation metrics help evaluate the model's performance on unseen datasets. In supervised learning, the task may be classification or regression, and each task requires different performance evaluation metrics. Therefore understanding the various metrics is of paramount importance. The following metrics are used for classification tasks:

- Accuracy: One of the most common classification metrics to implement is the accuracy metric. It can be estimated using Eq. (1.1); A true negative (TN) occurs when a model correctly predicts a negative outcome (i.e., a value of 0) and the actual outcome is also negative. A true positive (TP) is when a model correctly predicts a positive outcome (i.e., a value of 1), and the actual outcome is also positive. A false negative (FN) occurs when a model incorrectly predicts a negative outcome, but the actual outcome is positive. This is also known as a Type-II error. A false positive (FP) occurs when a model incorrectly predicts a positive outcome, but the actual outcome is negative. This is also known as a Type-I error.

$$\text{Accuracy} = (\text{Number of correct predictions})/(\text{Total number of prediction}) \tag{1.1}$$

- Precision: Precision is frequently utilized to address the limitation of accuracy. It measures the proportion of positive predictions that were accurate. Precision can be calculated by using the following using Eq. (1.2);

$$\text{Precision} = \text{TP}/(\text{TP} + \text{FP}) \tag{1.2}$$

- Recall/sensitivity: This is comparable to precision, but it aims to determine the proportion of actual positives identified correctly. It measures the number of correct positive predictions out of all the possible positive predictions that could have been made. Recall can be calculated by using the following using Eq. (1.3):

$$\text{Recall} = \text{TP}/(\text{TP} + \text{FN}) \tag{1.3}$$

- F-score: F-score or F1 score is a metric used to evaluate a binary classification model based on predictions made for the positive class. It is calculated using precision and recall. It is a single score that considers both precision and recall and gives equal weight to both. The F1 score can be calculated as the harmonic mean of precision and recall. The formula for calculating the F1 score is provided using Eq. (1.4):

$$\text{F1score} = 2^{*}(\text{precision}^{*}\text{recall})/(\text{precision} + \text{recall}) \tag{1.4}$$

- Confusion matrix: A confusion matrix represents the predicted outcomes in a binary classification task. This further shows the model performance on the test set when true values are known. Though the confusion matrix's implementation is straightforward, some of the terminology included may be confusing, as shown in Fig. 1.9.
- Area under the curve (AUC)-receiver operating characteristic (ROC) curve: The AUC-ROC is a tool that visualizes the performance of a classification model by plotting the true positive rate (TPR) against the false positive rate (FPR) on a graph. The ROC curve can help to understand how a classification

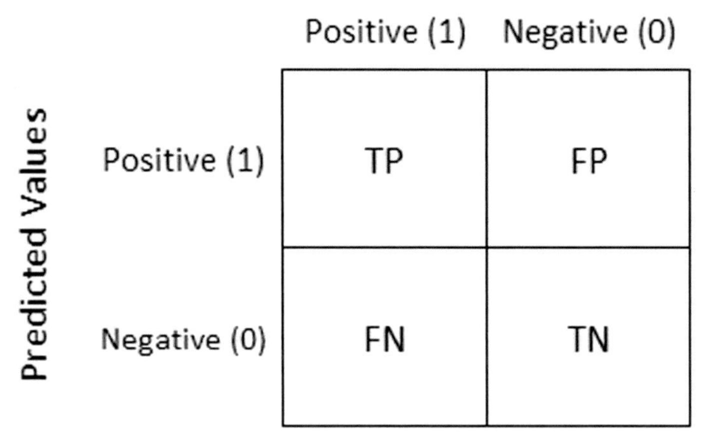

FIGURE 1.9

Confusion matrix in machine learning.

From Narkhede, S. (2018). The knowledge hub of ML and data science. https://www.sarangnarkhede.com/.

model's performance changes as the threshold is varied, and it is a good way to evaluate the overall performance of a model. TPR can be calculated using Eq. (1.5):

$$\text{TPR} = \text{TP}/(\text{TP} + \text{FN}) \tag{1.5}$$

FPR can be calculated using Eq. (1.6):

$$\text{TPR} = \text{FP}/(\text{FP} + \text{TN}) \tag{1.6}$$

AUC estimates the two-dimensional area under the entire ROC curve, as shown in Fig. 1.10.

AUC is a measure that evaluates the performance of a classification model across all thresholds by providing an aggregate score. The range of the AUC score is from 0 to 1, where a model that makes 100% incorrect predictions will have an AUC score of 0.0, and a model that makes 100% correct predictions will have an AUC score of 1.0.

In the case of regression tasks, the following evaluation metrics are used:

1. Mean absolute error: This is a statistical measure of how far off actual values are from predicted values.
2. Mean squared error (MSE): The MSE estimates the average squared deviation from the model's prediction and the actual value. When comparing alternative regression measures, MSE is often preferred since it can be differentiable and optimized more effectively.

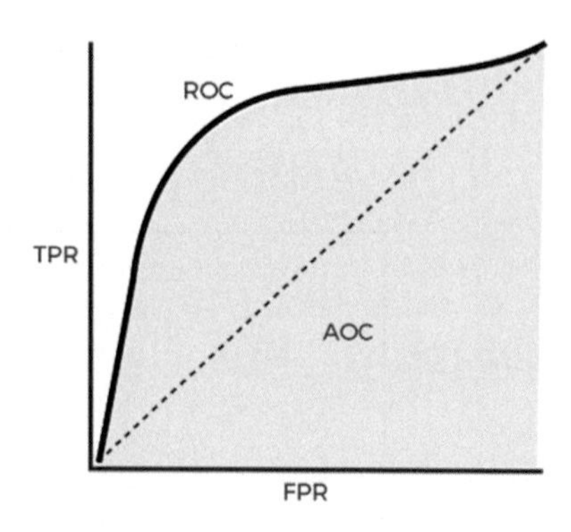

FIGURE 1.10

Receiver operating characteristic curve.

From Sarang, N. (2018). Understanding AUC - ROC curve. https://towardsdatascience.com/understanding-auc-roc-curve-68b2303cc9c5.

3. R squared score: The performance of a regression model can be evaluated by comparing its predicted value with a constant baseline using *R* squared error, also known as the coefficient of determination. Selecting the fixed baseline requires averaging the data and drawing the line at the midpoint.
4. Adjusted *R* squared: Adjusted *R* squared is improvement to the *R* squared error. Data scientists can be misled by *R* square's limitation of a score growing as the number of terms increases, even when the model needs to improve. An adjusted *R* square is employed to circumvent the problem posed by the *R* square, which always displays a lower value than R^2. It does so because it adjusts the increasing predictors and only indicates an improvement if one exists.

References

Adams, M. (2017) "Cost function (machine learning)," Radiopaedia.org [Preprint]. Available at: https://doi.org/10.53347/rid-56153.

Barcelos, G. (2022) Understanding bias in machine learning models, Arize AI. Available from https://arize.com/blog/understanding-bias-in-mL-models/. Accessed 05.11.22.

Boucher, P. (2020). Artificial intelligence: How does it work, why does it matter, and what can we do about it? European Parliament. Available from https://www.europarl.europa.eu/RegData/etudes/STUD/2020/641547/EPRS_STU(2020)641547_EN.pdf. Accessed 11.05.22.

Brownlee, J. (2019) Gentle introduction to the bias-variance trade-off in machine learning, machine learning mastery. Available from https://machinelearningmastery.com/gentle-introduction-to-the-bias-variance-trade-off-in-machine-learning/#:~:text=Bias%20Error,-Bias%20are%20the&;text=Generally%2C%20linear%20algorithms%20have%20a,assumptions%20of%20the%20algorithms%20bias. Accessed 06.11.22.

Burns, E., Laskowski, N., & Tucci, L. (2022). What is artificial intelligence (AI)? Definition, benefits, and use cases, SearchEnterpriseAI. TechTarget. Available from https://www.techtarget.com/searchenterpriseai/definition/AI-Artificial-Intelligence. Accessed 11.05.22.

Catrambone, R. (2022) Generalization by studying examples versus generalization by applying examples to problems, in: Proceedings of the twentieth annual conference of the cognitive science society, pp. 214–219. Available at: https://doi.org/10.4324/9781315782416-47.

Cost function (2017) Encyclopedia of machine learning and data mining, pp. 285–285. Available at: https://doi.org/10.1007/978-1-4899-7687-1_100091.

French, R. M. (2000). The turing test: The first 50 years. *Trends in Cognitive Sciences*, *4* (3), 115–122. Available from https://doi.org/10.1016/s1364-6613(00)01453-4.

Gong, D. (2022) Top 6 machine learning algorithms for classification, Top 6 Machine Learning Algorithms for Classification. Available from https://towardsdatascience.com/top-machine-learning-algorithms-for-classification-2197870ff501. Accessed 11.05.22.

Grau, M. M., Tajtakova, M., & Aranda, D. A. (2009). Machine learning methods for the market segmentation of the performing arts audiences. *International Journal of Business Environment*, *2*(3), 356. Available from https://doi.org/10.1504/ijbe.2009.023796.

Guyon, C., Bouwmans, T. and Zahzah, E.-h (2012) Robust principal component analysis for background subtraction: Systematic evaluation and comparative analysis, Principal component analysis [Preprint]. Available at: https://doi.org/10.5772/38267.

Hussin Adam Khatir, A. A., & Bee, M. (2022). Machine learning models and data-balancing techniques for credit scoring: What is the best combination? *Risks*, *10*(9), 169. Available from https://doi.org/10.3390/risks10090169.

Iacovazzi, A. and Raza, S. (2022) Ensemble of random and isolation forests for graph-based intrusion detection in containers, in: Proceedings of the IEEE International Conference on Cyber Security and Resilience (CSR) [Preprint]. Available at: https://doi.org/10.1109/csr54599.2022.9850307.

Isinkaye, F. O., Folajimi, Y. O., & Ojokoh, B. A. (2015). Recommendation systems: Principles, methods and evaluation. *Egyptian Informatics Journal*, *16*(3), 261–273. Available from https://doi.org/10.1016/j.eij.2015.06.005.

Katayama, H., Mori, Y. and Kuroda, M. (2022) Variable selection in nonlinear principal component analysis, Advances in principal component analysis [Preprint]. Available at: https://doi.org/10.5772/intechopen.103758.

Kucheryavskiy, S. et al. (2020) Procrustes cross-validation—A bridge between cross-validation and independent validation set. Available at: https://doi.org/10.26434/chemrxiv.12327803.

Kumar, Y. et al. (2022) Artificial Intelligence in disease diagnosis: A systematic literature review, synthesizing framework, and future research agenda, Journal of Ambient Intelligence and Humanized Computing. Available at: https://doi.org/10.1007/s12652-021-03612-z.

Kurama, V. (2019) Regression in machine learning: What it is and examples of different models, Built in. Available from https://builtin.com/data-science/regression-machine-learning. Accessed 11.05.22.

Li, Q., Yan, M. and Xu, J. (2021) Optimizing convolutional neural network performance by mitigating underfitting and overfitting, in: Proceedings of the IEEE/ACIS 19th international conference on computer and information science (ICIS) [Preprint]. Available at: https://doi.org/10.1109/icis51600.2021.9516868.

Logistic regression in machine learning – javatpoint. (no date). Available from: http://www.javatpoint.com, https://www.javatpoint.com/logistic-regression-in-machine-learning. Accessed 11.05.22.

Maulud, D., & Abdulazeez, A. M. (2020). A review on linear regression comprehensive in machine learning. *Journal of Applied Science and Technology Trends*, *1*(4), 140–147. Available from https://doi.org/10.38094/jastt1457.

Moshayedi, A. J., et al. (2022). Deep learning application pros and cons over algorithm. *EAI Endorsed Transactions on AI and Robotics*, *1*, 1–13. Available from https://doi.org/10.4108/airo.v1i.19.

Mustapha, M. T., et al. (2022). Breast cancer screening based on supervised learning and multi-criteria decision-making. *Diagnostics*, *12*(6), 1326. Available from https://doi.org/10.3390/diagnostics12061326.

Ozsahin, D.U. et al. (2022) Impact of feature scaling on machine learning models for the diagnosis of diabetes, in: Proceedings of the international conference on artificial intelligence in everything (AIE) [Preprint]. Available at: https://doi.org/10.1109/aie57029.2022.00024.

Ozsahin, I., et al. (2020). Review on diagnosis of COVID-19 from chest CT images using artificial intelligence. *Computational and Mathematical Methods in Medicine*, *2020*, 1–10. Available from https://doi.org/10.1155/2020/9756518.

Ray, S. (2021) SVM: Support vector machine algorithm in machine learning, Analytics Vidhya. Available from https://www.analyticsvidhya.com/blog/2017/09/understaing-support-vector-machine-example-code/#:~:text=%E2%80%9CSupport%20Vector%20Machine%E2%80%9D%20(SVM,mostly%20used%20in%20classification%20problems. Accessed 11.05.22.

Sarker, I. H. (2021a). Machine learning: Algorithms, real-world applications and research directions. *SN Computer Science*, *2*(3). Available from https://doi.org/10.1007/s42979-021-00592-x.

Sarker, I. H. (2021b). Deep learning: A comprehensive overview on techniques, taxonomy, applications and research directions. *SN Computer Science*, *2*(6). Available from https://doi.org/10.1007/s42979-021-00815-1.

Seng, J. K., et al. (2022). Artificial intelligence (AI) and machine learning for multimedia and edge information processing. *Electronics*, *11*(14), 2239. Available from https://doi.org/10.3390/electronics11142239.

Seyer Cagatan, A., et al. (2022). An alternative diagnostic method for C. Neoformans: Preliminary results of deep-learning based detection model. *Diagnostics*, *13*(1), 81. Available from https://doi.org/10.3390/diagnostics13010081, 2022.

Silvestrini, S., & Lavagna, M. (2022). Deep learning and artificial neural networks for spacecraft dynamics, navigation, and Control. *Drones*, *6*(10), 270. Available from https://doi.org/10.3390/drones6100270.

Singh, V. (2022) Difference between Type 1 and Type 2 Error, Naukri.com. Available from https://www.naukri.com/learning/articles/difference-between-type-1-and-type-2-error/. Accessed 11.05.22.

Terra, J. (2023, August 11). Artificial Intelligence and machine learning job trends in 2023: Simplilearn. Simplilearn.com. Available from https://www.simplilearn.com/rise-of-ai-and-machine-learning-job-trends-article.

Uzun Ozsahin, D. et al. (2022a) Impact of outliers and dimensionality reduction on the performance of predictive models for medical disease diagnosis, in: Proceedings of the international conference on artificial intelligence in everything (AIE) [Preprint]. Available at: https://doi.org/10.1109/aie57029.2022.00023.

Uzun Ozsahin, D., et al. (2022b). Computer-aided detection and classification of Monkeypox and chickenpox lesions in human subjects using deep learning framework. *Diagnostics*, *13* (2), 292. Available from https://doi.org/10.3390/diagnostics 13020292.

Varghese, D. (2018) Comparative study on classic machine learning algorithms, Towards Data Science. Available from https://towardsdatascience.com/comparative-study-on-classic-machine-learning-algorithms-24f9ff6ab222. Accessed 06.11.22.

Waseem, M. (2022). Classification in machine learning: Classification algorithms, Edureka. Available from: https://www.edureka.co/blog/classification-in-machine-learning/. Accessed 11.05.22.

Washizawa, Y. (2012) Subset basis approximation of kernel principal component analysis, Principal component analysis [Preprint]. Available at: https://doi.org/10.5772/37051.

Yan, J. et al. (2016) Machine learning in brain imaging genomics, Machine learning and medical imaging, pp. 411–434. Available at: https://doi.org/10.1016/b978-0-12-804076-8.00014-1.

Zhang, S. (2022). Challenges in KNN classification. *IEEE Transactions on Knowledge and Data Engineering*, *34*(10), 4663–4675. Available from https://doi.org/10.1109/tkde.2021.3049250.

Zhu, T. (2020). Analysis on the applicability of the random forest. *Journal of Physics: Conference Series*, *1607*(1), 012123. Available from https://doi.org/10.1088/1742-6596/1607/1/012123.

CHAPTER

Convolution neural network and deep learning

2

Mubarak Taiwo Mustapha[1,2], Ilker Ozsahin[1,3] and Dilber Uzun Ozsahin[1,4,5]

[1]*Operational Research Centre in Healthcare, Near East University, Nicosia/TRNC, Mersin-10, Turkey*

[2]*Department of Biomedical Engineering, Near East University, Nicosia/TRNC, Mersin-10, Turkey*

[3]*Brain Health Imaging Institute, Department of Radiology, Weill Cornell Medicine, New York, NY, United States*

[4]*Department of Medical Diagnostic Imaging, College of Health Science, University of Sharjah, Sharjah, United Arab Emirates*

[5]*Research Institute for Medical and Health Sciences, University of Sharjah, Sharjah, United Arab Emirates*

Abbreviations

BPTT	back-propagation through time
CNN	convolutional neural network
DBN	deep belief networks
GAN	generative adversarial network
GPUs	graphics processing units
GRU	gated recurrent unit
LSTM	long short-term memory
MNIST	Modified National Institute of Standards and Technology
MBConv	mobile inverted residual bottleneck
NLP	natural language processing
RBM	restricted Boltzmann machine
ReLu	rectified linear unit
RNNs	recurrent neural networks
SGD	stochastic gradient descent
Tanh	hyperbolic tangent function
VGG	visual geometry group

2.1 Brief history of deep learning

In 1943, Walter Pitts and Warren McCulloch created the initial mathematical representation of a neural network, showing how it mimics the human brain's cognitive function (Lettvin et al., 1959). This marked the beginning of exploring deep neural networks and deep learning. Frank Rosenblatt further advanced this field in 1957 with

Artificial Intelligence and Image Processing in Medical Imaging. DOI: https://doi.org/10.1016/B978-0-323-95462-4.00002-9

his publication "The Perceptron: A Perceiving and Recognizing Automaton," which presented a pattern recognition technique using a neural network with two layers (Rosenblatt, 1957). Alexey Ivakhnenko and V.G. Lapa created the first functional neural network in 1965 (Eliseeva et al., 2020). Alexey Ivakhnenko created the first eight-layer deep neural network in 1971, exhibited in the Alpha computer identification system. This was the introduction to deep learning in its entirety (Eliseeva et al., 2020). In 1980, Kunihiko Fukushima created the "Neocognitron," an artificial deep neural network with many convolutional layers for pattern recognition (Dettmers, 2022).

In 1985, Terry Sejnowski built NETtalk, which taught itself to speak English words (Adamson & Damper, 1996). Yann LeCun created a system in 1989 that used a convolutional deep neural network to read handwritten digits (LeCun et al., 1989). Due to the enormous amount of labeled data required to train deep learning models in supervised learning, Fei-Fei Li founded ImageNet in 2009 (Deng et al., 2009). ImageNet is a comprehensive collection of captioned images. Google Brain undertook a "Cat Experiment" in 2012, and the experiment's findings were made public. This experiment used unsupervised learning, in which the deep neural network recognized patterns and features in photos of cats using unlabeled input, and it accurately identified only 15% of images (Markoff, 2012). Facebook created DeepFace in 2014 as a deep learning system for recognizing and tagging people's faces in photographs. The important point to note about any deep learning model is that it requires a vast amount of data and a significant number of computational resources to train. At the time, this was a severe disadvantage for convolutional neural networks (CNN). As a result, CNNs remained confined to the postal sector and never made their way into machine learning (Dettmers, 2022).

2.2 Deep learning

Deep learning is a machine learning technique that uses artificial deep neural networks. The human brain comprises nerve cells, or neurons, processing information via signal transmission and reception. Deep neural network learning uses layers of "neurons" to communicate and process data. In deep learning, the term "deep" refers to the number of layers in a neural network, with a higher number of layers indicating a deeper network. These networks can learn from labeled and unlabeled data (Lee et al., 2017), allowing for supervised and unsupervised learning. However, training the deep neural network requires significant data, resulting in more accurate findings. Deep learning has established itself as a strong tool over the last few decades due to its capacity to manage massive volumes of data (Najafabadi et al., 2015).

Additionally, deep neural networks can detect and fix output errors or losses without human interaction (Sarker, 2021). The popularity of hidden layers has overtaken more traditional techniques, particularly pattern recognition. A CNN comprises three layers: the input, the hidden, and the output. The input layer in a deep neural network, the first layer, holds the parameters required for processing input data. It simply transfers these parameters to the next layer without performing any computations at this layer. An essential computation takes place in the hidden layer, and the result is passed on to the next layer. The number of layers and neurons in each layer of the deep neural network must be carefully chosen to enhance the network's efficacy. A deeper network is characterized by having more hidden layers (Andreas et al., 2015). After receiving the results from the previous layers, the output layer produces the final output. There is a weight attached to each connection between neurons in the next layer.

As the name implies, weights assign a numerical value to a particular feature. Some features may be more important than others in obtaining the desired output. These weights are used to determine the neuron's weighted sum. In addition to the weights, each hidden layer contains an activation function. The activation functions determine whether or not to activate a neuron based on its weighted sum. Activation functions, such as sigmoid and hyperbolic tangent function (tanh), are employed to introduce nonlinearity into calculations, thereby allowing for more sophisticated tasks.

The absence of an activation function would turn a deep neural network into a simple linear regression model. Rectified linear unit (ReLu), SoftMax, and tanh are all examples of activation functions. A bias neuron is also included in each layer to permit the activation function to be shifted left or right on the x-axis, resulting in a better fit of the activation function. When the input is absolute zero, the bias, a constant term, also acts as an output.

2.2.1 Convolution neural network

A CNN is a method for deep learning that can analyze an input image, recognize and differentiate various objects in it, and give them corresponding values. It comprises an input layer, one or more hidden layers, and an output layer, as illustrated in Fig. 2.1.

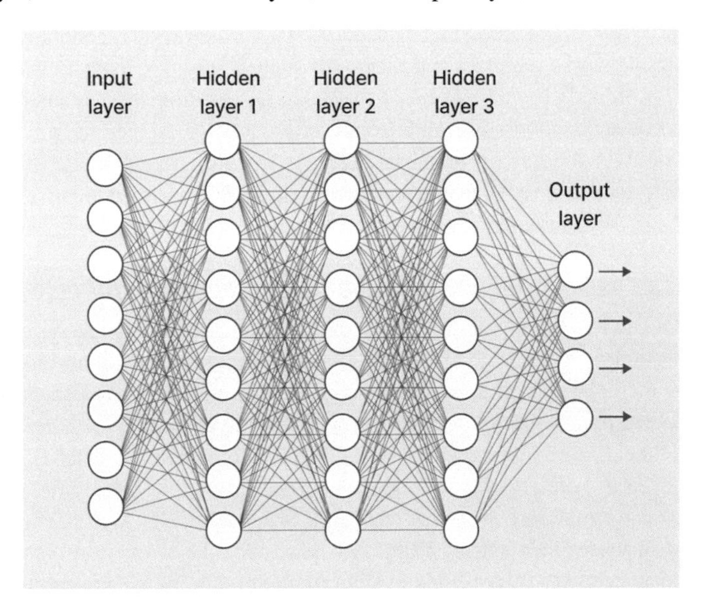

FIGURE 2.1

The convolutional neural network framework.

From Baheti, P. (2023). Convolutional neural networks: Architectures, types & examples. Convolutional Neural Networks: Architectures, Types & Examples. https://www.v7labs.com/blog/convolutional-neural-networks-guide.

Each node has a weight and threshold associated with it. Nodes are activated and begin sending information to the next layer of the network when their output exceeds a certain threshold. If this is not the case, no data will be sent to the next network layer.

CNNs are a powerful tool for computer vision, allowing a computer to "see" an image. They use a matrix of pixel values to represent the input image, which enables image recognition and classification. CNNs understand images as matrices of numbers and use feature maps to simplify the process. They are particularly useful in deep learning applications that involve large amounts of image data as they employ mathematical techniques to quickly and effectively understand the content of an image. CNNs are commonly used in image and video recognition and predictive analysis and can efficiently process large datasets without losing important details. Other neural networks, such as multilayer perceptron and recurrent neural networks (RNNs), can also be used for image processing. Still, CNNs are better suited for large datasets and preserving fine details. The unique mathematical principles used in CNN make them an ideal choice for deep learning applications that involve large amounts of data.

Different neural nets are available for various applications and data types. RNNs are frequently used in speech recognition and natural language processing (NLP) (Goldberg, 2016). CNN is increasingly used for classification and computer vision applications (Bhatt et al., 2021). Images were manually identified using feature extraction methods before the advent of CNN. CNN utilize techniques from linear algebra, such as matrix multiplication, to spot patterns in images, making them more efficient for image classification and object recognition. Despite this, they are computationally intensive, needing graphic processing units (GPUs) to train models. CNN may appear intricate but intriguing and highly useful in various technological areas. These neural networks require a significant amount of complex mathematics, so gaining a more profound comprehension of them can be accomplished by studying linear algebra and its applications. The architecture of a CNN is designed to mimic the neural connections in the human brain, taking cues from the organization of the visual cortex. Each neuron in a CNN responds to a specific, localized input known as a receptive field. When combined, these receptive fields cover the entire visual field.

2.2.1.1 The basic architecture of the convolutional neural network

While CNN can be complex in its design and function, a fundamental system governs its structure and operation. CNN consists of multiple layers, including hidden layers, which play a crucial role in assisting the computer in creating a representation of the image and determining how to classify it. The first layer in a CNN is referred to as the convolutional layer. As shown in Fig. 2.2, the fully-connected layer is always the last layer, even if there are additional convolutional or pooling layers after the convolutional layers. Complexity increases with each layer of a CNN. Colors and edges are highlighted in the earliest stages. With each successive layer of the CNN algorithm, the image data becomes increasingly evident in identifying the target object.

2.2.1.1.1 Convolutional layer

A convolution layer is a fundamental building block in deep learning neural networks, particularly CNN (Yamashita et al., 2018). The primary purpose of this layer is to extract meaningful features from the input data, such as images or audio (Seyer Cagatan et al., 2022). Convolution layers accomplish this by applying a set of filters known as kernels to the input data, as shown in Fig. 2.3. These filters are typically small, square matrices used to scan across the input data, performing a mathematical operation known as a

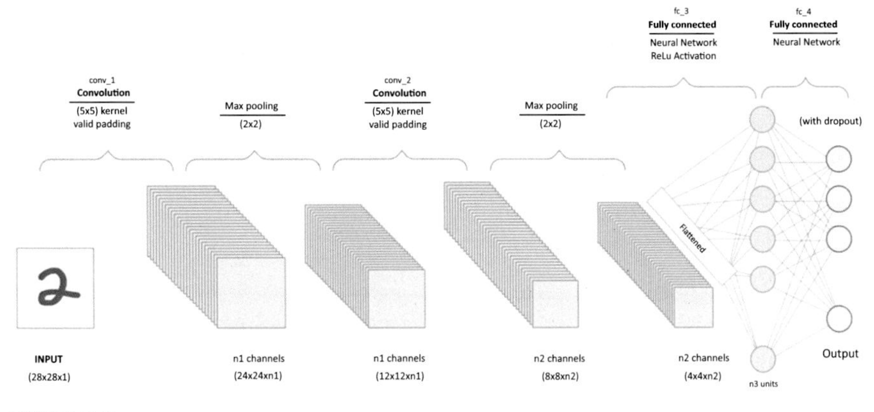

FIGURE 2.2

The architecture of the convolutional neural network.

From Cagatan, A., Mustapha, M., Bagkur, C., Sanlidag, T. & Uzun Ozsahin, D. (2022). An alternative diagnostic method for C. neoformans: Preliminary results of deep-learning based detection model. Diagnostics, 13(1), 81. https://doi.org/10.3390/diagnostics13010081.

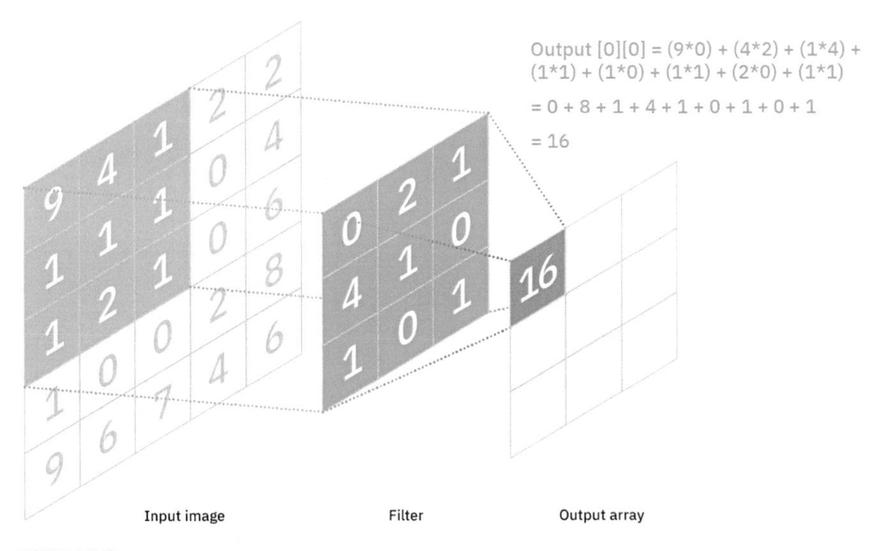

FIGURE 2.3

Convolutional layer.

From Kumar, A. (2021). Real-World Applications of Convolutional Neural Networks. https://vitalflux.com/real-world-applications-of-convolutional-neural-networks/.

convolution. The convolution operation is performed by element-wise multiplication of the filter with a small portion of the input data, known as a receptive field. The result of this multiplication is then summed up and passed through a nonlinear activation function, such as a ReLU, to introduce nonlinearity into the network. The process is repeated for each receptive field in the input data, resulting in feature maps representing the convolution layer's output. One of the key advantages of convolution layers is their ability to learn spatial hierarchies of features (Shi et al., 2022). As the filters are applied to different portions of the input data, they learn to detect different patterns and features. For example, the first convolutional layers might learn to detect edges and simple shapes in image classification. In contrast, deeper layers learn to detect more complex features, such as parts of objects. This hierarchical feature extraction makes CNN particularly effective at the image and video processing tasks.

2.2.1.1.2 Pooling layer

A pooling layer is commonly used in CNN to reduce the spatial dimensionality of the input data (Zafar et al., 2022). Pooling layers are typically inserted between consecutive convolution layers and summarize the features learned by the convolution layers, as shown in Fig. 2.4. The most common type of pooling is called max pooling, where the maximum value within a small region of the input data is selected as the output for that region. Max pooling is often used in CNNs because it effectively reduces the spatial resolution of the input data while maintaining the most important features learned by the convolution layers. This can help reduce the network's computational complexity and prevent overfitting by reducing the number of parameters in the network. Additionally, max pooling has the added benefit of making the network more robust to small translations or distortions in the input data. There are also other types of pooling layers, such as average pooling, which replaces the maximum value with the average value within a small region of the input data. This type of pooling can be less effective than max pooling in some cases but can also be useful in certain situations.

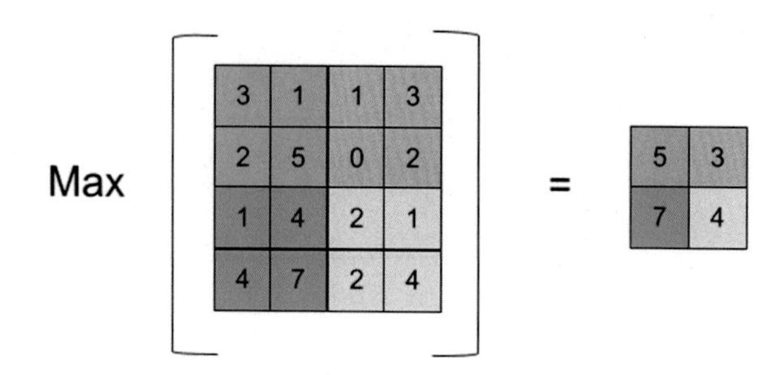

FIGURE 2.4

Pooling layer.

From Explain pooling layers: Max Pooling, average pooling, global average pooling, and Global Max Pooling. Explain Pooling layers: Max Pooling, Average Pooling, Global Average Pooling, and Global Max pooling. (2021). https://androidkt.com/explain-pooling-layers-max-pooling-average-pooling-global-average-pooling-and-global-max-pooling/.

2.2.1.1.3 Fully connected layer

The fully connected layer is the final layer in convolutional networks. Each neuron in this layer is connected to other neurons in a different layer, as shown in Fig. 2.5. This layer classifies the features and filters extracted from the previous layer. While convolutional and pooling layers often use the ReLU function to classify inputs efficiently, the fully connected layer typically uses the SoftMax activation function to produce a probability between 0 and 1.

2.2.2 Common convolutional neural network models

2.2.2.1 AlexNet

In 2012, a team consisting of Alex Krizhevsky, Ilya Sutskever, and Geoffery Hinton, widely considered the father of deep learning, introduced the AlexNet CNN in a research paper (Zhao & Shang, 2021). Since then, the use of deep CNN has surged, with many machine-learning solutions relying on them (Krizhevsky et al., 2017). Their study aimed to demonstrate that deep CNNs, efficient computation resources, and common CNN implementation techniques can be used to solve the relatively simple task of image classification. The paper showed that a deep

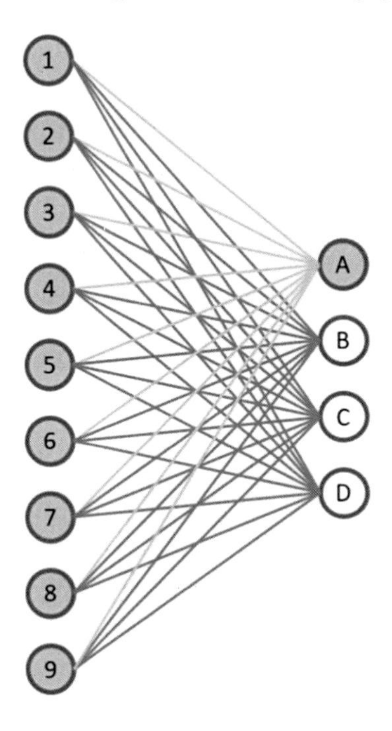

FIGURE 2.5

Fully connected layer.

From Unzueta, D. (2022). Fully connected layer vs. Convolutional layer: Explained. https://builtin.com/machine-learning/fully-connected-layer.

CNN consisting of 5 convolutional layers and three fully connected layers could classify images accurately and efficiently. Before the introduction of AlexNet, traditional neural networks, and CNNs performed well on datasets such as the Modified National Institute of Standards and Technology (MNIST) handwritten character dataset. Still, they needed help with image classification of everyday objects, which requires a larger dataset to account for the diversity of things. The introduction of large datasets such as ImageNet, which contains 22,000 classes across 15 million high-resolution images, solved this problem. Another limitation prior to AlexNet was the need for more sufficient computer resources to train such networks. However, the availability of optimized GPUs made it possible to train deep CNNs. The specific GPU used to train the AlexNet CNN architecture was the NVIDIA GTX 580 3GB GPU (Krizhevsky et al., 2017).

The AlexNet CNN architecture has eight layers of learnable parameters, as shown in Fig. 2.6. It comprises five convolutional layers and three fully connected layers. It was standard practice to use either tanh or sigmoid nonlinearity to train neurons within a neural network. However, the AlexNet model utilized ReLU, which Nair and Hinton introduced in previous research (Salakhutdinov et al., 2007). ReLU is applied after every convolutional and fully connected layer. ReLU ensures that positive values within the neurons are maintained while negative values are set to zero. Additionally, it accelerates the training process as gradient descent optimization occurs faster than standard nonlinearity techniques. Finally, dropout is applied before the first and second fully connected layers.

2.2.2.2 VGG16

VGG16 is a CNN model trained on the ImageNet dataset and developed in 2014 by the Visual Geometry Group (VGG) at the University of Oxford (Jiang et al., 2021). VGG16 is a deep model with 16 layers, as shown in Fig. 2.7, making it an effective tool for image classification tasks (Guan et al., 2019). The architecture of VGG16 is made up of a series of convolutional and max pooling layers, followed

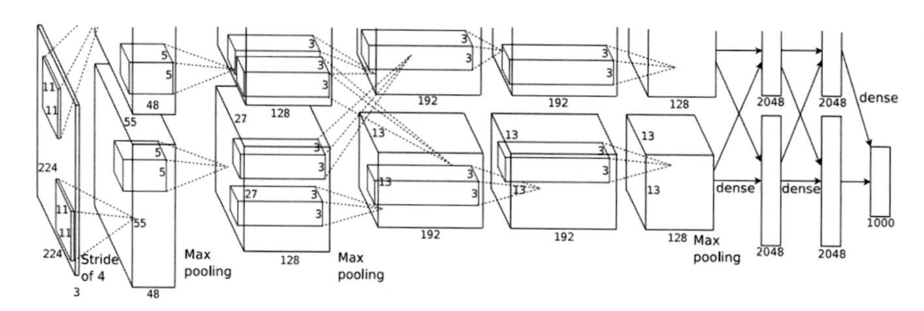

FIGURE 2.6

The basic architecture of the AlexNet model.

From Wei, J. (2019). AlexNet: The architecture that challenged CNNs. https://towardsdatascience.com/alexnet-the-architecture-that-challenged-cnns-e406d5297951.

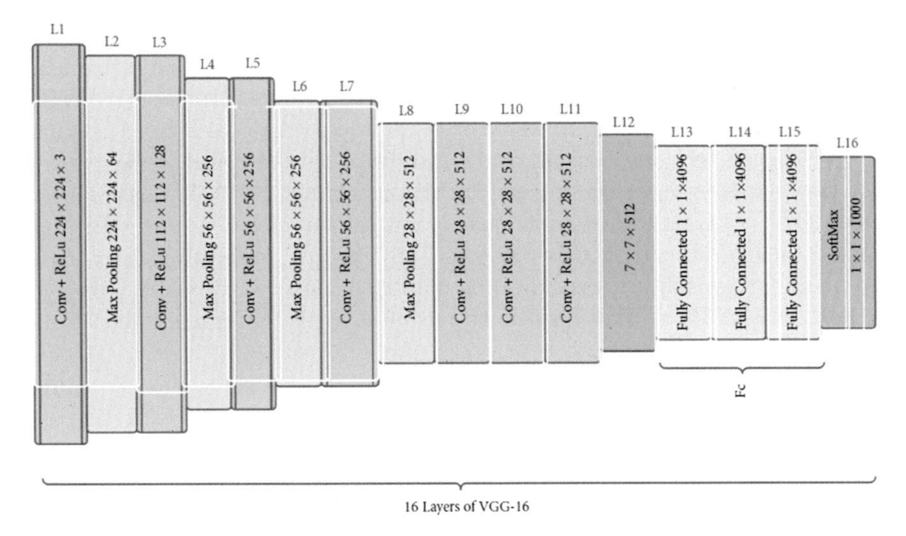

FIGURE 2.7

The basic architecture of the VGG16 model.

From Hasan, M., Fatemi, M., & Khan, M. (2021). Comparative analysis of skin cancer (Benign vs. Malignant) detection using convolutional neural networks. Journal of Healthcare Engineering, 1–17. https://doi.org/10.1155/2021/5895156.

by a few fully connected layers (Ang et al., 2022). The convolutional layers are responsible for identifying features from the input image. In contrast, the max-pooling layers decrease the spatial resolution of the feature maps, making the model more resistant to slight variations in the input image. The fully connected layers are used to classify the image based on the features extracted by the convolutional layers. One of the key features of VGG16 is the use of small convolutional filters with a size of 3×3 pixels. This enables the model to identify fine details in the image and decrease the number of parameters in the model. Also, VGG16 employs many convolutional layers, which aids the model in learning a hierarchy of features, from basic edges and textures to more intricate shapes and objects.

VGG16 has been exposed to a vast dataset of images and applied in multiple image classification tasks, including object recognition and scene comprehension. It has also been used as a feature extractor for other tasks, such as object detection and image segmentation. Despite its success, VGG16 has some drawbacks. It requires a large amount of computational resources and memory, making it difficult to use on resource-constrained devices (Ding et al., 2023). Additionally, the large number of parameters in the model can make it prone to overfitting on small datasets (Ding et al., 2023).

2.2.2.3 ResNet50

ResNet50 is a CNN model trained on the ImageNet dataset and was developed by Microsoft Research in 2015 (Shafiq & Gu, 2022). ResNet50, a variant of the

ResNet architecture, is a deep model with 50 layers, making it a very powerful tool for image classification tasks (Uzun Ozsahin et al., 2023). One of the main innovations of the ResNet architecture is the use of residual connections, which allows for training much deeper models than was previously possible. Traditional CNN struggles to learn valuable representations when the model is too deep, as the gradients tend to vanish or explode (Marquez et al., 2018). Residual connections allow gradients to flow more quickly through the model, enabling the training of much deeper models. Another key aspect of ResNet50 is batch normalization, which helps stabilize the training process by normalizing the activations of each layer. This enables the model to reach convergence quicker and with fewer training examples.

ResNet50 has been exposed to a vast dataset of images and applied in multiple image classification tasks (Uzun Ozsahin et al., 2023). It has also been used as a feature extractor for other tasks, such as object detection and image segmentation (Stateczny et al., 2022). ResNet50 has been widely used in industry and research and is considered one of the best models for image classification tasks. The model has excellent performance on the ImageNet dataset and has also been used to achieve state-of-the-art performance on other datasets. Despite its success, ResNet50 has some drawbacks. Like VGG16, it requires a large amount of computational resources and memory, making it difficult to use on resource-constrained devices (Gou et al., 2020). Additionally, the model has many parameters, which can make it prone to overfitting on small datasets (Uzun Ozsahin et al., 2022).

2.2.2.4 InceptionV3

InceptionV3 is a CNN model trained on the ImageNet dataset. The model was developed by Google in 2015 and is a variation of the Inception architecture (Ahmed et al., 2023; Uzun Ozsahin et al., 2022). InceptionV3 is a deep model with over 140 layers, which makes it a very powerful tool for image classification tasks (Uzun Ozsahin et al., 2022). The architecture of InceptionV3 is based on the concept of "Inception modules," which are designed to learn various features from the input image at different scales. An Inception module comprises a series of parallel convolutional layers with varying filter sizes, followed by a concatenation layer that combines the outputs of all the convolutional layers. This enables the model to learn features at different scales and complexities, which enhances the accuracy of the model. Another crucial aspect of InceptionV3 is the use of a global average pooling layer, which substitutes the usual fully connected layers at the end of the model. This helps to reduce the number of parameters in the model and improves its ability to generalize to new data.

The inceptionV3 model has been used as a feature extractor for object detection and image segmentation (Hoeser & Kuenzer, 2020). The inceptionV3 model has been widely used in industry and research and is considered one of the best models for image classification tasks (Cao et al., 2021). The model has excellent performance on the ImageNet dataset and has also been used to achieve state-of-the-art performance on other datasets. Despite its success, InceptionV3 has some drawbacks. It requires a large amount of computational resources and memory, making it challenging to use on resource-constrained devices. Additionally, the large number of parameters in the model can make it prone to overfitting on small datasets (Ahmad et al., 2020).

2.2.2.5 EfficientNet

EfficientNet is a CNN model trained on the ImageNet dataset and created by Google in 2019 (Rahhal et al., 2022). It was intended to be more efficient in terms of the number of parameters and computations needed to attain a certain level of performance. One of the main innovations of EfficientNet is the use of a compound scaling method, which scales the model's width, depth, and resolution in a systematic and consistent way. This allows the model to perform better on image classification tasks with fewer parameters and computation. Another important feature of EfficientNet is the use of a specialized convolutional layer called a mobile inverted residual bottleneck (MBConv) layer, designed to be more efficient in computation and memory usage.

The MBConv layer uses depth-wise convolutions, which reduces the number of parameters and computation needed while still enabling the model to identify useful features from the input image. EfficientNet has been widely used in industry and research, and it is considered one of the best models for image classification tasks in terms of efficiency. The model has been demonstrated to attain state-of-the-art performance on the ImageNet dataset and other datasets, with a significantly lower number of parameters and computation than other models. Despite its success, EfficientNet has some drawbacks. The model's architecture and scaling method can be difficult to understand and replicate, making it challenging for researchers and practitioners. Additionally, it may only be suitable for some tasks where accuracy is more important than efficiency (Rahhal et al., 2022).

2.2.3 Applications of convolutional neural networks

CNN has various applications in various technologies in our daily lives. Some of the most frequent uses include;

- Facial recognition: CNN allows computers to identify individuals by recognizing their faces in pictures, regardless of lighting or stance. It can detect unique features and match them with a name. This advancement empowers users of devices like smartphones or social media platforms such as Facebook to swiftly recognize the individuals depicted in photos.
- Document analysis: CNN is used to analyze and interpret handwriting by comparing it to a database of handwriting samples. They can read and understand written documents that are crucial in the banking and finance industry, as well as classify documents for museum archives.

- Genetics: Medical professionals can learn about new therapies and alternatives using CNNs that can analyze images of cells.
- Satellite images: Satellite images may be classified using convolutional layers and networks, saving humans time by classifying and identifying them automatically. Humans can't do this as quickly as computers can.
- Healthcare: In radiology, computer vision has been integrated to help clinicians detect cancerous tumors in healthy tissue more accurately.
- Retail: Visual search technology has been adopted by certain e-commerce websites, enabling brands to suggest complementary items to a customer's current wardrobe.
- Marketing: Social media platforms now simplify identifying and tagging friends in shared photo albums by suggesting who might be present in a photo on a user's profile.
- Automotive: Although fully autonomous cars are yet to be fully realized, the technology behind them is beginning to be integrated into vehicles, increasing the safety of drivers and passengers through features such as lane line detection and adaptive cruise control.

2.3 Common terminologies in deep learning

2.3.1 Neural network

A neural network is a machine-learning model that imitates the organization and operation of the human brain (Mustapha et al., 2022). It comprises layers of interconnected nodes, known as artificial neurons, built to process and analyze vast data. The artificial neuron is the basic building block of a neural network, which receives inputs, performs a calculation on them, and produces an output. The calculation performed by the neuron is typically a linear combination of the inputs followed by a nonlinear activation function. Neural networks can be utilized for various tasks, such as image recognition, NLP, and prediction (Mustapha et al., 2022; Stateczny et al., 2022; Uzun Ozsahin et al., 2022, 2023). They are particularly effective for handling large amounts of data and complex nonlinear relationships (Stateczny et al., 2022). Neural networks are highly adaptable and can be tailored to different types of data and tasks. They can be trained to improve their performance over time.

2.3.2 Recurrent neural network

The RNN is a neural network architecture tailored to process sequential data, such as time series, text, or speech (Yu et al., 2022). Unlike traditional feedforward neural networks, RNNs have a feedback connection that allows them to maintain a hidden state, which captures information about the previous inputs. This hidden state is used to inform the current input, allowing the network to learn the dependencies between the inputs over time. This makes RNNs particularly useful for tasks that involve sequential data, such as language translation, speech recognition, and time series forecasting (Sharma & Jain, 2021; Shewalkar et al., 2019; Yu et al., 2022). There are different RNNs, such as Long Short-Term Memory (LSTM), basic RNN, and Gated Recurrent Unit (GRU). These variations have slightly different architectures, but they all can maintain a hidden state to capture temporal dependencies. LSTM and GRU are more powerful than basic RNNs, can handle longer sequences, and are less prone to the vanishing gradient problem (Su & Kuo, 2019). RNNs can be used in unidirectional or bidirectional mode, meaning that they can process the input sequence in a forward or backward direction (Schuster & Paliwal, 1997). Bidirectional RNNs employ two distinct hidden states, one for the forward pass and one for the backward pass, enabling them to consider both past and future context.

2.3.3 Generative adversarial network

A generative adversarial network (GAN) is a deep learning model that can produce new data similar to the data it is trained on. It comprises many elements: a generator and a discriminator (Brownlee, 2019a). The generator creates new data, while the discriminator determines whether the generated data is real or fake. GANs are composed of two neural networks trained simultaneously in a competitive manner. The generator network attempts to generate data similar to the original data, and the discriminator tries to differentiate the generated data from the real data. The generator network improves its ability to replicate the real data in attempting to trick the discriminator network. In contrast, the discriminator network improves its ability to identify the generated data by trying to correctly classify the generator's output as fake. The two components are trained iteratively, with the generator improving at producing realistic data and the discriminator becoming better at identifying fake data. GANs have been used in many applications, such as image synthesis, text-to-speech, and video generation. GANs are also used in unsupervised learning, where they can be used to learn the underlying structure of the data (Wali et al., 2022).

2.3.4 Back-propagation

Back-propagation is an algorithm used for training artificial neural networks, specifically for supervised learning. It is a technique used to adjust the neural network weights by calculating the gradient of the loss function with respect to the weights. This is done by forwarding the input data through the network to produce an output and then comparing it to the actual output to calculate the error or loss. The error is then propagated backward through the network, starting from the output layer to adjust the weights. The chain rule of calculus is used to calculate the gradient of the loss function with respect to the weights, which is then used to update the weights in the direction of the negative gradient to minimize the loss function.

This is done by breaking down the loss function into a series of simpler functions, each of which depends on a single weight, and calculating the derivative of each of these simpler functions with respect to the weights. Back-propagation is a powerful algorithm widely used in neural network training, and it is the backbone of most deep learning techniques (Buscema, 1998). Back-propagation is an iterative method of training a neural network. It can be used with various optimization techniques, such as gradient descent and its variants, to minimize the loss function and improve the network's overall performance (Alake, 2022).

2.3.5 Gradient descent

Gradient descent is a widely used optimization algorithm in machine learning, especially in training neural networks. It is used to find the optimal values for a model's parameters (weights) that minimize a cost function (Yamashita et al., 2018). The algorithm starts with an initial set of parameter values. It iteratively updates them in the direction of the negative gradient of the cost function, with the goal of finding the global minimum of the function (Dabbura, 2022). The gradient descent algorithm is an iterative method for minimizing a cost function (Arefin & Asadujjaman, 2016). It first works by computing the cost function's gradient with respect to the parameters. The gradient points in the direction of the steepest descent, and the algorithm updates the parameters in the opposite direction with a step size determined by the learning rate. Different variations of gradient descent include-momentum-based gradient descent, stochastic gradient descent (SGD), and mini-batch gradient descent (Du, 2019). These variations differ in how they are calculated and the update is performed on the parameters. Still, they all share the same principle of using the gradient to update the parameters.

2.3.6 Activation function

An activation function is a nonlinear mathematical function that is applied to the input of a neuron in an artificial neural network. It is used to introduce into the network, allowing it to learn complex relationships between inputs and outputs (Kiliçarslan et al., 2021). Without an activation function, a neural network would be nothing more than a linear model and unable to learn nonlinear relationships. Activation functions are applied element-wise to a neuron's input, meaning that each input value is transformed by the function separately. Commonly used activation functions include sigmoid, ReLU, tanh (hyperbolic tangent), and Leaky ReLU (Ratnawati et al., 2020). The sigmoid function is a smooth, S-shaped function that maps any input value to a value between 0 and 1. In the hidden layers of a neural network, the ReLU function is commonly used. It maps all negative input values to 0 and all positive input values to the same value. The tanh function is similar to the sigmoid function but maps input values to a range between -1 and 1. The Leaky ReLU is a variant of the ReLU function; it allows a small, nonzero gradient for negative input values to avoid the vanishing gradient problem (Bai, 2022).

2.3.7 Overfitting

Overfitting is a common problem in machine learning, particularly deep learning, where a model becomes too complex and performs well on the training data but poorly on unseen data (Ying, 2019). It occurs when a model is trained too well on the training data and memorizes the noise and random variations in the training data instead of learning the underlying patterns. Overfitting can be caused by various factors, such as having too many parameters in a model, having too little training data, or using a too complex model for the given problem (Montesinos López et al., 2022; Ying, 2019). When a model overfits, it will perform well on the training data but poorly on unseen data, such as validation or test data. This is because the model has learned the noise and random variations in the training data, which do not generalize to new data. Several techniques can be used to prevent overfitting, such as regularization, early stopping, and cross-validation (Ying, 2019). Regularization is a technique that adds a penalty term to the loss function, reducing the model's complexity. Early stopping is a technique that stops training a model before it starts overfitting. Cross-validation is a technique that divides the data into training, validation, and test sets; it is used to evaluate the model performance on unseen data (Refaeilzadeh et al., 2009).

2.3.8 Batch normalization

Batch normalization is a method used to improve the performance and stability of neural networks by normalizing the inputs of each layer (Shen et al., 2021). It normalizes the activations of the neurons in each layer by scaling and shifting them. Batch normalization is a technique used to minimize the internal covariate shift, which is the change in the distribution of the inputs to each layer as the model parameters are updated during training (Brownlee, 2019b). It is done by normalizing the activations of each layer by calculating the mean and standard deviation and then subtracting the mean and dividing by the standard deviation. The values obtained are then used to shift and scale the activations to have zero mean and unit variance. This normalization process is performed for each minibatch during training. Batch normalization has several benefits, such as improving the stability of the model, reducing the number of epochs required to train the model, and reducing the model's sensitivity to the initial values of the parameters (Bilal et al., 2022). It also allows the model to use higher learning rates, which makes the training process faster. It also helps to prevent overfitting by adding some form of regularization to the model (Oppermann, 2020).

2.3.9 Transfer learning

Transfer learning is a method in machine learning that enables the use of a pretrained model for a new task rather than training a model from scratch. It is based on the idea that knowledge learned by a model while solving one task can be used to improve the performance of a model on a different but related task. Transfer learning is useful in situations with limited labeled data for a specific task, but a large amount of labeled data is available for a related task. In this case, a model trained on the related task can be used as a starting point for the new task and fine-tuned to adapt to the new task using the limited labeled data. There are two main types of transfer learning: feature-based and fine-tuning. Feature-based transfer learning is the process of using the features learned by a model on one task as input to a new model on a different task. Fine-tuning, on the other hand, is the process of retraining some or all of the layers of a pretrained model on a new task.

2.3.10 Autoencoder

An autoencoder is a neural network designed for unsupervised learning (Pattanayak, 2023). It comprises two main components: an encoder and a decoder (Addo et al., 2022). The encoder compresses the input data into a lower-dimensional representation known as the latent or bottleneck representation. The decoder then reconstructs the input data from the compressed representation. The main objective of an autoencoder is to learn a compressed representation of the input data that captures the most significant features (Mehrotra & Musolesi, 2018). This is achieved by training the network to reconstruct the input data from the compressed representation. The network is trained by minimizing the reconstruction error between the input data and the output of the decoder. Autoencoders can be used for dimensionality reduction, data compression, anomaly detection, and generative modeling (Sakurada & Yairi, 2014). They can also be used to learn a compact representation of input data and use it as a feature for other tasks (Hsieh et al., 2021). Finally, autoencoders can also be used for anomaly detection by training the network on a normal dataset and then using it to detect anomalies in the test data (Eswaran & Faloutsos, 2018).

2.3.11 Restricted Boltzmann machine

A restricted Boltzmann machine (RBM) is a generative stochastic artificial neural network capable of learning the input data's probability distribution. It is composed of two types of units, visible units that represent the input data and hidden units that represent the underlying structure of the data (Dabelow & Ueda, 2022). The visible and hidden units can be binary or continuous, and their connections are not directed (Tomczak & Gonczarek, 2016). RBMs are trained using a variant of the contrastive divergence algorithm, an unsupervised learning algorithm. During the training process, the RBM learns the probability distribution of the input data by adjusting the weights between the visible and hidden units. Once the RBM is trained, it can generate new samples similar to the input data by sampling from the learned probability distribution. RBMs have been used in many applications, such as image and speech recognition, NLP, collaborative filtering, and dimensionality reduction. They can also be used as building blocks for more complex models such as deep belief networks and GAN (Peng et al., 2018).

2.3.12 Convolutional layer

A convolutional layer is a fundamental building block in CNN used to extract local features from the input data (Yamashita et al., 2018). It applies a convolution operation to the input data, which involves a set of learnable filters known as kernels or weights. These filters are used to scan over the input data, capturing local patterns or features. Filters often come in small sizes, usually 3×3 or 5×5, and slide over the input data with a fixed step size called stride (Deshpande, 2019). The output of the convolution operation is a multidimensional array, also known as the feature map, which contains the filtered values of the input data. The feature map is then passed through a nonlinear activation function, such as ReLU or sigmoid, to introduce nonlinearity into the network. Convolutional layers can be stacked together, forming a deep network, which allows the network to learn more complex features from the input data. The deeper the network, the more abstract the features it can learn (Chen et al., 2016). The number of filters in each convolutional layer can be increased, which allows the network to capture more variations of the same feature.

2.3.13 Pooling layer

A pooling layer is a component of a CNN that is used to reduce the spatial dimensions of the feature maps while preserving important information (Zafar et al., 2022). Pooling layers are typically applied after one or more convolutional layers. They help reduce the amount of computation required by the network and make it more robust to small translations of the input data (Brownlee, 2019a). There are two main types of pooling: max pooling and average pooling. Max pooling selects the maximum value from each pooling window, while average pooling computes the average value (Basavarajaiah, 2019). Max pooling is more commonly used as it helps retain the input data's most important features (Zafar et al., 2022). The pooling operation is performed by applying a pooling window of a fixed size, usually 2×2 or 3×3, to the feature map and taking the maximum or average value within the window. The pooling window is moved across the feature map with a fixed stride, and the output of the pooling operation is a new feature map with reduced dimensions.

2.3.14 Fully connected layer

A fully connected layer, also known as a dense layer, is a layer in a neural network where all the neurons in one layer are connected to all the neurons in the next layer. This layer is typically used as the final layer of a neural network, and its primary purpose is to produce the network's final output. The weights of these connections are learned during the training process. Each neuron in the fully connected layer calculates a linear combination of the inputs received from all the neurons in the previous layer using a set of learned weights. The result is then passed through a nonlinear activation function such as ReLU or sigmoid to introduce nonlinearity into the network. This layer can learn a nonlinear mapping between the input and output, enabling it to perform more complex computations. It can also be used to reduce the dimensionality of the data and extract a more compact representation of the input (Yamashita et al., 2018).

2.3.15 Embedding layer

An embedding layer is a neural network layer that represents discrete data such as words, characters, or categorical variables as continuous dense vectors in a high-dimensional space (Liang et al., 2019). These embeddings are learned during the training process and can capture the underlying semantic or syntactic relationships between the discrete data. Embedding layers are commonly used in NLP tasks such as language translation, sentiment analysis, and text generation (Li & Gong, 2021). In these tasks, the embedding layer converts words in a text into dense vectors that the rest of the network can process. The embeddings are learned such that semantically, similar words have similar representations in the high-dimensional space. Embedding layers can also be used in other domains, such as computer vision and speech recognition. They can also be used to convert images or audio data into dense vectors that can be processed by the network (Aldarmaki et al., 2022).

2.3.16 Cross-entropy

Cross-entropy is a commonly used loss function in neural network training, particularly in multiclass classification problems (Demirkaya et al., 2020). The function measures the dissimilarity or difference between the predicted probability distribution and the true distribution of the target classes. It trains the network to produce more accurate predictions by minimizing the difference between the predicted and true distributions (Ribeiro & Moniz, 2020). The cross-entropy loss is calculated by taking the negative logarithm of the predicted probability of the true class. It is a measure of how well the network's predictions are. The lower the loss, the better the predictions are (Brownlee, 2020). The loss is calculated for each sample, and the network's parameters are updated to minimize the overall loss across all samples.

2.3.17 Optimizer

An optimizer is an algorithm applied to modify the parameters of a neural network to minimize the loss function (Nwankpa, 2020). It is a key component in the training process of a neural network. The objective of the optimizer is to identify the values of the network's parameters that minimize the loss function, which enables the network's predictions to be more precise. Several types of optimizers are available such as SGD, Adam, Adagrad, and Adadelta (Zaheer & Shaziya, 2019). The selection of an optimizer depends on the problem at hand and the nature of the data; each optimizer has its pros and cons. For example, SGD is a simple and widely used optimizer that updates the network's parameters by taking the gradient of the loss function with respect to the parameters. Adam optimizer, on the other hand, is a more advanced optimizer that adapts the learning rate for each parameter based on historical gradient information (Ruder, 2020).

2.3.18 Back-propagation through time

Back-propagation through time (BPTT) trains RNNs that handle sequential data. RNNs can maintain an internal state and handle input sequences of variable lengths, unlike feed-forward networks (Zhang et al., 2021). However, the traditional back-propagation algorithm cannot be applied to RNNs directly because of the temporal dependency of the hidden states. BPTT is an extension of back-propagation used to compute the gradients for the network weights. The basic idea of BPTT is to unroll the recurrent connections in time and treat the network as a feed-forward network with additional connections between layers (Bellec et al., 2020). The network's weights are updated by computing the gradients of the loss function with respect to the weights at each time step. The gradients are then back-propagated through the unrolled network, and the weights are updated using an optimizer. BPTT is a computationally intensive process and can be challenging to implement in practice due to the large number of time steps (Sutskever & Hinton, 2010). Other methods have been proposed to address this issue, such as truncated back-propagation through time and real-time recurrent learning (Tang & Glass, 2018).

2.3.19 Batch size

In deep learning, the batch size is a hyperparameter used to regulate the number of samples employed in one forward/backward pass. It determines the number of samples processed by the network before updating the weights. Typically, the larger the batch size, the more computationally efficient the training process is, but it also requires more memory (Yao et al., 2019). The batch size can significantly affect the performance of the model. A small batch size can result in a noisy and unstable model, while a large batch size can cause the model to converge more slowly (Hoffer et al., 2020). The optimal batch size depends on the specific problem and the characteristics of the data, and it is often determined through experimentation. It is worth mentioning that there is also another approach called SGD in which the batch size is set to 1. It is less computationally efficient but can be useful in scenarios where the data set is too big to fit into memory.

2.3.20 Epoch

An epoch is a complete iteration of a neural network's entire training dataset. In other words, it is the number of times the network will see the whole training set. The number of epochs is a hyperparameter that controls the times the training process will iterate over the training dataset. The number of epochs is an important hyperparameter to tune as it affects the model's performance. The number of epochs during training is an important hyperparameter that determines how often the model will see the training data. If the number of epochs is too low, the model may not have enough time to converge to an optimal solution, but if it is too high, the model may start overfitting the training data and perform poorly on unseen data. To avoid this, the model's performance is monitored on a validation set during the training process, and the training process is stopped when the performance on the validation set stops improving or starts to degrade. This approach is called early stopping (Brownlee, 2022).

2.3.21 Vanishing gradient

The vanishing gradient problem is a common issue that can occur in deep neural networks during the training process. The problem occurs when the gradients of the weights, which are used to update the network's parameters during training, become very small. This can happen when the network has many layers and the gradients are passed through multiple layers during back-propagation. When the gradients are very small, the weights in the network are updated very slowly or not at all. This can cause the training process to converge very slowly or not at all. This problem is particularly pronounced in deep architectures and affects the training of the lower layers the most. Several techniques can be used to address the vanishing gradient problem. One popular technique is to use activation functions more robust to the vanishing gradient problem, such as ReLU (Hu et al., 2021). Additionally, batch normalization techniques can mitigate the problem (Awais et al., 2021).

2.3.22 Strides

The stride is a hyperparameter that determines the step size of the convolution operation in a CNN. It controls how the kernel (filter) moves over the input image and can affect the spatial dimensions of the output and the number of parameters in the network. A stride of 1 means that the kernel moves one pixel at a time, while a stride of 2 means that the kernel moves two pixels simultaneously. Larger strides can reduce the dimensions of the output but also may lead to loss of information. The stride can also be applied differently in the height and width dimension; this is referred to as a different stride value for the height and width dimension. A stride of (1,1) in the height and width dimension would mean that the kernel moves one pixel at a time in the height and width dimension (Brownlee, 2019c).

2.3.23 Padding

Padding is a technique used in CNN to control the spatial dimensions of the output from the convolutional layers. The padding operation is applied to the input image before the convolution operation. It consists of adding extra pixels around the edges of the image, which are set to a specific value (usually zero). The purpose of padding in CNN is to maintain the spatial dimensions of the input image, avoiding the reduction of the spatial dimensions of the output image. Padding prevents a decrease in spatial dimensions, which can result in a loss of information if it is not used. Padding can be applied symmetrically around the image by adding the same number of pixels on each side. This approach is called "valid" padding. Alternatively, padding can be applied asymmetrically by adding more pixels on one side of the image than the other. This approach is called "same" padding (Yamashita et al., 2018).

2.3.24 Hyperparameter

A CNN's hyperparameters are values set before the training process and cannot be learned from the data. They control the architecture and behavior of the network, and their values can significantly impact the network performance (Radhakrishnan, 2017). Examples of hyperparameters in a CNN include the number of layers, the number of filters, the size of the filters, the dropout rate, and the learning rate (Ang et al., 2022). The number of filters in a CNN is a hyperparameter that controls the number of features extracted from the image. Increasing the number of filters will increase the number of extracted features and the computational complexity of the network. On the other hand, decreasing the number of filters will reduce the number of extracted features, but it will also make the network less complex (Brownlee, 2019d).

The size of the filters is another hyperparameter that can be adjusted (Hinz et al., 2018). Smaller filters, such as 3×3 or 5×5, can extract fine-grained features, while larger filters, such as 7×7 or 11×11, can extract more coarse-grained features (Hinz et al., 2018). The number of layers in a CNN is also a hyperparameter that can be adjusted (Neary, 2018). More layers can extract more complex features, but they also increase the computational complexity of the network. The learning rate is another hyperparameter that controls the speed of the training process (Nabi, 2019). A high learning rate will cause the network to converge quickly, but it may not find the optimal solution (Park et al., 2020). A low learning rate will cause the network to converge slowly, but it will find a more accurate solution (Park et al., 2020). The dropout rate is another hyperparameter used to prevent overfitting by randomly dropping out a certain percentage of neurons during the training process (Sumera et al., 2022). Common dropout rates are between 0.2 and 0.5, meaning that 20%–50% of the neurons are dropped out during each iteration (Baldi & Sadowski, 2014).

2.3.25 Filters

A filter in a CNN is a set of weights used to extract features from an image. The filter is applied to the image using convolution (Yamashita et al., 2018). During convolution, the filter slides over the image, element-wise multiplying the values of the image at each position with the values of the filter and summing the results. The output of this process is called a feature map. Filters extract different features from an image, such as edges, textures, and patterns (Hung et al., 2019). For example, a filter sensitive to horizontal edges will highlight horizontal lines in the image, while a filter sensitive to vertical edges will highlight vertical lines. By applying multiple filters to an image, a CNN can extract a wide range of features, which can be used to classify the image.

2.3.26 Dropout

Dropout is a regularization technique commonly used in CNN to prevent overfitting (Jabir & Falih, 2021). Dropout helps to combat overfitting by randomly dropping out (or setting to zero) a certain number of neurons during the training process. This forces the network to learn multiple independent data representations and reduces the dependence on any specific feature, making the model more robust. Dropout is applied to the hidden layers of a CNN during the training process, with a common dropout rate of between 0.2 and 0.5. The dropout rate is set to zero during the testing phase, allowing the model to use all the information learned during the training process to make predictions. Dropout is a simple yet powerful technique that has been shown to improve the performance of CNNs on various tasks, such as image classification, object detection, and NLP. It is also computationally efficient and can be used in conjunction with other regularization techniques to further improve the CNN performance (Sanjar et al., 2020).

References

Adamson, M. J., & Damper, R. I. (1996). A recurrent network that learns to pronounce English text. In *Proceedings of the fourth international conference on spoken language processing (ICSLP 1996)*. Available from https://doi.org/10.21437/icslp.1996-433.

Addo, D., Zhou, S., Jackson, J. K., Nneji, G. U., Monday, H. N., Sarpong, K., Patamia, R. A., Ekong, F., & Owusu-Agyei, C. A. (2022). EVAE-net: An ensemble variational autoencoder deep learning network for COVID-19 classification based on chest X-ray images. *Diagnostics*, *12*(11), 2569. Available from https://doi.org/10.3390/diagnostics12112569.

Ahmad, J., Jan, B., Farman, H., Ahmad, W., & Ullah, A. (2020). Disease detection in plum using convolutional neural network under true field conditions. *Sensors*, *20*(19), 5569. Available from https://doi.org/10.3390/s20195569.

Ahmed, M., Afreen, N., Ahmed, M., Sameer, M., & Ahamed, J. (2023). An inception V3 approach for malware classification using machine learning and transfer learning. *International Journal of Intelligent Networks*, *4*, 11–18. Available from https://doi.org/10.1016/j.ijin.2022.11.005.

Alake, R. (2022, March 17). An introduction to gradient descent and backpropagation in machine learning algorithms. Medium. Retrieved January 31, 2023. Available from https://towardsdatascience.com/an-introduction-to-gradient-descent-and-backpropagation-in-machine-learning-algorithms-a14727be70e9.

Aldarmaki, H., Ullah, A., Ram, S., & Zaki, N. (2022). Unsupervised automatic speech recognition: A Review. *Speech Communication*, *139*, 76–91. Available from https://doi.org/10.1016/j.specom.2022.02.005.

Andreas, Purnomo, M. H., & Hariadi, M. (2015). Controlling the hidden layers' output to optimizing the training process in the deep neural network algorithm. In *Proceedings of the IEEE international conference on cyber technology in automation, control, and intelligent systems (CYBER)*. Available from https://doi.org/10.1109/cyber.2015.7288086.

Ang, K. M., El-kenawy, E.-S. M., Abdelhamid, A. A., Ibrahim, A., Alharbi, A. H., Khafaga, D. S., Tiang, S. S., & Lim, W. H. (2022). Optimal design of convolutional neural network architectures using teaching–learning-based optimization for image classification. *Symmetry*, *14*(11), 2323. Available from https://doi.org/10.3390/sym14112323.

Arefin, M. R., & Asadujjaman, M. (2016). Minimizing average of loss functions using gradient descent and stochastic gradient descent. *Dhaka University Journal of Science*, *64*(2), 141–145. Available from https://doi.org/10.3329/dujs.v64i2.54490.

Awais, M., Bin Iqbal, M. T., & Bae, S.-H. (2021). Revisiting internal covariate shift for batch normalization. *IEEE Transactions on Neural Networks and Learning Systems*, *32* (11), 5082–5092. Available from https://doi.org/10.1109/tnnls.2020.3026784.

Bai, Y. (2022). Relu-function and derived function review. *SHS Web of Conferences*, *144*, 02006. Available from https://doi.org/10.1051/shsconf/202214402006.

Baldi, P., & Sadowski, P. (2014). The dropout learning algorithm. *Artificial Intelligence*, *210*, 78–122. Available from https://doi.org/10.1016/j.artint.2014.02.004.

Basavarajaiah, M. (2019, August 22). Which pooling method is better? Max-pooling vs Minpooling vs average pooling. Medium. Retrieved January 31, 2023. Available from https://medium.com/@bdhuma/which-pooling-method-is-better-maxpooling-vs-minpooling-vs-average-pooling-95fb03f45a9.

Bellec, G., Scherr, F., Subramoney, A., Hajek, E., Salaj, D., Legenstein, R., & Maass, W. (2020). A solution to the learning dilemma for recurrent networks of spiking neurons. *Nature Communications*, *11*(1). Available from https://doi.org/10.1038/s41467-020-17236-y.

Bhatt, D., Patel, C., Talsania, H., Patel, J., Vaghela, R., Pandya, S., Modi, K., & Ghayvat, H. (2021). CNN variants for computer vision: History, architecture, application, challenges and future scope. *Electronics*, *10*(20), 2470. Available from https://doi.org/10.3390/electronics10202470.

Bilal, M. A., Ji, Y., Wang, Y., Akhter, M. P., & Yaqub, M. (2022). Early earthquake detection using batch normalization graph Convolutional Neural Network (BNGCNN. *Applied Sciences*, *12*(15), 7548. Available from https://doi.org/10.3390/app12157548.

Brownlee, J. (2019a, July 19). A gentle introduction to generative adversarial networks (Gans). MachineLearningMastery.com. Retrieved January 31, 2023. Available from https://machinelearningmastery.com/what-are-generative-adversarial-networks-gans/.

Brownlee, J. (2019b, December 3). A gentle introduction to batch normalization for Deep Neural Networks. MachineLearningMastery.com. Retrieved January 31, 2023. Available from https://machinelearningmastery.com/batch-normalization-for-training-of-deep-neural-networks/.

Brownlee, J. (2019c, August 15). A gentle introduction to padding and stride for Convolutional Neural Networks. MachineLearningMastery.com. Retrieved January 31, 2023. Available from https://machinelearningmastery.com/padding-and-stride-for-convolutional-neural-networks/.

Brownlee, J. (2019d, July 5). A gentle introduction to 1x1 convolutions to manage model complexity. MachineLearningMastery.com. Retrieved January 31, 2023. Available from https://machinelearningmastery.com/introduction-to-1x1-convolutions-to-reduce-the-complexity-of-convolutional-neural-networks/.

Brownlee, J. (2020, December 22). A gentle introduction to cross-entropy for Machine Learning. MachineLearningMastery.com. Retrieved January 31, 2023. Available from https://machinelearningmastery.com/cross-entropy-for-machine-learning/.

Brownlee, J. (2022, August 15). Difference between a batch and an epoch in a neural network. MachineLearningMastery.com. Retrieved January 31, 2023. Available from https://machinelearningmastery.com/difference-between-a-batch-and-an-epoch/.

Buscema, M. (1998). Back propagation neural networks. *Substance Use & Misuse*, *33*(2), 233–270. Available from https://doi.org/10.3109/10826089809115863.

Cao, J., Yan, M., Jia, Y., Tian, X., & Zhang, Z. (2021). Application of a modified inception-V3 model in the dynasty-based classification of ancient murals. *EURASIP Journal on Advances in Signal Processing*, *2021*(1). Available from https://doi.org/10.1186/s13634-021-00740-8.

Chen, Y., Jiang, H., Li, C., Jia, X., & Ghamisi, P. (2016). Deep feature extraction and classification of hyperspectral images based on convolutional neural networks. *IEEE Transactions on Geoscience and Remote Sensing*, *54*(10), 6232–6251. Available from https://doi.org/10.1109/tgrs.2016.2584107.

Dabbura, I. (2022, September 27). Gradient descent algorithm and its variants. Medium. Retrieved January 31, 2023. Available from https://towardsdatascience.com/gradient-descent-algorithm-and-its-variants-10f652806a3.

Dabelow, L., & Ueda, M. (2022). Three learning stages and accuracy–efficiency tradeoff of restricted boltzmann machines. *Nature Communications*, *13*(1). Available from https://doi.org/10.1038/s41467-022-33126-x.

Demirkaya, A., Chen, J., & Oymak, S. (2020). Exploring the role of loss functions in multiclass classification. In *Proceedings of the 54th annual conference on information sciences and systems (CISS)*. Available from https://doi.org/10.1109/ciss48834.2020.1570627167.

Deng, J., Dong, W., Socher, R., Li, L.-J., Kai, Li, & Li, F.-F. (2009). ImageNet: A large-scale hierarchical image database. *2009 IEEE Conference on Computer Vision and Pattern Recognition*. Available from https://doi.org/10.1109/cvpr.2009.5206848.

Deshpande, A. (2019). A beginner's guide to understanding convolutional neural networks part 2. A beginner's guide to understanding convolutional neural networks part 2—Adit Deshpande—Engineering at forward | UCLA CS '19. Retrieved January 31, 2023. Available from https://adeshpande3.github.io/A-Beginner's-Guide-To-Understanding-Convolutional-Neural-Networks-Part-2/.

Dettmers, T. (2022, October 10). Deep learning in a nutshell: History and training. NVIDIA Technical Blog. Retrieved January 31, 2023. Available from https://developer.nvidia.com/blog/deep-learning-nutshell-history-training/.

Ding, Y., Fang, W., Liu, M., Wang, M., Cheng, Y., & Xiong, N. (2023). JMDC: A joint model and data compression system for deep neural networks collaborative computing in edge-cloud networks. *Journal of Parallel and Distributed Computing*, *173*, 83–93. Available from https://doi.org/10.1016/j.jpdc.2022.11.008.

Du, J. (2019). The frontier of SGD and its variants in machine learning. *Journal of Physics: Conference Series*, *1229*(1), 012046. Available from https://doi.org/10.1088/1742-6596/1229/1/012046.

Eliseeva, D. Y., Fedosov, A. Y., Agaltsova, D. V., Mnatsakanyan, O. L., & Kuchmezov, K. K. (2020). The evolution of Artificial Intelligence and the possibility of its application in Cyber Games. *Revista Amazonia Investiga*, *9*(28), 123–129. Available from https://doi.org/10.34069/ai/2020.28.04.15.

Eswaran, D., & Faloutsos, C. (2018). Sedanspot: Detecting anomalies in edge streams. In *Proceedings of the IEEE international conference on data mining (ICDM)*. Available from https://doi.org/10.1109/icdm.2018.00117.

Goldberg, Y. (2016). A Primer on neural network models for Natural Language Processing. *Journal of Artificial Intelligence Research*, *57*, 345–420. Available from https://doi.org/10.1613/jair.4992.

Gou, X., Qing, L., Wang, Y., Xin, M., & Wang, X. (2020). Re-training and parameter sharing with the hash trick for compressing convolutional neural networks. *Applied Soft Computing*, *97*, 106783. Available from https://doi.org/10.1016/j.asoc.2020.106783.

Guan, Q., Wang, Y., Ping, B., Li, D., Du, J., Qin, Y., Lu, H., Wan, X., & Xiang, J. (2019). Deep convolutional neural network VGG-16 model for differential diagnosing of papillary thyroid carcinomas in cytological images: A pilot study. *Journal of Cancer*, *10*(20), 4876–4882. Available from https://doi.org/10.7150/jca.28769.

Hinz, T., Navarro-Guerrero, N., Magg, S., & Wermter, S. (2018). Speeding up the hyperparameter optimization of deep convolutional neural networks. *International Journal of Computational Intelligence and Applications*, *17*(2), 1850008. Available from https://doi.org/10.1142/s1469026818500086.

Hoeser, T., & Kuenzer, C. (2020). Object detection and image segmentation with deep learning on earth observation data: A review-part I: Evolution and recent trends. *Remote Sensing*, *12*(10), 1667. Available from https://doi.org/10.3390/rs12101667.

Hoffer, E., Ben-Nun, T., Hubara, I., Giladi, N., Hoefler, T., & Soudry, D. (2020). Augment your batch: Improving generalization through instance repetition. In *Proceedings of the IEEE/CVF conference on computer vision and pattern recognition (CVPR)*. Available from https://doi.org/10.1109/cvpr42600.2020.00815.

Hsieh, T.-Y., Sun, Y., Wang, S., & Honavar, V. (2021). Functional autoencoders for functional data representation learning. In *Proceedings of the SIAM international conference on data mining (SDM)*, 666–674. Available from https://doi.org/10.1137/1.9781611976700.75.

Hu, Z., Zhang, J., & Ge, Y. (2021). Handling vanishing gradient problem using artificial derivative. *IEEE Access*, *9*, 22371–22377. Available from https://doi.org/10.1109/access.2021.3054915.

Hung, C.-C., Song, E., & Lan, Y. (2019). Image texture, texture features, and image texture classification and segmentation. *Image Texture Analysis*, 3–14. Available from https://doi.org/10.1007/978-3-030-13773-1_1.

Jabir, B., & Falih, N. (2021). Dropout, a basic and effective regularization method for a deep learning model: A case study. *Indonesian Journal of Electrical Engineering and Computer Science*, *24*(2), 1009. Available from https://doi.org/10.11591/ijeecs.v24.i2.pp1009-1016.

Jiang, Z.-P., Liu, Y.-Y., Shao, Z.-E., & Huang, K.-W. (2021). An improved VGG16 model for pneumonia image classification. *Applied Sciences*, *11*(23), 11185. Available from https://doi.org/10.3390/app112311185.

Kiliçarslan, S., Adem, K., & Çelik, M. (2021). An overview of the activation functions used in deep learning algorithms. *Journal of New Results in Science*, *10*(3), 75–88. Available from https://doi.org/10.54187/jnrs.1011739.

Krizhevsky, A., Sutskever, I., & Hinton, G. E. (2017). ImageNet classification with deep convolutional neural networks. *Communications of the ACM*, *60*(6), 84–90. Available from https://doi.org/10.1145/3065386.

LeCun, Y., Boser, B., Denker, J. S., Henderson, D., Howard, R. E., Hubbard, W., & Jackel, L. D. (1989). Backpropagation applied to handwritten zip code recognition. *Neural Computation*, *1*(4), 541–551. Available from https://doi.org/10.1162/neco.1989.1.4.541.

Lee, H.-W., Kim, N.-ri, & Lee, J.-H. (2017). Deep neural network self-training based on unsupervised learning and dropout. *The International Journal of Fuzzy Logic and Intelligent Systems*, *17*(1), 1–9. Available from https://doi.org/10.5391/ijfis.2017.17.1.1.

Lettvin, J., Maturana, H., McCulloch, W., & Pitts, W. (1959). What the frog's eye tells the frog's brain. *Proceedings of the IRE*, *47*(11), 1940–1951. Available from https://doi.org/10.1109/jrproc.1959.287207.

Li, S., & Gong, B. (2021). Word embedding and text classification based on deep learning methods. *MATEC Web of Conferences*, *336*, 06022. Available from https://doi.org/10.1051/matecconf/202133606022.

Liang, X., Min, M. R., Guo, H., & Wang, G. (2019). Learning K-way d-dimensional discrete embedding for hierarchical data visualization and retrieval. In *Proceedings of the twenty-eighth international joint conference on artificial intelligence*. Available from https://doi.org/10.24963/ijcai.2019/411.

Markoff, J. (2012, June 25). How many computers to identify a cat? 16,000. The New York Times. Retrieved January 31, 2023. Available from https://www.nytimes.com/2012/06/26/technology/in-a-big-network-of-computers-evidence-of-machine-learning.html.

Marquez, E. S., Hare, J. S., & Niranjan, M. (2018). Deep cascade learning. *IEEE Transactions on Neural Networks and Learning Systems*, *29*(11), 5475–5485. Available from https://doi.org/10.1109/tnnls.2018.2805098.

Mehrotra, A., & Musolesi, M. (2018). Using autoencoders to automatically extract mobility features for predicting depressive states. *Proceedings of the ACM on Interactive, Mobile, Wearable and Ubiquitous Technologies*, *2*(3), 1–20. Available from https://doi.org/10.1145/3264937.

Montesinos López, O. A., Montesinos López, A., & Crossa, J. (2022). Overfitting, model tuning, and evaluation of prediction performance. *Multivariate Statistical Machine Learning Methods for Genomic Prediction*, 109–139. Available from https://doi.org/10.1007/978-3-030-89010-0_4.

Mustapha, M. T., Ozsahin, D. U., Ozsahin, I., & Uzun, B. (2022). Breast cancer screening based on supervised learning and multi-criteria decision-making. *Diagnostics*, *12*(6), 1326. Available from https://doi.org/10.3390/diagnostics12061326.

Nabi, J. (2019, March 16). Hyper-parameter tuning techniques in deep learning. Medium. Retrieved January 31, 2023. Available from https://towardsdatascience.com/hyper-parameter-tuning-techniques-in-deep-learning-4dad592c63c8.

Najafabadi, M. M., Villanustre, F., Khoshgoftaar, T. M., Seliya, N., Wald, R., & Muharemagic, E. (2015). Deep learning applications and challenges in big data analytics. *Journal of Big Data*, *2*(1). Available from https://doi.org/10.1186/s40537-014-0007-7.

Neary, P. (2018). Automatic hyperparameter tuning in deep convolutional neural networks using asynchronous reinforcement learning. In *Proceedings of the IEEE international conference on cognitive computing (ICCC)*. Available from https://doi.org/10.1109/iccc.2018.00017.

Nwankpa, C. E. (2020). Advances in optimisation algorithms and techniques for Deep Learning. *Advances in Science, Technology and Engineering Systems Journal*, *5*(5), 563–577. Available from https://doi.org/10.25046/aj050570.

Oppermann, A. (2020, August 11). Regularization in deep learning-L1, L2, and dropout. Medium. Retrieved January 31, 2023. Available from https://towardsdatascience.com/regularization-in-deep-learning-l1-l2-and-dropout-377e75acc036.

Park, J., Yi, D., & Ji, S. (2020). A novel learning rate schedule in optimization for neural networks and its convergence. *Symmetry*, *12*(4), 660. Available from https://doi.org/10.3390/sym12040660.

Pattanayak, S. (2023). Unsupervised learning with restricted boltzmann machines and auto-encoders. *Pro Deep Learning with TensorFlow 2.0*, 407–510. Available from https://doi.org/10.1007/978-1-4842-8931-0_5.

Peng, X., Gao, X., & Li, X. (2018). On better training the infinite restricted boltzmann machines. *Machine Learning*, *107*(6), 943–968. Available from https://doi.org/10.1007/s10994-018-5696-2.

Radhakrishnan, P. (2017, October 18). What are hyperparameters? and how to tune the hyperparameters in a deep neural network? Medium. Retrieved January 31, 2023. Available from https://towardsdatascience.com/what-are-hyperparameters-and-how-to-tune-the-hyperparameters-in-a-deep-neural-network-d0604917584a.

Rahhal, M. M., Bazi, Y., Jomaa, R. M., Zuair, M., & Melgani, F. (2022). Contrasting EfficientNet, VIT, and GMLP for COVID-19 detection in ultrasound imagery. *Journal of Personalized Medicine*, *12*(10), 1707. Available from https://doi.org/10.3390/jpm12101707.

Ratnawati, D. E., Marjono, Widodo, & Anam, S. (2020). Comparison of activation function on Extreme Learning Machine (ELM) performance for classifying the active compound. In *Proceedings of the symposium on biomathematics 2019 (SYMOMATH 2019)*. Available from https://doi.org/10.1063/5.0023872.

Refaeilzadeh, P., Tang, L., & Liu, H. (2009). Cross-validation. *Encyclopedia of Database Systems*, 532–538. Available from https://doi.org/10.1007/978-0-387-39940-9_565.

Ribeiro, R. P., & Moniz, N. (2020). Imbalanced regression and extreme value prediction. *Machine Learning*, *109*(9–10), 1803–1835. Available from https://doi.org/10.1007/s10994-020-05900-9.

Rosenblatt, F. (1957). The perceptron: A perceiving and recognizing automaton, Report 85-60-1, Cornell Aeronautical Laboratory, Buffalo, New York.

Ruder, S. (2020, March 20). An overview of gradient descent optimization algorithms. ruder.io. Retrieved January 31, 2023. Available from https://www.ruder.io/optimizing-gradient-descent/.

Sakurada, M., & Yairi, T. (2014). Anomaly detection using autoencoders with nonlinear dimensionality reduction. In *Proceedings of the MLSDA 2014 2nd workshop on machine learning for sensory data analysis*. Available from https://doi.org/10.1145/2689746.2689747.

Salakhutdinov, R., Mnih, A., & Hinton, G. (2007). Restricted boltzmann machines for collaborative filtering. In *Proceedings of the 24th international conference on machine learning*. Available from https://doi.org/10.1145/1273496.1273596.

Sanjar, K., Rehman, A., Paul, A., & JeongHong, K. (2020). Weight dropout for preventing neural networks from overfitting. In *Proceedings of the 8th international conference on orange technology (ICOT)*. Available from https://doi.org/10.1109/icot51877.2020.9468799.

Sarker, I. H. (2021). Deep learning: A comprehensive overview on techniques, taxonomy, applications and research directions. *SN Computer Science*, *2*(6). Available from https://doi.org/10.1007/s42979-021-00815-1.

Schuster, M., & Paliwal, K. K. (1997). Bidirectional recurrent neural networks. *IEEE Transactions on Signal Processing*, *45*(11), 2673–2681. Available from https://doi.org/10.1109/78.650093.

Seyer Cagatan, A., Taiwo Mustapha, M., Bagkur, C., Sanlidag, T., & Ozsahin, D. U. (2022). An alternative diagnostic method for C. Neoformans: Preliminary results of deep-learning based detection model. *Diagnostics*, *13*(1), 81. Available from https://doi.org/10.3390/diagnostics13010081.

Shafiq, M., & Gu, Z. (2022). Deep residual learning for image recognition: A survey. *Applied Sciences*, *12*(18), 8972. Available from https://doi.org/10.3390/app12188972.

Sharma, A., & Jain, S. K. (2021). Deep learning approaches to time series forecasting. *Recent Advances in Time Series Forecasting*, 91–97. Available from https://doi.org/10.1201/9781003102281-6.

Shen, Y., Wang, J., & Navlakha, S. (2021). A correspondence between normalization strategies in artificial and biological neural networks. *Neural Computation*, *33*(12), 3179–3203. Available from https://doi.org/10.1162/neco_a_01439.

Shewalkar, A., Nyavanandi, D., & Ludwig, S. A. (2019). Performance evaluation of deep neural networks applied to speech recognition: RNN, LSTM and GRU. *Journal of Artificial Intelligence and Soft Computing Research*, *9*(4), 235–245. Available from https://doi.org/10.2478/jaiscr-2019-0006.

Shi, C., Zhang, X., Wang, T., & Wang, L. (2022). A lightweight convolutional neural network based on hierarchical-wise convolution fusion for remote-sensing scene image classification. *Remote Sensing*, *14*(13), 3184. Available from https://doi.org/10.3390/rs14133184.

Stateczny, A., Uday Kiran, G., Bindu, G., Ravi Chythanya, K., & Ayyappa Swamy, K. (2022). Spiral search grasshopper features selection with VGG19-resnet50 for remote sensing object detection. *Remote Sensing*, *14*(21), 5398. Available from https://doi.org/10.3390/rs14215398.

Su, Y., & Kuo, C.-C. J. (2019). On extended long short-term memory and dependent bidirectional recurrent neural network. *Neurocomputing*, *356*, 151–161. Available from https://doi.org/10.1016/j.neucom.2019.04.044.

Sumera, S., Sirisha, R., Anjum, N., & Vaidehi, K. (2022). Implementation of CNN and ANN for fashion-MNIST-dataset using different optimizers. *Indian Journal of Science and Technology*, *15*(47), 2639–2645. Available from https://doi.org/10.17485/ijst/v15i47.1821.

Sutskever, I., & Hinton, G. (2010). Temporal-kernel recurrent neural networks. *Neural Networks*, *23*(2), 239–243. Available from https://doi.org/10.1016/j.neunet.2009.10.009.

Tang, H., & Glass, J. (2018). On training recurrent networks with truncated backpropagation through time in speech recognition. In *Proceedings of the IEEE spoken language technology workshop (SLT)*. Available from https://doi.org/10.1109/slt.2018.8639517.

Tomczak, J. M., & Gonczarek, A. (2016). Learning invariant features using subspace restricted Boltzmann machine. *Neural Processing Letters*, *45*(1), 173–182. Available from https://doi.org/10.1007/s11063-016-9519-9.

Uzun Ozsahin, D., Mustapha, M. T., Bartholomew Duwa, B., & Ozsahin, I. (2022). Evaluating the performance of deep learning frameworks for malaria parasite detection using microscopic images of peripheral blood smears. *Diagnostics*, *12*(11), 2702. Available from https://doi.org/10.3390/diagnostics12112702.

Uzun Ozsahin, D., Mustapha, M. T., Uzun, B., Duwa, B., & Ozsahin, I. (2023). Computer-aided detection and classification of Monkeypox and chickenpox lesion in human subjects using Deep Learning Framework. *Diagnostics*, *13*(2), 292. Available from https://doi.org/10.3390/diagnostics13020292.

Wali, A., Alamgir, Z., Karim, S., Fawaz, A., Ali, M. B., Adan, M., & Mujtaba, M. (2022). Generative adversarial networks for speech processing: A review. *Computer Speech & Language*, *72*, 101308. Available from https://doi.org/10.1016/j.csl.2021.101308.

Yamashita, R., Nishio, M., Do, R. K., & Togashi, K. (2018). Convolutional neural networks: An overview and application in Radiology. *Insights into Imaging*, *9*(4), 611–629. Available from https://doi.org/10.1007/s13244-018-0639-9.

Yao, Z., Gholami, A., Xu, P., Keutzer, K., & Mahoney, M. W. (2019). Trust region based adversarial attack on neural networks. In *Proceedings of the IEEE/CVF conference on computer vision and pattern recognition (CVPR)*. Available from https://doi.org/10.1109/cvpr.2019.01161.

Ying, X. (2019). An overview of overfitting and its solutions. *Journal of Physics: Conference Series*, *1168*, 022022. Available from https://doi.org/10.1088/1742-6596/1168/2/022022.

Yu, J., de Antonio, A., & Villalba-Mora, E. (2022). Deep learning (CNN, RNN) applications for smart homes: A systematic review. *Computers*, *11*(2), 26. Available from https://doi.org/10.3390/computers11020026.

Zafar, A., Aamir, M., Mohd Nawi, N., Arshad, A., Riaz, S., Alruban, A., Dutta, A. K., & Almotairi, S. (2022). A comparison of pooling methods for convolutional neural networks. *Applied Sciences*, *12*(17), 8643. Available from https://doi.org/10.3390/app12178643.

Zaheer, R., & Shaziya, H. (2019). A study of the optimization algorithms in deep learning. In *Proceedings of the third international conference on inventive systems and control (ICISC)*. Available from https://doi.org/10.1109/icisc44355.2019.9036442.

Zhang, J., Zeng, Y., & Starly, B. (2021). Recurrent neural networks with long term temporal dependencies in machine tool wear diagnosis and prognosis. *SN Applied Sciences*, *3*(4). Available from https://doi.org/10.1007/s42452-021-04427-5.

Zhao, Q., & Shang, Z. (2021). Deep learning and its development. *Journal of Physics: Conference Series*, *1948*(1), 012023. Available from https://doi.org/10.1088/1742-6596/1948/1/012023.

CHAPTER

Image preprocessing phase with artificial intelligence methods on medical images

3

Kamil Dimililer[1,2,3], Binnur Demir Erdem[4], Devrim Kayali[1] and Oluwaseun Priscilla Olawale[5]

[1]*Department of Electrical and Electronic Engineering, Near East University, Nicosia, Northern Cyprus, Mersin-10, Turkey*

[2]*Applied Artificial Intelligence Research Centre, Near East University, Nicosia, Northern Cyprus, Mersin-10, Turkey*

[3]*Center for Science Technology and Engineering, Near East University, Nicosia, Northern Cyprus, Mersin-10, Turkey*

[4]*Department of Automotive Engineering, Near East University, Nicosia, Northern Cyprus, Mersin-10, Turkey*

[5]*Department of Software Engineering, Near East University, Nicosia, Northern Cyprus, Mersin-10, Turkey*

3.1 Introduction

Medical imaging is a vast growing domain in medicine. New algorithms built on technologies such as artificial intelligence (AI), machine learning (ML), and deep learning (DL) have made accurate image analysis possible (Tiwari et al., 2019). In medical analysis, "detection of image" refers to the direct analysis and treatment of images using intelligent, self-driven algorithms. For instance, supervised learning techniques are employed to categorize tumor sizes as benign or malignant in the detection model for breast cancer. For image detection in medical imaging, convolutional neural network (CNN) techniques are used due to their great accuracy and dependability. The advancement of AI in the area of medical image analysis is crucial for the detection and diagnosis of diseases. To address the need for healthcare, the findings from basic research and clinical investigations are put into practice. The research on the fundamentals of AI and medical imaging creates new techniques that improve image data and data science, which in turn enables the successful analysis of AI applications in medical imaging.

Computer vision, digital video processing, biometric verification, optical character and face recognition, remote sensing, as well as medical imaging, largely depend on image processing (IP) techniques (Olawale & Dimililer, 2020). IP is a prominent subfield of AI that aims to discover techniques to extract differences or similarities from a group of visuals. In terms of identifying features from a set of images, IP is a

Artificial Intelligence and Image Processing in Medical Imaging. DOI: https://doi.org/10.1016/B978-0-323-95462-4.00003-0

rapidly expanding field. The advancement of computer vision and IP systems has paved the way for any technician to collect statistical data sets from visual imaging outputs. IP completely eradicates the subjective nature of conventional analysis by accurately measuring and quantifying the image. Digital Image Processing (DIP) is an effective, low cost, and fast approach that is fascinating and currently used by several scientists in almost all fields (Soosai et al., 2021). It centers on applications utilizing IP techniques to extract significant data, characteristics, and attributes from a set of images. IP is mainly concerned with retrieving data from complex systems not excluding IP in medicine or medical IP, machine and robotic vision, factory or production IP, and pattern recognition (Velan & Poluru, 2020).

IP is divided into various stages; the first phase includes capturing the image. The required images are captured using the appropriate image-capturing device and then digitized. The acquired image is then preprocessed in the second stage. The frequency and spatial domain techniques are the two types of image preprocessing techniques. The spatial domain analysis is more computationally intensive because it works with the direct modification of pixels contained in an image. While the frequency domain analysis involves taking the Fourier transforms of the particular image that needs to be improved, multiplying the output with a particular filter value, and finally transforming it back to the spatial domain (Soosai et al., 2021). IP approaches may be classified into—image enhancement, image compression, image segmentation, image restoration, and image resizing.

Preprocessing is a crucial part of many image applications. In addition to eliminating noise, preprocessing reduces distortions in the input images. Before the main analysis and extraction of information, radiometric and geometric corrections are usually required as part of the preprocessing operation (Sann et al., 2021). Different conditions, such as differences in illumination, camera orientation, and noise, affect the authentic images received from the photo acquisition system. As a result, the use of immediate authentic images for irregular detection is inefficient. As a result, the authentic images need to be preprocessed at the start of the IP process (Yan et al., 2019). Noise removal, deblurring, and contrast enhancement are all part of traditional preprocessing. Preprocessing techniques aim to improve image quality, which aids in subsequent processing (Deepa & Jyothi, 2018). Histogram equalization, power law filter, log filter, median filter, mean filter, and Gaussian filter are some other preprocessing techniques available (Kavipriya & Hiremath, 2021).

3.2 Medical imaging

Medical imaging is the technique and process of creating visual representations of the interior of a body for clinical analysis and medical intervention. The images produced by medical imaging techniques such as X-ray, computed tomography (CT), magnetic resonance imaging (MRI), ultrasound, and nuclear medicine can be used to diagnose and treat a wide variety of medical conditions. These images can be used to view internal organs, bones, and other tissues and can help physicians make a diagnosis and plan a course of treatment.

There are several different techniques used in medical imaging, including:

- X-ray imaging: uses electromagnetic radiation to create images of bones and internal organs.
- CT scans: uses X-rays to create detailed, cross-sectional images of the body.
- MRI: uses a magnetic field and radiowaves to create detailed images of soft tissue and organs.
- Ultrasound: uses high-frequency sound waves to create images of internal organs and structures.
- Positron emission tomography scans: uses a small amount of radioactive material to create images of organ function and metabolism.
- Nuclear medicine: uses small amounts of radioactive material to create images of organs and bones.
- Fluoroscopy: uses X-rays to create real-time images of internal organs and structures during a medical procedure.
- Radiography: uses electromagnetic radiation to create images of internal structures, similar to X-ray but with more detailed images.
- Mammography: This technique uses X-rays to create images of breast tissue.

These are some of the most common imaging techniques used in medicine, each with its own unique characteristics and indications for use.

3.3 Image processing

To optimize image preprocessing methods, maximum image preprocessing literature research is presented primarily based on those three processes, that is, image graying, image geometric changing, and image enhancement. Image denoising and enhancement strategies together with image length normalization, median filtering, and image enhancement are other methods that may be used. Yan, Wen, and Ga (2019) stated a technique, from the research work of another author, that bills for all skew, blur, and broken images as a result of different factors and produces higher image effects than the conventional preprocessing method.

The discrete cosine transform (DCT) is a technique for compressing images and it may also be employed as a preprocessing method to classify problems. During the compression process, minimizing the number of irrelevant pixels in the image focuses on effective and relevant feature extraction (Dimililer & Sekeroglu, 2021). The key advantages of IP are its consistency, processability, easy methods, and efficient approach.

3.4 Histogram equalization

Histogram equalization is a popular approach for enhancing the advent of images. Histogram equalization normally improves the overall contrast of images, particularly while the image's usable records are represented with the aid of using near contrast values (Sann et al., 2021). The detail(s) in the histogram is geared toward other IP applications like image compression and segmentation. The histogram shows the frequency with which each gray level from 0 to 255 occurs. In other words, it represents the rate of the image's gray level. It is responsible for distributing the values of a single pixel in an image.

3.5 Power-law transformation

A power law governs the behavior of a variety of image capture, display, and printing devices. The exponent in the power-law model is referred to as gamma. Gamma correction is the method used to correct these power-law response characteristics. If presenting a picture precisely on a computer display is an issue, gamma correction is necessary. Images that haven't been properly adjusted can appear bleached out or, more commonly, overly dark. Trying to replicate colors accurately also necessitates some understanding of gamma correction, because changing the gamma correction value affects not only the brightness but also the red-green-blue ratios (Dhawan et al., 2013). Mathematically, the power-law transform can be given as:

$$s = cr^i \tag{3.1}$$

where s = pixel value output

r = pixel value input
i, c = real numbers

According to Partner (2021), different levels of enhancement can be obtained for different values of i. Changing the parameters produces a full family of curves, where i and c are fixed values. With decimal values of i power-law curves transfer a small set of dark input values into a greater set of output values, while the transpose is correct for higher input levels and the gray levels must be in the range [0.0, 1.0], when we set c to 1. 0.

3.6 Linear transformation

Identity and negative transformations are both included in the linear transformation. In the negative transformation, negative images are confined to boost gray or white features in the dark regions. Dark values are converted to light values, and light values are converted to dark values (Kavipriya & Hiremath, 2021). When every variable is replaced with a single log, this is an example of log transformation (x). The log transformation reduces or removes the skew in our original data, improving data accuracy. To be published, original data must follow or approximate log-normal distributions. Log transformations, in contrast to other types of transformations that modify skewed data to conform to normality to be consistent with their conclusions, are undoubtedly the most popular. The distribution of log-normal distributed data is generally centered or near the curve, hence there should be no regression with logs (icsid.org, 2022). The linear transformation is given by:

$$s = L - 1 - r \tag{3.2}$$

where $L-1$ = the maximum value of the pixel

r = input pixel value
s = output pixel value

3.7 Log transformation

The log transformation increases the dark pixel values while compressing the light pixel values. It converts a small amount of low gray-level intensities into a larger value of the output set. Similarly, it converts a wide variety of high gray-level intensities into limited values of high-quality output (icsid.org, 2022). Mathematically, it may be obtained via the given formula:

$$s = c \log(r + 1) \tag{3.3}$$

where c = constant

r = input pixel value
s = output pixel value

3.8 Mean filter

The mean filter (also known as the average filter) is a concise, convenient, and logical technique of smoothing images, that is, minimizing the degree of intensity variance between pixels. This is a common technique for reducing image noise.

The mean filter substitutes every value with the mean value of nearby pixels, including itself, as it passes through the image pixel by pixel (Partner, 2021). Fig. 3.1 shows the mean filtering process. Some issues associated with the mean filter may be summarized into the following:

- One pixel with an out-of-the-ordinary value will have a positive impact on the mean value of all pixels nearby.
- When a filter region reaches a boundary, it interpolates new values for pixels on the edge, blurring it. This could be an issue if the output requires clean regions.

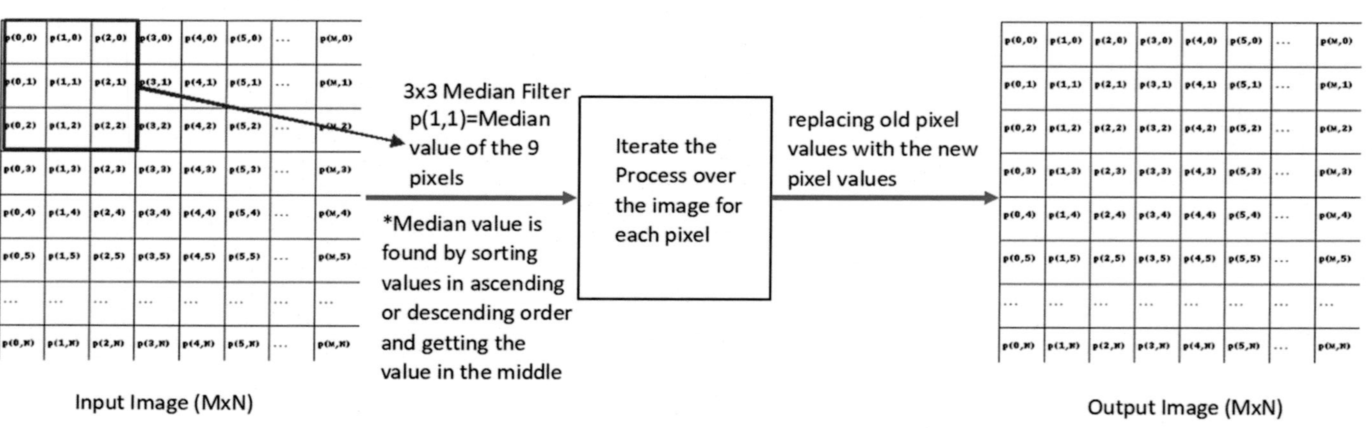

FIGURE 3.1

Mean filtering process of a 3 × 3 filter.

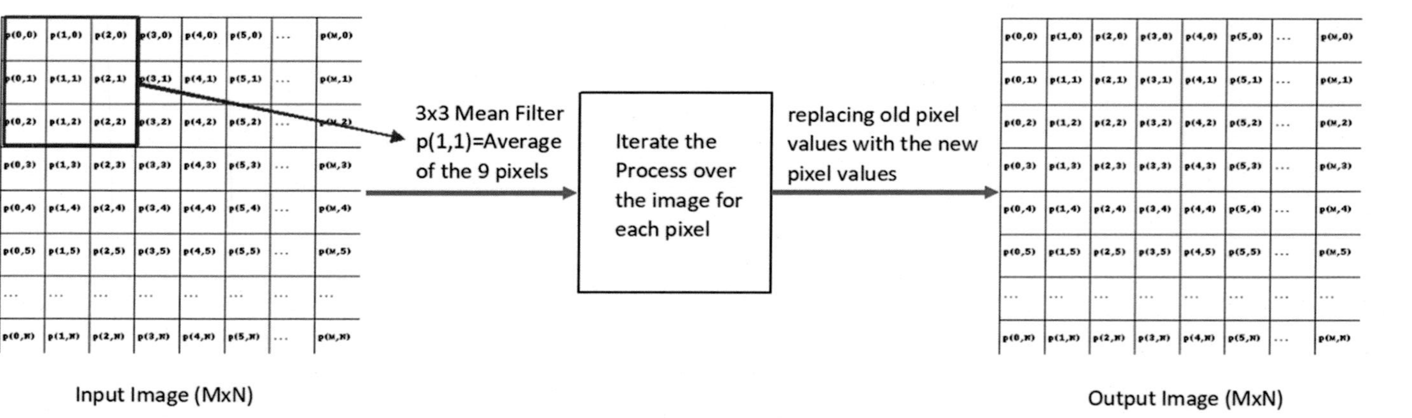

FIGURE 3.2

Median filtering process of a 3 × 3 filter.

3.9 Median filter

The median filter in IP, like the mean filter, is typically used to reduce noise in an image. However, it frequently outperforms the mean filter in terms of maintaining significant data in the image. The median filter looks at each pixel in the image separately and evaluates it to the surrounding pixels to see if it is representative of its region. It uses the median of surrounding pixel values instead of the mean values to replace the pixel value. The median is derived by numerically ranking all of the pixel values in the immediate surrounding before replacing the pixel in question with the middle pixel value.

After computing the median of all the pixels beneath the kernel window in ascending or descending order, the median filter replaces this result with the central pixel (Fig. 3.2). This filter can be described as nonlinear digital filtering that preserves edges while reducing noise (Kadhim, 2021).

3.10 Gaussian filter

By definition (icsid.org, 2022), the Gaussian filter is a weighted windowed linear filter. A smoothing operator that is similar to the mean filter is the Gaussian blur, except that it employs various kernels, giving distant pixels less weight. The image is blurred and the noise is reduced. The dimension of the fixed window specified by the user measures the distance between each pixel in the window and the center pixel (Wan et al., 2018).

3.11 Image compression

Image compression is necessary for complex work such as digital data communication systems and web transactions. Additionally, for every compression technique, two essential factors play a significant role: the compression determinant and the standard of the decompressed image. Image compression is an IP technology for reducing the amount of image data. The main goals are to enable additional images to be stored on the same storage device and to send more images at the same terminal (Al-khassaweneh & AlShorman, 2020).

An image may be compressed using either a lossy or lossless image compression approach. According to Khashman and Dimililer (2008), a suitable image compression framework will always achieve the maximum quality of compressed images with the optimum compression ratio, while preserving the least amount of time. Block diagram of the compression process in shown in Fig. 3.3.

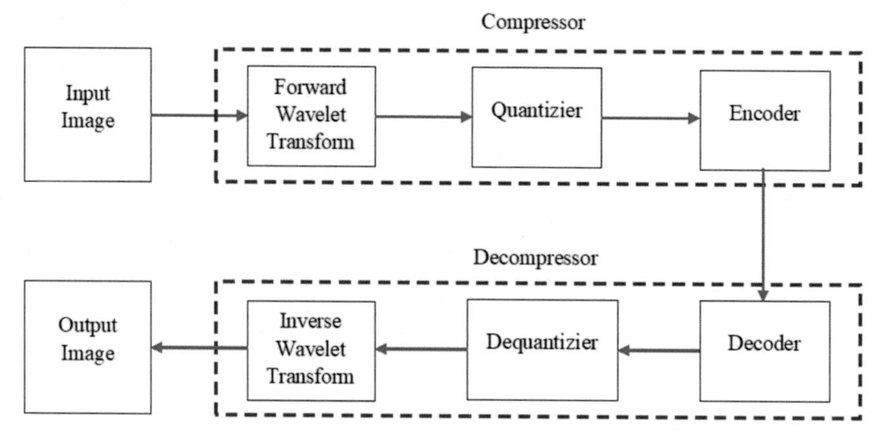

FIGURE 3.3

Block diagram of the compression process.

3.12 Image enhancement

The term "image enhancement" refers to a method of manipulating image pixels to achieve a concise perception of a particular image, in a particular situation. Sometimes, image enhancement requires the application of one or more IP methods, to obtain important information. Although image enhancement appears to be a difficult task, the primary goal of IP is to produce an image that is significantly greater in quality than its original (Olawale & Dimililer, 2020).

3.13 Image resizing

Conventional methods for resizing images usually work in pixel space and use a variety of important units of measurement. The challenge is to change the shape of the image while preserving important content. Most image resizing methods start with a prominent classification procedure and determine the most important region in an image. The image is then generated using a scaling operator when conserving these regions, in the hopes of avoiding artifacts. These two steps are still difficult. Current measures of importance primarily consider low-level features and ignore important higher-level semantic aspects. Secondly, the present scaling operators rarely take into account the alternative quality criterion, namely the preservation of the natural appearance of the resulting image (Danon et al., 2021).

3.14 Image restoration

According to Rasti et al. (2021), image restoration is about recovering the original, unseen part of an image from a degraded, visible image, and it is a burgeoning field of remote sensing. Image restoration is the process to retrieve missing data in an image from a broken copy. The image that was retrieved will further aid in the processing of future remote-sensing images. The fundamental method of image restoration processing involves taking a picture of a deteriorated or defiled image and using some prior knowledge of the degradation phenomenon to rebuild or restore original images (Bao, 2021).

3.15 Image segmentation

It refers to the classification of an image into various segments. This method is often applied to segregate the abnormal regions in an image, for a particular situation (Sowmya et al., 2020). According to Haque and Neubert (2020), all image segmentation algorithms can be grouped into manual, semiautomatic, and fully automatic segmentation algorithms.

- Manual segmentation algorithms: These techniques require professionals to initially define the segmented area, after which the edges surrounding the exact areas are sketched to accurately label the image pixel.
- Semiautomatic segmentation algorithms: These techniques rely on minimal user interaction with computerized methods to obtain appropriate segmentation output.
- Fully automatic segmentation algorithms: These techniques do not necessitate any human engagement. Nevertheless, the majority of these methods rely solely on the supervised learning algorithms that need training dataset.

The three basic steps of IP are as follows:

- Importing the image using image capture software or hardware
- Analyzing, preprocessing, and then processing the image using the desired IP technique
- Output, which may be a modified image or a report based on the image analysis

Thanks to technological advances, large volumes of high-quality photos can now be obtained at a very low price. As a result, there has been significant growth in the usage of IP techniques. Automatic IP or evaluation techniques have been developed to obtain important data (Haque & Neubert, 2020). Other disciplines that may be combined with IP include data mining, databases (for accessing datasets), programming with different languages and/or frameworks, and eventually AI. Fig. 3.4 represents the disciplines associated with image preprocessing.

Data mining methods are also integrated with IP methods to improve image interoperability, which is notably useful in the medical domain for early illness diagnosis (Olawale et al., 2021). Data mining refers to a set of techniques for data collected from various sources, such as detecting completely undiscovered correlations and identifying underlying patterns. Data mining is a method of analyzing data using statistics, for analyzing massive records, discovering anomalies, collecting relevant information, and estimating outcomes. According to Lv (2021), the kind of feature to

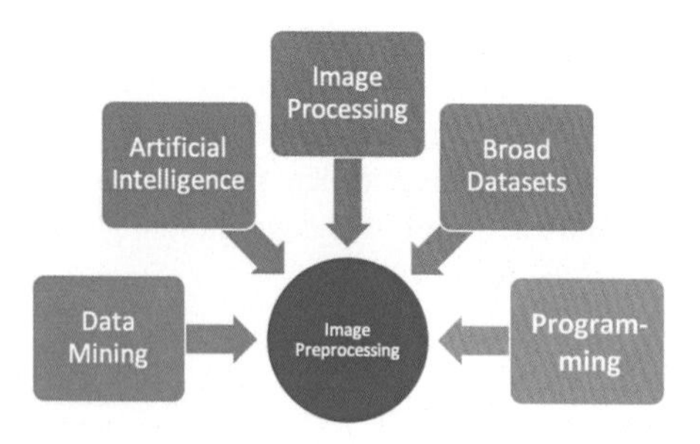

FIGURE 3.4

Disciplines associated with image preprocessing.

be identified for the completion of the mining process is provided based on the data mining parameters to be achieved. A mining model is an abstract description of a set of facts. According to the role of the model, it is commonly separated into predictive and descriptive models. Predictive modeling refers to how specific data characteristics are extracted from known data collection to perform classification analysis on unknown data sources; descriptive modeling refers to reporting the principles, regulations, and information features in the data set without classifying the data.

MATLAB®, Java, C++, and Python are common frameworks for achieving AI and IP results. Even though each of these programming languages has a distinct use, Python language is the most extensively used because it works well with ML and DL (Olawale, Dimililer & Al-Turjman, 2020).

AI on the other hand comprises both DL and ML, both having comparable characteristics. One major distinction is that DL uses artificial neural networks (ANN) to solve problems. DL develops numerous connected networks regardless a human agent is available, whereas ML requires an agent to be present for training to take place (Olawale, Dimililer & Al-Turjman, 2020). DL requires no human engagement because the computer obtains data and distributes it simultaneously to prevent big or severe losses (Dimililer et al., 2021). ML is low-cost, uses less power, and provides reliable data. ML algorithms are used to complete tasks that humans find difficult to completely execute (Dimililer, 2021).

In virtually every sector, DL has made significant advances in the last decade. The neural networks serve as the foundation for many nonlinear operations that abstract data. The principal idea behind profound learning is to attempt to demonstrate information with exceptionally preoccupied data by lessening the information aspects with different handling layers. DLCNN architectures, include AlexNet, VGGNet, GoogLeNet, ResNet, and Xception. There is a plethora of currently accessible algorithms, each targeting a specific aspect of machine vision. Nevertheless, many of these approaches may not be used on medical image data instantly. (Samira Masoudi et al., 2021). The important preprocessing processes ideally improve the performance of the DL algorithm in prediction or classification. Due to the high level of accuracy of DL in other practical systems, it already offers effective promising alternatives for diagnostic imaging and is seen as a critical tool for future healthcare applications (Razzak, 2018). AI enables more secure, precise, and effective imaging solutions than traditional imaging systems that rely significantly on professional effort (Shi et al., 2021). The developed medical diagnostic systems are based on AI and, in particular, DL implementations' ability to examine even the tiniest characteristics of visuals that are undetectable to medical professionals and take decisions based on them (Dimililer & Sekeroglu, 2021).

3.16 Artificial intelligence in medical imaging

AI is one of the fundamental sciences of the future that always make life easier, offer solutions to numerous issues, and enhance the overall quality of life. Numerous studies and practical studies have produced in-depth thoughts in this field. Machine intelligence is a new development that will benefit humanity in many ways and will be used in all areas of theoretical and practical research, including astronomy, aviation, engineering, and medicine. Nowadays, many businesses use AI to update their systems. Machine intelligence made significant contributions to improving human lives and advancing medical radiology (Daoud & Otair, 2020). AI-powered image acquisition can greatly assist in automating the scanning process and reshaping the workflow, offering the best protection for the imaging technicians. Additionally, by precisely identifying infections in X-ray and CT images and easing subsequent quantification, AI can increase job efficiency. The computer-aided platforms also assist radiologists in clinical decisions, such as disease diagnosis, tracking, and prognosis (Shi et al., 2021). With the advent of the Coronavirus disease 2019 (COVID-19) pandemic, for example, to make the deployment of a contactless scanning workflow easier, contemporary X-ray and CT devices were used in the medical imaging industry. Through a live video transmission from a camera, technicians could keep an eye on the patient from the control room. AI shows off its innate capacity to combine data from several sources to execute accurate and effective diagnosis, analysis, and follow-up.

In terms of precision and speed, AI is efficient in many areas. AI has the potential to revolutionize the field of medical imaging. AI algorithms can be trained to analyze medical images and identify patterns, anomalies, and diagnostic information that may be difficult for human radiologists to detect. This can lead to earlier and more accurate diagnoses, as well as improved patient outcomes.

One of the key areas where AI is being used in medical imaging is in the detection of cancer. AI algorithms can be trained to analyze mammograms, CT scans, and other imaging modalities to identify signs of breast cancer, lung cancer, and other types of cancer. This can help radiologists to more accurately detect and diagnose cancer at an earlier stage, when it is more likely to be treatable.

Another area where AI is being used in medical imaging is in the analysis of medical images for the diagnosis and treatment of neurological disorders. AI algorithms can be trained to analyze brain scans and identify patterns that are indicative of conditions such as Alzheimer's disease, multiple sclerosis, and other neurological disorders. This can help neurologists to more accurately diagnose and treat these conditions.

AI is also being used in medical imaging to improve the accuracy of diagnostic imaging. AI algorithms can be trained to identify features in images that are indicative of certain conditions, such as abnormalities in the blood vessels, bones, or organs. This can help radiologists to more accurately diagnose and treat a wide range of conditions, such as heart disease, osteoporosis, and other conditions.

Overall, AI has the potential to greatly improve the field of medical imaging. By using AI algorithms to analyze medical images, radiologists, and other medical professionals can more accurately detect and diagnose a wide range of conditions, leading to improved patient outcomes. However, more research is needed to develop and validate these AI-based imaging systems, and to ensure they are safe and effective for use in clinical practice.

Dagli et al. (2021) made a review on AI in the field of surgery, where they discussed current applications in the field, practical examples, and challenges. They also mention the utility of computer vision and natural language processing to improve patient care, research, and surgical outcomes.

Sideraki and Drigas (2021) examined the role of AI in autism. AI is actively used for assisting people when diagnosing autism and various disorders more easily. Besides diagnosis, AI is also used for intervention and treatment purposes for children who have difficulties with socialization and speech. Virtual reality, chatbots, and robots with AI are some examples of these kinds of applications.

Diabetes diagnosis is another subject where AI has achieved accurate results. Nithyalakshmi et al. (2021) compared the K-nearest neighborhood (KNN) classifier and backpropagation neural network algorithm using the diabetes database. For evaluation, 10-fold cross-validation was used, and the backpropagation neural network algorithm had better results with 85% accuracy and 89% AUC scores than KNN with an accuracy of 79% and 72% AUC scores. In another research by Fitriati and Murtako (2017), fast adaptive histogram equalization was used as a preprocessing approach, and extreme learning machine was used as a classification method. The area of blood vessels was the feature employed. DiaretDB0 attained a 97.5% response rate in the earlier study. Then, using RSCM data, they used these insights. However, for evaluating the accuracy, it only achieved 52% in the experiment.

Hasanzad et al. (2022) reviewed AI in endocrine diseases. AI can help make better clinical decisions and it is called augmented intelligence. In diabetes management, AI is used for the management, diagnosis, and detection of diabetic problems. Some other endocrine diseases that are assisted by AI include osteoporosis and thyroid cancer.

Dental panoramic radiographs (DPRs) were frequently utilized in the past to diagnose and comprehend patient dental conditions. Many ML and DL techniques have recently been used to solve medical picture recognition issues. Additionally, data augmentation and image preprocessing techniques can be effective when used with DL techniques. In order to create a novel and useful two-phase DPR detection and classification approach to aid dentists in diagnosis, we combined data augmentation and data preprocessing methods with advanced DL methods. Expediting and saving priceless physician manpower costs and time would aid in enhancing the medical quality of dental services. Without image preprocessing or augmentation, the trained network correctly detected the dental position numbering with an accuracy of 90.93% and the dental condition with an accuracy of 93.33%, according to the experimental data. After data augmentation, tooth numbering accuracy can be enhanced to 95.62%, and dental condition accuracy may be increased to 98.33%.

The most effective and popular approach for finding breast cancer is a mammogram. Boudouh & Bouakkaz (2022) developed a reliable breast tumor detection model that distinguishes between normal and abnormal mammography pictures (tumor). Based on the results, the preprocessing filters that were used were successful. According to earlier studies, the results were highly optimal. AlexNet and VGG16 (with trainable layers) achieved an accuracy of 99.5% and 100% respectively, better than the current breast tumor detection models.

To verify AI-assisted medical treatment feasibility, Ma et al. (2022) used deep models for unsupervised detection and classification of cardiovascular diseases on diagnostic imaging. Their objective was to obtain a thorough cardiac diagnosis as well as the location of the susceptible lesion. They built a network model that consists of both classification and regression nodes. To improve the imbalance of data between the classes, a weighted loss function was used. They also mentioned that even under weakly supervised conditions, the suggested system reveals the likelihood of tumor detection.

3.17 Applications of image preprocessing in medical imaging

In medical image analysis, demarcating organs using the image segmentation technique is a crucial process; however, it is difficult to work with due to the little labeled data, poor contrast, and nonhomogenous textures. Compared to natural images, body parts and tissues in clinical image datasets include distinct anatomical information (such as organ size), which can be used to improve segmentation accuracy. In their study, Huang et al. (2021) presented a novel segmentation approach that incorporates the anatomical prior from medical images into DL models through loss. The proposed prior loss function, known as the "deep atlas prior," (DAP) is based on a probabilistic atlas. It contains previous information about the position and shape of the organs, which is crucial for the proper segmentation of the organs.

Huang et al. (2021) combined a DAP loss with conventional losses (dice, focal) to form an adaptive Bayesian loss for improved segmentation. The adaptive loss adjusts the ratio of DAP and likelihood losses in each training epoch. The proposed loss function can be applied to a variety of existing models, as proven by the researchers' experiments on public (LiTS 2017 Challenge) and private datasets for liver and spleen segmentation using both fully- and semisupervised models.

Kavipriya and Hiremath (2021) used a variety of preliminary processing approaches to increase the quality of the angiographic images and minimize the noise in the images. The angiographic data employed in the study was collected with the consent of the competent authorities from the specified hospital. The histogram equalization method is applied to the initial image to be preprocessed and control and improve the contrast of the image. Massive medical systems, such as MRI and CT systems, are used to examine tumors. There have been numerous studies in the past 20 years on brain tumor detection using MRI. This system detects the location of the tumor as well as its precise size.

Dimililer & İlhan (2016) applied IP techniques, such as contrast enhancement, erosion, and median filtering, on MRI to conserve image data. The goal of the study was to create an IP technique for detecting tumors in MRI-generated scans. The effect of categorization on backpropagation neural networks were compared using initial and transformed image visuals. To resolve the limitations of current techniques for automatically detecting brain-related illness using MRI scans, Renjith and Wagaj (2020) suggested a novel DL algorithm executing tasks such as preprocessing, hybrid feature learning landmark estimation, normalization, and classification. The raw brain MRI image is first preprocessed with noise-removing filters and contrast enhancement functions before being used to estimate landmarks.

Dimililer & Sekeroglu (2021) implemented the use of a deep CNN and a DCT on COVID-19 research using X-ray pictures. Two separate datasets comprising COVID-19 X-ray pictures were investigated, and DCT was applied to these images to undertake several studies. The preliminary results of the two datasets demonstrated that using DCT in the preprocessing phase of deep CNNs can improve classification rates by 10% and 5%, respectively, in terms of sensitivity and ROC AUC score. Also, Albahri et al. (2020) provided a comprehensive methodology for evaluating and benchmarking AI techniques used in all classification tasks of COVID-19 medical images; this methodology was presented in three steps. The identification procedure for the construction of four decision matrices, namely binary, multiclass, multilevel, and hierarchical, is described based on the intersection of performance standards for each classification task and AI classification approaches. The multicriteria approach for evaluating AI classification systems is created using the integrated analytical hierarchy process and the VlseKriterijumska Optimizacija I Kompromisno Resenje method. Finally, the proposed benchmarking solutions were validated using both objective and subjective methods.

For chronic respiratory diagnosis, Subramaniam et al. (2021) developed an effective preprocessing and classification technique. In the proposed technique, the best features were extracted, and the lungs in X-ray images were separated into left and right by using local binary pattern, Haar transform, and histogram of oriented gradients algorithm. The accuracy of results in COVID-19 detection algorithms or other machine/DL approaches can be improved by segmenting lungs from X-rays. To compare the algorithms, the segmented lungs are validated using intersection over union (IoU) scores. Preprocessed X-ray images provide higher classification accuracy than unprocessed raw images in all three classes (normal/COVID-19/pneumonia).

Gupta et al. (2021) described a fog-cloud system for lung cancer illness early detection and management. At the fog layer, an IP-based preprocessing phase has been added to detect various anomalies using feature extraction from CT-scanned pictures. For the classification of different retrieved features, a backpropagation-based ANN was used.

Beeravolu et al. (2021) proposed efficient image preprocessing methods for preprocessing breast cancer to create datasets that reduce computational time for the neural network while improving accuracy and classification rates. To that end, the methods proposed in this study include pectoral muscle removal, background removal, image noise addition, and image enhancement. Adding noise to images without affecting their detail quality makes the input images for the neural network more representative, which may improve the neural network model's performance.

Although promising results have been obtained in the field of Alzheimer's disease diagnosis using hippocampal atrophy analysis, the majority of these solutions produce good results with various MRI preprocessing techniques. Furthermore, most recent works use classification for Alzheimer's disease, and this causes a research gap to use object detection. Fong et al. (2020) made two contributions to the solution of this problem.

Due to storage capacity and bandwidth constraints, medical images must be efficiently compressed before transmission and storage. An ideal image compression system should achieve a high compression ratio while maintaining decent image quality.

Some problems associated with IP can be given as the removal of reflections, following the trend of real data collection. Since most previous models capture images of postcards, static objects, or color plates, the capture of real data suffers from a lack of scene diversity (Zhou et al., 2021). High-performance segmentation models based on CNNs have made remarkable advances in the medical imaging literature. Despite the new high performance, large, representative, and well-annotated datasets, more are still required for the new advanced segmentation models. However, we rarely have a perfect training dataset, especially in medical imaging where both data and annotations are expensive to obtain. Much recent research has addressed the problem of segmenting medical images with imperfect datasets (Tajbakhsh et al., 2020).

Sengoz et al. (2022) used both IP and DL on histopathology images taken from animals with paratuberculosis and intact intestine. By using a dataset of 520 histopathology images, they made a comparative study on the importance of preprocessing using contrast limited adaptive histogram equalization (CLAHE). Without preprocessing the F1 score was 93% while using CLAHE as preprocessing improved the result and the F1 score was 98%.

Oh et al. (2022) made a study on endoscopic ultrasonography (EUS) images that were used with DL for automatic segmentation of the pancreatic cyst lesion (PCL). Attention U-Net was employed and compared with residual U-Net, basic U-Net, and U-Net++ models. According to internal test results, the attention U-Net had a better IoU score and dice similarity coefficient (DSC) than the other models. For the external test, there wasn't any significant difference between the models, but basic U-Net had higher DSC and IoU scores. Results in this study show that DL has successful implementation of automatically segmented PCL using EUS images.

Lin et al. (2022) developed an AI-based system to detect fetal intracranial abnormalities in different patterns using ultrasound images. The system is called Prenatal Ultrasound Diagnosis Artificial Intelligence Conduct System. YOLOv3 was used as a real-time CNN algorithm to build the system. They evaluated the system using an image dataset from the third tertiary hospital and also compared the results with the diagnoses of different experts. Accuracy, specificity, sensitivity, ROC, and AUC metrics were used to calculate the performance of the system. Results show that both in real-time scan and the image dataset the performance was as good as that of the medical experts and had shorter time consumption for recognition of various fetal CNS abnormalities.

Saidnassim et al. (2021) used transformer-based networks for breast cancer diagnosis. Bootstrap Your Own Latent (BYOL), a self-supervised learning algorithm, is proposed for processing diagnostic mammogram images. BYOL can also handle unlabeled image data and enhance the DL mechanism for data analytics. UNet networks with Xavier initializations were used to compare their performance when implemented with and without BYOL. Overall network performance improves when trained using the proposed BYOL algorithm.

Sun and Yuan (2022) studied methods to increase segmentation accuracy from shape-changeable adaptive capabilities and boundary recognition perspective. They studied the active contour model based on boundary constraints and for the automatic realization of the CT cutting algorithm they proposed a superpixel boundary-aware convolution network. According to experimental results, accuracy and sensitivity were increased by about 12% when compared to the traditional CT algorithm. When the cut images of the malignant tumors are compared, the cutting effect is 34% higher.

Janjua et al. (2022) made a study on the AI-based diagnosis of common thoracic lung diseases using a chest X-ray dataset built on the data of multimodal radiation diagnostics. The framework in this study is based on fuzzy-empowered DL for object detection and classification. Preprocessing, image segmentation for highlighting diagnostic objects, and classification are the stages used to determine the benign and malignant objects.

Zhang et al. (2021) made image segmentation on spine medical images using DL. In this study, they improved the U-shaped network (BN-U-Net) algorithm and applied image segmentation of 22 research objects to the spinal MRI images. Accuracy, sensitivity, and specificity were used for evaluation. U-shaped network (BN-U-Net) had 94.54% ± 3.56% accuracy, 88.76% ± 2.67% sensitivity, and 86.27% ± 6.23%, which were significantly better when compared to a fully convolutional network (FCN) algorithm. Also, the processing time for the U-Net and the FCN algorithms was more than 6 minutes while the processing time of the patients' BN-U-Net algorithm was 5−10 seconds.

(Ben Ali Kaddour and Abdulaziz (2021) researched developing an AI model that can perform pathological reports automatically. Unlike previous studies, they aimed for the diagnosis of multiple diseases from a medical image. Feature extraction is done by morphological properties and ML and DL with CNN architectures are applied for image classification. The resulting model had 90.74% accuracy and 83.32% F1-score respectively and diagnosed 12 medical disorders with an overall of 29 diagnostic cases.

Salama et al. (2021) studied lung cancer detection from CT images by using preprocessing, segmentation, and classification. For preprocessing phase, the Wiener filter, and image size-dependent normalization technique were used for noise suppression and enhancement. For image segmentation, U-Net was used prior to classification to increase the system performance. At the classification phase, the ResNet50 model was used to classify malignant or benign lung CT images. The proposed framework was applied to National Lung Screening Trial, and Lung Nodule Analysis 2016 (Luna 16) datasets. The computational time of the network was 1.9876 second, and according to classification results, 98.98% accuracy, 98.43% precision, 98.86% F1-score, 98.99% sensitivity, and 98.65% AUC score were obtained.

Sun et al. (2021) made a comparative study where they compared two-dimensional (2D) and three-dimensional (3D) semantic segmentation models prediction performance for metastatic tumors having a volume greater or equal to 0.3 mL. Postcontrast T1 whole-brain MRI images were used. The advantage of the 3D segmentation model is that it fully uses the spatial information between neighboring slices while 2D image segmentation models do not. The 2D model had 81.5% of AUC while the 3D model had 87.6% on the prediction of intracranial metastatic tumors.

3.18 Simplified: applications of artificial intelligence and image preprocessing in medical imaging

AI and image preprocessing have been discussed in the previous sections. In this particular section, we will examine specific medical images and techniques

Table 3.1 Applications of artificial intelligence and image preprocessing in medical imaging.

Authors	Medical images (cases)	Image preprocessing technique	AI technique
Dimililer et al. (2017)	Lung cancer—CT image	Gaussian filter and discrete wavelet transform	–
Khastavaneh and Haron (2015)	MRI	Image segmentation	Massive training artificial neural network (MTANN)
Lu et al. (2019)	Mammograms	Median filter, contrast-limited adaptive histogram equalization, and data augmentation	Convolutional neural network (CNN)
Temiz and Bilge (2020)	Ultrasound	Super-resolution	CNN
Lin and Chang (2021)	Dental panoramic radiograph	Edge detection	CNNs
Escorcia-Gutierrez et al. (2022)	Breast cancer	Gaussian filter and Tsallis entropy-based method	ResNet34 and wavelet neural network (WNN)
Kavitha et al. (2022)	Breast cancer	Adaptive fuzzy based median filtering and OKMT-SGO	CapsNet based and back-propagation neural network
Jiang et al. (2021)	Synthetic digital mammography		Gradient guided conditional generative adversarial networks
Punithavathi (2021)	Mammogram	Filtering and segmentation	Hybrid modified SVM and KNN classifier
Malebary and Hashmi (2021)	Mammogram	–	Combination of K-means, recurrent neural network, CNNs, random forest, and boosting techniques
Song et al. (2022)	Panoramic radiographs	–	U-Net
Yang et al. (2022)	Ultrasound	–	Deep CNN

AI, *artificial intelligence;* CT, *computed tomography;* KNN, *K-nearest neighborhood;* MRI, *magnetic resonance imaging;* OKMT-SGO, *optimal Kapur's-based multilevel thresholding with shell game optimization algorithm;* SVM, *support vector machine.*

(comprising both image preprocessing and AI), alongside some medical cases. Table 3.1 below provides information at a glance regarding the applications of AI and image preprocessing in medical imaging, after which they are discussed based on the findings of other researchers.

3.19 Medical cases

3.19.1 Lung cancer—computed tomography scans

According to Dimililer et al. (2016), images from CT scans give necessary electronic densities for tissues of interest. Good spatial resolution and a contrast between soft and hard tissues are required for specific target identification. X-ray and MRI images are not as desirable as CT procedures. Utilizing pictures from CT has begun to gain popularity in IP approaches.

The development of a tumor, in the respiratory system's airways, is known as lung cancer. Proper timely detection of lung cancer opens the door to all treatment options, lowers the chance of pernicious surgery, and increases survival rates. The enhancement of lung tumor images for early identification and treatment phases makes great use of IP techniques recently. This is because finding the anomaly problems is important in target images and takes time (Dimililer et al., 2017).

3.19.2 Mammogram

Breast cancer incidence has steadily increased over the last few decades. The current breast cancer screening strategy is based on traditional X-ray imaging. The sensitivity and specificity of the diagnosis are largely dependent on the radiologists' experiences, and uncertain diagnosis is quite common due to resolution limitations and concerns about lawsuits arising from incorrect diagnosis or undetected lesions (Lu et al., 2019).

3.19.3 Dental panoramic radiograph

A DPR is a scanned wide-angle X-ray radiograph of the patient's upper and lower jaws. Medical personnel frequently used DPR as an important reference diagnostic basis for understanding patients' dental conditions. It aided dentists in providing the most immediate medical services to patients. The DPR is now interpreted and marked by trained medical workers, which indirectly increases their effort. The medical quality of dental services will be improved if the image recognition method of DL can automatically assess the patient's panoramic X-ray images. This is especially true when deciding the content of a substantial percentage of panoramic X-ray images (Lin & Chang, 2021). Oyelade et al. (2022), who proposed a U-Net-based segmentation model for automatic identification of mandibular canal in panoramic radiographs. The results showed that the model outperformed traditional methods and achieved an average DSC of 0.88.

3.19.4 Ultrasound

Ultrasound imaging (USI) is risk-free, noninvasive, and inexpensive. It is almost often the method of choice for imaging the cardiovascular system, the abdomen, urology, the vascular system, obstetrics, gynecology, and other areas of medicine. USI, on the other hand, experiences artifacts that happen as sound waves travel across the medium. Artifacts are frequently caused by the rapid transmission of ultrasound signals. As a result, different types of noise have an impact on the signals. Due to its inherent imaging properties, it produces images with lower resolution (LR) and lower quality when compared to other medical imaging modalities. A variety of image enhancement approaches have been thoroughly investigated to address these issues. Super-resolution (SR) is one of these techniques that aim to obtain high-resolution (HR) images while enlarging LR photos (Temiz & Bilge, 2020).

Yang et al. (2022) collected ultrasound images from patients to prepare a train and test dataset for locating the interscalene brachial plexus. The images were annotated by three senior anesthesiologists who are experts in regional anesthesia. These annotated images were used as ground truth to develop a deep CNN. Their experimental results showed that the mean distances of the predictions from the obtained model were shorter than the predictions of five nonexpert anesthesiologists.

3.19.5 Magnetic resonance imaging

Soft tissues are sensitive to MRI; therefore, each tissue has a unique intensity value. This is also true of dead or ill tissues, which have distinctive intensity values. Because they have different intensity levels, they seem distinct from normal tissues (Khastavaneh & Haron, 2015).

3.19.6 Melanoma

One of the main reasons for an increase in cancer mortality rates is the lethal pattern of skin cancer disorders known as melanoma. Due to the apparent resemblance between benign and malignant skin lesions, it is difficult for medical experts to identify skin lesions in dermoscopic graphs. In order to help with the analysis of amelanotic lesions, dermoscopy is a noninvasive technology for the visual evaluation of a substructure of skin.

Kaplan et al. (2019) developed a clinical decision support system to help make clinical decisions and increase diagnostic accuracy. The proposed system uses pretrained deep-learning models to predict melanoma using dermoscopic images.

3.20 Image processing techniques

3.20.1 Image segmentation

Object identification, recognition tasks, video surveillance, medical imaging, and content-based image retrieval are just a few of the many uses for image segmentation. In computer-assisted detection and computer-assisted diagnostics systems, segmentation is necessary for detecting tumors and diseases, analyzing anatomical structures, measuring tissue volume, diagnosing colon and breast cancer, and spotting lengthy nodules (Khastavaneh & Haron, 2015).

3.20.2 Super-resolution

SR is the practice of enhancing resolution. The goal of the SR image is to provide finer details that are absent from the LR image while scaling it up to create a HR image (Temiz & Bilge, 2020). With SR, resolution can be raised while keeping or even enhancing image details. It can be used with either a single image or a collection of images. Multiframe SR techniques first register image sequences before estimating relative motion. After that, an interpolation technique is used to project the pertinent scene sequences into a coordinate system. To enhance image quality, several methods for image augmentation interpolate finer details. In contrast, a single image SR technique aims to produce an HR image from a single LR image.

3.20.3 Data augmentation

The overfitting issue frequently arises during the DL model training phase as a result of data issues. However, data augmentation techniques can potentially address the issue of overfitting in addition to dropout. Data augmentation addresses the issue of overfitting by enhancing the diversity of image data using techniques like random pixel panning, random pixel enlargement and reduction, or random pixel color adjustment. The likelihood of overfitting decreases as more data are ingested by the model during training from these processed inputs. Data augmentation can be helpful in the realm of medical research in addition to resolving the overfitting issue.

3.20.4 Artificial intelligence techniques

DL algorithms are frequently employed in the analysis of images. CNNs can be taught to assess and categorize medical images in the context of mammograms in order to find probable breast cancer. Multiple layers of filters are employed throughout the process, and they are trained to recognize particular traits in the mammography pictures. These characteristics may consist of mass size and shape, tissue density, and other traits that are diagnostic of cancer. The network's succeeding layers then process the filter outputs to create a final prediction concerning the existence of cancer. The benefit of utilizing CNNs for mammography analysis is that they can automatically pick up on the characteristics that are crucial for classification, eliminating the need for manual feature engineering and enabling more precise predictions. A clustering algorithm called K-means (MacQueen, 1967) divides data into groups according to similarity. Recurrent neural networks (RNNs) are a subclass of DL neural networks used for processing sequential data, including time series data or natural language. Random forest

(Breiman, 2001) is an ensemble learning technique for classification and regression that combines different decision trees to create predictions that are more accurate. Boosting (Freund & Schapire, 1997) is an ensemble learning strategy that joins several weak models to create a stronger model.

Gradient-guided conditional generative adversarial networks (GGGAN) are a type of DL algorithm that has been recently applied in the field of synthetic digital mammography (SDM). GGGANs are a combination of generative adversarial networks (GANs) and gradient-based optimization methods that generate high-quality synthetic mammography images (Oyelade et al., 2022). The GGGAN architecture consists of two neural networks, the generator and the discriminator, that work together to produce synthetic mammography images, which are indistinguishable from real images. The generator network generates synthetic images, while the discriminator network assesses the quality of the generated images. The gradient information from the discriminator network is used to guide the generator network, improving the quality of the generated images over time. In the field of SDM, GGGANs have been used to generate synthetic mammography images for data augmentation and to overcome the limitations of real mammography data, such as small sample sizes and limited diversity. GGGANs have shown promising results in various SDM applications, including image classification, breast density assessment, and lesion detection.

DL architecture CapsNet, commonly referred to as the Capsule Network, was unveiled in 2017 by Sabour et al. (2017). Each capsule in the CapsNet architecture is a tiny neural network that represents a particular object or component of an image. Each capsule produces a vector as its final product, which is then utilized to illustrate how an image's component pieces relate to its overall composition. On the other hand, back-propagation neural network (BPNN) employs backpropagation as its learning algorithm and is a conventional neural network architecture. The weights of the nodes in BPNN's numerous layers of interconnected nodes are modified throughout the training process. Learning the characteristics of the images and predicting the class labels connected to them are the two goals of BPNN. Both CapsNet and BPNN can be used for image preprocessing to extract features from images, decrease the dimensionality of the data, and improve the quality of the images. While BPNN is used to discover the underlying patterns and features of an image, CapsNet is used to determine the relationships between various components of an image. The outcomes of these preprocessing stages are then used to do the final classification of the images using a classification model, such as a support vector machine (SVM) or a CNN.

While BPNN is better suited for activities that demand learning the characteristics of an image, CapsNet is more suited for applications that need capturing the relationships between various components of an image.

In order to perform classification tasks, a hybrid modified SVM and K-nearest neighbors (KNN) classifier combines the strengths of the two well-known ML methods. The goal of this hybrid strategy is to enhance the classifier's overall performance by utilizing the advantages of both algorithms. SVM is a linear classifier that maximizes the margin between classes to choose the best line dividing them. With small datasets and high dimensional data, it performs effectively. Large datasets or complicated nonlinear correlations between features and class labels could make it less useful. On the other hand, KNN is a nonlinear classifier (Jain & Singh, 2012) that assigns a sample to one of its KNN depending on the majority class. Large datasets can be handled effectively, although it can be computationally expensive and may be sensitive to the choice of k (Hasan & Yusoff, 2010). The SVM is first utilized in the hybrid strategy to categorize the data and find the samples that are challenging to classify. The KNN classifier is then given these samples to examine further. Based on the findings of both classifiers combined, the ultimate choice is chosen. ResNet34 is a deep residual network that was developed to overcome the vanishing gradient problem in deep neural networks. The network consists of residual blocks with convolutional layers, activation functions, and skip connections. In the context of breast cancer diagnosis, ResNet34 has been used to classify mammogram images as benign or malignant. A study showed that ResNet34 achieved a high accuracy rate of 94.3% in detecting breast cancer from mammogram images.

Wavelet neural network (WNN) is a type of deep neural network that combines the benefits of wavelet transform and neural networks. Wavelet transform is a mathematical tool that helps to extract features from images, while neural networks are used to perform classification tasks. WNN has been used to diagnose breast cancer by analyzing mammogram images. A study showed that WNN achieved an accuracy rate of 95.2% in detecting breast cancer from mammogram images, outperforming traditional machine-learning models.

Both ResNet34 and WNN have demonstrated high accuracy rates in detecting breast cancer from mammogram images. The choice of which model to use would depend on the specific requirements and limitations of the application. However, these DL models have the potential to revolutionize the way breast cancer is diagnosed, making the process faster, more accurate, and less dependent on human error (Escorcia-Gutierrez et al., 2022).

Over 9000 mammograms were processed using a median filter, contrast-limited adaptive histogram equalization, and data augmentation. A CNN was then used to train a categorized model. The findings of the experiment showed that the model with preprocessed image dataset performed much better in terms of accuracy than the model without preprocessed images (Lu et al., 2019). Jiang et al. (2021) proposed a deep CNN with gradient guided conditional generative adversarial networks (GGGAN) that transform digital breast tomosynthesis to SDM. They stated that it can be used instead of full-field digital mammography, which reduces the dose of radiation for breast cancer screening. Song et al. (2022) made a study to evaluate the deep CNN performance for the segmentation of apical lesions using panoramic radiographs. They stated that deep CNNs using U-Net can achieve high performance for apical lesion detection. According to their results, 0.828, 0.815, and 0.742 F1-scores were obtained with 0.3, 0.4, and 0.5 IoU thresholds, respectively.

A method for training an ANN is called MTANN. In this training method, subregions of the training image are extracted and given to an ANN. These subregion pixel sets are linked with a likelihood distribution map, which serves as the teacher image. By contrasting the network output with the teacher image, the network parameters are found. The likelihood distribution map that results from the combination of all the subregional outputs can be used to spot abnormalities. Massive training is a technique for training an ANN with a large number of subregions on a small number of images. To find anomalies like nodules on lung CT images, MTANN was frequently used (Khastavaneh & Haron, 2015).

Escorcia-Gutierrez et al. (2022) mentioned that there are increased false positive results in breast cancer identification and proposed an automated DL-based breast cancer diagnosis model to solve this. Gaussian filter and Tsallis entropy-based methods were used for preprocessing and image segmentation purposes, respectively. For feature extraction, ResNet34 was used while tuning its hyperparameters with the chimp optimization algorithm. And for the detection of breast cancer in digital mammograms, a WNN was used.

A DL model was developed by Kavitha et al. (2022) which is called optimal multilevel thresholding-based segmentation with DL enabled capsule network. This model uses mammograms for breast cancer diagnosis. Adaptive fuzzy-based median filtering was used for noise removal, and optimal Kapur's-based multilevel thresholding with shell game optimization algorithm was applied for segmentation. Their feature extractor was CapsNet based and BPNN was used as a classifier to detect breast cancer. Digital database for screening mammography (DDSM) and mini-Mammographic Image Analysis Society (MIAS) datasets were used to evaluate the proposed model. 97.55% and 98.50% accuracies were obtained on the DDSM and mini-MIAS datasets, respectively.

Mittal et al. (2022) used patch extraction with DL for image SR. In the proposed method the low-resolution image is divided into patches and then networks like residual dense networks and enhanced SR generative adversarial networks are applied to these image patches. PSNR values of each image patch are compared and the one with the highest PSNR value is picked and the SR image is reconstructed as output.

To improve the classification accuracy on low-contrast images, Malebary and Hashmi (2021) proposed a breast mass classification system that classifies normal, malignant, and benign breast masses. Their architecture is a combination of K-means, RNN, CNN, random forest, and boosting techniques. They used two public datasets DDSM and MIAS for evaluation and the resulting accuracies were 96% and 95%, respectively. Punithavathi (2021) made a study on the classification of three class mammogram images, which are normal, benign, and malignant. Filtering is applied to denoise the images and then segmentation is applied to extract features from that region of the image. The study states that the hybrid modified SVM and KNN classifier obtained higher performance when compared to the KNN and SVM individually.

3.21 Conclusion

In this chapter, different IP techniques and their use in various applications were discussed. AI is at the forefront of these applications. The image preprocessing phase takes place in most of the applications since it has been observed that the results obtained are more optimal when the IP step is included. Image preprocessing in medical imaging was also included and discussed in this chapter, which was related to AI. In systems related to medical imaging, more accurate results become more important because it is a sensitive area where health is at stake. To give more specific applications, information about medical cases and their applications was also indicated in Section 3.19. After mentioning each medical case, commonly used image processing techniques and artificial intelligence techniques related to these cases were also discussed in the last section. As reviewed in the sections of this chapter, besides their successful applications, IP and AI are very powerful when used together and can make contributions to humanity in many different areas such as security, education, assisted technology for disabled persons, and business.

References

Albahri, O. S., Zaidan, A. A., Albahri, A. S., Zaidan, B. B., Abdulkareem, K. H., Al-qaysi, Z. T., Alamoodi, A. H., Aleesa, A. M., Chyad, M. A., Alesa, R. M., Kem, L. C., Lakulu, M. M., Ibrahim, A. B., & Rashid, N. A. (2020). Systematic review of artificial intelligence techniques in the detection and classification of COVID-19 medical images in terms of evaluation and benchmarking: Taxonomy analysis, challenges, future solutions and methodological aspects. *Journal of Infection and Public Health*, *10*, 1381–1396. Available from https://doi.org/10.1016/j.jiph.2020.06.028, http://www.elsevier.com/wps/find/journaldescription.cws_home/716388/descriptio.

Al-khassaweneh, M., & AlShorman, O. (2020). Frei-Chen bases based lossy digital image compression technique. *Applied Computing and Informatics*. Available from https://doi.org/10.1016/j.aci.2019.120.004.

Bao, S. (2021). An improved non-local mean filtering algorithm based on medical image restoration, Proceedings – 2021 International Conference on Computer Engineering and Artificial Intelligence, ICCEAI 2021, Institute of Electrical and Electronics Engineers Inc. China, 43–47. Available from https://doi.org/10.1109/ICCEAI52939.2021.00008, http://ieeexplore.ieee.org/xpl/mostRecentIssue.jsp?punumber = 9544086.

Beeravolu, A. R., Azam, S., Jonkman, M., Shanmugam, B., Kannoorpatti, K., & Anwar, A. (2021). Preprocessing of breast cancer images to create datasets for deep-CNN. *IEEE Access*, 33438–33463. Available from https://doi.org/10.1109/ACCESS.2021.3058773, http://ieeexplore.ieee.org/xpl/RecentIssue.jsp?punumber = 6287639.

Ben Ali Kaddour, A., & Abdulaziz, N. (2021). Artificial Intelligence Pathologist: The use of Artificial Intelligence in Digital Healthcare, IEEE Global Conference on Artificial Intelligence and Internet of Things, GCAIoT 2021, Institute of Electrical and

Electronics Engineers Inc. United Arab Emirates, 31–36. Available from https://doi.org/10.1109/GCAIoT53516.2021.9693090, http://ieeexplore.ieee.org/xpl/mostRecentIssue.jsp?punumber = 9692905.

Boudouh, S. S., & Bouakkaz, M. (2022). Breast tumor detection in mammogram images using convolutional neural networks. In *2022 IEEE 9th International Conference on Sciences of Electronics, Technologies of Information and Telecommunications (SETIT)*, (pp. 330–336). Hammamet, Tunisia: IEEE. Available from https://doi.org/10.1109/SETIT54465.2022.9875567.

Breiman, L. (2001). Random forests. *Machine Learning*, *1*, 5–32. Available from https://doi.org/10.1023/A:1010933404324.

Dagli, M. M., Rajesh, A., Asaad, M., & Butler, C. E. (2021). The use of artificial intelligence and machine learning in surgery: A comprehensive literature review. *American Surgeon*. Available from https://doi.org/10.1177/00031348211065101, https://journals.sagepub.com/home/asua.

Danon, D., Arar, M., Cohen-Or, D., & Shamir, A. (2021). Image resizing by reconstruction from deep features. *Computational Visual Media*, *4*, 453–466. Available from https://doi.org/10.1007/s41095-021-0216-x, http://link.springer.com/journal/41095.

Daoud, M. K. & Otair, M. (2020). The role of artificial intelligence and the internet of things in the development of medical radiology (an experimental study on magnetic resonance imaging). Proceedings – 2020 International Conference on Intelligent Computing and Human-Computer Interaction, ICHCI 2020, Institute of Electrical and Electronics Engineers Inc. Malaysia, 17–20. Available from https://doi.org/10.1109/ICHCI51889.2020.00011, http://ieeexplore.ieee.org/xpl/mostRecentIssue.jsp?punumber = 9424768.

Deepa Jyothi, K. (2018). A robust and efficient pre processing techniques for stereo images 2018. International Conference on Electrical, Electronics, Communication Computer Technologies and Optimization Techniques, ICEECCOT 2017, Institute of Electrical and Electronics Engineers Inc. India, 89–92. Available from https://doi.org/10.1109/ICEECCOT.2017.8284645.

Dhawan, V., Sethi, G., Lather, V. S., & Sohal, K. (2013). Power law transformation and adaptive gamma correction: A comparative study. *International Journal of Electronics & Communication Technology*, 2, 118–123.

Dimililer, K., Dindar, H., & Al-Turjman, F. (2021). Deep learning, machine learning and internet of things in geophysical engineering applications: An overview. *Microprocessors and Microsystems*. Available from https://doi.org/10.1016/j.micpro.2020.103613, https://www.journals.elsevier.com/microprocessors-and-microsystems.

Dimililer, K., Ever, Y. K. & Ugur, B. (2016), ILTDS: Intelligent lung tumor detection system on CT images. Advances in Intelligent Systems and Computing, 530 Springer Verlag, Cyprus, 225–235. Available from http://www.springer.com/series/11156, https://doi.org/10.1007/978-3-319-47952-1_17 21945357.

Dimililer, K. & Sekeroglu, B. (2021). The effect of discrete cosine transform on COVID-19 differentiation from chest X-Ray images: A preliminary study. International Conference on INnovations in Intelligent SysTems and Applications, INISTA 2021 - Proceedings, Institute of Electrical and Electronics Engineers Inc. Cyprus. Available from https://doi.org/10.1109/INISTA52262.2021.9548637, http://ieeexplore.ieee.org/xpl/mostRecentIssue.jsp?punumber = 9548323.

Dimililer, K. (2021). DCT-based medical image compression using machine learning 2021. Available from https://doi.org/10.1007/s11760-021-01951-0.

Dimililer, K., Hesri, A., Ever, Y. K., & Yang, H. (2017). Lung lesion segmentation using gaussian filter and discrete wavelet transform. ITM Web of Conferences. 01018. Available from https://doi.org/10.1051/itmconf/20171101018.

Dimililer, K., & İlhan, A. (2016). Effect of image enhancement on MRI brain images with neural networks. *Procedia Computer Science*, *102*, 39–44.

Escorcia-Gutierrez, J., Mansour, R. F., Beleño, K., Jiménez-Cabas, J., Pérez, M., Madera, N., & Velasquez, K. (2022). Automated Deep Learning Empowered Breast Cancer Diagnosis Using Biomedical Mammogram Images. *Computers, Materials and Continua*, *2*, 4221–4235. Available from https://doi.org/10.32604/cmc.2022.022322, https://www.techscience.com/cmc/v71n2/46353.

Fitriati, D. & Murtako, A. (2017), Implementation of Diabetic Retinopathy screening using realtime data. International Conference on Informatics and Computing, ICIC 2016, Institute of Electrical and Electronics Engineers Inc. Indonesia, 198–203. Available from https://doi.org/10.1109/IAC.2016.7905715.

Fong, J. X., Shapiai, M. I., Tiew, Y. Y., Batool, U. & Fauzi, H. (2020). Bypassing MRI Pre-processing in Alzheimer's Disease Diagnosis using Deep Learning Detection Network. Proceedings –2020 16th IEEE International Colloquium on Signal Processing and its Applications, CSPA 2020, Institute of Electrical and Electronics Engineers Inc. Malaysia, 219–224. Available from https://doi.org/10.1109/CSPA48992.2020.9068680, http://ieeexplore.ieee.org/xpl/mostRecentIssue.jsp?punumber = 9052211.

Freund, Y., & Schapire, R. E. (1997). A decision-theoretic generalization of on-line learning and an application to boosting. *Journal of Computer and System Sciences*, *1*, 119–139. Available from https://doi.org/10.1006/jcss.1997.1504.

Gupta, A., Jain, V., & Hussain, W. (2021). A fog–cloud computing-inspired image processing-based framework for lung cancer diagnosis using deep learning. *Springer Science and Business Media LLC*, 19–27. Available from https://doi.org/10.1007/978-981-16-2248-9_3.

Haque, I., & Neubert, J. (2020). Deep learning approaches to biomedical image segmentation. Informatics in Medicine Unlocked. Available from https://doi.org/10.1016/j.imu.2020.100297.

Hasan, M. A., & Yusoff, M. R. M. (2010). Hybrid of SVM and KNN for improving the accuracy of text categorization. *Journal of Theoretical and Applied Information Technology*, *1*, 46–50.

Hasanzad, M., Aghaei Meybodi, H. R., Sarhangi, N., & Larijani, B. (2022). Artificial intelligence perspective in the future of endocrine diseases. *Journal of Diabetes and Metabolic Disorders*. Available from http://www.jdmdonline.com/, https://doi.org/10.1007/s40200-021-00949-2.

Huang, H., Zheng, H., Lin, L., Cai, M., Hu, H., Zhang, Q., Chen, Q., Iwamoto, Y., Han, X., Chen, Y. W., & Tong, R. (2021). Medical image segmentation with deep atlas prior. *IEEE Transactions on Medical Imaging*, *12*, 3519–3530. Available from https://doi.org/10.1109/TMI.2021.3089661, http://ieeexplore.ieee.org/xpl/RecentIssue.jsp?punumber = 42.

Jain, P. K., & Singh, A. K. (2012). Hybrid SVM and KNN classifier for improved classification of mammogram images. *Journal of Medical Systems*, *3*, 953–961.

Janjua, J. I., Khan, T. A., & Nadeem, M. (2022). Chest X-Ray Anomalous Object Detection and Classification Framework for Medical Diagnosis. International Conference on Information Networking, 2022, IEEE Computer Society, Pakistan, 158–163. Available from https://doi.org/10.1109/ICOIN53446.2022.9687110, http://www.icoin.org/.

Jiang, G., Wei, J., Xu, Y., He, Z., Zeng, H., Wu, J., Qin, G., Chen, W., & Lu, Y. (2021). Synthesis of mammogram from digital breast tomosynthesis using deep convolutional neural network with gradient guided cGANs. *IEEE Transactions on Medical Imaging*, *8*, 2080–2091. Available from https://doi.org/10.1109/TMI.2021.3071544, http://ieeexplore.ieee.org/xpl/RecentIssue.jsp?punumber = 42.

Kadhim, M. A. (2021). Restoration Medical Images from Speckle Noise Using Multifilters. 2021 7th International Conference on Advanced Computing and Communication Systems, ICACCS 2021, Institute of Electrical and Electronics Engineers Inc. Iraq, 1958–1963. Available from https://doi.org/10.1109/ICACCS51430.2021.9441814, http://ieeexplore.ieee.org/xpl/mostRecentIssue.jsp?punumber = 9441490.

Kaplan, A., Guldogan, E., Colak, C., & Arslan, A. K. (2019). Prediction of melanoma from dermoscopic images using deep learning-based artificial intelligence techniques. International Conference on Artificial Intelligence and Data Processing Symposium, IDAP 2019. Institute of Electrical and Electronics Engineers Inc. Turkey. Available from https://doi.org/10.1109/IDAP.2019.8875970, http://ieeexplore.ieee.org/xpl/most RecentIssue.jsp?punumber = 8864030 9781728129327.

Kavipriya, K., & Hiremath, M. (2021). Analysis of benchmark image pre-processing techniques for coronary angiogram images. International Conference on Innovative Trends in Information Technology, ICITIIT 2021, Institute of Electrical and Electronics Engineers Inc. India. Available from https://doi.org/10.1109/ICITIIT51526.2021.9399602, http://ieeexplore.ieee.org/xpl/mostRecentIssue.jsp?punumber = 9399548.

Kavitha, T., Mathai, P. P., Karthikeyan, C., Ashok, M., Kohar, R., Avanija, J., & Neelakandan, S. (2022). Deep learning based capsule neural network model for breast cancer diagnosis using mammogram images. *Interdisciplinary Sciences – Computational Life Sciences*, *1*, 113–129. Available from https://doi.org/10.1007/s12539-021-00467-y, http://www.springer.com/life + sciences/systems + biology + and + bioinformatics/journal/12539.

Khashman, A., & Dimililer, K. (2008). Image compression using neural networks and Haar wavelet. WSEAS Transactions on Signal Processing. (5), 330–339 Available from http://www.wseas.us/e-library/transactions/signal/2008/27-363.pdf. Cyprus.

Khastavaneh, H., & Haron, H. (2015). A conceptual model for segmentation of multiple scleroses lesions in magnetic resonance images using massive training artificial neural network. Proceedings - International Conference on Intelligent Systems, Modelling and Simulation, ISMS, 2015, IEEE Computer Society, Malaysia, 273–278. Available from https://doi.org/10.1109/ISMS.2014.53, http://ieeexplore.ieee.org/xpl/mostRecentIssue.jsp?punumber = 6168459.

Lin, M., He, X., Guo, H., He, M., Zhang, L., Xian, J., Lei, T., Xu, Q., Zheng, J., Feng, J., Hao, C., Yang, Y., Wang, N., & Xie, H. (2022). Use of real-time artificial intelligence in detection of abnormal image patterns in standard sonographic reference planes in screening for fetal intracranial malformations. *Ultrasound in Obstetrics and Gynecology*, *3*, 304–316. Available from https://doi.org/10.1002/uog.24843, http://obgyn.onlinelibrary.wiley.com/hub/journal/0.10.1002/(ISSN)1469-0705/.

Lin, S. Y., & Chang, H. Y. (2021). Tooth numbering and condition recognition on dental panoramic radiograph images using CNNs. *IEEE Access*, 166008–166026. Available from https://doi.org/10.1109/ACCESS.2021.3136026, http://ieeexplore.ieee.org/xpl/RecentIssue.jsp?punumber = 6287639.

Lu H. C., Loh E. W. & Huang S. C. (2019). The Classification of Mammogram Using Convolutional Neural Network with Specific Image Preprocessing for Breast Cancer Detection. 2nd International Conference on Artificial Intelligence and Big Data,

ICAIBD 2019, Institute of Electrical and Electronics Engineers Inc. Taiwan, 9–12. Available from https://doi.org/10.1109/ICAIBD.2019.8837000, http://ieeexplore.ieee.org/xpl/mostRecentIssue.jsp?punumber = 8826551.

Lv, F. (2021). Data Preprocessing and Apriori Algorithm Improvement in Medical Data Mining. Proceedings of the 6th International Conference on Communication and Electronics Systems, ICCES 2021, Institute of Electrical and Electronics Engineers Inc. China, 1205–1208. Available from https://doi.org/10.1109/ICCES51350.2021.9489242, http://ieeexplore.ieee.org/xpl/mostRecentIssue.jsp?punumber = 9488658.

Ma, P., Li, Q., & Li, J. (2022). Application of artificial intelligence in cardiovascular imaging. *Journal of Healthcare Engineering*. Available from https://doi.org/10.1155/2022/7988880, http://www.hindawi.com/journals/jhe/contents/.

MacQueen J. (1967). 1967 Proceedings of 5th Berkeley symposium on mathematical statistics and probability some methods for classification and analysis of multivariate observations icsid.org 2022 What Is Power Law Transformation In Image Processing? Available from https://www.icsid.org/ 2022.

Malebary, S. J., & Hashmi, A. (2021). Automated breast mass classification system using deep learning and ensemble learning in digital mammogram. *IEEE Access*, 55312–55328. Available from https://doi.org/10.1109/ACCESS.2021.3071297, http://ieeexplore.ieee.org/xpl/RecentIssue.jsp?punumber = 6287639.

Mittal, H., Rai, V., Sonawane, S., & Mhatre, S. (2022). Image Resolution Enhancer using Deep Learning. Proceedings - International Conference on Applied Artificial Intelligence and Computing, ICAAIC 2022, Institute of Electrical and Electronics Engineers Inc. India, 578–586. Available from https://doi.org/10.1109/ICAAIC53929.2022.9792975, http://ieeexplore.ieee.org/xpl/mostRecentIssue.jsp?punumber = 9792618.

Nithyalakshmi, V., Sivakumar, Dr. R., & Sivaramakrishnan, Dr. A. (2021). Automatic detection and classification of diabetes using artificial intelligence. *International Academic Journal of Innovative Research*, *1*, 01–05. Available from https://doi.org/10.9756/iajir/v8i1/iajir0801.

Oh, S., Kim, Y. J., Park, Y. T., & Kim, K. G. (2022). Automatic pancreatic cyst lesion segmentation on eus images using a deep-learning approach. *Sensors*, *1*. Available from https://doi.org/10.3390/s22010245, https://www.mdpi.com/1424-8220/22/1/245/pdf.

Olawale, O., Dimililer, K., & Al-Turjman, F. (2020). *Chapter Six - AI simulations and programming environments for drones: An overview* (pp. 93–106). Elsevier. Available from https://doi.org/10.1016/B978-0-12-819972-5.00006-9.

Olawale, O. P., & Dimililer, K. (2020). Individual Eye Gaze Prediction with the Effect of Image Enhancement Using Deep Neural Networks. 4th International Symposium on Multidisciplinary Studies and Innovative Technologies, ISMSIT 2020 – Proceedings, Institute of Electrical and Electronics Engineers Inc. Turkey. Available from https://doi.org/10.1109/ISMSIT50672.2020.9254786, http://ieeexplore.ieee.org/xpl/mostRecentIssue.jsp?punumber = 9254214.

Olawale, O. P., Ozdamli, F., & Dimililer, K. (2021). Data Mining Techniques for the Classification of Medical Cases: A Survey. ISMSIT 2021—5th International Symposium on Multidisciplinary Studies and Innovative Technologies, Proceedings, Institute of Electrical and Electronics Engineers Inc. Turkey, 68–73. Available from https://doi.org/10.1109/ISMSIT52890.2021.9604724, http://ieeexplore.ieee.org/xpl/mostRecentIssue.jsp?punumber = 9604521.

Oyelade, O. N., Ezugwu, A. E., Almutairi, M. S., Saha, A. K., Abualigah, L., & Chiroma, H. (2022). A generative adversarial network for synthetization of regions of interest based on digital mammograms. *Scientific Reports* (1). Available from https://doi.org/10.1038/s41598-022-09929-9, http://www.nature.com/srep/index.html.

Partner, B. (2021). Digital image processing. Available from https://benchpartner.com/.

Punithavathi, V. (2021). A hybrid algorithm with modified SVM and KNN for classification of mammogram images using medical image processing with data mining techniques.

Rasti, B., Chang, Y., Dalsasso, E., Denis, L., & Ghamisi, P. (2021). Image restoration for remote sensing: Overview and toolbox. *IEEE Geoscience and Remote Sensing Magazine*. Available from https://doi.org/10.1109/MGRS.2021.3121761, http://ieeexplore.ieee.org/xpl/RecentIssue.jsp?punumber = 6245518.

Razzak, M. (2018). Deep learning for medical image processing: Overview, challenges and the future. *Classification in BioApps*, 323–350.

Renjith, C. V. & Wagaj, S. C. (2020). MRI Brain Disease Detection using Enhanced Landmark based Deep Feature Learning. IEEE, Greater Noida, India 2020. Available from https://doi.org/10.1109/ICACCCN51052.2020.9362863.

Sabour, S., Frosst, N., & Hinton, G. E. (2017). Dynamic routing between capsules. Advances in Neural Information Processing Systems, Neural information processing systems foundation Canada 2017, 3857–3867.

Saidnassim, N., Abdikenov, B., Kelesbekov, R., Akhtar, M. T. & Jamwal, P. (2021). Self-supervised Visual Transformers for Breast. Cancer Diagnosis. 2021 Asia-Pacific Signal and Information Processing Association Annual Summit and Conference, APSIPA ASC 2021 – Proceedings, Institute of Electrical and Electronics Engineers Inc. Kazakhstan, 423–427. Available from http://ieeexplore.ieee.org/xpl/mostRecentIssue.jsp?punumber = 9688568.

Salama, W. M., Aly, M. H. & Elbagoury, A. M. (2021). Lung Images Segmentation and Classification Based on Deep Learning: A New Automated CNN Approach. Journal of Physics: Conference Series, 2128, 1, IOP Publishing Ltd Egypt. Available from https://doi.org/10.1088/1742-6596/2128/1/012011, http://iopscience.iop.org/journal/1742-6596.

Samira Masoudi, S., Raviprakash, H., Bagci, U., Choyke, P., & Turkbey, B. (2021). Quick guide on radiology image pre-processing for deep learning applications in prostate cancer research. *Journal of Medical Imaging*, *1*. Available from https://doi.org/10.1117/1.JMI.8.1.010901.

Sann, S. S., Win, S. S., & Thant, Z. M. (2021). An analysis of various image pre-processing techniques in butterfly image. *International Journal for Advance Research and Development*, *1*, 1–4.

Sengoz, N., Yigit, T., Ozmen, O., & Işık, A. H. (2022). Importance of preprocessing in histopathology image classification using deep convolutional neural network.

Shi, F., Wang, J., Shi, J., Wu, Z., Wang, Q., Tang, Z., He, K., Shi, Y., & Shen, D. (2021). Review of artificial intelligence techniques in imaging data acquisition, segmentation, and diagnosis for COVID-19. *IEEE Reviews in Biomedical Engineering*, 4–15. Available from https://doi.org/10.1109/RBME.2020.2987975, http://ieeexplore.ieee.org/xpl/RecentIssue.jsp?punumber = 4664312.

Sideraki, A., & Drigas, A. (2021). Artificial intelligence (AI) in autism. *Technium Social Sciences Journal*, 262–277. Available from https://doi.org/10.47577/tssj.v26i1.5208.

Song, I.-S., Shin, H.-K., Kang, J.-H., Kim, J.-E., Huh, K.-H., Yi, W.-J., Lee, S.-S., & Heo, M.-S. (2022). Deep learning-based apical lesion segmentation from panoramic radiographs. *Imaging Science in Dentistry*, *4*, 351. Available from https://doi.org/10.5624/isd.20220078.

Soosai, M. R., Joshya, Y. C., Kumar, R. S., Moorthy, I. G., Karthikumar, S., Chi, N. T. L., & Pugazhendhi, A. (2021). Versatile image processing technique for fuel science: A review. *Science of the Total Environment*. Available from https://doi.org/10.1016/j.scitotenv.2021.146469, http://www.elsevier.com/locate/scitotenv.

Sowmya, B. J., Shetty, Chetan, Seema, S., & Srinivasa, K. G. (2020). Chapter 7 - Utility system for premature plant disease detection using machine learning. *Hybrid Computational Intelligence for Pattern Analysis and Understanding*, 149–172. Available from https://doi.org/10.1016/B978-0-12-818699-2.00008-1.

Subramaniam, U., Subashini, M. M., Almakhles, D., Karthick, A., & Manoharan, S. (2021). An expert system for COVID-19 infection tracking in lungs using image processing and deep learning techniques. *BioMed Research International*. Available from https://doi.org/10.1155/2021/1896762, http://www.hindawi.com/journals/biomed/.

Sun, J., & Yuan, X. (2022). Application of artificial intelligence nuclear medicine automated images based on deep learning in tumor diagnosis. *Journal of Healthcare Engineering*. Available from https://doi.org/10.1155/2022/7247549, http://www.hindawi.com/journals/jhe/contents/.

Sun, Y. C., Hsieh, A. T., Fang, S. T., Wu, H. M., Kao, L. W., Chung, W. Y., Chen, H. H., Liou, K. D., Lin, Y. S., Guo, W. Y., & Lu, H. H. S. (2021). Can 3D artificial intelligence models outshine 2D ones in the detection of intracranial metastatic tumors on magnetic resonance images? *Journal of the Chinese Medical Association: JCMA*, *10*, 956–962. Available from https://doi.org/10.1097/JCMA.0000000000000614.

Tajbakhsh, N., Jeyaseelan, L., Li, Q., Chiang, J. N., Wu, Z., & Ding, X. (2020). Embracing imperfect datasets: A review of deep learning solutions for medical image segmentation. *Medical Image Analysis*. Available from https://doi.org/10.1016/j.media.2020.101693, http://www.elsevier.com/inca/publications/store/6/2/0/9/8/3/index.htt.

Temiz, H., & Bilge, H. S. (2020). Super resolution of B-mode ultrasound images with deep learning. *IEEE Access*, 78808–78820. Available from https://doi.org/10.1109/ACCESS.2020.2990344, http://ieeexplore.ieee.org/xpl/RecentIssue.jsp?punumber = 6287639.

Tiwari, A., Chaudhari, M., & Rai, A. (2019). Multidisciplinary Approach of Artificial Intelligence over Medical Imaging: A Review, Challenges, Recent Opportunities for Research. Proceedings of the 3rd International Conference on I-SMAC IoT in Social, Mobile, Analytics and Cloud, I-SMAC 2019, Institute of Electrical and Electronics Engineers Inc. India. 237–242. Available from https://doi.org/10.1109/I-SMAC47947.2019.9032566, http://ieeexplore.ieee.org/xpl/mostRecentIssue.jsp?punumber = 9018087.

Velan, S., & Poluru, V. (2020). Application of Digital Image Processing Techniques in Determining the Quality of ARC and MIG Welding of Steel Joints. 1320–1325. Available from https://doi.org/10.1109/ICRITO48877.2020.9198021.

Wan, C., Ye, M., Yao, C., & Wu, C. (2018). Brain MR image segmentation based on Gaussian filtering and improved FCM clustering algorithm. Proceedings – 2017 10th International Congress on Image and Signal Processing, BioMedical Engineering and Informatics, CISP-BMEI 2017, Institute of Electrical and Electronics Engineers Inc. China 2018, 1–5. Available from https://doi.org/10.1109/CISP-BMEI.2017.8301978.

Yan, X., Wen, L., & Ga, L. (2019). A fast and effective image preprocessing method for hot round steel surface. *Open Access*. Available from https://doi.org/10.1155/2019/9457826.

Yang, X. Y., Wang, L. T., Li, G. D., Yu, Z. K., Li, D. L., Guan, Q. L., Zhang, Q. R., Guo, T., Wang, H. L., & Wang, Y. W. (2022). Artificial intelligence using deep neural network learning for automatic location of the interscalene brachial plexus in ultrasound images. *European Journal of Anaesthesiology*, *9*, 758–765. Available from https://doi.org/10.1097/EJA.0000000000001720, http://journals.lww.com/ejanaesthesiology/pages/default.aspx.

Zhang, Q., Du, Y., Wei, Z., Liu, H., Yang, X., & Zhao, D. (2021). Spine medical image segmentation based on deep learning. *Journal of Healthcare Engineering*. Available from https://doi.org/10.1155/2021/1917946, http://www.hindawi.com/journals/jhe/contents/.

Zhou, Y., Ren, D., Emerton, N., Lim, S., & Large, T. (2021). Image Restoration for Under-Display Camera. Proceedings of the IEEE Computer Society Conference on Computer Vision and Pattern Recognition, IEEE Computer Society undefined, 9175–9184. Available from https://doi.org/10.1109/CVPR46437.2021.00906.

CHAPTER

Artificial intelligence in mammography: advances and challenges

4

Salam Dhou[1], Khaldoon Alhusari[1] and Mohanad Alkhodari[2,3]

[1]*Department of Computer Science and Engineering, American University of Sharjah, Sharjah, United Arab Emirates*

[2]*Cardiovascular Clinical Research Facility (CCRF), Radcliffe Department of Medicine, University of Oxford, Oxford, United Kingdom*

[3]*Healthcare Engineering Innovation Center (HEIC), Department of Biomedical Engineering, Khalifa University, Abu Dhabi, United Arab Emirates*

4.1 Introduction to mammography as early breast cancer detection tool

Female breast cancer has surpassed lung cancer as the most commonly diagnosed cancer worldwide according to the GLOBOCAN 2020 estimates of cancer incidence and mortality produced by the International Agency for Research on Cancer (Sung et al., 2021). It is the leading cause of cancer death among women worldwide.

Early detection of breast cancer can significantly increase the chances of successful treatment and survival. Breast imaging is a specific type of radiology imaging that focuses solely on breast health. It plays a vital role in the early detection of breast cancer because it can show early signs of cancer in the breast a long time before the cancer is clinically evident. The most commonly used breast imaging modalities are mammography, ultrasound, breast magnetic resonance imaging, and breast tomosynthesis (3D mammography) (Sechopoulos et al., 2021). Other imaging modalities such as computed tomography (CT) or positron emission tomography can be used to detect the spread of breast cancer. Microwave imaging is a promising imaging modality for early breast cancer detection both as a standalone or complementary technique. However, it is still facing challenges that need to be addressed to unlock its full potential and translate it to clinical settings (AlSawaftah et al., 2022).

Screening mammography is the principal modality for early breast cancer detection, because it is the only method consistently found to decrease mortality rates caused by breast cancer. Mammography is a low-dose X-ray exam that produces images of the breast called mammograms. In its first implementation,

Artificial Intelligence and Image Processing in Medical Imaging. DOI: https://doi.org/10.1016/B978-0-323-95462-4.00004-2

screen-film mammography (SFM) was used for regular breast screening. Since the early 2000′s, the use of film was replaced with a digital X-ray detector, which resulted in the development and introduction of digital mammography (DM) in clinical use. Using DM, the resulting mammograms are stored and read in digital format which facilitates the enhancement, magnification, or manipulation of the exams for further evaluation (Sechopoulos et al., 2021). Mammography exams tend to vary in the number of views performed. Standard views which comprise routine screening mammography are bilateral craniocaudal (CC), medio-lateral oblique (MLO), medio-lateral, and lateromedially, with CC and MLO being the most commonly used for breast cancer detection and diagnosis.

Mammography can reveal early signs of breast cancer up to 4 years before a cancer becomes clinically evident (Marmot et al., 2013). Despite its efficacy, studies showed that 20% of newly diagnosed cancers and almost one-third on interval cancers are evident retrospectively on previous mammograms (Hoff et al., 2011). Interval cancers can be defined as those diagnosed within 1 year of a mammogram reported as normal an annual screening program or within 2 years of a mammogram reported as normal. Misinterpretations of perceived abnormalities and overlooked abnormalities are of the most common reason for missed breast cancers (Majid et al., 2003). To address this problem, strategies such as double reading have been selectively used to increase the cancer detection rate. However, this is labor intensive hence costly and therefore is not widely used, except when mandated by medical or government agencies.

Traditional computer-aided detection (CAD) systems were developed and used to increase the detection of breast cancer by reducing overlooked abnormalities without the need of a second observer (i.e., radiologist). CAD algorithms search for the same features that a radiologist looks for during case evaluation. In breast screening mammograms, CAD algorithms search for microcalcifications and masses (Castellino, 2005). Although CAD systems can be helpful in screening mammography, subsequent large-scale studies showed no benefit of CAD tools in improving radiologist diagnostic performance (Katzen & Dodelzon, 2018; Lehman & Topol, 2021). One of the largest studies to date, which included approximately 500,000 mammograms interpreted by 271 radiologists, found that screening performance was not improved with CAD on any metric (Lehman et al., 2015).

Artificial intelligence (AI) methods have been recently applied on mammography images to improve early detection of breast cancer. These techniques offer promise over the traditional CAD systems (Kohli et al., 2017) because of their ability to discover the features that indicate abnormalities in the images by the AI algorithms themselves during the training rather than being defined by the human programmer (Sechopoulos et al., 2021). Thus, unlike traditional CAD systems, AI algorithms have the potential to detect features and relationships that are unnoticeable by humans or currently unknown (Bahl, 2019).

4.2 Artificial intelligence applications in mammography

With recent advances in computer processing power and storage capabilities in addition to the vast amount of available digital health data, there is growing interest in applying AI and its applications such as machine learning to automate healthcare tasks and enhance accuracy in clinical medicine (Trister et al., 2017). AI technologies have the potential to support healthcare professionals in their repetitive and time-consuming healthcare tasks by automating these tasks and performing them entirely or partially. Studies expect that AI applications in healthcare will grow in the next decade (Cardoso et al., 2022).

Machine learning is a branch of AI that incorporate a variety of statistical and optimization techniques to give the computers the ability to learn from examples and to detect and categorize hard-to-identify patterns from massive, noisy, or complex datasets (Houssami et al., 2017). Applying these techniques to medical images, such as screening mammography, have the potential to enhance the efficacy of the breast cancer screening and improve the interpretive performance of radiologists (Trister et al., 2017). Mammography has attracted great attention for the use of AI in healthcare (Lehman & Topol, 2021). To date, AI has been used to analyze mammography images for different purposes. The following sections 4.2.1–4.2.5 present principle applications of AI in mammography (Lei et al., 2021). In each of the following sections, selected studies for each of these AI applications are discussed.

4.2.1 Artificial intelligence in the detection and classification of breast masses

Masses are one of the most common symptoms of breast cancer. Detecting and diagnosing masses can be difficult because of their variation in the shape, size, and margins. Having dense breasts makes it even harder to detect breast masses. Thus, mass detection and classification is a major step in computer-aided diagnosis (CAD) systems. Several AI-based methods have been developed to detect, segment, analyze, and classify breast masses in mammography images.

Rodriguez-Ruiz and colleagues (2019) compared the performance of a commercially available AI system to that of radiologists for breast cancer detection in DM and digital breast tomosynthesis (DBT). The AI system uses deep learning convolutional neural networks, feature classifiers, and image analysis algorithms to detect masses, calcifications, and other breast abnormalities. It was found that the AI system was as accurate in screening mammography interpretation as the radiologist (had an AUC-ROC higher than 61.4% of the radiologists). In another study by McKinney and colleagues (McKinney et al., 2020), the performance of an AI system was compared to that of human radiologists on large representative datasets from the UK and the USA. The study showed that the AI system outperformed all of human readers (AUC-ROC for the AI system was greater than the AUC-ROC for the average radiologist by an absolute margin of 11.5%).

Transfer learning in deep convolutional neural networks (DCNNs) is being used extensively in several medical imaging applications to solve medical imaging tasks. The generalization capabilities of the transfer learning techniques facilitate translating the "knowledge" learned from nonmedical images to medical diagnostic tasks. Transfer learning has been used in screening mammography for the purpose of breast mass detection and classification.

Samala and colleagues (2017) proposed a multitask transfer learning DCNN for the purpose of classifying breast masses as benign or malignant. A dataset consisting of SFMs and DMs collected from the University of Michigan Health System along with additional SFMs from the public digital database for screening mammography (DDSM) database were used in this study. This study showed that the multitask transfer learning DCNN model, which was trained on both SFMs and DMs and tested on SFMs, was found to have significantly ($P = .007$) higher lesion-based performance ($AUC = 0.82 \pm 0.02$) compared to the single-task transfer learning DCNN that was trained and tested using the SFMs only ($AUC = 0.78 \pm 0.02$). This study demonstrates that multitask transfer learning may be an effective approach for training DCNN in medical imaging applications when training samples from a single modality are limited.

Carneiro and colleagues (2017) developed a system that is based on deep learning models to classify both unregistered mammogram views and respective segmentation maps of breast lesions (i.e., masses and microcalcifications). They adopted two approaches in validating their system; the semiautomated approach (using manually defined mass and microcalcification segmentation maps) and the fully-automated approach (using segmentation maps generated by automated mass and microcalcification detection systems). Two publicly available datasets (INbreast and DDSM) were used to validate the system. While the results for the two approaches (i.e., the fully-automated and the semiautomated) were accurate, the semiautomated approach produced better results.

Al-masni and colleagues (2018) proposed a CAD system based on DCNN and fully connected neural network called you only look once (YOLO). Their proposed system aims to detect and classify breast masses into benign or malignant in one framework. The study used 600 mammograms from DDSM and their augmented mammograms of 2400 to train and test the system. The overall accuracy of breast mass detection was 99.7% while the accuracy of classifying the detected masses as benign or malignant was 97%. The proposed system was able to detect and classify masses even in challenging cases where masses were located in dense regions of the breast or over the pectoral muscles.

Al-antari and colleagues (2018) developed a fully integrated CAD where three deep learning models were used for different tasks. The first deep learning model used is YOLO. It was utilized for mass detection from the entire mammograms. The second deep learning model used was based on full resolution convolutional network (FrCN). It was adopted to segment the detected masses. The third deep learning model is a CNN one that was used to classify the masses as benign or malignant. The authors used INbreast dataset to validate their method. The CNN model used for classifying the detected breast masses achieved an overall accuracy of 95.64%, an AUC of 94.78%, a Matthews correlation coefficient (MCC) of 89.91%, and an F1-score of 96.84%

Ribli and colleagues (2018) proposed a CAD system based on faster region-based convolutional neural networks (R-CNN) to detect and classify breast lesions on mammograms as benign or malignant. Their method was tested on the public INbreast database and achieved an AUC of 95%.

Salama and Aly (2021) developed a framework for breast image segmentation and classification. They utilized the trained modified U-Net model to segment the breast area from the mammogram images. Then, several transfer learning models were adopted to classify the images as benign or malignant, including InceptionV3, DenseNet121, ResNet50, VGG16, and MobileNetV2. Three publicly available datasets were used to test their model, namely, MIAS, DDSM, and the curated breast imaging subset of DDSM (CBIS-DDSM). Data augmentation was applied and achieved the best results.

Tsochatzidis and colleagues (2021) proposed a method that integrates the mammographic mass segmentation information into a CNN to improve the diagnosis of breast cancer. In their proposed method, each convolutional layer of the CNN was modified to incorporate the segmentation maps that correspond to the input images. Two mammographic mass datasets, namely DDSM-400 and CBIS-DDSM were used to evaluate the proposed method. This study found that integrating segmentation information into a CNN improved the performance of breast cancer diagnosis of mammographic masses. Table 4.1 summarizes the machine learning studies for mass detection and classification.

4.2.2 Artificial intelligence in the detection and classification of microcalcifications

Breast calcifications are small calcium deposits in breast tissue. They appear as bright specks, lines or dots on mammograms because calcium readily absorbs the X-rays from mammography. Thus, mammography is commonly used to detect calcifications (Nalawade, 2009). There are two types of calcifications: microcalcifications and macrocalcifications. Macrocalcifications are large with diameter greater than 0.5 mm, typically well-defined, and are mostly benign and age-related. Microcalcifications are smaller in size with diameters usually less than 0.5 mm, with or without visible masses (Cruz-Bernal et al., 2018; Logullo et al., 2022). The existence of microcalcifications is common in breast cancer patients and they can be the early and only presenting sign of breast cancer. Certain features of microcalcifications can help the radiologist judge whether they are likely benign or malignant. These features are mostly related to their

Table 4.1 Summary of machine learning and deep learning studies for breast mass detection and classification.

References	Purpose	Dataset(s)	Method(s)	Results
Samala et al. (2017)	Classification of breast masses as benign or malignant	Private dataset from the University of Michigan Health System (1655 SFMs and 310 DMs), DDSM (277 SFMs)	Lesion-based multitask transfer learning DCNN model	AUC: 0.82 ± 0.02
Carneiro et al. (2017)	Classification of unregistered mammograms and respective segmentation maps of breast lesions (i.e., masses and microcalcifications)	INbreast and DDSM	CNN and CNN-F Chatfield et al. (2014) that is a simplified version of the AlexNet model	INbreast: manually-detected lesions: specificity: 0.92, sensitivity: 0.69, AUC: 0.91. INbreast: auto-detected lesions: specificity: 0.66, sensitivity: 0.69, AUC: 0.86 DDSM: manually-detected lesions: specificity: 0.97, sensitivity: 0.94, AUC: 0.99.
Al-masni et al. (2018)	Detection and classification of breast masses as benign or malignant	DDSM (600 mammograms)	CAD system based on DCNN and FC-NNs called You Only Look Once (YOLO)	Mass detection accuracy: 99.7%. Classification accuracy: 97%, AUC: 96.45%, sensitivity: 100%, and specificity: 94.0%

Al-Antari et al. (2018)	Detection, segmentation, and classification of breast masses as benign or malignant	INbreast (410 mammograms) from 115 patients	A completely integrated CAD system (YOLO) to detect masses, FrCN to segment masses, and the deep CNN to classify breast masses as benign or malignant	YOLO: mass detection accuracy: 98.96%, Matthews correlation coefficient (MCC): 97.62%, and F1-score: 99.24% FrCN: mass segmentation accuracy: 92.97%, MCC: 85.93%, and Dice (F1-score): 92.69%, and Jaccard similarity coefficient metrics: 86.37%. Deep CNN: classification accuracy: 95.64%, AUC: 94.78%, MCC: 89.91%, and F1-score: 96.84%
Ribli et al. (2018)	Detection and classification of breast lesions as benign or malignant	Trained on DDSM (2620 FSMs) and a dataset from Semmelweis University in Budapest (847 FFDM images), tested on INbreast (115 FFDM cases)	Faster R-CNN (based on VGG16)	AUC: 0.95

(*Continued*)

Table 4.1 Summary of machine learning and deep learning studies for breast mass detection and classification. *Continued*

References	Purpose	Dataset(s)	Method(s)	Results
Chougrad et al. (2018)	Classification of breast lesions as benign or malignant	DDSM + BCDR + InBreast (6116 images)	InceptionV3, ResNet50, VGG16	Accuracy: InceptionV3: 0.989, ResNet50: 0.987, VGG16: 0.986
Ragab et al. (2019)	Classification of breast lesions as benign or malignant	DDSM (2,620 cases), CBIS-DDSM (753 microcalcification cases and 891 mass cases)	DCNN: (AlexNet + SVM Classifier)	Sensitivity: 0.899, Accuracy: 0.872, AUC: 0.94
Salama and Aly (2021)	Image segmentation and classification as benign and malignant	MIAS (302 images), DDSM (534 images), and CBIS-DDSM (300 images)	InceptionV3, DenseNet121, ResNet50, VGG16 and MobileNetV2 for Classification	Accuracy: 98.87%, AUC: 98.88%, Sensitivity: 98.98%, Precision: 98.79%, F1-Score: 97.99%.
Tsochatzidis et al. (2021)	Diagnosis of segmented masses as benign or malignant	DDSM (400 images), CBIS-DDSM (1837 images)	Segmentation: U-Net and U-Net++, Diagnosis: ResNet-50	Using ground-truth segmentation maps, AUC of 0.898 for DDSM and 0.862 for CBIS-DDSM Using U-Net automatic segmentation, AUC of 0.880 for DDSM and 0.860 for CBIS-DDSM

appearance and distribution in the breast. However, microcalcifications are often missed or wrongly categorized during screening because of their small sizes and indirect scattering in mammogram images (Leong et al., 2022).

Several CAD systems have been developed to detect and categorize calcifications in mammography images. Guo et al., (2016) proposed a method for the detection of microcalcifications clusters. Their method achieves this goal by preprocessing the mammograms first to remove labels, noise, and unwanted structures (i.e., pectoral muscle), and enhance microcalcifications clusters. Then nonlinking simplified pulse-coupled neural network was used to detect microcalcifications clusters. Mammograms from publicly available datasets, namely MIAS and Japanese Society of Medical Imaging Technology were used to validate the method in addition to 20 images from the People's Hospital of Gansu Province, China. The detection results reveal that this method can detect the calcifications using mammogram images and hence can be used in clinical applications related to breast cancer diagnosis.

Suhail and colleagues (2018) developed a method for the classification of microcalcifications into benign and malignant classes. Their method was based on an improved Fisher linear discriminant analysis (LDA) approach for the linear transformation of segmented microcalcification data in combination with an SVM variant for the classification between the two classes. To validate their method, they used the DDSM, which has 288 regions of interest (ROIs) containing identified microcalcifications. Their classification method achieved an average accuracy of 96%.

Cai and colleagues (2019) developed a CNN model for the detection, analysis, and classification of microcalcifications in mammography images. They tested their algorithm on datasets collected from two medical institutions. Their results showed that using features extracted by the CNN model outperformed handcrafted features.

In another study, Rehman and colleagues (2021) proposed a CAD system that is based on FC-DSCNN for the detection and classification of microcalcification clusters into malignant and benign classes from mammograms. Their method was evaluated using DDSM dataset and another dataset collected from a local hospital (PINUM). Experimental results demonstrated that the proposed method achieved high performance in detecting and classifying microcalcifications.

Leong and colleagues (2022) proposed an adaptive transfer learning DCNN method that aims to segment breast mammogram images with microcalcification cases for early breast cancer diagnosis. Several deep neural network models, namely, ResNet50, ResNet34, VGG16, and AlexNet, were utilized and trained on mammogram images of breast microcalcifications. Results of this study showed that ResNet50 achieved the highest accuracy with a value of 97.58%. Table 4.2 summarizes the machine learning studies for the detection and classification of microcalcifications.

Table 4.2 Summary of machine learning and deep learning studies for the detection and classification of microcalcifications.

References	Purpose	Dataset(s)	Method(s)	Results
Guo et al. (2016)	Detection of microcalcification	118 mammograms from MIAS and Japanese Society of Medical Imaging Technology (JSMIT), and 20 mammograms from the People's Hospital of Gansu Province	SPCNN	Specificity: 94.7%, Sensitivity: 96.3%, AUC: 97.0%, Accuracy: 95.8%, MCC: 90.4%
Suhail et al. (2018)	Classification of microcalcifications as benign or malignant	DDSM (288 ROIs containing identified microcalcifications)	LDA approach combined with an SVM variant	Accuracy: 96%, AUC: 95%
Cai et al. (2019)	Detection, analysis, and classification of microcalcifications	Total of 990 images from the Sun Yat-Sen University Cancer Center (SYSUCC) and Nanhai Affiliated Hospital of Southern Medical University (Foshan, China)	A deep CNN with the same five convolutional layers as AlexNet	Accuracy: 89.32%, Sensitivity: 86.89%
Rehman et al. (2021)	Detection and classification of microcalcification as benign or malignant	DDSM (3568 images) and a dataset from PINUM hospital in Pakistan (2885 images)	FC-DSCNN	Using PINUM dataset: Sensitivity: 99%, specificity: 82%, accuracy: 90%, F1_score: 85%, precision: 89%, recall: 82% Using DDSM dataset: Sensitivity: 97%, Specificity: 83%, Accuracy: 87%, F1_score: 83%, precision: 85%, recall: 83%

Heenaye-Mamode Khan et al. (2021, n.d.)	Detection and classification of various types of breast abnormalities: calcifications, masses, asymmetry and carcinomas	CBIS-DDSM (2620 images)	ResNet50 and DCNN	Accuracy: DCNN: 88%
Leong et al. (2022)	Classification of microcalcifications as benign or malignant	CIBS-DDSM	ResNet50 compared with ResNet34, VGG16, and AlexNet	Accuracy: ResNet50: 97.58% (highest). ResNet34: 97.35%, VGG16: 96.97%, AlexNet: 83.06%

4.2.3 Artificial intelligence in mammographic breast density estimation

4.2.3.1 Importance of studying mammographic breast density

Early detection of breast cancer using known risk factors is crucial to saving patients' lives. One such strong risk factor is known as mammographic breast density. In simple terms, breast density refers to a measurement (can be quantitative or qualitative) of the proportion of radio-dense, non-fatty tissue—often called "fibroglandular" tissue—within a breast (Kallenberg et al., 2016). This measurement is a useful metric because it has been noted that women with high breast density are far more likely to develop breast cancer (Boyd et al., 2010).

Breast density can be ascertained from mammograms. This is because, within a mammogram, dense tissue appears light, whereas, in contrast, fatty tissue appears dark (Boyd et al., 2010). Samples of mammograms with differing densities are provided in Fig. 4.1. Potentially harmful growths and tumors also appear light within mammograms. This contrast makes it easier for radiologists to correctly identify tumors in low-density (fatty) breasts, as tumors will naturally stand out. However, in the case of highly dense breasts, mammogram sensitivity falls short, as dense tissue can mask potential tumors (Birdwell Robyn, 2009).

There are several widely used systems for measuring and classifying breast density. One such system, an industry standard, is the breast imaging-reporting

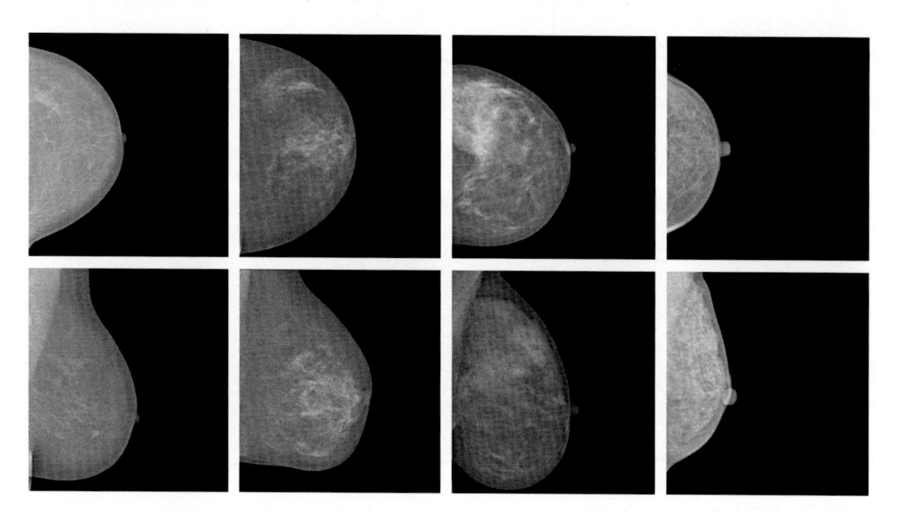

FIGURE 4.1 Samples of medio-lateral oblique (MLO) and craniocaudal (CC) view mammograms with differing densities.

The images are compiled from the INbreast dataset. The top row depicts CC mammograms and the bottom row depicts MLO mammograms. The mammograms are ordered in terms of increasing mammographic breast density from left to right, with the leftmost breast being almost entirely fatty, and the rightmost breast being almost entirely dense.

and data system (BI-RADS), which categorizes breasts into four indistinct density classes: fatty (BI-RADS I), scattered areas (BI-RADS II), heterogeneously dense (BI-RADS III), and extremely dense (BI-RADS IV) (Sickles, 2013). Fatty breasts have almost no fibroglandular tissue and are virtually entirely composed of fat. Fatty breasts make for highly sensitive mammograms, within which tumors are easily identifiable. Breasts with scattered areas of density are also mostly composed of fat but do contain a relatively small amount of fibro-glandular tissue. Growths within such breasts are also not too difficult to identify. In heterogeneously dense breasts, the moderate majority of the breast tissue (i.e., >50%) is fibroglandular. The density in such breasts may be concentrated in certain areas while other areas appear entirely fatty. It is possible for tumors to be hidden by the dense areas of heterogeneously dense breasts. Extremely dense breasts are almost entirely made up of fibro-glandular tissue, which lowers mammogram sensitivity, and consequently makes tumor identification very difficult and highly subjective.

The efficacy of mammograms has, since their inception, relied solely on the interpretation of the attending radiologist. In terms of breast density alone, this is not considered an issue, as radiologists tend to agree on the majority of cases (Redondo et al., 2012). However, there is a noticeable degree of variability in radiologist assessments, especially with regard to dense breasts. A study of the assessments of 83 radiologists noted that the percentage of breasts that were dense according to each radiologist's interpretation ranged from 6.3% to 84.5% (Sprague et al., 2016). This amount of subjectivity may dangerously influence a radiologist's decision as to whether harmful lesions are present within a mammogram.

In addition, it is rather common for cancers to be missed in mammograms only to later be found in retrospective reviews (Birdwell Robyn, 2009). This can sometimes be due to misinterpretations of true breast density, commonly caused by issues such as the assumption of asymmetry of fibroglandular tissue—in some cases, dense tissue is only visible in one mammographic view of the breast and not the other, which can cause a radiologist to mistake a lesion for dense tissue. Further, misjudgments of breast density can cause radiologists to make biased decisions. For example, they could assume a higher risk factor if they thought the breast was extremely dense, in which case it is possible for them to suggest more invasive procedures, such as DBT, which requires twice as much radiation as mammography (Nazari & Mukherjee, 2018). This is the case because mammogram sensitivity drops significantly for extremely dense breasts—the work in (Lamb et al., 2020) suggests it can be as low as 40%. Radiologists also commonly miss cancers in mammograms due to cognitive biases, misinterpretations, and other factors detailed in (Birdwell Robyn, 2009). As such, radiologist interpretation, while generally sound, can be problematic.

Considering the aforementioned issues, the need for a less subjective, more accurate, and easily integrable method of breast density estimation becomes clear. Various studies exist in the literature for automatic breast density estimation such the segmentation-based approaches and commercial software tools including

Quantra (Hartman et al., 2008) and Volpara (Highnam et al., 2010). However, these segmentation-based approaches and software tools are not discussed in this chapter. In this chapter, AI-based methods, specifically the ones that employ machine learning and deep learning methods for breast density estimation, are exclusively discussed.

4.2.3.2 Breast density estimation using artificial intelligence

Traditional machine learning and deep learning can both be used to automate breast density assessment (Destounis et al., 2017). The difference lies in that, for traditional machine leaning methods to work as intended, they must be preceded by a hand-crafted process of feature extraction. In this process, numerical data is derived from raw mammographic images. This data is consequently used to build a machine learning model that can predict the density class of an input mammogram. In contrast, deep learning techniques, can learn features directly from input mammograms, and do not require a designed feature extraction step. However, for such methods, the architecture of the model used requires scrutiny.

The work published in (Gong et al., 2019) details a method in which several gray-level co-occurrence matrices (GLCMs) were calculated for each input mammogram in order to create a feature vector of 528 dimensions. Additionally, the mean, skewness, and kurtosis were extracted from each image. Interestingly, (Gong et al., 2019) also calculates a density ratio based on pixel sums and combines it with the previously extracted features to create a 532-dimensional feature vector. This feature vector is then used to train a one-against-one support vector machine (SVM) model as the main model, as well as an extreme learning machine (ELM) model—a feedforward neural network detailed in (Huang et al., 2006)—for comparison. The models in (Gong et al., 2019) were trained and tested on three datasets—mini-MIAS (Suckling et al., 1994), DDSM (Heath et al., 2001), and a combination of mini-MIAS and a private dataset—and validated using 10-fold cross-validation. The best performing model was a linear kernel SVM, which achieved an average accuracy of 95.85% on the three datasets. ELM faded in comparison, getting an average accuracy of 82.90% for the three datasets.

In another prominent study, (Kumar et al., 2017), 11 first-order statistical features—including energy, uniformity, and entropy—were computed from histograms of regions of interest extracted from mammograms. In addition, 13 GLCM features, 5 gray-level difference statistics (GLDS) features, 11 gray-level run length matrix (GLRLM) features, and 210 laws' texture energy features were extracted. Furthermore, (Kumar et al., 2017) implemented 2D Gabor wavelet transform (2D GWT) to produce features that measure Gabor filtered statistical details including the mean and standard deviation—this is done for 21 images to produce 42 features. All of the features were then fed into six hierarchical models with varying architectures. The idea behind the models used in (Kumar et al., 2017) was to break the four-class problem down into three binary classification problems. As such, the first part of each of the models separates BI-RADS I breasts from the rest, the second part distinguishes BI-RADS II breasts from the rest, and the third part classifies breasts as either BI-RADS III or BI-RADS IV. In implementing this architecture, K-nearest neighbors (KNN), probabilistic neural network (PNN), artificial neural network (ANN), neuro fuzzy classifier (NFC), and SVM were used. In every model architecture aside from the sixth, a hierarchy of three blocks of the algorithm was implemented, with each block being preceded by probabilistic principle component analysis (PCA) for dimensionality reduction. The sixth model was created following the same hierarchy using the 3 best performing algorithms (one for each step): SVM, NFC, and KNN. The models were trained and tested using the DDSM database (Heath et al., 2001). Out of the first five models, the best performing model was PCA-NFC with an accuracy of 80.41%. It was followed by PCA-SVM and PCA-KNN whose accuracies were 78.33% and 72.50%, respectively. The hybrid model, consisting of a combination of all three, performed best overall, with an accuracy of 84.17%.

Also worth noting is the work in (Liu et al., 2011), where mammographic images were divided into subregions, and histogram moments were used to produce features that measure the variance, skewness, kurtosis, and means of each of the subregions. The features were then fed into an SVM, which was used to separate the subregions into high-density and low-density categories. Based on the output of the SVM, the density was estimated as the quotient of the number of high-density subregions and the total number of subregions. This numeric density estimate was then used to classify the mammogram into one of four density categories. The model was trained and tested on a private dataset and achieved an accuracy of 86.40%.

Uniquely, (Saffari et al., 2020) made use of a conditional generative adversarial network (cGAN) to generate a binary mask of input mammograms. The encoder of the cGAN's generator learns features about the input mammograms, and the decoder of the generator produces the binary mask. The mask is then fed to a convolutional neural network (CNN) consisting of three convolutional layers and two fully connected layers. The CNN is used to classify a given mammogram into one of four density classes. The model also estimates a numerical density value based on a ratio of the number of dense tissue pixels and the number of all breast tissue pixels. The model was trained and tested on the INbreast database (Moreira et al., 2012), on balanced and imbalanced input data of different sizes (128×128 and 64×64 pixels). The best performance was accrued when using a balanced dataset and an input size of 128×128 pixels. The accuracy was 98.75%, with a precision of 97.50% and a sensitivity of 97.50%. For the imbalanced dataset, the model achieved better results with an input size of 64×64 pixels. Its accuracy in this case was 94.95%, with a precision of 94.17% and a sensitivity of 94.17%.

The work done in (Ciritsis et al., 2019) implemented a deep CNN (DCNN) with 13 convolutional layers, with four dense layers and a fully connected SoftMax classification layer. The model was trained and tested on MLO and CC mammograms from a private database. For MLO mammograms, the model achieved an overall accuracy of 92.2%, and an area under the receiver operating characteristic curve (AUC) of 98%. For CC mammograms, the model's overall accuracy was 87.4%, and the AUC was 97%.

In (Shi et al., 2019), a lightweight CNN architecture was employed to extract deep features from mammographic images. The CNN consisted of three convolutional layers (where max-pooling is also implemented) and three connected layers. The model was trained and tested on the mini-MIAS database (Suckling et al., 1994) and was validated using five-fold cross-validation. The authors also experimented with different kinds of image preprocessing, as well as different numbers of convolutional layers. The main three-layer model achieved the best reported accuracy of 83.9% and the lowest loss at 0.52. In terms of preprocessing procedures, the best performing model was one where pectoral muscles were extracted alongside the breast region. The accuracy of this model was 83.6% and its loss was 0.52.

Another interesting study, (Zhao et al., 2021), proposed an architecture dubbed BASCNet, wherein residual networks (ResNets) from (He et al., 2016) were employed in combination with an adaptive spatial attention module (ASAM) and an adaptive channel attention module (ACAM). In the model detailed in (Zhao et al., 2021), ASAM helps learn distinctive features, and ACAM highlights informative channels. The model was trained on CC mammograms from the DDSM database (Heath et al., 2001) as well as on MLO and CC mammograms from the INbreast database (Moreira et al., 2012). For the DDSM database, the accuracy of the model was in the range of 85.10%, and the F1 Score was in the region of 73.92%. For the INbreast database, the accuracy was in the range of 90.51%, and the F1Score was in the region of 78.11%.

The work done in (Xu et al., 2018) details another CNN-based method, this time using residual learning. The proposed CNN in this study has a total of 70 convolutional layers with two fully connected layers and a SoftMax layer for classification. The model was trained on an augmented version of the INbreast dataset (Moreira et al., 2012). It achieved a reported accuracy of 96.8%, suggesting that the method proposed has great potential.

Another approach was taken in (Ionescu et al., 2019), where a CNN-based architecture was implemented to predict visual analog scale (VAS) scores from mammograms. VAS a subjective method that has been used to visually assess breast density of patients. Essentially, it is a 10 cm scale, on which the zero point represents 0% density, and the 10 cm point represents 100% density. Attending radiologists can leave a mark on the scale and then calculate the distance in millimeters from the zero point to the mark to attain the speculated density percentage. In this study's case, a CNN consisting of six groups of two convolutional layers is trained on mammograms from the PROCAS dataset (Evans et al., 2012) labeled with the averages of the VAS scores estimated by two different readers. The predictions made by the model showed high correlation with the preassigned VAS scores.

Table 4.3 summarizes the studies mentioned in this chapter; it notes the datasets used, image preprocessing steps taken, feature extraction methods employed, machine learning models implemented, and results achieved in each work.

Table 4.3 Summary of machine learning and deep learning methods for breast density assessment.

References	Dataset(s)	Preprocessing & feature extraction	Model(s)	Results
Gong et al. (2019)	Mini-MIAS, DDSM, and a private database	Preprocessing: denoising and removal of labels and pectoral muscle Statistical feature extraction, GLCM, and area-based density estimation	SVM	Accuracy: MIAS: 96.19% DDSM: 96.35% MIAS + private DB (mixed): 95.01%
Kumar et al. (2017)	DDSM	Preprocessing: ROI extraction from central breast region Statistical feature extraction, GLDS, GLCM, GLRLM, Laws' texture energy, and 2D GWT	PCA-KNN, PCA-PNN, PCA-ANN, PCA-NFC, PCA-SVM, and a Hybrid Hierarchical Framework	Accuracy: PCA-KNN: 72.50% PCA-PNN: 68.33% PCA-ANN: 50.41% PCA-NFC: 80.41% PCA-SVM: 78.33% Hybrid: 84.17%
Liu et al. (2011)	Private database	Preprocessing: removal of artefacts, pectoral muscle, and contour line, as well as enhancement of images through dyadic wavelet transform Statistical feature extraction	SVM	Accuracy: 86.40%
Saffari et al. (2020)	INbreast	Removal of pectoral muscle and rescaling	cGAN-CNN	Accuracy: 98.75% Precision: 97.50% Sensitivity: 97.50% Specificity: 99.16%
Ciritsis et al. (2019)	Private Database	Rescaling	DCNN	MLO Accuracy: 92.20% MLO AUC: 0.980 CC Accuracy: 87.40% CC AUC: 0.970

(*Continued*)

Table 4.3 Summary of machine learning and deep learning methods for breast density assessment. *Continued*

References	Dataset(s)	Preprocessing & feature extraction	Model(s)	Results
Shi et al. (2019)	Mini-MIAS	Rescaling and pectoral muscle removal	CNN	Accuracy: 83.9% Loss: 0.52
Zhao et al. (2021)	DDSM and INbreast	Denoising, breast area extraction, and rescaling	BASCNet: ResNet + ASAM + ACAM	DDSM (CC): Accuracy: 85.10% ± 2.50% F1 Score: 73.92% ± 3.82% AUC: 91.54% ± 0.88% INbreast (CC & MLO): Accuracy: 90.51% ± 5.08% F1 Score: 78.11% ± 10.30% AUC: 99.09% ± 1.20%
Xu et al. (2018)	INbreast	Augmentation through rotation, flipping, and rescaling	CNN	Accuracy: 96.8%
Ionescu et al. (2019)	PROCAS	Padding to equalize image sizes, reorientation of breasts, cropping and downscaling through bicubic interpolation, and pixel inversion and histogram equalization	CNN	Highest Pearson's correlation (*P*) of .851 between predicted and reader VAS scores

4.2.4 Artificial intelligence in breast cancer risk assessment

There are several risk factors for breast cancer including age, family history, reproductive factors, breast density, estrogen's levels, and lifestyle choices (Sun et al., 2017). Of course, AI has been extensively implemented in the literature with the express purpose of assessing cancer risk in patients. Several studies in the literature has reported remarkable results, which suggests that AI can not only assist, but often outperform expert assessment of breast cancer risk.

A 2018 meta-analysis by Nindrea et al. examined the diagnostic accuracy of varying machine learning algorithms including KNN, ANN, and SVM, when applied to breast cancer risk assessment (Nindrea et al., 2018). The authors reviewed a total of 1879 articles and selected 11 papers for their review. They assessed the methods employed in those papers by comparing the measures of the ROC AUC for each method. The study found that the AUC for SVM in particular was over 90%, and as such concluded that SVM is able to predict breast cancer risk better than other machine learning algorithms.

Another study, (Sepandi et al., 2018), focused solely on building a capable ANN for breast cancer risk assessment. In particular, they applied ANN on a dataset comprised of mammographic images, as well as risk factors and clinical findings. The algorithm was accompanied by a cytopathological diagnosis to improve on the overall accuracy. To evaluate the method, metrics such as accuracy and ROC AUC were calculated. The outlined method performed well, achieving an AUC of 95.5%. With this result in mind, the authors concluded that the described ANN method may be applicable as a decision-support tool to aid doctors when deciding whether a biopsy is necessary.

A separate study by He et al. (2019), also focuses primarily on providing decision assistance when it comes to biopsy recommendations, in order to reduce unnecessary biopsies for patients. Also of note is that this study focused only on patients with a BI-RADS IV breast density class, since those patients are the ones who are most likely to be recommended a biopsy, as they are the most at risk. Initially, the authors compiled a dataset that included ultrasound images, mammographic reports, patient demographics, and pathology results for over 5000 patients. Then, they applied a natural language processing model to extract relevant cancer-risk information from the clinical reports. Lastly, a deep learning model was built and trained using the extracted cancer risk information and the data collected prior. The model works by predicting malignancy and outputting a biopsy-recommendation index measure that can aid attending physicians in making informed decisions. According to the authors, the model achieved a total accuracy of 81% and an overall AUC of 93%.

In (Yala et al., 2019), the authors hypothesized that better cancer risk assessment performance may be achieved if traditional risk factor information, gathered using questionnaires and medical records of patients, mammograms, and breast density were combined as input into a deep learning model. To test this theory, three models were created: a logistic regression model trained on traditional risk factor information alone, a deep learning model trained on mammograms alone, and a hybrid deep learning model trained on both. The models were assessed using AUC as a metric, and the results were compared to results obtained using the Tyrer-Cuzick model—a previous standard method that also incorporates breast density. The hybrid deep learning model achieved the best results overall, with an AUC of 70%.

A key follow-up study, (Yala et al., 2021), introduced an encoder-aggregator deep learning model, dubbed "Mirai," which predicts the breast cancer risk for patients across 5-years into the future. The model was trained and evaluated on multiple datasets from varying demographics using C-indices and ROC AUC as metrics. On one particular dataset, the Massachusetts General Hospital (MGH) dataset, a C-index of 0.76 and a 5-year AUC of 76% were achieved, outperforming the hybrid deep learning model proposed in (Yala et al., 2019) and the Tyrer-Cuzick model. In addition, Mirai was able to accurately predict 41.5% of the patients in the MGH dataset who would go on to develop breast cancer in a 5-year window, as compared to only 22.9% correctly predicted by the Tyrer-Cuzick model. Later, A. Yala et al. published another paper, (Yala et al., 2022), that continued to evaluate the performance of Mirai, this time using several datasets from a wider variety of populations. They found that Mirai was able to maintain its performance across the different test sets. According to Yala et al., this makes Mirai the most broadly evaluated AI cancer-risk assessment model to date, which consequently suggests that Mirai may have a place in active clinics in the near future.

Table 4.4 summarizes the studies detailed in this section, making note of the datasets used, the methods employed, and the results of each study.

Table 4.4 Summary of machine learning and deep learning methods for breast cancer risk assessment.

References	Datasets	Method	Best Results
Nindrea et al. (2018)	Wisconsin breast cancer dataset Wu et al. (2019), Mini-MIAS Suckling et al. (1994), and several private databases	SVM	Accuracy: 97.53%, AUC: >90%
Sepandi et al. (2018)	Private database	ANN	AUC: 95.5%
He et al. (2019)	Private database	Autoencoder-multilayer perceptron	Accuracy: 81%, AUC: 93%
Yala et al. (2019)	Private database	Hybrid deep learning model	AUC: 70%
Yala et al. (2021)	Private database	Encoder-aggregator deep learning model	AUC: 76%

4.2.5 Artificial intelligence for mammographic image improvement

The quality of an image can be the deciding factor for whether it proves useful or not, and the case is no different for mammograms. In fact, the quality of a mammographic image may influence the presiding doctor's decisions rather dramatically. Raw mammographic images are often not accurately representative. This is because these images are subject to noise and human error (Ekpo et al., 2018; Joseph et al., 2017). The quality of the images—acquired through X-ray—is also subject to deterioration due to factors such as the size of the breast or the movement of the patient (Lee et al., 2019). This means that the entire architecture of breast parenchyma may be misshaped within a mammogram. Furthermore, certain radiographic features may be difficult to perceive correctly within mammograms, which serves to increase errors in breast cancer detection (Ekpo et al., 2018).

In their efforts to reduce mammographic errors, researchers have come up with methods of improving mammographic image quality. Some of these methods focus on denoising mammograms, while others try to enhance the overall contrast and quality of the images. Recently, a number of AI-based techniques for image denoising and contrast enhancement have been applied to mammograms in the literature, and the results are rather promising.

In an attempt to restore important structural features from noisy mammographic images, (Ghosh et al., 2020) employs a deep convolutional autoencoder. This denoising autoencoder works by minimizing a total-variational multinorm loss function. The model proposed can restore image data in the feature space by reducing the dimensionality of mammograms, while preserving the key features. The output of the model is a mammogram with decreased noise and highlighted structural features. The authors contend that in terms of speed and performance, the suggested method outperforms previous algorithms after a fine-tuning process.

The work in (Singh et al., 2019) describes a deep CNN with 17 layers capable of denoising mammograms afflicted by Gaussian noise. The authors report exceptional performance, with a structural similarity index (SSIM) of 0.98 and a peak signal-to-noise (PSNR) ratio of 41.53 dB.

Another CNN-based approach was proposed by Eckert et al. (2020). In this approach, Anscombe Transformation is first applied to mammograms with Poisson noise in order to obtain mammograms with gaussian noise, in a process of data augmentation. Following this, a deep CNN is trained on the noisy images, allowing it to learn the noise maps of the images. Afterwards, the network can be used to denoise images with both types of noise. The authors report a performance better than state-of-the-art approaches such as BM3D and DNCNN.

As for image improvement, several studies simply precede an AI-based detection or classification task with an image-processing stage wherein previously established, non-AI-based image improvement algorithms are implemented. Examples include (Shenbagavalli & Thangarajan, 2018), where an artificial neural network is trained on mammograms improved through Shearlet transform, and

Table 4.5 Summary of machine learning and deep learning methods for mammographic image improvement.

References	Dataset(s)	Method	Best Results
Ghosh et al. (2020)	MedPix and CBIS-DDSM Lee et al. (2016)	Convolutional Autoencoder	SSIM: 0.891, mean squared error (MSE): 0.0188
Singh et al. (2019)	Mini-MIAS Suckling et al. (1994)	CNN	SSIM: 0.98, PSNR: 41.53 dB
Eckert et al. (2020)	Private database	CNN	SSIM: 0.841, PSNR: 36.18 dB
Shenbagavalli and Thangarajan (2018)	DDSM Heath et al. (2001)	Shearlet transform + ANN	Improved classification accuracy: 93.45%.
Teare et al. (2017)	DDSM Heath et al. (2001) and a Private Database	CLAHE + DCNN	Improved classification performance (AUC: 0.922)

(Teare et al., 2017) where contrast-limited adaptive histogram equalization (CLAHE) precedes dual deep CNNs and a random forest gating network. Both studies report improvements in their prediction tasks following the application of image improvement techniques.

Table 4.5 summarizes the studies detailed in this section, making note of the datasets used, the methods employed, and the results of each study.

4.3 Conclusions, challenges, and future directions

4.3.1 Early breast cancer detection

As discussed in Section 4.2.1 and 4.2.2, AI holds an incredible promise for breast cancer detection. Many initial AI studies demonstrated remarkable improvement in accuracy over that of radiologists (McKinney et al., 2020; Rodriguez-Ruiz et al., 2019), but a recent systematic review showed that there is insufficient scientific evidence to support these findings. Freeman and colleagues (2021) evaluated the AI algorithms that are used to detect cancer in digital mammograms alone or in combination with radiologists. In their review, they found that 34 out of 36 evaluated AI systems were less accurate than a single radiologist, and all AI systems were less accurate than the consensus of two or more radiologists. Thus, as an emerging technology, several steps need to be taken before adopting AI systems in clinical settings. For example, larger and more representative datasets of screening populations are needed to robustly validate the AI systems and truly prove their efficacy before considering them in mammography screening and interpretation (Rodriguez-Ruiz et al., 2019).

Moreover, the feasibility and potential of considering AI systems in the clinic still need to be investigated. As missed breast cancers on screening mammography remain the most litigious situation for medical malpractice lawsuits (Arleo et al., 2016), it is still uncertain if patients and physicians would accept the use of AI systems alone for mammography interpretation. For example, in the case of missed breast cancer by the AI system, it is not yet clear who would take the responsibility (Houssami et al., 2017). In this regard, Arleo and colleagues (2016) expect that radiologists will be tasked with reviewing mammography exams that are annotated by AI algorithms that point out suspicious regions and abnormalities in these mammograms. The radiologists will still likely be responsible for the rest of the tasks required in the mammography revision workflow such as ordering and interpreting any subsequent diagnostic imaging or tissue biopsies to confirm malignancy if needed. However, the details of where and how AI systems can help in the workflow remains unknown (Arleo et al., 2016).

The likelihood of using AI as a replacement for human radiologists has been discussed. For example, in a systematic review conducted by Freeman and colleagues (2021), seven studies reporting the test accuracy of AI as a standalone system used in breast cancer detection from screening mammography were identified and discussed. Three of these studies evaluated AI as a replacement for one or all radiologists (McKinney et al., 2020; Salim et al., 2020; Schaffter et al., 2020). Four of these studies evaluated AI as a triage tool to identify women whose mammograms show no signs of malignancy so do not need radiologist review (Christiana et al., 2020; Dembrower et al., 2020; Lång et al., 2021; Raya-Povedano et al., 2021) and one of the four studies evaluated AI to detect missed cancers after screening in women who were not recalled for further tests by the radiologist (Dembrower et al., 2020). These studies indicate promise for using AI systems in the triage process and after radiologist review but AI algorithms are found to be not yet specific enough to replace a human radiologist (Taylor-Phillips & Freeman, 2022). Thus, future research should focus on designing solutions to use of AI as a decision support tool rather than simply designing the AI system to directly replace humans (Taylor-Phillips & Freeman, 2022).

As discussed earlier, several studies have shown the potential of deep learning models in breast cancer detection and diagnosis. However, these studies were based on retrospective and/or in-silico data resources that had high risk of bias and may not be representative of real-world clinical practice (Freeman et al., 2021; Lehman & Topol, 2021). Thus, to leverage the power of AI to solve the current challenges in early and accurate breast cancer detection and diagnosis, research can be directed to focus on several perspectives (Lehman & Topol, 2021). For instance, future studies should consider designing AI systems best suited for the precise clinical applications. Furthermore, the fusion of the knowledge provided by the AI system and abilities of a human radiologist need to be emphasized. Moreover, prospective studies should be designed and applied to large cohorts of subjects to prove the superiority of AI systems in early breast cancer detection and diagnosis.

4.3.2 Breast density estimation

In terms of breast density estimation approaches that rely on AI—excluding methods that work through segmentation—multiple challenges were observed as follows. First, at present, the image datasets that are publicly available do not host enough images to make a proper evaluation of the models proposed possible, as has been noted in (Shi et al., 2019). The majority of studies covered note that any future work must encompass a revaluation with more images. The biggest issue here is the fact that the standardization of images—which would be required for optimal AI performance—is a difficult, tedious, and costly task. As such, even newer studies with no access to private datasets will opt to use the same small-scale, public datasets simply for a lack of an alternative. Another issue of note is that expert opinions, which have been shown to be highly variable when it comes to breast density estimation, are what is currently being used as ground truth for the machine learning models (Saffari et al., 2020). As such, even models that perform extremely well on test and validation sets will not outperform practicing physicians in distinguishing between different density classes. Perhaps one way to reduce the impact of variability in the labels would be to introduce measures that deal with label noise (Karimi et al., 2020). However, ultimately, the quality of the model will depend on the quality of the data it is trained on. This, along with the aforementioned issue, highlight the need for a sizeable, high quality, public dataset with attested density labels. Lastly, it appears as though state-of-the-art deep learning techniques like transformers are seldom used by researchers to solve the problem. It is possible that such methods are not well suited to this particular task, but more experimentation is required to prove or disprove their efficacy.

4.3.3 Breast cancer risk assessment

For breast cancer risk assessment, there is enough evidence in the literature to attest to the efficacy of several methods (Lei et al., 2021). In fact, a few methods are currently being tested on a variety of populations as a possible prelude to being incorporated into active clinics as decision-support tools (Yala et al., 2021). However, more research need to be done to determine which risk factors contribute most to the accuracy of the models. It can be noted, for example, that methods that incorporate breast density perform better than methods that do not (Yala et al., 2019). Knowing this will allow researchers to come up with better, more cost-effective models for cancer risk assessment.

4.3.4 Mammographic image improvement

When it comes to mammographic image improvement, AI performs well, but is perhaps not the most cost-effective option. While deep learning techniques such as autoencoders have decent denoising capabilities, so too do general, non-AI-based techniques such as filtering is several times less costly to implement.

However, such techniques are not capable of generalizing to each and every image, therefore, AI has the potential to outperform them, given enough time and experimentation (Ghosh et al., 2021).

References

Al-Antari, M. A., Al-Masni, M. A., Choi, M. T., Han, S. M., & Kim, T. S. (2018). A fully integrated computer-aided diagnosis system for digital X-ray mammograms via deep learning detection, segmentation, and classification. *International Journal of Medical Informatics*, *117*, 44–54. Available from https://doi.org/10.1016/j.ijmedinf.2018.06.003.

Al-masni, M. A., Al-antari, M. A., Park, J. M., Gi, G., Kim, T. Y., Rivera, P., Valarezo, E., Choi, M. T., Han, S. M., & Kim, T. S. (2018). Simultaneous detection and classification of breast masses in digital mammograms via a deep learning YOLO-based CAD system. *Computer Methods and Programs in Biomedicine*, *157*, 85–94. Available from https://doi.org/10.1016/j.cmpb.2018.01.017.

AlSawaftah, N., El-Abed, S., Dhou, S., & Zakaria, A. (2022). Microwave imaging for early breast cancer detection: Current state, challenges, and future directions. *Journal of Imaging*, *8*(5), 123. Available from https://doi.org/10.3390/jimaging8050123.

Arleo, E. K., Saleh, M., & Rosenblatt, R. (2016). Lessons learned from reviewing breast imaging malpractice cases. *Journal of the American College of Radiology*, *13*(11), R58–R60. Available from https://doi.org/10.1016/j.jacr.2016.09.028.

Bahl, M. (2019). Detecting breast cancers with mammography: Will AI succeed where traditional CAD failed? *Radiology*, *290*(2), 315–316. Available from https://doi.org/10.1148/radiol.2018182404.

Birdwell Robyn, L. (2009). The preponderance of evidence supports computer-aided detection for screening mammography. *Radiology*, *253*(1). Available from https://doi.org/10.1148/radiol.2531090611.

Boyd, N. F., Martin, L. J., Bronskill, M., Yaffe, M. J., Duric, N., & Minkin, S. (2010). Breast tissue composition and susceptibility to breast cancer. *The Author. Journal of the National Cancer Institute*, *102*(16). Available from https://doi.org/10.1093/jnci/djq239.

Cai, H., Huang, Q., Rong, W., Song, Y., Li, J., Wang, J., Chen, J., & Li, L. (2019). Breast microcalcification diagnosis using deep convolutional neural network from digital mammograms. *Computational and Mathematical Methods in Medicine*, *2019*, 1–10. Available from https://doi.org/10.1155/2019/2717454.

Cardoso, M. J., Houssami, N., Pozzi, G., & Séroussi, B. (2022). Artificial intelligence (AI) in breast cancer care - leveraging multidisciplinary skills to improve care. *Artificial Intelligence in Medicine* (123, p. 102215). Elsevier B.V. Available from https://doi.org/10.1016/j.artmed.2021.102215.

Carneiro, G., Nascimento, J., & Bradley, A. P. (2017). Automated analysis of unregistered multi-view mammograms with deep learning. *IEEE Transactions on Medical Imaging*, *36*(11), 2355–2365. Available from https://doi.org/10.1109/TMI.2017.2751523.

Castellino, R. A. (2005). Computer aided detection (CAD): An overview. *Cancer Imaging: The Official Publication of the International Cancer Imaging Society*, *5*(1), 17–19. Available from https://doi.org/10.1102/1470-7330.2005.0018.

Chatfield, K., Simonyan, K., Vedaldi, A., & Zisserman, A. (2014). Return of the devil in the details: Delving deep into convolutional nets.

Chougrad, H., Zouaki, H., & Alheyane, O. (2018). Deep convolutional neural networks for breast cancer screening. *Computer Methods and Programs in Biomedicine*, *157*, 19–30. Available from https://doi.org/10.1016/j.cmpb.2018.01.011.

Christiana, B., Alejandro, R.-R., Christoph, M., Nico, K., & Sylvia, H.-K. (2020). Going from double to single reading for screening exams labeled as likely normal by AI: What is the impact? In: *15th International Workshop on Breast Imaging (IWBI2020)*, 66-66 SPIE. Available from https://doi.org/10.1117/12.2564179 9781510638310.

Ciritsis, A., Rossi, C., De Martini, I. V., Eberhard, M., Marcon, M., Becker, A. S., Berger, N., & Boss, A. (2019). Determination of mammographic breast density using a deep convolutional neural network. *British Journal of Radiology*, *92*(1093). Available from https://doi.org/10.1259/bjr.20180691.

Cruz-Bernal, A., Flores-Barranco, M. M., Almanza-Ojeda, D. L., Ledesma, S., & Ibarra-Manzano, M. A. (2018). Analysis of the cluster prominence feature for detecting calcifications in mammograms. *Journal of Healthcare Engineering*, *2018*, 1–11. Available from https://doi.org/10.1155/2018/2849567.

Dembrower, K., Wåhlin, E., Liu, Y., Salim, M., Smith, K., Lindholm, P., Eklund, M., & Strand, F. (2020). Effect of artificial intelligence-based triaging of breast cancer screening mammograms on cancer detection and radiologist workload: A retrospective simulation study. *The Lancet Digital Health*, *2*(9), e468–e474. Available from https://doi.org/10.1016/S2589-7500(20)30185-0.

Destounis, S., Arieno, A., Morgan, R., Roberts, C., & Chan, A. (2017). Qualitative versus quantitative mammographic breast density assessment: Applications for the US and abroad. *Diagnostics*, *7*(2). Available from https://doi.org/10.3390/diagnostics7020030.

Eckert, D., Vesal, S., Ritschl, L., Kappler, S., & Maier, A. (2020). Deep learning-based denoising of mammographic images using physics-driven data augmentation. Informatik aktuell. Available from https://doi.org/10.1007/978-3-658-29267-6_21.

Ekpo, E. U., Alakhras, M., & Brennan, P. (2018). Errors in mammography cannot be solved through technology alone. *Asian Pacific Journal of Cancer Prevention*, *19*(2). Available from https://doi.org/10.22034/APJCP.2018.19.2.291.

Evans, D. G. R., Warwick, J., Astley, S. M., Stavrinos, P., Sahin, S., Ingham, S., McBurney, H., Eckersley, B., Harvie, M., Wilson, M., Beetles, U., Warren, R., Hufton, A., Sergeant, J. C., Newman, W. G., Buchan, I., Cuzick, J., & Howell, A. (2012). Assessing individual breast cancer risk within the U.K. National Health Service Breast Screening Program: A new paradigm for cancer prevention. *Cancer Prevention Research*, *5*(7). Available from https://doi.org/10.1158/1940-6207.CAPR-11-0458.

Freeman, K., Geppert, J., Stinton, C., Todkill, D., Johnson, S., Clarke, A., & Taylor-Phillips, S. (2021). Use of artificial intelligence for image analysis in breast cancer screening programmes: Systematic review of test accuracy. *BMJ (Clinical Research ed.)*, *374*, n1872. Available from https://doi.org/10.1136/bmj.n1872.

Ghosh, S. K., Biswas, B., & Ghosh, A. (2020). Restoration of mammograms by using deep convolutional denoising auto-encoders. *Advances in Intelligent Systems and Computing*, *990*. Available from https://doi.org/10.1007/978-981-13-8676-3_38.

Ghosh, S. K., Biswas, B., & Ghosh, A. (2021). A novel stacked sparse denoising autoencoder for mammography restoration to visual interpretation of breast lesion. *Evolutionary Intelligence*, *14*(1). Available from https://doi.org/10.1007/s12065-019-00344-0.

Gong, X., Yang, Z., Wang, D., Qi, Y., Guo, Y., & Ma, Y. (2019). Breast density analysis based on glandular tissue segmentation and mixed feature extraction. *Multimedia Tools and Applications*, *78*(22). Available from https://doi.org/10.1007/s11042-019-07917-2.

Guo, Y., Dong, M., Yang, Z., Gao, X., Wang, K., Luo, C., Ma, Y., & Zhang, J. (2016). A new method of detecting micro-calcification clusters in mammograms using contourlet transform and non-linking simplified PCNN. *Computer Methods and Programs in Biomedicine*, *130*, 31–45. Available from https://doi.org/10.1016/j.cmpb.2016.02.019.

Hartman, K., Highnam, R., Warren, R., & Jackson, V. (2008). Volumetric assessment of breast tissue composition from FFDM images 5116 LNCS. Lecture Notes in Computer Science (including subseries Lecture Notes in Artificial Intelligence and Lecture Notes in Bioinformatics). Springer-Verlag Berlin Heidelberg. Available from https://doi.org/10.1007/978-3-540-70538-3_5.

Heath, M. D., Bowyer, K., Kopans, D., Moore, R. H., & Philip, K. W. (2001). Medical physics the digital database for screening mammography. International Workshop on Digital Mammography, 212–218.

Heenaye-Mamode Khan, M., Boodoo-Jahangeer, N., Dullull, W., Nathire, S., Gao, X., Sinha, G. R., & Nagwanshi, K. K. (2021). Multi- class classification of breast cancer abnormalities using Deep Convolutional Neural Network (CNN). *PLoS One*, *16*(8), e0256500. Available from https://doi.org/10.1371/journal.pone.0256500.

He, T., Puppala, M., Ezeana, C. F., Huang, Y. S., Chou, P. H., Yu, X., Chen, S., Wang, L., Yin, Z., Danforth, R. L., Ensor, J., Chang, J., Patel, T., & Wong, S. T. C. (2019). A deep learning–based decision support tool for precision risk assessment of breast cancer. *JCO Clinical Cancer Informatics*, *3*. Available from https://doi.org/10.1200/cci.18.00121.

He, K., Zhang, X., Ren, S., & Sun. J. (2016). Deep residual learning for image recognition. In: *Proceedings of the IEEE Computer Society Conference on Computer Vision and Pattern Recognition*, 2016-December. Available from https://doi.org/10.1109/CVPR.2016.90.

Highnam, R., Brady, M., Yaffe, M. J., Karssemeijer, N., & Harvey, J. (2010). Robust breast composition measurement - Volpara™ 6136 LNCS. Lecture notes in computer science (including subseries lecture notes in artificial intelligence and lecture notes in bioinformatics). Springer-Verlag. Available from https://doi.org/10.1007/978-3-642-13666-5_46.

Hoff, S. R., Samset, J. H., Abrahamsen, A.-L., Vigeland, E., Klepp, O., & Hofvind, S. (2011). Missed and true interval and screen-detected breast cancers in a population based screening program. *Academic Radiology*, *18*(4), 454–460. Available from https://doi.org/10.1016/j.acra.2010.11.014.

Houssami, N., Lee, C. I., Buist, D. S. M., & Tao, D. (2017). Artificial intelligence for breast cancer screening: Opportunity or hype? *Breast (Edinburgh, Scotland)*, *36*, 31–33. Available from https://doi.org/10.1016/j.breast.2017.09.003.

Huang, G. B., Zhu, Q. Y., & Siew, C. K. (2006). Extreme learning machine: Theory and applications. *Neurocomputing*, *70*(1–3). Available from https://doi.org/10.1016/j.neucom.2005.12.126.

Ionescu, G. V., Fergie, M., Berks, M., Harkness, E. F., Hulleman, J., Brentnall, A. R., Cuzick, J., Evans, D. G., & Astley, S. M. (2019). Prediction of reader estimates of mammographic density using convolutional neural networks. *Journal of Medical Imaging*, *6*(3). Available from https://doi.org/10.1117/1.jmi.6.3.031405.

Joseph, A. M., John, M. G., & Dhas, A. S. (2017). Mammogram image denoising filters: A comparative study. In: Conference on Emerging Devices and Smart Systems, ICEDSS 2017. Available from https://doi.org/10.1109/ICEDSS.2017.8073679.

Kallenberg, M., Petersen, K., Nielsen, M., Ng, A. Y., Diao, P., Igel, C., Vachon, C. M., Holland, K., Winkel, R. R., Karssemeijer, N., & Lillholm, M. (2016). Unsupervised deep learning applied to breast density segmentation and mammographic risk scoring.

IEEE Transactions on Medical Imaging, *35*(5). Available from https://doi.org/10.1109/TMI.2016.2532122.

Karimi, D., Dou, H., Warfield, S. K., & Gholipour, A. (2020). Deep learning with noisy labels: Exploring techniques and remedies in medical image analysis. *Medical Image Analysis*, *65*. Available from https://doi.org/10.1016/j.media.2020.101759.

Katzen, J., & Dodelzon, K. (2018). A review of computer aided detection in mammography. *Clinical Imaging*, *52*, 305–309. Available from https://doi.org/10.1016/j.clinimag.2018.08.014.

Kohli, M., Prevedello, L. M., Filice, R. W., & Geis, J. R. (2017). Implementing machine learning in radiology practice and research. *American Journal of Roentgenology*, *208*(4), 754–760. Available from https://doi.org/10.2214/AJR.16.17224.

Kumar, I., Bhadauria, H. S., Virmani, J., & Thakur, S. (2017). A hybrid hierarchical framework for classification of breast density using digitized film screen mammograms. *Multimedia Tools and Applications*, *76*(18). Available from https://doi.org/10.1007/s11042-016-4340-z.

Lamb, L. R., Fonseca, M. M., Verma, R., & Seely, J. M. (2020). Missed breast cancer: Effects of subconscious bias and lesion characteristics. *Radiographics: A Review Publication of the Radiological Society of North America, Inc*, *40*(4). Available from https://doi.org/10.1148/rg.2020190090.

Lee, R. S., Gimenez, F., Hoogi, A., & Rubin, D. (2016). Curated breast imaging subset of DDSM. *The Cancer Imaging Archive*, *8*.

Lee, S., Jin Park, S., Jeon, J. M., Lee, M. H., Ryu, D. Y., Lee, E., Kang, S. H., & Lee, Y. (2019). Noise removal in medical mammography images using fast non-local means denoising algorithm for early breast cancer detection: A phantom study. *Optik*, *180*. Available from https://doi.org/10.1016/j.ijleo.2018.11.167.

Lehman, C. D., & Topol, E. J. (2021). Digital medicine Readiness for mammography and artificial intelligence 398. http://www.thelancet.com.

Lehman, C. D., Wellman, R. D., Buist, D. S. M., Kerlikowske, K., Tosteson, A. N. A., & Miglioretti, D. L. (2015). Diagnostic accuracy of digital screening mammography with and without computer-aided detection. *JAMA Internal Medicine*, *175*(11), 1828. Available from https://doi.org/10.1001/jamainternmed.2015.5231.

Lei, Y. M., Yin, M., Yu, M. H., Yu, J., Zeng, S. E., Lv, W. Z., Li, J., Ye, H. R., Cui, X. W., & Dietrich, C. F. (2021). Artificial intelligence in medical imaging of the breast. *Frontiers in Oncology*, *11*. Available from https://doi.org/10.3389/fonc.2021.600557, Is artificial intelligence about to transform the mammogram?.

Leong, Y. S., Hasikin, K., Lai, K. W., Mohd Zain, N., & Azizan, M. M. (2022). Microcalcification discrimination in mammography using deep convolutional neural network: Towards rapid and early breast cancer diagnosis. *Frontiers in Public Health*, *10*. Available from https://doi.org/10.3389/fpubh.2022.875305.

Liu, Q., Liu, L., Tan, Y., Wang, J., Ma, X., & Ni, H. (2011). Mammogram density estimation using sub-region classification. In: *Proceedings - 2011 4th International Conference on Biomedical Engineering and Informatics*, 1, BMEI 2011. IEEE. Available from https://doi.org/10.1109/BMEI.2011.6098327.

Logullo, A., Prigenzi, K., Nimir, C., Franco, A., & Campos, M. (2022). Breast microcalcifications: Past, present and future (Review. *Molecular and Clinical Oncology*, *16*(4), 81. Available from https://doi.org/10.3892/mco.2022.2514.

Lång, K., Dustler, M., Dahlblom, V., Åkesson, A., Andersson, I., & Zackrisson, S. (2021). Identifying normal mammograms in a large screening population using artificial intelligence. *European Radiology*, *31*(3), 1687–1692. Available from https://doi.org/10.1007/s00330-020-07165-1.

Majid, A. S., de Paredes, E. S., Doherty, R. D., Sharma, N. R., & Salvador, X. (2003). Missed breast carcinoma: Pitfalls and pearls. *Radiographics: A Review Publication of the Radiological Society of North America, Inc*, *23*(4), 881–895. Available from https://doi.org/10.1148/rg.234025083.

Marmot, M. G., Altman, D. G., Cameron, D. A., Dewar, J. A., Thompson, S. G., & Wilcox, M. (2013). The benefits and harms of breast cancer screening: An independent review. *British Journal of Cancer*, *108*(11), 2205–2240. Available from https://doi.org/10.1038/bjc.2013.177.

McKinney, S. M., Sieniek, M., Godbole, V., Godwin, J., Antropova, N., Ashrafian, H., Back, T., Chesus, M., Corrado, G. S., Darzi, A., Etemadi, M., Garcia-Vicente, F., Gilbert, F. J., Halling-Brown, M., Hassabis, D., Jansen, S., Karthikesalingam, A., Kelly, C. J., King, D., . . . Shetty, S. (2020). International evaluation of an AI system for breast cancer screening. *Nature*, *577*(7788), 89–94. Available from https://doi.org/10.1038/s41586-019-1799-6.

Moreira, I. C., Amaral, I., Domingues, I., Cardoso, A., Cardoso, M. J., & Cardoso, J. S. (2012). INbreast: Toward a full-field digital mammographic database. *Academic Radiology*, *19*(2). Available from https://doi.org/10.1016/j.acra.2011.09.014.

Nalawade, Y. V. (2009). Evaluation of breast calcifications. *Indian Journal of Radiology and Imaging*, *19*(4), 282–286. Available from https://doi.org/10.4103/0971-3026.57208.

Nazari, S. S., & Mukherjee, P. (2018). An overview of mammographic density and its association with breast cancer. *Breast Cancer (Tokyo, Japan)*, *25*(3). Available from https://doi.org/10.1007/s12282-018-0857-5.

Nindrea, R. D., Aryandono, T., Lazuardi, L., & Dwiprahasto, I. (2018). Diagnostic accuracy of different machine learning algorithms for breast cancer risk calculation: A meta-analysis. *Asian Pacific Journal of Cancer Prevention*, *19*(7). Available from https://doi.org/10.22034/APJCP.2018.19.7.1747.

Ragab, D. A., Sharkas, M., Marshall, S., & Ren, J. (2019). Breast cancer detection using deep convolutional neural networks and support vector machines. *PeerJ*, *7*, e6201. Available from https://doi.org/10.7717/peerj.6201.

Raya-Povedano, J. L., Romero-Martín, S., Elías-Cabot, E., Gubern-Mérida, A., Rodríguez-Ruiz, A., & Álvarez-Benito, M. (2021). AI-based strategies to reduce workload in breast cancer screening with mammography and tomosynthesis: A retrospective evaluation. *Radiology*, *300*(1), 57–65. Available from https://doi.org/10.1148/radiol.2021203555.

Redondo, A., Comas, M., Macià, F., Ferrer, F., Murta-Nascimento, C., Maristany, M. T., Molins, E., Sala, M., & Castells, X. (2012). Inter- and intraradiologist variability in the BI-RADS assessment and breast density categories for screening mammograms. *British Journal of Radiology*, *1019*(85). Available from https://doi.org/10.1259/bjr/21256379.

Rehman, K., Li, J., Pei, Y., Yasin, A., Ali, S., & Mahmood, T. (2021). Computer vision-based microcalcification detection in digital mammograms using fully connected depthwise separable convolutional neural network. *Sensors*, *21*(14), 4854. Available from https://doi.org/10.3390/s21144854.

Ribli, D., Horváth, A., Unger, Z., Pollner, P., & Csabai, I. (2018). Detecting and classifying lesions in mammograms with deep learning. *Scientific Reports*, *8*(1), 4165. Available from https://doi.org/10.1038/s41598-018-22437-z.

Rodriguez-Ruiz, A., Lång, K., Gubern-Merida, A., Broeders, M., Gennaro, G., Clauser, P., Helbich, T. H., Chevalier, M., Tan, T., Mertelmeier, T., Wallis, M. G., Andersson, I., Zackrisson, S., Mann, R. M., & Sechopoulos, I. (2019). Stand-alone artificial intelligence for breast cancer detection in mammography: Comparison with 101 radiologists. *JNCI: Journal of the National Cancer Institute*, *111*(9), 916–922. Available from https://doi.org/10.1093/jnci/djy222.

Saffari, N., Rashwan, H. A., Abdel-Nasser, M., Singh, V. K., Arenas, M., Mangina, E., Herrera, B., & Puig, D. (2020). Fully automated breast density segmentation and classification using deep learning. *Diagnostics*, *10*(11). Available from https://doi.org/10.3390/diagnostics10110988.

Salama, W. M., & Aly, M. H. (2021). Deep learning in mammography images segmentation and classification: Automated CNN approach. *Alexandria Engineering Journal*, *60*(5), 4701–4709. Available from https://doi.org/10.1016/j.aej.2021.03.048.

Salim, M., Wåhlin, E., Dembrower, K., Azavedo, E., Foukakis, T., Liu, Y., Smith, K., Eklund, M., & Strand, F. (2020). External evaluation of 3 commercial artificial intelligence algorithms for independent assessment of screening mammograms. *JAMA Oncology*, *6*(10), 1581. Available from https://doi.org/10.1001/jamaoncol.2020.3321.

Samala, R. K., Chan, H. P., Hadjiiski, L. M., Helvie, M. A., Cha, K. H., & Richter, C. D. (2017). Multi-task transfer learning deep convolutional neural network: Application to computer-aided diagnosis of breast cancer on mammograms. *Physics in Medicine & Biology*, *62*(23), 8894–8908. Available from https://doi.org/10.1088/1361-6560/aa93d4.

Schaffter, T., Buist, D. S. M., Lee, C. I., Nikulin, Y., Ribli, D., Guan, Y., Lotter, W., Jie, Z., Du, H., Wang, S., Feng, J., Feng, M., Kim, H. E., Albiol, F., Albiol, A., Morrell, S., Wojna, Z., Ahsen, M. E., Asif, U., ... Jung, H. (2020). Evaluation of combined artificial intelligence and radiologist assessment to interpret screening mammograms. *JAMA Network Open*, *3*(3), e200265. Available from https://doi.org/10.1001/jamanetworkopen.2020.0265.

Sechopoulos, I., Teuwen, J., & Mann, R. (2021). Artificial intelligence for breast cancer detection in mammography and digital breast tomosynthesis: State of the art. *Seminars in Cancer Biology*, *72*, 214–225. Available from https://doi.org/10.1016/j.semcancer.2020.06.002.

Sepandi, M., Taghdir, M., Rezaianzadeh, A., & Rahimikazerooni, S. (2018). Assessing breast cancer risk with an artificial neural network. *Asian Pacific Journal of Cancer Prevention*, *19*(4). Available from https://doi.org/10.22034/APJCP.2018.19.4.1017.

Shenbagavalli, P., & Thangarajan, R. (2018). Aiding the digital mammogram for detecting the breast cancer using Shearlet transform and neural network. *Asian Pacific Journal of Cancer Prevention*, *19*(9). Available from https://doi.org/10.22034/APJCP.2018.19.9.2665.

Shi, P., Wu, C., Zhong, J., & Wang, H. (2019). Deep learning from small dataset for bi-rads density classification of mammography images. In: *Proceedings - 10th International Conference on Information Technology in Medicine and Education*, ITME 2019. Available from https://doi.org/10.1109/ITME.2019.00034.

Sickles. (2013). ACR BI-RADS® Mammography. In: *ACR BI-RADS® Atlas, Breast Imaging Reporting and Data System*. Reston, VA: American College of Radiology.

Singh, G., Mittal, A., & Aggarwal, N. (2019). Deep convolution neural network based denoiser for mammographic images. *Communications in Computer and Information Science*, *1045*. Available from https://doi.org/10.1007/978-981-13-9939-8_16.

Sprague, B. L., Conant, E. F., Onega, T., Garcia, M. P., Beaber, E. F., Herschorn, S. D., Lehman, C. D., Tosteson, A. N. A., Lacson, R., Schnall, M. D., Kontos, D., Haas, J. S., Weaver, D. L., & Barlow, W. E. (2016). Variation in mammographic breast density assessments among radiologists in clinical practice: A multicenter observational study. *Annals of Internal Medicine*, *165*(7). Available from https://doi.org/10.7326/M15-2934.

Suckling, J., Parker, J., Dance, D., Astley, S., Hutt, I., Boggis, C., Ricketts, I., Stamatakis, E., Cerneaz, N., Kok, S., Taylor, P., Betal, D., & Savage, J. (1994). The mammographic image analysis society digital mammogram database. *Experta Medica, International Congress Series*, *1069*(JANUARY 1994).

Suhail, Z., Denton, E. R. E., & Zwiggelaar, R. (2018). Classification of micro-calcification in mammograms using scalable linear Fisher discriminant analysis. *Medical & Biological Engineering & Computing*, *56*(8), 1475–1485. Available from https://doi.org/10.1007/s11517-017-1774-z.

Sung, H., Ferlay, J., Siegel, R. L., Laversanne, M., Soerjomataram, I., Jemal, A., & Bray, F. (2021). Global cancer statistics 2020: GLOBOCAN estimates of incidence and mortality worldwide for 36 cancers in 185 countries. *CA: A Cancer Journal for Clinicians*, *71*(3). Available from https://doi.org/10.3322/caac.21660.

Sun, Y. S., Zhao, Z., Yang, Z.-N., Xu, F., Lu, H.-J., Zhu, Z.-Y., Shi, W., Jiang, J., Yao, P.-P., & Zhu, H.-P. (2017). Risk factors and preventions of breast cancer. These findings represent a small step in the long fight against breast cancer. *International Journal of Biological Sciences*, *11*(3). Available from https://doi.org/10.7150/ijbs.21635.

Taylor-Phillips, S., & Freeman, K. (2022). Artificial intelligence to complement rather than replace radiologists in breast screening. *The Lancet Digital Health*, *4*(7), e478–e479. Available from https://doi.org/10.1016/S2589-7500(22)00094-2.

Teare, P., Fishman, M., Benzaquen, O., Toledano, E., & Elnekave, E. (2017). Malignancy detection on mammography using dual deep convolutional neural networks and genetically discovered false color input enhancement. *Journal of Digital Imaging*, *30*(4). Available from https://doi.org/10.1007/s10278-017-9993-2.

Trister, A. D., Buist, D. S. M., & Lee, C. I. (2017). Will machine learning tip the balance in breast cancer screening? *JAMA Oncology*, *3*(11), 1463. Available from https://doi.org/10.1001/jamaoncol.2017.0473.

Tsochatzidis, L., Koutla, P., Costaridou, L., & Pratikakis, I. (2021). Integrating segmentation information into CNN for breast cancer diagnosis of mammographic masses. *Computer Methods and Programs in Biomedicine*, *200*, 105913. Available from https://doi.org/10.1016/j.cmpb.2020.105913.

Wu, N., Phang, J., Park, J., Shen, Y., Kim, S. G., Heacock, L., Moy, L., Cho, K., Geras, K. J., Worrall, D. E., Welling, M., Moreira, I. C., Amaral, I., Domingues, I., Cardoso, A., Cardoso, M. J., Cardoso, J. S., Lee, R. S., Gimenez, F., … Ara, T. (2019). Breast cancer wisconsin (diagnostic) data set | Kaggle. *Kaggle*, *4*(November).

Xu, J., Li, C., Zhou, Y., Mou, L., Zheng, H., & Wang, S. (2018). Classifying mammographic breast density by residual learning.

Yala, A., Lehman, C., Schuster, T., Portnoi, T., & Barzilay, R. (2019). A deep learning mammography-based model for improved breast cancer risk prediction. *Radiology*, *292*(1). Available from https://doi.org/10.1148/radiol.2019182716.

Yala, A., Mikhael, P. G., Strand, F., Lin, G., Satuluru, S., Kim, T., Banerjee, I., Gichoya, J., Trivedi, H., Lehman, C. D., Hughes, K., Sheedy, D. J., Matthis, L. M., Karunakaran, B., Hegarty, K. E., Sabino, S., Silva, T. B., Evangelista, M. C., Caron, R. F., ... Barzilay, R. (2022). Multi-institutional validation of a mammography-based breast cancer risk model. *Journal of Clinical Oncology*, *40*(16). Available from https://doi.org/10.1200/JCO.21.01337.

Yala, A., Mikhael, P. G., Strand, F., Lin, G., Smith, K., Wan, Y. L., Lamb, L., Hughes, K., Lehman, C., & Barzilay, R. (2021). Toward robust mammography-based models for breast cancer risk. *Science Translational Medicine*, *13*(578). Available from https://doi.org/10.1126/scitranslmed.aba4373.

Zhao, W., Wang, R., Qi, Y., Lou, M., Wang, Y., Yang, Y., Deng, X., & Ma, Y. (2021). BASCNet: Bilateral adaptive spatial and channel attention network for breast density classification in the mammogram. *Biomedical Signal Processing and Control*, *70*. Available from https://doi.org/10.1016/j.bspc.2021.103073.

CHAPTER

5 Segmentation of breast tissue structures in mammographic images

Bashar Rajoub[1], Hani Qusa[2], Hussein Abdul-Rahman[1] and Heba Mohamed[3]
[1]*Faculty of Engineering Technology and Science, Higher Colleges of Technology, Dubai, United Arab Emirates*
[2]*Faculty of Computer Information Science, Higher Colleges of Technology, Dubai, United Arab Emirates*
[3]*Faculty of Health Sciences, Higher Colleges of Technology, Dubai, United Arab Emirates*

5.1 Introduction

Mammographic screening is a challenging task for humans: It relies on double-reading by expert radiologists to identify abnormalities in the appearance of breast tissue [e.g., lesions, lumps, cysts, and dense structures, which may correspond to calcifications or masses (Population screening programmes, 2021)]. Suspicious areas in mammograms are examined to get more information using ultrasound and magnetic resonance imaging scans. This provides clinicians with a more detailed information such as the size and location of existing abnormalities and if additional irregularities are present. Diagnosis of breast cancer is confirmed by analyzing biopsies of the breast's tissue.

Missed breast cancers during initial mammogram screenings can be as high as 30% (Elter & Horsch, 2009). Also, only 15%–30% of patients referred to biopsy after initial diagnosis will have cancer (Hadjiiski et al., 2006). As such, mammography has high sensitivity but low specificity. A false positive (FP) mammography diagnosis can result in unnecessary operations, anxiety, and pain for the patients, while, on the one hand, missed detections (false negatives) will lead to delaying life-saving treatments. This therefore emphasizes the need to minimize the number of incorrect predictions.

Computer aided diagnosis (CAD) integrates contributions from image processing, computer vision, and machine learning to detect the abnormalities present in images. Using CAD for mammograms had spread quickly after it was approved by the U.S. Food and Drug Administration in June 1998. CAD systems are known to produce high sensitivity/specificity for some types of abnormalities (e.g., lesions) but low sensitivity/specificity for the more subtle abnormalities (e.g., calcified deposits).

CAD offers a pathway to improve and automate mammographic screening. CAD systems help in detecting early signs of cancer and provide clinicians a roadmap of suspicious regions, saving clinicians' time and improving the

Artificial Intelligence and Image Processing in Medical Imaging. DOI: https://doi.org/10.1016/B978-0-323-95462-4.00005-4

reliability of the diagnoses. CAD can be introduced in the decision making pipeline and can serve as a second opinion tool and reduce oversight errors (Delogu et al., 2007; Guliato et al., 2008; Shi et al., 2008; Wei et al., 2005).

Researchers in the medical imaging field have focused on combining imaging modalities and computer algorithms to identify early signs of cancer (Fenton et al., 2007). Studies in (Freer & Ulissey, 2001; Nishikawa, 2007; Warren Burhenne et al., 2000) showed that CAD systems can detect lesions that were missed by radiologists. Both studies indicated CAD increases the detection rate of early-stage cancers (reduce the false negative rate) without excessive effect on the recall rate or positive predictive value for biopsy.

Breast cancer CAD systems rely on image processing and machine learning. In the initial stages of CAD system development, mammogram images are processed through a pipeline involving segmentation, feature extraction, feature selection, and classification steps. CAD systems are based on extracting specific hand-engineered image features, such as shape, morphology, cluster-based features, intensity, or texture based on gray-level appearance, and topological features (Chen, Denton, et al., 2012; Cheng et al., 2003; Elter & Horsch, 2009).

5.2 Image-based feature analysis for breast cancer classification and diagnosis

Finding good clinical/image-based features impact classification performance of breast cancer. Such features are expected to discriminate between benign and malignant findings. In the literature, several image features were tested. The features are expressed in terms of image descriptors, for example, morphological descriptors, intensity and texture descriptors, and distribution and location of pathological lesions (Breast Imaging Reporting & Data System, 2009).

Lesions in mammograms are among the most common symptoms of breast cancer (DeSantis et al., 2011). Common lesions include masses, architectural distortion (AD), and bilateral asymmetry of the breast (bilateral asymmetry refers to areas of high density appearing only in one breast as compared to the same areas in the other). It has been shown that the sensitivity of CAD systems significantly decreases as the density of the breast increases (Ho & Lam, 2003). Human recall for identifying lesions in mammographic images is estimated to be between 0.75 and 0.92 (Gur et al., 2004). Beam et al. (1996) show that there is a wide variability in the accuracy of mammogram interpretation of 108 radiologists who reviewed 45 cancerous mammograms (from 79 women). The radiologists' reading sensitivity and specificity varied from 47% to 100% and 36% to 99%. Similar results in (Berg et al., 2000) showed that the intra and interobserver variability in the assessment of lesions was highly variable, where radiologists agreed on only 55% of the total 86 lesions.

In the literature, several techniques that rely on the extraction of low-level image features, such as shapes, texture, and local key-point descriptors, were used to aid in the detection of mammographic patterns. Common image segmentation methods include gray-level thresholding, morphological filtering, active contours (ACs), and region-based segmentation techniques (e.g., region growing, split and merge, and watershed methods).

Unsupervised algorithms group image regions by measuring their similarity. This generates groupings (clusters) of pixels in mammograms, such as dense vs. translucent regions. In (Lou et al., 2000), mammography images were processed using an adaptation of the c-means clustering algorithm. An initial breast region is extracted based on the assumption that the gray values of the skin-air trace is a monotonic decreasing function. A refinement boundary position for each initial boundary point is estimated with an extrapolation method. The final breast boundary is obtained by linking all true boundary points.

5.3 Breast region and pectoral muscle segmentation

The goal of breast region segmentation is to identify the breast boundary. This requires separating the breast tissue from the background, which contains annotations, and also separates the breast region from the pectoral muscle in mediolateral oblique (MLO) views. A good breast region segmentation algorithm must reduce the loss of breast tissue. This helps in modeling mammographic parenchymal patterns as irrelevant features from unwanted regions are ignored (Sheba et al., 2018).

Segmentation of the breast region requires accurate detection of the pectoral muscle contour. The pectoral muscle is a large dense area of mass tissue that supports the breast body, which can be seen in the MLO. It appears as a bright triangular-shaped region in the top right or left corner of the mammogram. The pectoral muscle may have regions with similar appearance and intensity to mammographic parenchyma and does not provide any diagnostic information on the presence or absence of cancer in the breast region. Segmenting and removing the pectoral muscle from a mammogram helps increase the accuracy of breast cancer detection while reducing the number of FPs. It also helps to estimate cancer risk category, as it can produce more accurate breast tissue density distribution.

Image preprocessing of mammograms aims to enhance the structures within the mammographic image, emphasize the breast profile from the background, and remove noise, artefacts/labels. A median filter can be used for noise filtering. Zhu and Huang (2012) proposed a filtering algorithm that uses the correlation of the image to resize the mask according to the noise levels of the mask. In (Petrick et al., 1996), an adaptive density-weighted contrast enhancement filter using Laplacian of Gaussian (LoG) filter was used to detect the boundaries of structures more accurately.

Ferrari et al. (2004) used a multiresolution approach using Gabor filter bank to enhance the directional piece-wise linear structures in the ROI containing the pectoral muscle. The magnitude and phase for each pixel in the ROI was used to produce candidate edge boundaries. Finally, the pectoral muscle boundary is obtained through an edge-flow postprocessing procedure by connecting disjoint boundary segments and removing false edges. Wavelets have also been used in the context of image denoising (Mencattini et al., 2008). For low contrast mammograms, contrast enhancement (Kim et al., 2012) can be used as a preprocessing step. Enhancement of microcalcifications and suspicious masses, especially in low-contrast and noisy mammograms, can improve the detection of early signs of breast cancer. Other approaches performed denoising using multiscale adaptive gain and a wavelet shrinkage with adaptive threshold (Mencattini et al., 2008). In (Sheba & Raj, 2017), adaptive fuzzy logic based bi-histogram equalization is used for contrast enhancement, where all the parameters are computed based on the characteristics of the mammographic images.

Breast region segmentation can be considered as a pixel-wise classification (supervised) problem where the classes are breast or nonbreast region. Segmentation as a classification problem relies on features extracted from the labeled mammogram images (breast region and background) and using these features to train a machine learning classifier. Supervised algorithms use features to learn a mapping to class labels (e.g., malignant vs. benign mass).

Segmentation of the pectoral muscle can be challenging because of the variations in pectoral muscle sizes, shapes, and due to the low contrast between the tissue and the muscle in some images. The pectoral muscle also shares similarities in intensity and texture with other breast tissue. The presence of unsegmented regions from the image can lead to increased classifier false-negative and FP rates (Liu et al., 2012; Mughal et al., 2017; Wei et al., 2016).

5.3.1 Thresholding and edge-based techniques

The aim of edge-based segmentation is to find discontinuities in intensity values across the image. Edge detection can be achieved using a wide range of image processing and math operator functions. Edge detection generates an edge map comprising a disconnected set of edge fragments. In order to reveal the underlying structures and shapes, these edges are refined and combined into meaningful edges that represent the shapes of existing objects. Prior knowledge on the expected shape can be used to refine the raw edge map. For example, for pectoral muscle detection, the Hough transform can be used to find the edges that are part of a macro line segment.

Edge detection can be performed using first-order derivatives of the image, for example, using gradient-based operators such as Sobel, Prewitt, and Roberts. Second-order derivatives (e.g., the Laplacian) use the zero-crossings information for edge detection. Edge detection based on the gradient or Laplacian methods can be sensitive to noise, this will cause spikes in the derivative function. It is recommended that images are smoothed first before applying edge detection. The Canny detector uses a multistage algorithm to detect the edges in the image. It is considered a powerful and robust edge detector when compared to gradient and Laplacian-based detectors (Sonka et al., 2014).

MCndez et al. (n.d.) used gradient-based algorithm to detect the breast border and several algorithms to estimate the position of the nipple (maximum height of the breast border, maximum gradient, and maximum second derivative of the gray levels across the median-top section of the breast). First, a two-level thresholding was used to get an approximate breast region, which is then refined using an average filter. The breast region is then divided into three regions, which form the seeds for a tracking algorithm, which uses gray-level gradients in order to search for edge pixels (boundary).

In (Zhou et al., 2001), the breast region is first segmented from the surrounding image background (as well as the pectoral muscle in MLO view mammograms) by a gradient-based boundary detection. Gray-level thresholding (a pixel classification problem) was performed by combining the discriminant analysis method by Otsu's method (Otsu, 1979) and the maximum entropy principle to estimate an optimal threshold (Wong & Sahoo, 1989).

In (Mughal et al., 2018), the contours of the breast region were estimated using differential operators, such as isotropic, Sobel, and Prewitt operators. Differential operators can detect sharp changes in image intensity by computing the horizontal and vertical differences within the image. This stage produces rough boundaries in the image, which need to be refined further. The accumulative histogram at each pixel location is then used to filter and keep the pixels with the largest gradients based on some threshold value. The edges of a broken boundary are joined using morphological closing using a disk shaped structuring element. A convex hull function generates a topographic map to segment out the pectoral muscle. The authors established that the Prewitt filter is very suitable for enhancing high frequency and low frequency within the edges of the images edge detection and is the best option for pectoral muscle segmentation.

Thresholding is the simplest method for image segmentation. It can be used to segment mammographic images into distinct regions that share similar characteristics. Threshold segmentation mainly extracts foreground based on gray value information. Selecting a suitable threshold plays a major role in the final segmentation result.

Cheng et al. (2006) showed that image regions with abnormalities impose extra peaks on a histogram compared with healthy region, which has only a single peak. Therefore, a global threshold can be extracted from the image by analyzing the gray level distribution (histograms) of individual pixels (Brzakovic et al., 1990).

There are various thresholding schemes: global thresholding, local/adaptive thresholding, bilevel/multilevel thresholding, interactive/automatic thresholding. Global thresholding is based on a single threshold used for the whole image. Local/adaptive thresholding is based on a threshold that varies over the image depending on the local characteristics of neighboring subregions (pixels) within the image. Bi-level thresholding produces a binary image, where pixels with value 1 correspond to foreground, and pixels with value 0 correspond to the background. In multithresholding, the resulting image consists of a set of distinct categories corresponding to independent objects. Interactive thresholding select optimal threshold based on the visual assessment of expert operators of the segmentation result. Automatic threshold selection select optimal thresholds based on histogram analysis (Sonka et al., 2014), minimizing the total misclassification error, maximizing a measure of class separability, and based on co-occurrence matrix analysis.

Global thresholding methods rarely produce good results since masses are often superimposed on the tissue of the same intensity level. Also, global thresholding methods do not incorporate the spatial information of neighboring pixels or local texture information. Local and adaptive thresholding techniques can perform better than global thresholding. In (Yin et al., 1994), a semiautomatic segmentation method was applied to remove nontissue areas. Then the median of all the pixel values within the breast region is used as a threshold to partition the breast region into more dense and less dense tissue.

5.3.2 Morphological-based techniques

The pectoral muscle is triangular and appears in the upper left/right corner of the mammogram. A bounding box for the breast contour in the mammogram can be used to create an upper triangle with one third the width of the bounding box. The pectoral muscle is then removed by setting all the pixels in the upper triangle to zeros (Sheba et al., 2018). Also, the regions of interest (ROI) are automatically detected and segmented from mammograms using global thresholding bands on Otsu's method and morphological operations. Shape, texture, and gray-level features are extracted from the ROIs. A classifier and regression tree (CART) produced optimal features that were used to train feed-forward artificial neural networks model. Experimental results show that the proposed method achieved an accuracy of 96% with 83% sensitivity and 98% specificity (Sheba et al., 2018).

Considering the pectoral muscle contour to follow a straight-line with angles ranging between 45° and 90°, early methods for automatic identification of the pectoral muscle used image processing techniques. In (Karssemeijer, 1998), the pectoral muscle is segmented by first dividing the breast area into regions in which the distance of pixels to the skin line is approximately equal. The distance of a pixel to the skin line is represented by the number of erosions (using a circular structuring element) needed to exclude the pixel from the breast mask. The gradient image was obtained from the ROIs and edge detection based on the Sobel operator is performed. The Hough transform is then used to detect the boundary of the pectoral muscle. Breast tissue was separated from the background based on a threshold value which was determined by performing peak detection on the histogram.

In (Kwok et al., 2004), the Hough transform is used to approximate the pectoral muscle boundary as a straight line. The intensity profiles from search paths perpendicular to the straight line were then iteratively used to extract cliff locations and fit a more accurate curve. A closed region is then formed to segment the pectoral muscle. Chakraborty et al. (2012) initialize the boundary by a straight lines approximation. The image profile along each horizontal line is used to find the maximum discontinuity points based on the weighted average gradient. An adaptive search is then applied to identify the most probable band containing the pectoral edge points. A straight line is fitted on the resulting pectoral edge points. Results show that the average gradient produced very few FPs and false negatives of the pectoral muscle segmentation.

Straight lines approximation of the pectoral muscle might be an oversimplification. In (Wang et al., 2010), the stochastic nature of the pectoral muscle boundary is captured with a Markov chain model to produce a coarse segmentation and the final boundary is estimated using an active AC model. In (Muhammad, 2017), mammograms are decomposed in four sub-bands using multilevel 2D wavelet decomposition in combination with morphological operators on the approximation sub-band to refine the contours.

5.3.3 Region growing techniques

Region growing segmentation groups pixels into regions based on a growing criterion. Starting with a marker/seed a point(s), a region is iteratively grown by adding similar neighboring pixels until no more neighboring pixels satisfy the growing criterion. The seed point can be manually selected by an operator or automatically according to specific criteria. A commonly used growing criterion is homogeneity. The basic idea is to divide an image into zones of maximum homogeneity; the criteria for homogeneity can be based on gray-level, color, texture, and shape of the grown region (Sonka et al., 2014).

In (Raba et al., 2005), a constrained region growing algorithm based on an intensity criterion was used to segment the pectoral muscle. The algorithm was initialized with a seed point, which was determined based on an estimate of breast curvature (the point was placed at the first pixel of the noncurved side). Split and merge segmentation recursively splits the image until all regions satisfy a splitting criterion (e.g., homogeneity, cluster analysis, pixel classification, etc.). Adjacent regions that satisfy a predefined constraints are then merged to get the final segmentation. Although this seems to be dual to region growing, region splitting does not result in the same segmentation result even if the same homogeneity criteria are used (Sonka et al., 2014).

Watershed segmentation is another region-based segmentation that utilizes image morphology. It treats the image as a topological surface, where the surface height is represented by pixel intensity. The process starts by selecting at least one seed point belonging to each object in the image, including the background as a separate object. The markers are chosen by an operator or are provided by an automatic procedure. The watershed transform segments the image by finding the catchment basins and ridges on the corresponding surface. The intuition behind this algorithm can be explained as follows: if water falls on a 3D surface, it will be collected into catchment basins (local minima), two adjacent basins can be thought of as being separated by a ridge (watershed); as more water falls, the water level of each catchment basin will rise until reaching the ridge, which then causes the two basins to merge.

In (Camilus et al., 2010), graph cuts segmentation technique was used to segment and detect the pectoral muscle edge. The segmentation method is initialized by specifying a rectangular ROI, which contains the whole pectoral muscle. Results indicate that this method produced excellent results comported with earlier methods. Pectoral muscle segmentation based on normalized graph cuts was also proposed in (Abdellatif et al., 2012). The segmentation process is initialized using knowledge of the shape and location of the pectoral muscle. A refinement stage of the pectoral muscle counters smoothens the muscle edge using a Bezier curve. Region-based segmentation techniques perform better than edge-based methods. However, a combination of edge-based methods and region-based methods could provide better segmentation results.

5.4 Characterization and classification of microcalcifications and microcalcification clusters

The radiological definition of microcalcification clusters (MCCs) is that at least three microcalcifications are located in an area within a 1 cm^2 (Nalawade, 2009). The presence of MCCs in mammograms is a primary risk factor of breast cancer. MCCs can be benign, but there are types that are associated with a high probability of malignancy (Wilkinson et al., n.d).

Two classes of calcifications exist: macrocalcifications and microcalcifications. Macrocalcifications are large bright dots within the breast area, while microcalcifications are tiny deposits of calcium embedded in the breast tissue, which appear as bright (dense) spots in mammograms (Recent Advances in Breast Imaging, Mammography, & Computer-Aided Diagnosis of Breast Cancer, 2006). The presence of microcalcifications might show a high risk of malignancy, especially if forming dense clusters, while scattered microcalcifications are usually benign. Macrocalcifications are usually not linked with the development of the disease.

Different types of calcifications have been identified (e.g., punctate and amorphous). Mammography enables the detection of microcalcifications at an early stage. Detection of microcalcifications requires high spatial resolution scans and sharp displays, as they tend to be small (e.g., 100 μm). The presence of microcalcifications may indicate an increased breast cancer risk. Benign microcalcifications are usually easily seen on a mammogram and appear as large, diffusely distributed, and often round with smooth margins (Breast Imaging Reporting & Data System, 2009), while malignant microcalcifications tend to be numerous, clustered, small, and irregularly shaped.

Burnside et al. (2007) showed that microcalcification descriptors and categories in breast imaging-reporting and data system (BI-RADS) 4th edition help predict the risk of malignancy for suspicious microcalcifications. Data consists of images obtained from 115 women who consecutively underwent image-guided biopsy of microcalcifications. It was found that calcification descriptors helped to stratify the probability of malignancy as follows: coarse heterogeneous, one of 14 (7%); amorphous, four of 30 (13%); fine pleomorphic, 10 of 34 (29%); and fine linear, 10 of 19 (53%).

5.4.1 Morphological descriptors

Several feature descriptors exist for the characterization and classification of breast tissue. Morphological descriptors extract features such as size, contrast, shape of microcalcifications and their variations within clusters. Examples of shape features used in the literature include (Elter & Horsch, 2009; Leichter et al., 2004; Veldkamp et al., 2000; Zhang et al., 2001): MCC area; MCC eccentricity (cluster geometry feature); MCC compactness; MCC moment; MCC Fourier descriptors; MCC region size; MCC shape rate; shape factor and number of neighbors (i.e. shape of the individual micro calcifications); perimeters, circularities, rectangularity; orientations; morphology of individual microcalcifications using the means and standard deviations of: area of calcification, compactness, moments, Fourier descriptor, eccentricity, spread and number of calcifications in a cluster (Kallergi et al., 2005). These shape features were also found to be useful for characterization and classification of masses. In (Surendiran & Vadivel, 2012), the distribution eccentricity, elongatedness, circularity 1/circularity 2 (circularity 1/2 measure how much a mass resembles a circle and ellipse), compactness, standard deviation, and dispersion.

Summary statistics can be further applied on the descriptors computed by calculating the mean, standard deviations, minimum, maximum, median, and normalized central moments of the shape descriptor. In (Kallergi et al., 2005), the authors employed 13 image-based descriptors extracted from the morphology of individual micro calcifications (using the means) and the distribution of the clusters (using standard deviation) of: area of calcification, compactness, moments, Fourier descriptor, eccentricity, spread and number of calcifications in cluster, combined with patients age to build an artificial neural network. The classification performance of the algorithm reached a 100% sensitivity for a specificity of 85% and receiver operating characteristic area index $A_z = 0.98$).

Shape descriptors can represent individual microcalcification in a cluster and capture properties of the whole microcalcifications cluster. Shape descriptors extract features such as the areas, perimeters, circularities, rectangularities, orientations, and eccentricities. This can be applied to individual microcalficiations

(Fu et al., 2005; Kallergi et al., 2005; Papadopoulos et al., 2005; Veldkamp et al., 2000). To capture cluster shape or margin or even shapes of mammographic masses, we need first to segment the mass/cluster, compute contour/centroids and get the convex hull, from which, we can extract the shape descriptors stated above.

Morphological descriptors extract features such as size, contrast, shape of micro calcifications and their variations within clusters. For example, morphological features, such as focus/foci features, can describe lesions smaller than 5 mm. Also, mass/legion shape (e.g., round, oval lobulated and irregular), the margin (e.g., obscured, irregular and speculated), and the internal mass characteristics (e.g., homogeneous, heterogeneous, rim enhancement, dark internal septations, enhancing internal septatios, and central enhancement). Nonmass structures can be characterized by various distribution patterns (e.g., focal, linear, ductal, segmental, regional, multiple regions and diffuse, homogeneous, heterogeneous, stippled/punctate, clumped and reticular/dendritic, and between breasts symmetry) (Pérez, n.d.).

Ciecholewski (2017) used morphological transformations to segment microcalcifications in mammograms. The images are first enhanced in contrast and denoised. Morphological image transformations were used to detect micro calcifications, and then the watershed segmentation is used to extract the shape micro calcifications is used. The method produced a similarity index of 80.5% on a set of 200 ROIs 512×512 pixels in size, taken from mammograms in the digital database for screening mammography (DDSM) (images contain 100 malignant and 100 benign lesions).

Zhang et al. (2013) used morphological image processing and wavelet transform to detect MC. Morphological processing using top-hat transform enhances the appearance of micro calcifications and produces candidates of microcalcifications. A refinement step based on multilevel wavelets is then applied. Finally, MCs are detected using connected components by labeling an image block as MC if the total number of connected components is larger than three. The proposed method detected 92.9% of true MC cluster per image and 0.08 false MC cluster per image.

Leichter et al. (2004) analyzed the influence of two types of features; eccentricity (cluster geometry feature) and the shape factor and number of neighbors (i.e. shape of the individual micro calcifications); for differentiating between benign and malignant clusters. Classification results on 324 biopsy-proven MCCs using a linear discriminant analysis show that the cluster geometry feature was more effective in differentiating benign from malignant clusters (AUC value of 0.87) than was the shape of individual micro calcification (Leichter et al., 2004).

5.4.2 Image and texture-based descriptors

In the literature, the mean and variance of gray values and the contrast of individual microcalcifications have been employed for describing the optical density of microcalcifications (Leichter et al., 1999; Papadopoulos et al., 2005). Texture-based features have also been successfully used for the characterization of tissue surrounding microcalcifications (Chan et al., 1998; Dhawan et al., 1996; Elter & Horsch, 2009; Fu et al., 2005; Soltanian-Zadeh et al., 2004) and also for the characterization of masses (Gradient & texture analysis for the classification of mammographic masses, 2021; Mavroforakis et al., 2006; Varela et al., 2006).

Texture descriptors have been successfully used for describing microcalcifications. Texture descriptors are extracted from the gray-level co-occurrence matrices (GLCM) (GLCM encode second-order statistics of the gray levels in a given ROI). In (Haralick et al., 1973), the Haralick's texture descriptors were extracted from the gray-level co-occurrence matrices. Quintanilla-Domínguez et al. (2016) used texture analysis to detect MCCs in digitized mammograms. ROI in mammograms with dense tissue and existing MCs were selected for feature extraction. The top-hat transform was used to increase the contrast between the MCs clusters and background. Window-based features such as mean and standard deviation were applied. Two clustering algorithms, k-means and self-organizing maps, were used in order to segment the image into different areas (background and MC).

Soltanian-Zadeh et al. (2001) extracted 17 shape features of individual microcalcifications and microcalcifications clusters. Also, 44 texture features based on the co-occurrence method of Haralick were extracted. Feature selection based on genetic algorithm and maximizing the area under the ROC curve criterion. Using a k-nearest neighbor (kNN) classifier, shape features produced a ROC area of 0.82 ($k = 7$), while Haralick features produce a ROC area of 0.72 ($k = 9$). Soltanian-Zadeh et al. (2004) compared the performance of various texture and shape features sets such as GLCM based, shape, Haralick, wavelet, and multiwavelet features for microcalcification classification in mammograms. Other derived features include the symmetry, orthogonality, short support, and a higher number of vanishing moments. The authors concluded that the multiwavelet features outperformed the other features, achieving an AUC of 0.89 [based on Nijmegen database (Karssemeijer et al., 1991)].

Malar et al. (2012) compared the performance of various features such as wavelet-based texture features, gray level spatial dependence matrix, and Gabor filter bank features for microcalcification detection in digitized mammograms. The extreme learning machine (ELM) classifier was used and validated on 120 ROIs containing normal and microcalcification images (images were extracted from 55 mammogram images from the mini-MIAS database). The best classification accuracy (94%) was achieved when using wavelet features.

Yu et al. (2006) proposed a microcalcification classifier combining texture features and hand crafted auxiliary features. First, suspicious MCs were detected by thresholding the output of a wavelet filter based on the mean pixel value of the image. Texture features from the neighborhood of every candidate calcification are computed using a Derin–Elliott Markov random field model. Three additional features were extracted: the mean pixel value, the gray level variance, and a measure of edge density. Bayes and Feed-Forward Back Propagation neural network classifiers were trained and evaluated on the MiniMammographic Database (Suckling et al., 2015) using 20 mammograms containing 25 areas of clustered microcalficiations marked by radiologists. The proposed classifier identified and removed 1341 FPs out of 1356 (98.9%). The sensitivity (true positives rate) is 92%, with 0.75 FPs per image (Yu et al., 2006).

Tiedeu et al. developed a method for detecting MCCs using texture analysis (Tiedeu et al., 2012). Input images were smoothed using Gaussian filters and subtracted from their contrast enhanced versions. This suppresses the background tissue. MCs are detected in two stages: first, a segmentation step designed to find as many MCs as possible using a multithreshold algorithm. In the second stage, a selection step was used to reduce FPs from the segmentation and keeping only MCs belonging to clusters (a cluster of MCs was defined based on a minimal density of 5 MCs/cm^3. The detection method showed 85.65% of success rate for individual MCs with 2.50 FP image and an area under ROC the curve of 96.8%.

AbuBaker et al. examined the properties of true microcalcifications compared to falsely detected MCCs in mammogram images obtained from the University of South Florida mammogram image database (USF) (USF Digital Mammography Home Page, 2022). First order texture features (i.e. mean, entropy, standard deviation, skewness, kurtosis) and second order texture features (i.e. angular second moment, contrast, absolute value, inverse difference, entropy, maximum probability) were extracted. The authors concluded that the collective usage of several first and second order feature extraction techniques has failed to provide efficient discrimination between TP and FP clusters. They have also found that the entropy is the most effective individual feature to reduce the FP ratio without affecting the TP ratio. Entropy reduces the ratio of FPs by 18% without significantly affecting the ratio of true positives (only 0.3% TP cases missed) (AbuBaker et al., 2007).

Dhawan et al. (1996) presented gray-level image structure feature-based approach to classify “difficult- to-diagnose” mammographic microcalcifications. Two sets of texture features extracted from GLCM (histogram statistics for representing global texture) and wavelet features representing local texture from ROIs containing microcalcification lesions. They reported an AUC of 0.86 for the classification of 191 mammographic microcalcifications using the neural network classifier and the feature set selected through the GA search method.

Also, the texture analysis based on the wavelet descriptors has demonstrated to be important for Micro calcifications characterization (Soltanian-Zadeh et al., 2004). Ramos et al. (Texture extraction, 2022) extracted texture features from co-occurrence matrices, wavelet and ridgelet transform of mammogram images. The extracted features include entropy, energy, sum average, sum variance, and cluster tendency are fed to a feature selection stage based on genetic algorithms. The best classification rates of masses were obtained with the wavelet-based feature extraction and Random Forests classifier, giving an AUC of 0.90.

Haindl and Remeš (2019) used a two-dimensional adaptive causal autoregressive texture model to enhance the visualization of breast tissue abnormalities, such as micro calcifications and masses. They extracted 200 local textural features from different frequency bands and used the Karhunen–Loeve transform to extract the top three most informative projections that retain 99% of the overall energy. The three components are then used as the red, green and blue color channels of the resulting enhanced image. The results indicate that the algorithm works well both for small and bigger findings. State-of-the-art results where barely visible abnormalities, with the same average gray level as their surroundings, are highlighted very well their algorithm.

5.4.3 Spatial descriptors

Spatial descriptors encode the spatial distribution of microcalficiations inside a cluster. Example of spatial features are the mean, standard deviation of the distances between individual microcalcifications, and the eccentricity and the normalized central moments of the microcalcification centroids (Leichter et al., 1999; Schmidt et al., 1999).

The location of the MCCs within the breast region can be used to infer the probability of breast cancer. Location-based descriptors require a robust segmentation and detection methods for breast structures such as the pectoral muscle, breast boundary, and the location of the MCCs within the breast region (Breast Imaging Reporting & Data System, 2009). The coordinates can be based on clinical orientation quadrants. The side (left/right) of the lesion is given first, second the location is specified (e.g., upper outer quadrant, upper inner quadrant, lower outer quadrant, and lower inner quadrant in the left or right or both sides), and last depth is described (e.g., anterior, middle, posterior, or subareolar) (Breast Imaging Reporting & Data System, 2009; Pérez, n.d.). Location-based descriptors require a robust segmentation and detection methods for breast structures such as the pectoral muscle, breast boundary, and the nipples (Elter & Horsch, 2009; Pérez, n.d.). In (Veldkamp et al., 2000), the relative distance of a cluster to the pectoral muscle and the breast edge in the mammogram was used as a location feature.

In (Veldkamp et al., 2000), a set of 16 intensity and shape descriptors extracted from microcalcifications based on the distribution of individual microcalcifications within a cluster, cluster area and eccentricity, and cluster location features. For classification, the kNN method was used in a leave-one-patient-out procedure. Results showed that the method's performance was considerably higher than that of the radiologists.

Zhang et al. (2001) extracted 11 features from both spatial and morphology domains in order to describe the MCCs. Two stages allow feature generation from different perspectives. The first stage detects coarse visual features, while the second stage facilitates finer analysis and classification. The features are grouped into three categories: gray-level description, shape description (average gray level of the foreground and background, standard deviation of the gray level of the foreground and background, compactness, moment, and Fourier descriptors), and MCC features (cluster region size and cluster shape rate). Experimental results using a back-propagation neural network showed that the false detection rate of calcifications in the image is reduced by 42%.

5.4.4 Fractal geometry

Modeling mammograms by deterministic fractal coding produced visually similar patterns to the original image containing microcalcifications. In (Sankar & Thomas, 2010), a fast fractal modeling approach was used to detect the presence of microcalcifications. Fractal image modeling is achieved by searching for a matching domain based on mean and variance, dynamic range of the image blocks, and mass center features. A normal mammogram detection score of 89.09% is obtained for mean variance methods.

The application of fractal geometry for breast mass classification in mammography images has been also reported in the literature. Nguyen and Rangayyan (2005) observed that the fractal dimension (FD) of benign masses is, in general, less than the FD of malignant tumors. They also showed that the computation of the FD of microcalficiations contours based on the box grid method provides discrimination between benign and malignant clusters.

Fractal geometry analysis views the microcalficiations cluster as a fractal normal background superimposed by a nonfractal foreground. In (Huang & Yu, 2007), a method for microcalcification classification using a combination of 4th order Markov random field (MRF) and a deterministic fractal model (FM). Fifty 88×88 ROI samples with 25 samples belonging to the class MCs and another 25 samples to class normal were used to train and test a three-layer neural network. Combining the 4th order MRF and FM modeling produced a 0.90 AUC performance (LOO validation).

5.5 Segmentation and classification of masses

A mass or lesion is defined as a 3D density structure that can be seen in at least two different projections (Breast Imaging Reporting & Data System, 2009). The literature indicates that masses with speculated boundaries, irregular shapes, and margins are indicative of carcinoma (DiSaia et al., 2017; Kamal et al., 2016). Nonmass structures can be characterized by various distribution patterns (e.g., focal, linear, ductal, segmental, regional, multiple regions and diffuse, homogeneous, heterogeneous, stippled/punctate, clumped and reticular/dendritic, and between breasts symmetry) (Breast Imaging Reporting & Data System, 2009; Pérez, n.d.).

Asymmetric densities (AD) is a type of abnormality indicating distortion of breast tissue, such as tethering or indentation of breast tissue region, the appearance of straight thin lines, speculated radiating lines, or focal retraction (Dilhuydy, 2007). AD is caused due to postsurgical scars or soft tissue damage is benign; however, if masses are present within AD, this becomes highly suspicious of malignancy (Banik et al., 2013).

Asymmetric densities (or asymmetries for short) are density structures caused by the superposition of normal breast tissues and are normally seen only in one projection and do not confirm to the definition of mass (Burnside et al., 2009; Dilhuydy, 2007).

Choi et al. (2014) proposed a method for the detection of breast masses by integrating features gained from unsupervised and supervised algorithms. Unsupervised detection recognizes that masses have usually dense core regions, while supervised detection can learn the subtle textural and boundary variation of masses. In order to maximize detection sensitivity, an effective ROI combination solution that combines supervised and unsupervised detections is proposed, along with an ensemble classification algorithm. This significantly reduces the number of FP detections as much as 70% at only a cost of 4.6% in sensitivity loss.

Masses are abnormal regions in breast tissue that can be visually identified in mammograms in at least two different projections. Visual descriptors of masses adopted in BI-RADS include shape, margin, and density are used to characterize masses. Early development of CAD schemes was presented in (Doi, 2007). Research on improving CAD is key to promote advances in the ability to detect lesions and predict the risk of cancer. CAD systems deploy image processing and computer vision techniques to improve the diagnostic consistency by radiologists. It provides radiologists with tools to detect suspicious regions in a given image and assists in making correct diagnoses in clinical settings.

Clinical features of pathological lesions in mammography images are described in the ACR (BI-RADS) standard based on morphological and localization properties (Breast Cancer Digital Repository, 2021; Burnside et al., 2009). BI-RADS offers characterization of masses according to density, margin type (circumscribed, microlobular, obscured, indistinct, and speculated) and shape (round, oval, lobular, irregular). Benign lesions/masses tend to be fatty (i.e., low density), homogeneous, or mixed-density circumscribed (i.e., well-defined edges). They normally appear smooth with possible macrolobulations, regularly shaped (such as round and oval) with very well-defined margins (i.e., well-defined transition between the lesion and the surrounding tissue separated by a fine radiolucent line [fatty halo]). Isodense to high-density masses can be benign or malignant. Isodense masses with lobular shape and microlubulated margins are moderately suspicious (Atlas of Mammography, 2009; Dilhuydy, 2007; Found a Breast Mass?, 2019).

Malignant lesions exhibit ill-defined, rough, or irregular-shaped contours with microlobulations, spiculations, and concavities. Varela et al. (2006) showed that the sharpness features perform better than microlobulation features for masses classification. Speculated masses appear as lines radiating from the margins of the mass, which may show infiltration by the lesion of the healthy breast tissue (Atlas of Mammography, 2009; Breast Imaging Reporting & Data System, 2009; Burnside et al., 2009).

Classification of masses can be performed according to its circumscribed margin. Obscured or partially obscured margins occur when superimposed breast tissues hide margins. If the margin is at least 75% circumscribed and 25% obscured, the mass can be classified on the basis of its circumscribed margin, similarly, if the margin is partially circumscribed, the mass can be classified on the basis of its indistinct margins (Dilhuydy, 2007; Pérez, n.d.).

Oliver et al. (2010) studied the performance of seven mass detection methods for the detection and segmentation of masses. The authors combined four algorithms with the best performance using addition and multiplication of the probability images. They indicated that fusing the predictions of multiple algorithms can improve detection compared with using the algorithms individually.

5.5.1 Shape features

According to the BI-RADS system, benign and malignant masses can be differentiated using their shape, size, and density features. Shape-based features can be categorized based on both the shape and margin of the mass. Surendiran and Vadivel (2012) extracted 17 shapes and geometrical features from the ROIs of images containing all types of masses: lobular, circular, oval, AD, irregular, and more. The features include the circularity of the lesions, their irregularity, and their margin characteristics. Shape and margin features include area, perimeter, eccentricity, equivalent diameter, compactness, thinness ratio, circularity 1, circularity 2, elongatedness, dispersion, shape index, Euler number, maximum radius, minimum radius, and SD of the edge. These features were found to be effective in discriminating benign and malignant masses using a CART classifier. Unlike statistical-based measures, which are based on gray value of the masses, geometric shape, and margin, features tend to be robust against the presence of noise or overenhancement (Surendiran & Vadivel, 2012).

Since masses usually have greater intensity than the surrounding tissue, a global threshold value could be used to initialize mass detection algorithms. Li et al. (1999) used local adaptive thresholding and multiscale processing methods to segment mammographic image into parts belonging to the same classes. A high mass detection performance with 93% sensitivity and 3.1 FP rate per image is achieved using a combined "hard" and "soft" decision-making strategy.

In (Raguso et al., 2010), differences in shape complexity of breast mass contours are represented by the FD. FD was found to be a discriminating feature for classification of breast masses with a ROC score of 0.97. In (Kamal et al., 2016), morphological descriptors of BI-RADS were used to discriminate between breast lesion types The study was carried out on a total of 261 enhancing breast lesions identified on contrast-enhanced spectral mammography in 239 patients (26.1% benign lesions and 73.9% malignant lesions). The study showed that the most suggestive morphological descriptors of malignant masses are irregular-shaped mass lesions with speculated and irregular margins.

Sheba et al. (2018) experimentally observed that masses, whether cancerous or noncancerous, have areas ranging from 400 to 21,000 pixels. Multithresholding based on a modified Otsu's method is used to extract a pixel label matrix (Chen, Zhao et al., 2012). Pixel regions with label sizes from 400 to 21,000 pixels are chosen as candidate masses. The contours of the pixel regions are refined using morphological operations (Zhang et al., 2016).

In (Moura & Guevara López, 2013), a shape descriptor based on histograms of gradient divergence (HGD) was compared against other image descriptors (intensity statistics, histogram measures, invariant moments, Zernike moments, Haralick features, gray-level run length, gray-level difference matrix, Gabor filter banks, histogram of oriented gradients, wavelets, and curvelets). Different machine learning classifiers were tested and results indicated that the HGD was best descriptor for predicting breast masses.

Rojas Domínguez and Nandi (2009) obtained initial mass contours by segmented masses using dynamic programming for boundary tracking and constrained region growing. The initial mass contours are fitted to ellipses and used to obtain features to describe the mass and its margin. The features include: contrast between foreground and background regions, edge strength, fuzziness of mass margins, relative gradient orientation, and edge-signature information. The results show improved segmentation and classification performance.

In (Liu et al., 2012), an initial mass contour made by radiologists is refined by a level set segmentation method. Features are then extracted using shape analysis and Fourier descriptor of normalized accumulative angle. LDA and SVM classifiers are then trained and tested on 292 mass ROIs obtained from DDSM mammogram images. Rangayyan et al. (2000) used an iterative boundary segmentation approach for the classification of manually segmented mammographic masses into benign/malignant. They used two shape factor features: the first is fractional concavity, which describes portions of the boundary as concave or convex segments; the second factor is the speculation index that is based on the concavity fraction of a mass boundary and the degree of the narrowness of spicules. The final boundary is refined using a global shape feature of compactness. A classification accuracy of 81.5% was reported.

Rangayyan et al. (2008) obtained shape features with the ability to capture important details of spicules, lobulations, and shape roughness. The features are based on the turning angle functions of contours produced high classification accuracies in discriminating between benign breast masses and malignant tumors. In (Rangayyan et al., 2000), image gradient in a region surrounding the mass contour is used to estimate the sharpness of a mass contour. Similarly, Shi et al. (2008) used rubberband-straightening transform to map a band of pixels surrounding the mass onto a rectangular image and then applied line detection to measure margin sharpness.

In (Varela et al., 2006), the sharpness of the margin and the presence of microlobulations was done for interior, border, and outer mass segment. Eltonsy et al. (2007) modeled the growth of a mass as a set of multiple concentric layers around focal areas in breast tissue. The morphological concentric layer model assigns bright intensity at the focal activity layer and lower brightness levels for the evolving concentric layers. The focal layer morphological features are indicative of the presence of a potentially malignant mass. The probability of malignant abnormality tends to increase if the relative incidence of focal areas with multiple concentric layers is low in the breast region.

Li et al. (2017) proposed a contour descriptor that translates the 2D contour of breast mass into 1D signature using the root mean square slope. The contour descriptor consists of four local features to capture both the contours and breast mass regularity. Classical machine-learning models using KNN, SVM, and ANN classifiers are finally trained to classify the masses into benign/malignant. The best classification accuracy of 99.66% is achieved using the SVM classifier.

Location descriptors represent lesion coordinates based on clinical orientation quadrants. Usually, the side (left/right) of the lesion is given first, second the location is specified (e.g., upper outer quadrant, upper inner quadrant, lower outer quadrant, and lower inner quadrant in the left or right or both sides), and last depth is described (e.g., anterior, middle, posterior, or subareolar) (Breast Imaging Reporting & Data System, 2009; Pérez, n.d.).

Sahiner et al. (2008) used computer vision techniques to emulate radiologists' learned-by-experience descriptors. An initial mass shape estimate within the ROI was obtained using a pixel-by-pixel K-means clustering algorithm, which is then refined using an active AC segmentation method. The study has identified two features that had high accuracy for characterizing mass margins according to BI-RADS. These are the spiculation measure for a pixel on the mass boundary and the circumscribed margin.

5.5.2 Image and texture-based features

Masses tend to be brighter and have higher contrast as compared to normal tissue. The statistical features can be useful in detecting suspicious masses. Sheba et al. (2018) extracted six statistical features from candidate ROIs; the mean, variance, skewness, kurtosis, energy, and entropy. In (Farhan & Kamil, 2020), mammographic texture analysis is used to generate features to discriminate breast masses from normal tissue. Features based on LBP, HOG, and GLCM techniques were extracted from specified ROIs. In addition, contrast, energy, correlation, and homogeneity were used as features properties. The LBP and logistic regression produced the best accuracy of 92.5% (HOG + SVM classifier performed at 90%, GLCM + KNN classifier performed at 89.3%).

Benign masses tend to have smooth texture, whereas malignant masses exhibit rough textures. Hence, GLCM features help in distinguishing between soft texture and hard texture. In (Sheba et al., 2018), 52 texture feature values were obtained from the GLCM matrix each at angles 0, 45, 90, and 135 degrees at distance $d = 1$. The features include energy, contrast, correlation, variance, homogeneity, entropy, sum average, sum entropy, sum variance, difference variance, difference entropy, first correlation measure, and second correlation measure. The method achieved an accuracy of 96%, 83% sensitivity, and 98% specificity.

Dominguez and Nandi (2007) used multiple threshold levels (30 levels with a step size of 0.025) to convert the images to binary images. Mammographic images are segmented and a set of features is computed from each of the segmented regions. The method achieved a sensitivity of 80% at 2.3 FPs per image on 57 images of masses from the mini-MIAS database.

Texture descriptors have been successfully used for describing masses (Mavroforakis et al., 2006; Rangayyan et al., 2000; Varela et al., 2006). Matos et al. (2019) extracted local features using SIFT, speed up robust feature, oriented fast and rotated BRIEF (ORB) to classify malignancy and benignity of masses. A bag of features model is used to provide new lower dimensional representations. The support vector machine (SVM), adaptive boosting (Adaboost), and random forests (RF) classifiers were tested and reported a 100% sensitivity, 99.65% accuracy, and 99.24% specificity for benign and malignant classification.

Image processing tools can also be used to derive descriptors for detecting mammogram lesions. Low-level image features, such as edges, linear structures, and intensity, besides higher level features such as shapes and texture descriptors, are discussed in this section. Edge detection searches for gray level discontinuities in the image by measuring the rate of change in the gray level using gradients. A discussion of various operators for edge detection, such as Prewitt operator, Sobel operator, Roberts operator, and LoG operator, can be found in (Rangayyan et al., 2008).

Biswas and Mukherjee (2011) used a multiscale-oriented filter bank to extract low-level rotation-invariant textural features at different scales. They proposed a generative model to capture discriminatory texture patterns (i.e., the AD specific textures and normal textures) using mixture of Gaussian texton distributions to learn the latent textural primitives. Recognizing AD is posed as a texture-classification problem.

Braz Junior et al. (2019) proposed a method to detect mass regions through mean shift segmentation, spatial diversity texture analysis, geostatistical indexes, geometric analysis, and concave geometry. The effectiveness of each feature is evaluated on the MIAS and DDSM databases using a SVM. The results demonstrate that concave geometry can capture inner texture distribution of masses and help reduce the FPs. Speculated structures, normally associated with malignant masses, exhibit margin abruptness. Texture analysis and border information describing the mass margins is therefore a strong candidate for classification of mammographic masses. Unlike texture analysis for microcalficiations where the entire ROI surrounding the microcalcifications is used, texture analysis for masses can be carried out in a band of pixels close to the mass margin where speculated structures are more prominent.

In (Wei et al., 2005) combined gradient field analysis and gray level information to identify candidate regions that might contain masses. A clustering-based region growing algorithm was then used for extracting the suspicious lesions. Features are generated based on a set of shape and texture descriptors from the extracted lesion regions and used to train ruled-based and LDA mass/nonmass classifiers.

In (Bellotti et al., 2006), dynamical threshold is used to segment suspicious lesions. Texture features are generated from segmented lesions based on the gray tone spatial dependence matrix (GTSDM) computed for different angles (GTSDM captures second-order spatial statistics on pixel intensity). The features are used to train a neural network mass classifier.

In Elmoufidi et al. (2018), candidate regions are detected based on a modified K-means algorithm that generates the number of segmentation classes automatically. Three types of texture features selected from ROIs were used: Haralick features based on gray level co-occurrence matrices, gray-level run-length matrices, and histograms of local binary patterns. In addition, shape features such as the equivalent circle of ROI and the bounding box, which is defined by the smallest rectangle containing the ROI, are used, which have the same area as the ROI and have the center of gravity of the ROI. Finally, a classifier based on multi-instance learning is used to classify masses.

Tai et al. (2014) extracted local and discrete texture features for mammographic mass detection. The texture features based on co-occurrence matrix and optical density transformation are extracted from suspicious areas using adaptive square ROI. A stepwise linear discriminant analysis is used to classify abnormal regions by selecting and rating the individual performance of each feature. The authors claimed a satisfactory detection sensitivity and FP rate and better performance for lower density mammograms.

Mohanty et al. (2013) extracted features from the run lengths such as energy, inertia, entropy, maxprob, inverse, short run emphasis (SRE), long run emphasis (LRE), gray level nonuniformity (GLN), run length nonuniformity (RLN), low gray level run emphasis (LGRE), high gray level run emphasis (HGRE), short run high gray level emphasis (SRLGE), and ARM. The method can distinguish malignant masses from benign masses with an accuracy of 96.7/% (decision tree classifier) and area under the receiver operating curve of 0.995. Sheba et al. (2018) also used the GLRLM method and extracted 44 GLRLM texture features at angles 0, 45, 135, and 90 degrees. The GLRLM features include SRE, LRE, GLN, run percentage (RP), RLN, LGRE, HGRE, short run low gray level emphasis (SRLGE), SRHGE, low run low gray level emphasis (LRLGE), and low run high gray level emphasis (LRHGE). Texture features based on GLCM resulted in higher accuracy than texture features based on GLRLM, but the best accuracy claimed was for texture features from the combined GLRLM and GLCM (Mohanty et al., 2013). Also, GLRLM features play an important role in differentiating benign lesions, which have smooth soft texture, from malignant lesions, which have a coarse texture.

5.5.3 Multiresolution features

Wang et al. (2014) segmented images using a modified wavelet transformation of the local modulus maxima algorithm. Five textural and five morphological features were extracted from the ROIs and used to train an ELM classifier for breast tumor detection. The ELM classifier produced better classification accuracy than SVM and improved training speed. de Lima et al. (2016) applied multiresolution wavelet analysis and extracted Zernike moments from the ROI of the images from each wavelet component. They combined them with texture and shape features. Considering BIRADS criteria, the classification accuracy of masses reached 94.11% (based on 355 images of fatty breast tissue from IRMA database, with 233 normal instances (no lesion), 72 benign, and 83 malignant cases and using support-vector/extreme-learning machines).

Varela et al. (2007) used a multiscale iris filter and an adaptive threshold to segment suspicious regions. Features based on the iris filter output, gray level, texture, contour-related, and morphological features were extracted and used to train a back-propagation neural network classifier. The method yielded a sensitivity of 88% and 1.02 FPs per image for lesion-based and case-based evaluation, respectively. Results suggest that the proposed method could help radiologists as a second reader in mammographic screening.

Hu et al. (2011) used multiresolution analysis and a combination of window-based adaptive thresholding and global thresholding to detect suspicious lesions in mammograms. Global thresholding produced a coarse segmentation, which is filtered using morphological enhancement. The window-based adaptive local thresholding method produced a finer segmentation. The algorithm has a sensitivity of 91.3% with 0.71 FPs per image.

Pereira et al. (2014) combined wavelet analysis and genetic algorithms to determine an appropriate number of threshold levels and values that can segment masses in the image (from CC and MLO views). The GA and wavelet combination decreases computational times for threshold selection. A true positive ROI prediction occurs if there is a mass in the ROI and the algorithm detects it. The reported sensitivity on a set of selected cases from the public DDSM was 95%.

Beura et al. (2015) implemented a method to classify breast tissues as normal, benign, or malignant using wavelet and GLCM to all the detailed coefficients from 2D-DWT of the ROI. Feature selection using *t*-test and *F*-test were used separately to find the most relevant features. The authors reported an accuracy of 98.0% for normal-abnormal and a 94.2% accuracy for benign-malignant classes from the MIAS database.

5.6 Conclusion

In conclusion, mammographic screening is a complex task that relies on expert radiologists to identify abnormalities in breast tissue. CAD systems have been developed to improve the accuracy of mammogram interpretation by integrating techniques from image processing, computer vision, and machine learning. Breast region segmentation, edge detection, and various thresholding schemes are utilized to identify breast boundaries and extract relevant features. These features are then used to train machine learning classifiers for pixel-wise classification. Several feature descriptors, such as morphological, spatial, and shape descriptors, have been employed in the literature to characterize and classify breast tissue abnormalities. Recent advancements in deep learning techniques, such as fully convolutional networks and U-Net architectures, have also shown promising results in mass segmentation and calcification detection. Ultimately, these approaches aim to improve the accuracy and efficiency of mammographic screening, ultimately enhancing early detection and treatment of breast cancer.

References

Abdellatif, H., Taha, T.E., Zahran, O.F., Al-Nauimy, W., & Abd El-Samie F.E. (2012). K2. Automatic pectoral muscle boundary detection in mammograms using eigenvectors segmentation. In *2012 29th National Radio Science Conference (NRSC)* (pp. 633–640). Available from https://doi.org/10.1109/NRSC.2012.6208576.

Abu Baker, A., Qahwaji, R., & Ipson, S. (2007). Texture-based feature extraction for the microcalcification from digital mammogram images. In *2007 IEEE international conference on signal processing and communications* (pp. 896–899). Available from https://doi.org/10.1109/ICSPC.2007.4728464.

Atlas of Mammography. (2009). 3rd ed. Radiology. 252(3), 663–663, https://pubs.rsna.org/doi/10.1148/radiol.2523092526. Available from https://doi.org/10.1148/radiol.2523092526.

Banik, S., Rangayyan, R. M., & Desautels, J. E. L. (2013). Measures of angular spread and entropy for the detection of architectural distortion in prior mammograms. *International Journal of Computer Assisted Radiology and Surgery*, *8*(1), 121–134. Available from https://doi.org/10.1007/s11548-012-0681-x, http://link.springer.com/10.1007/s11548-012-0681-x.

Beam, C. A., Layde, P. M., & Sullivan, D. C. (1996). Variability in the interpretation of screening mammograms by US radiologists: Findings from a national sample. *Archives of Internal Medicine*, *156*(2), 209–213. Available from https://doi.org/10.1001/archinte.1996.00440020119016.

Bellotti, R., De Carlo, F., Tangaro, S., Gargano, G., Maggipinto, G., Castellano, M., Massafra, R., Cascio, D., Fauci, F., Magro, R., Raso, G., Lauria, A., Forni, G., Bagnasco, S., Cerello, P., Zanon, E., Cheran, S. C., Lopez Torres, E., Bottigli, U., ... De Nunzio, G. (2006). A completely automated CAD system for mass detection in a large mammographic database. *Medical Physics*, *33*(8), 3066–3075. Available from https://doi.org/10.1118/1.2214177.

Berg, W. A., Campassi, C., Langenberg, P., & Sexton, M. J. (2000). Breast imaging reporting and data system inter- and intraobserver variability in feature analysis and final assessment. *American Journal of Roentgenology*, *174*(6), 1769–1777. Available from https://doi.org/10.2214/ajr.174.6.1741769, https://www.ajronline.org/doi/full/10.2214/ajr.174.6.1741769.

Beura, S., Majhi, B., & Dash, R. (2015). Mammogram classification using two dimensional discrete wavelet transform and gray-level co-occurrence matrix for detection of breast cancer. *Neurocomputing*, *154*, 1–14. Available from https://doi.org/10.1016/j.neucom.2014.12.032, https://www.sciencedirect.com/science/article/pii/S0925231214016968.

Biswas, S. K., & Mukherjee, D. P. (2011). Recognizing architectural distortion in mammogram: A multiscale texture modeling approach with GMM. *IEEE Transactions on Biomedical Engineering*, *58*(7), 2023–2030. Available from https://doi.org/10.1109/TBME.2011.2128870.

Braz Junior, G., da Rocha, S. V., de Almeida, J. D. S., de Paiva, A. C., Silva, A. C., & Gattass, M. (2019). Breast cancer detection in mammography using spatial diversity, geostatistics, and concave geometry. *Multimedia Tools and Applications*, *78*(10), 13005–13031. Available from https://doi.org/10.1007/s11042-018-6259-z, http://link.springer.com/10.1007/s11042-018-6259-z.

Breast Cancer Digital Repository. (2021). 6 30 2021/06/30/00:14:33. https://bcdr.eu/information/about.

Breast Imaging Reporting & Data System. (2009). A comprehensive guide providing standardized breast imaging terminology, report organization, assessment structure and a classification system for mammography, ultrasound and MRI of the breast https://www.acr.org/Clinical-Resources/Reporting-and-Data-Systems/Bi-Rads.

Brzakovic, D., Luo, X. M., & Brzakovic, P. (1990). An approach to automated detection of tumors in mammograms. *IEEE Transactions on Medical Imaging*, *9*(3), 233–241. Available from https://doi.org/10.1109/42.57760.

Burnside, E. S., Ochsner, J. E., Fowler, K. J., Fine, J. P., Salkowski, L. R., Rubin, D. L., & Sisney, G. A. (2007). Use of microcalcification descriptors in BI-RADS 4th edition to stratify risk of malignancy. *Radiology*, *242*(2), 388–395. Available from https://doi.org/10.1148/radiol.2422052130, http://pubs.rsna.org/doi/10.1148/radiol.2422052130.

Burnside, E. S., Sickles, E. A., Bassett, L. W., Rubin, D. L., Lee, C. H., Ikeda, D. M., Mendelson, E. B., Wilcox, P. A., Butler, P. F., & D'Orsi, C. J. (2009). The ACR BI-RADS® experience: Learning from history. *Journal of the American College of Radiology*, *6*(12), 851–860. Available from https://doi.org/10.1016/j.jacr.2009.07.023, https://www.jacr.org/article/S1546-1440(09)00390-1/abstract.

Camilus, K. S., Govindan, V. K., & Sathidevi, P. S. (2010). Computer-aided identification of the pectoral muscle in digitized mammograms. *Journal of Digital Imaging*, *23*(5), 562–580. Available from https://doi.org/10.1007/s10278-009-9240-6, https://www.ncbi.nlm.nih.gov/pmc/articles/PMC3046680/.

Chakraborty, J., Mukhopadhyay, S., Singla, V., Khandelwal, N., & Bhattacharyya, P. (2012). Automatic detection of pectoral muscle using average gradient and shape based feature. *Journal of Digital Imaging*, *25*(3), 387–399. Available from https://doi.org/10.1007/s10278-011-9421-y, http://link.springer.com/10.1007/s10278-011-9421-y.

Chan, H.-P., Sahiner, B., Lam, K. L., Petrick, N., Helvie, M. A., Goodsitt, M. M., & Adler, D. D. (1998). Computerized analysis of mammographic microcalcifications in morphological and texture feature spaces. *Medical Physics*, *25*(10), 2007–2019. Available from https://doi.org/10.1118/1.598389, https://aapm.onlinelibrary.wiley.com/doi/abs/10.1118/1.598389.

Cheng, H. D., Cai, X., Chen, X., Hu, L., & Lou, X. (2003). Computer-aided detection and classification of microcalcifications in mammograms: A survey. *Pattern Recognition*, *36*(12), 2967–2991. Available from https://doi.org/10.1016/S0031-3203(03)00192-4, https://www.sciencedirect.com/science/article/pii/S0031320303001924.

Cheng, H. D., Shi, X. J., Min, R., Hu, L. M., Cai, X. P., & Du, H. N. (2006). Approaches for automated detection and classification of masses in mammograms. *Pattern Recognition*, *39*

(4), 646–668. Available from https://doi.org/10.1016/j.patcog.2005.07.006, https://www.sciencedirect.com/science/article/pii/S0031320305002955.

Chen, Z., Denton, E. R. E., Zwiggelaar, R., Maidment, A. D. A., Bakic Predrag, R., & Gavenonis, S. (2012). *Image patch benign case boundary pixel connected subgraph true positive rate. Classification of microcalcification clusters based on morphological topology analysis* (pp. 521–528). Springer. Available from https://doi.org/10.1007/978-3-642-31271-7_67, 978-3-642-31271-7.

Chen, Q., Zhao, L., Lu, J., Kuang, G., Wang, N., & Jiang, Y. (2012). Modified two-dimensional Otsu image segmentation algorithm and fast realisation. *IET Image Processing*, *6*(4), 426. Available from https://doi.org/10.1049/iet-ipr.2010.0078, https://digital-library.theiet.org/content/journals/10.1049/iet-ipr.2010.0078.

Choi, J. Y., Kim, D. H., Plataniotis, K. N., & Ro, Y. M. (2014). Computer-aided detection (CAD) of breast masses in mammography: Combined detection and ensemble classification. *Physics in Medicine and Biology*, *59*(14), 3697–3719. Available from https://doi.org/10.1088/0031-9155/59/14/3697, https://iopscience.iop.org/article/10.1088/0031-9155/59/14/3697.

Ciecholewski, M. (2017). Microcalcification segmentation from mammograms: A morphological approach. *Journal of Digital Imaging*, *30*(2), 172–184. Available from https://doi.org/10.1007/s10278-016-9923-8, http://link.springer.com/10.1007/s10278-016-9923-8.

Delogu, P., Evelina Fantacci, M., Kasae, P., & Retico, A. (2007). Characterization of mammographic masses using a gradient-based segmentation algorithm and a neural classifier. *Computers in Biology and Medicine*, *37*(10), 1479–1491. Available from https://doi.org/10.1016/j.compbiomed.2007.01.009, https://www.sciencedirect.com/science/article/pii/S0010482507000248.

DeSantis, C., Siegel, R., & Jemal, A. (2011). Breast cancer facts & figures 2011–2012. 36.

Dhawan, A. P., Chitre, Y., & Kaiser-Bonasso, C. (1996). Analysis of mammographic microcalcifications using gray-level image structure features. *IEEE Transactions on Medical Imaging*, *15*(3), 246–259. Available from https://doi.org/10.1109/42.500063.

Dilhuydy, M. H. (2007). Breast imaging reporting and data system (BI-RADS) or French "classification ACR" What tool for what use? A point of view. *European Journal of Radiology*, *61*(2), 187–191. Available from https://doi.org/10.1016/j.ejrad.2006.08.032.

DiSaia, P. J., Creasman, W. T., Mannel, R. S., McMeekin, S., & Mutch, D. G. (2017). *Clinical gynecologic oncology E-book* (p. 893) Elsevier Health Sciences.

Doi, K. (2007). Computer-aided diagnosis in medical imaging: Historical review, current status and future potential. *Computerized Medical Imaging and Graphics: The Official Journal of the Computerized Medical Imaging Society*, *31*(4–5), 198–211. Available from https://doi.org/10.1016/j.compmedimag.2007.02.002, https://www.ncbi.nlm.nih.gov/pmc/articles/PMC1955762/.

Dominguez, A.R., & Nandi, A.K. (2007). IEEE enhanced multi-level thresholding segmentation and rank based region selection for detection of masses in mammograms. In *2007 IEEE international conference on acoustics, speech, and signal processing*. I-449-I-452. Available from https://ieeexplore.ieee.org/document/4217113/, https://doi.org/10.1109/ICASSP.2007.366713.

Elter, M., & Horsch, A. (2009). CADx of mammographic masses and clustered microcalcifications: A review. *Medical Physics*, *36*(6), 2052–2068. Available from https://doi.org/10.1118/1.3121511.

Eltonsy, N. H., Tourassi, G. D., & Elmaghraby, A. S. (2007). A concentric morphology model for the detection of masses in mammography. *IEEE Transactions on Medical Imaging*, *26*(6), 880–889. Available from https://doi.org/10.1109/TMI.2007.895460.

Farhan, A. H., & Kamil, M. Y. (2020). Texture analysis of breast cancer via LBP, HOG, and GLCM techniques. *IOP Conference Series: Materials Science and Engineering*, *928*(7)072098. Available from https://doi.org/10.1088/1757-899X/928/7/072098, https://iopscience.iop.org/article/10.1088/1757-899X/928/7/072098.

Fenton, J. J., Taplin, S. H., Carney, P. A., Abraham, L., Sickles, E. A., D'Orsi, C., Berns, E. A., Cutter, G., Hendrick, R. E., Barlow, W. E., & Elmore, J. G. (2007). Influence of computer-aided detection on performance of screening mammography. *The New England Journal of Medicine*, *356*(14), 1399–1409. Available from https://doi.org/10.1056/NEJMoa066099.

Ferrari, R. J., Rangayyan, R. M., Desautels, J. E. L., Borges, R. A., & Frère, A. F. (2004). Automatic identification of the pectoral muscle in mammograms. *IEEE Transactions on Medical Imaging*, *23*(2), 232–245. Available from https://doi.org/10.1109/tmi.2003.823062.

Found a Breast Mass? (2019). If a breast mass is found on Mammography what does this mean? All about the different features and characteristics of a breast lump and what they mean. Breast cancer – Moose and doc found a breast mass? What does it mean? – Moose and doc found a breast mass? https://breast-cancer.ca/mass-chars/.

Freer, T. W., & Ulissey, M. J. (2001). Screening mammography with computer-aided detection: Prospective study of 12,860 patients in a community breast center. *Radiology*, *220*(3), 781–786. Available from https://doi.org/10.1148/radiol.2203001282.

Fu, J. C., Lee, S. K., Wong, S. T. C., Yeh, J. Y., Wang, A. H., & Wu, H. K. (2005). Image segmentation feature selection and pattern classification for mammographic microcalcifications. *Computerized Medical Imaging and Graphics: The Official Journal of the Computerized Medical Imaging Society*, *29*(6), 419–429. Available from https://doi.org/10.1016/j.compmedimag.2005.03.002.

Gradient and texture analysis for the classification of mammographic masses | IEEE Journals & Magazine | IEEE Xplore. (2021). https://ieeexplore.ieee.org/document/887618.

Guliato, D., Rangayyan, R. M., Carvalho, J. D., & Santiago, S. A. (2008). Polygonal modeling of contours of breast tumors with the preservation of spicules. *IEEE Transactions on Biomedical Engineering*, *55*(1), 14–20. Available from https://doi.org/10.1109/TBME.2007.899310.

Gur, D., Sumkin, J. H., Rockette, H. E., Ganott, M., Hakim, C., Hardesty, L., Poller, W. R., Shah, R., & Wallace, L. (2004). Changes in breast cancer detection and mammography recall rates after the introduction of a computer-aided detection system. *Journal of the National Cancer Institute*, *96*(3), 185–190. Available from https://doi.org/10.1093/jnci/djh067.

Hadjiiski, L., Chan, H.-P., & Sahiner, B. (2006). Advances in CAD for diagnosis of breast cancer. *Current Opinion in Obstetrics & Gynecology*, *18*(1), 64.

Haindl, M., & Remeš, V. (2019). Pseudocolor enhancement of mammogram texture abnormalities. *Machine Vision and Applications*, *30*(4), 785–794. Available from https://doi.org/10.1007/s00138-019-01028-6, http://link.springer.com/10.1007/s00138-019-01028-6.

Haralick, R. M., Shanmugam, K., & Dinstein, I. (1973). Textural features for image classification. *IEEE Transactions on Systems, Man, and Cybernetics*, *SMC-3*(6), 610–621.

Available from https://doi.org/10.1109/TSMC.1973.4309314, http://ieeexplore.ieee.org/document/4309314/.

Ho, W. T., & Lam, P. W. T. (2003). Clinical performance of computer-assisted detection (CAD) system in detecting carcinoma in breasts of different densities. *Clinical Radiology*, *58*(2), 133–136. Available from https://doi.org/10.1053/crad.2002.1131, https://www.sciencedirect.com/science/article/pii/S0009926002911311.

Huang, Y.K., & Yu, S.N. (2007). 29th annual international conference of the IEEE engineering in medicine and biology society. In *IEEE recognition of microcalcifications in digital mammograms based on Markov random field and deterministic fractal modeling* (pp. 3922–3925). http://ieeexplore.ieee.org/document/4353191/. Available from https://doi.org/10.1109/IEMBS.2007.

Hu, K., Gao, X., & Li, F. (2011). Detection of suspicious lesions by adaptive thresholding based on multiresolution analysis in mammograms. *IEEE Transactions on Instrumentation and Measurement*, *60*(2), 462–472. Available from https://doi.org/10.1109/TIM.2010.2051060.

Kallergi, M., Heine, J. J., Tembey, M., Suri, J. S., Wilson, D. L., & Laxminarayan, S. (2005). *Computer-aided diagnosis of mammographic calcification clusters: Impact of segmentation handbook of biomedical image analysis: Volume II: Segmentation models part B* (pp. 707–751). Boston, MA: Springer US. Available from https://doi.org/10.1007/0-306-48606-7_13.

Kamal, R. M., Helal, M. H., Mansour, S. M., Haggag, M. A., Nada, O. M., Farahat, I. G., & Alieldin, N. H. (2016). Can we apply the MRI BI-RADS lexicon morphology descriptors on contrast-enhanced spectral mammography? *The British Journal of Radiology*, *89*(1064)20160157. Available from https://doi.org/10.1259/bjr.20160157.

Karssemeijer, N. (1998). Automated classification of parenchymal patterns in mammograms. *Physics in Medicine and Biology*, *43*(2), 365–378. Available from https://doi.org/10.1088/0031-9155/43/2/011, https://iopscience.iop.org/article/10.1088/0031-9155/43/2/011.

Karssemeijer, N., Colchester Alan, C. F., & Hawkes David, J. (1991). *A stochastic model for automated detection of calcifications in digital mammograms. Segmentation mammography image analysis pattern recognition* (pp. 227–238). Springer. Available from https://doi.org/10.1007/BFb0033756.

Kim, D., Hoe, C., Jae, Y., Choi, S., Hyeong, R., & Yong, M. (2012). SPIE Mammographic enhancement with combining local statistical measures and sliding band filter for improved mass segmentation in mammograms. *Medical Imaging 2012: Computer-Aided Diagnosis*, *8315*, 581–586. Available from https://doi.org/10.1117/12.911147, https://www.spiedigitallibrary.org/conference-proceedings-of-spie/8315/83151Z/Mammographic-enhancement-with-combining-local-statistical-measures-and-sliding-band/10.1117/12.911147.full.

Kwok, S. M., Chandrasekhar, R., Attikiouzel, Y., & Rickard, M. T. (2004). Automatic pectoral muscle segmentation on mediolateral oblique view mammograms. *IEEE Transactions on Medical Imaging*, *23*(9), 1129–1140. Available from https://doi.org/10.1109/TMI.2004.830529.

Leichter, I., Lederman, R., Bamberger, P., Novak, B., Fields, S., & Buchbinder, S. S. (1999). The use of an interactive software program for quantitative characterization of microcalcifications on digitized film-screen mammograms. *Investigative Radiology*, *34*(6), 394–400. Available from https://doi.org/10.1097/00004424-199906000-00002.

Leichter, I., Lederman, R., Buchbinder, S. S., Bamberger, P., Novak, B., & Fields, S. (2004). Computerized evaluation of mammographic lesions: What diagnostic role does the shape of the individual microcalcifications play compared with the geometry of the cluster? *American Journal of Roentgenology*, *182*(3), 705–712. Available from https://doi.org/10.2214/ajr.182.3.1820705, https://www.ajronline.org/doi/full/10.2214/ajr.182.3.1820705.

de Lima, S. M. L., da Silva-Filho, A. G., & dos Santos, W. P. (2016). Detection and classification of masses in mammographic images in a multi-kernel approach. *Computer Methods and Programs in Biomedicine*, *134*, 11–29. Available from https://doi.org/10.1016/j.cmpb.2016.04.029, https://www.sciencedirect.com/science/article/pii/S0169260716304242.

Liu, C. C., Tsai, C. Y., Liu, J., Yu, C. Y., & Yu, S. S. (2012). A pectoral muscle segmentation algorithm for digital mammograms using Otsu thresholding and multiple regression analysis. *Computers & Mathematics with Applications*, *64*(5), 1100–1107. Available from https://doi.org/10.1016/j.camwa.2012.03.028, https://linkinghub.elsevier.com/retrieve/pii/S0898122112002337.

Li, H., Meng, X., Wang, T., Tang, Y., & Yin, Y. (2017). Breast masses in mammography classification with local contour features. *Biomedical Engineering Online*, *16*(1)44. Available from https://doi.org/10.1186/s12938-017-0332-0, https://doi.org/10.1186/s12938-017-0332-0.

Li, L., Qian, W., Clarke, L.P., Clark, R.A., & Thomas, J.A. (1999). SPIE improving mass detection by adaptive and multiscale processing in digitized mammograms. In *Medical imaging 1999: Image processing* (pp. 490–498). https://www.spiedigitallibrary.org/conference-proceedings-of-spie/3661/0000/Improving-mass-detection-by-adaptive-and-multiscale-processing-in-digitized/10.1117/12.348604.full.3661, Available from https://doi.org/10.1117/12.348604.

Lou, S. L., Lin, H. D., Lin, K. P., & Hoogstrate, D. (2000). Automatic breast region extraction from digital mammograms for PACS and telemammography applications. *Computerized Medical Imaging and Graphics*, *24*(4), 205–220. Available from https://doi.org/10.1016/S0895-6111(00)00009-4, https://www.sciencedirect.com/science/article/pii/S0895611100000094.

Malar, E., Kandaswamy, A., Chakravarthy, D., & Giri Dharan, A. (2012). A novel approach for detection and classification of mammographic microcalcifications using wavelet analysis and extreme learning machine. *Computers in Biology and Medicine*, *42*(9), 898–905. Available from https://doi.org/10.1016/j.compbiomed.2012.07.001, https://linkinghub.elsevier.com/retrieve/pii/S0010482512001059.

Matos, C. E. F., Souza, J. C., Diniz, J. O. B., Junior, G. B., de Paiva, A. C., de Almeida, J. D. S., da Rocha, S. V., & Silva, A. C. (2019). Diagnosis of breast tissue in mammography images based local feature descriptors. *Multimedia Tools and Applications*, *78*(10), 12961–12986. Available from https://doi.org/10.1007/s11042-018-6390-x, http://link.springer.com/10.1007/s11042-018-6390-x.

Mavroforakis, M. E., Georgiou, H. V., Dimitropoulos, N., Cavouras, D., & Theodoridis, S. (2006). Mammographic masses characterization based on localized texture and dataset fractal analysis using linear, neural and support vector machine classifiers. *Artificial Intelligence in Medicine*, *37*(2), 145–162. Available from https://doi.org/10.1016/j.artmed.2006.03.002.

MCndez, A.J., Tahocesb, P.G., Lade, M.J., So, M., Correab, J.L., & Vidal, J.J. (n.d). Automatic detection of breast border and nipple i mammograms. 10.

Mencattini, A., Salmeri, M., Lojacono, R., Frigerio, M., & Caselli, F. (2008). Mammographic images enhancement and denoising for breast cancer detection using dyadic wavelet processing. *IEEE Transactions on Instrumentation and Measurement*, *57*(7), 1422–1430. Available from https://doi.org/10.1109/TIM.2007.915470.

Mohanty, A. K., Senapati, M. R., Beberta, S., & Lenka, S. K. (2013). Texture-based features for classification of mammograms using decision tree. *Neural Computing and Applications*, *23*(3–4), 1011–1017. Available from https://doi.org/10.1007/s00521-012-1025-z, http://link.springer.com/10.1007/s00521-012-1025-z.

Elmoufidi, A., El Fahssi, K., Jai-andaloussi, S., Sekkaki, A., Gwenole, Q., & Lamard, M. (2018). Anomaly classification in digital mammography based on multiple-instance learning. *IET Image Processing*, *12*(3), 320–328. Available from https://doi.org/10.1049/iet-ipr.2017.0536, https://onlinelibrary.wiley.com/doi/abs/10.1049/iet-ipr.2017.0536.

Moura, D. C., & Guevara López, M. A. (2013). An evaluation of image descriptors combined with clinical data for breast cancer diagnosis. *International Journal of Computer Assisted Radiology and Surgery*, *8*(4), 561–574. Available from https://doi.org/10.1007/s11548-013-0838-2, https://doi.org/10.1007/s11548-013-0838-2.

Mughal, B., Muhammad, N., Sharif, M., Rehman, A., & Saba, T. (2018). Removal of pectoral muscle based on topographic map and shape-shifting silhouette. *BMC Cancer*, *18* (1)778. Available from https://doi.org/10.1186/s12885-018-4638-5, https://bmccancer.biomedcentral.com/articles/10.1186/s12885-018-4638-5.

Mughal, B., Sharif, M., & Muhammad, N. (2017). Bi-model processing for early detection of breast tumor in CAD system. *The European Physical Journal Plus*, *132*(6), 266. Available from https://doi.org/10.1140/epjp/i2017-11523-8, http://link.springer.com/10.1140/epjp/i2017-11523-8.

Muhammad, N. (2017). Extraction of breast border and removal of pectoral muscle in wavelet domain. *Biomedical Research (Tokyo, Japan)*, *28*(11), 3.

Nalawade, Y. V. (2009). Evaluation of breast calcifications. *The Indian Journal of Radiology & Imaging*, *19*(4), 282–286. Available from https://doi.org/10.4103/0971-3026.57208, https://www.ncbi.nlm.nih.gov/pmc/articles/PMC2797739/.

Nguyen, T.M., & Rangayyan, R.M. (2005). IEEE shape analysis of breast masses in mammograms via the fractal dimension. In *2005 IEEE engineering in medicine and biology 27th annual conference* (pp. 3210–3213). http://ieeexplore.ieee.org/document/1617159/, Available from https://doi.org/10.1109/IEMBS.2005.1617159.

Nishikawa, R. M. (2007). Current status and future directions of computer-aided diagnosis in mammography. *Computerized Medical Imaging and Graphics*, *31*(4–5), 224–235. Available from https://doi.org/10.1016/j.compmedimag.2007.02.009, https://linkinghub.elsevier.com/retrieve/pii/S0895611107000171.

Oliver, A., Freixenet, J., Martí, J., Pérez, E., Pont, J., Denton, E. R. E., & Zwiggelaar, R. (2010). A review of automatic mass detection and segmentation in mammographic images. *Medical Image Analysis*, *14*(2), 87–110. Available from https://doi.org/10.1016/j.media.2009.12.005, https://linkinghub.elsevier.com/retrieve/pii/S1361841509001492.

Otsu, N. (1979). A threshold selection method from gray-level histograms. *IEEE Transactions on Systems, Man, and Cybernetics*, *9*(1), 62–66. Available from https://doi.org/10.1109/TSMC.1979.4310076.

Papadopoulos, A., Fotiadis, D., & Likas, A. (2005). Characterization of clustered microcalcifications in digitized mammograms using neural networks and support vector machines. *Artificial Intelligence in Medicine*, *34*, 141–150. Available from https://doi.org/10.1016/j.artmed.2004.10.001.

Pereira, D. C., Ramos, R. P., & Do Nascimento, M. Z. (2014). Segmentation and detection of breast cancer in mammograms combining wavelet analysis and genetic algorithm. *Computer Methods and Programs in Biomedicine*, *114*(1), 88–101. Available from https://doi.org/10.1016/j.cmpb.2014.01.014, https://www.sciencedirect.com/science/article/pii/S0169260714000261.

Petrick, N., Chan, H. P., Sahiner, B., & Wei, D. (1996). An adaptive density-weighted contrast enhancement filter for mammographic breast mass detection. *IEEE Transactions on Medical Imaging*, *15*(1), 59–67. Available from https://doi.org/10.1109/42.481441.

Population screening programmes. (2021). List of information about NHS breast screening (BSP) programme. Population screening programmes: NHS breast screening (BSP) programme – detailed information – GOV.UK. Population screening programmes. https://www.gov.uk/topic/population-screening-programmes/breast.

Pérez, N.P. (n.d). Improving variable selection and mammography-based machine learning classifiers for breast cancer CADx. 185.

Quintanilla-Domínguez, J., Barrón-Adame, J. M., Gordillo-Sosa, J. A., Lozano-Garcia, J. M., Estrada-García, H., & Guzmán-Cabrera, R. (2016). Analysis of mammograms using texture segmentation. *Research in Computing Science*, *123*(1), 119–126. Available from https://doi.org/10.13053/rcs-123-1-11, http://rcs.cic.ipn.mx/2016_123/Analysis%20of%20Mammograms%20using%20Texture%20Segmentation.pdf.

Raba, D., Oliver, A., Martí, J., Peracaula, M., Espunya, J., Marques, J. S., Pérez de la Blanca, N., & Pina, P. (2005). *Breast segmentation with pectoral muscle suppression on digital mammograms pattern recognition and image analysis* (pp. 471–478). Berlin, Heidelberg: Springer Berlin Heidelberg. Available from http://link.springer.com/10.1007/11492542_58.

Raguso, G., Ancona, A., Chieppa, L., L'Abbate, S., Pepe, M.L., Mangieri, F., De Palo, M., & Rangayyan, R.M. (2010). Breast shape fractals benign tumors malignant tumors complexity theory silicon. In *2010 Annual international conference of the IEEE engineering in medicine and biology. Application of fractal analysis to mammography* (pp. 3182–3185). Available from https://doi.org/10.1109/IEMBS.2010.5627180.

Rangayyan, R.M., Guliato, D., De Carvalho, J., & Santiago, S.A. (2008). Feature extraction from the turning angle function for the classification of contours of breast tumors.

Rangayyan, R. M., Mudigonda, N. R., & Desautels, J. E. L. (2000). Boundary modelling and shape analysis methods for classification of mammographic masses. *Medical and Biological Engineering and Computing*, *38*(5), 487–496. Available from https://doi.org/10.1007/BF02345742.

Recent Advances in Breast Imaging, Mammography, and Computer-Aided Diagnosis of Breast Cancer. (2006). 2021. Suri | Publications | Spie. https://spie.org/Publications/Book/651880?SSO = 1.

Rojas Domínguez, A., & Nandi, A. K. (2009). Toward breast cancer diagnosis based on automated segmentation of masses in mammograms. *Pattern Recognition*, *42*(6), 1138–1148. Available from https://doi.org/10.1016/j.patcog.2008.08.006, https://www.sciencedirect.com/science/article/pii/S0031320308003154.

Sahiner, B., Hadjiiski, L.M., Chan, H.P., Paramagul, C., Nees, A., Helvie, M., & Shi, J. (2008). Medical imaging 2008: Computer-aided diagnosis. SPIE Concordance of computer-extracted image features with BI-RADS descriptors for mammographic mass margin. 6915, 513–518. https://www.spiedigitallibrary.org/conference-proceedings-of-spie/6915/69151N/Concordance-of-computer-extracted-image-features-with-BI-RADS-descriptors/10.1117/12.770752.full. Available from https://doi.org/10.1117/12.770752.

Sankar, D., & Thomas, T. (2010). A new fast fractal modeling approach for the detection of microcalcifications in mammograms. *Journal of Digital Imaging*, *23*(5), 538–546. Available from https://doi.org/10.1007/s10278-009-9224-6, https://www.ncbi.nlm.nih.gov/pmc/articles/PMC3046673/.

Schmidt, F., Sorantin, E., Szepesvàri, C., Graif, E., Becker, M., Mayer, H., & Hartwagner, K. (1999). An automatic method for the identification and interpretation of clustered microcalcifications in mammograms. *Physics in Medicine and Biology*, *44*(5), 1231–1243. Available from https://doi.org/10.1088/0031-9155/44/5/011.

Sheba, K.U., & Raj, S.G. (2017). Adaptive fuzzy logic based Bi—Histogram equalization for contrast enhancement of mammograms. In *2017 International conference on intelligent computing, instrumentation and control technologies (ICICICT)* (pp. 156–161). Available from https://doi.org/10.1109/ICICICT1.2017.8342552.

Sheba, K. U., Raj, S. G., & Akhloufi, M. (2018). An approach for automatic lesion detection in mammograms. *Cogent Engineering*, *5*(1). Available from https://doi.org/10.1080/23311916.2018.1444320, https://www.cogentoa.com/article/10.1080/23311916.2018.1444320.

Shi, J., Sahiner, B., Chan, H. P., Ge, J., Hadjiiski, L., Helvie, M. A., Nees, A., Wu, Y. T., Wei, J., Zhou, C., Zhang, Y., & Cui, J. (2008). Characterization of mammographic masses based on level set segmentation with new image features and patient information. *Medical Physics*, *35*(1), 280–290. Available from https://doi.org/10.1118/1.2820630.

Soltanian-Zadeh, H., Pourabdollah-Nezhad, S., & Rafiee-Rad, F. (2001). SPIE Shape-based and texture-based feature extraction for classification of microcalcifications in mammograms. *Medical Imaging 2001: Image Processing*, *4322*, 301–310. Available from https://doi.org/10.1117/12.431100, https://www.spiedigitallibrary.org/conference-proceedings-of-spie/4322/0000/Shape-based-and-texture-based-feature-extraction-for-classification-of/10.1117/12.431100.full.

Soltanian-Zadeh, H., Rafiee-Rad, F., & Pourabdollah-Nejad D, S. (2004). Comparison of multiwavelet, wavelet, Haralick, and shape features for microcalcification classification in mammograms. *Pattern Recognition*, *37*(10), 1973–1986. Available from https://doi.org/10.1016/j.patcog.2003.03.001, https://www.sciencedirect.com/science/article/pii/S0031320304001323.

Sonka, M., Hlavac, V., & Boyle, R. (2014). *Image processing, analysis, and machine vision* (p. 930) Cengage Learning.

Suckling, J., Parker, J., Dance, D., Astley, S., Hutt, I., Boggis, C., Ricketts, I., Stamatakis, E., Cerneaz, N., Kok, S., Taylor, P., Betal, D., & Savage, J., (2015). Mammographic image analysis society (MIAS) database v1. 21. https://www.repository.cam.ac.uk/handle/1810/250394. doi: 10/250394.

Surendiran, B., & Vadivel, A. (2012). Mammogram mass classification using various geometric shape and margin features for early detection of breast cancer. *International Journal of Medical Engineering and Informatics*, *4*(1), 36–54. Available from https://doi.

org/10.1504/IJMEI.2012.045302, https://www.inderscienceonline.com/doi/abs/10.1504/IJMEI.2012.045302.

Tai, S.-C., Chen, Z.-S., & Tsai, W.-T. (2014). An automatic mass detection system in mammograms based on complex texture features. *IEEE Journal of Biomedical and Health Informatics*, *18*(2), 618–627. Available from https://doi.org/10.1109/JBHI.2013.2279097.

Texture extraction. (2022). Texture extraction: An evaluation of ridgelet, wavelet and co-occurrence based methods applied to mammograms | Elsevier Enhanced Reader Texture extraction. https://reader.elsevier.com/reader/sd/pii/S0957417412004903?token = 7BCD1D6D60BA5BB74DE80DB52B5B3E5ED7459C37E6EDAFA4803E86-A2484D9F21CB57B15E3E59F0EFE0B4E5CCA13B211E&originRegion = us-east-1&originCreation = 20221010070347.

Tiedeu, A., Daul, C., Kentsop, A., Graebling, P., & Wolf, D. (2012). Texture-based analysis of clustered microcalcifications detected on mammograms. *Digital Signal Processing*, *22*(1), 124–132. Available from https://doi.org/10.1016/j.dsp.2011.09.004, https://linkinghub.elsevier.com/retrieve/pii/S1051200411001254.

USF Digital Mammography Home Page. (2022). http://www.eng.usf.edu/cvprg/mammography/database.html.

Varela, C., Tahoces, P. G., Méndez, A. J., Souto, M., & Vidal, J. J. (2007). Computerized detection of breast masses in digitized mammograms. *Computers in Biology and Medicine*, *37*(2), 214–226. Available from https://doi.org/10.1016/j.compbiomed.2005.12.006, https://www.sciencedirect.com/science/article/pii/S0010482506000060.

Varela, C., Timp, S., & Karssemeijer, N. (2006). Use of border information in the classification of mammographic masses. *Physics in Medicine and Biology*, *51*(2), 425–441. Available from https://doi.org/10.1088/0031-9155/51/2/016.

Veldkamp, W. J. H., Karssemeijer, N., Otten, J. D. M., & Hendriks, J. H. C. L. (2000). Automated classification of clustered microcalcifications into malignant and benign types. *Medical Physics*, *27*(11), 2600–2608. Available from https://doi.org/10.1118/1.1318221, https://aapm.onlinelibrary.wiley.com/doi/abs/10.1118/1.1318221.

Wang, Z., Yu, G., Kang, Y., Zhao, Y., & Qu, Q. (2014). Breast tumor detection in digital mammography based on extreme learning machine. *Neurocomputing*, *128*, 175–184. Available from https://doi.org/10.1016/j.neucom.2013.05.053, https://www.sciencedirect.com/science/article/pii/S0925231213010163.

Wang, L., Zhu, M., Deng, L., & Yuan, X. (2010). Automatic pectoral muscle boundary detection in mammograms based on Markov chain and active contour model. *Journal of Zhejiang University Science C*, *11*(2), 111–118. Available from https://doi.org/10.1631/jzus.C0910025, http://link.springer.com/10.1631/jzus.C0910025.

Warren Burhenne, L. J., Wood, S. A., D'Orsi, C. J., Feig, S. A., Kopans, D. B., O'Shaughnessy, K. F., Sickles, E. A., Tabar, L., Vyborny, C. J., & Castellino, R. A. (2000). Potential contribution of computer-aided detection to the sensitivity of screening mammography. *Radiology*, *215*(2), 554–562. Available from https://doi.org/10.1148/radiology.215.2.r00ma15554.

Wei, C.-H., Gwo, C.-Y., & Huang, P. J. (2016). Identification and segmentation of obscure pectoral muscle in mediolateral oblique mammograms. *The British Journal of Radiology*, *89*(1062)20150802. Available from https://doi.org/10.1259/bjr.20150802, https://www.ncbi.nlm.nih.gov/pmc/articles/PMC5258151/.

Wei, J., Sahiner, B., Hadjiiski, L. M., Chan, H.-P., Petrick, N., Helvie, M. A., Roubidoux, M. A., Ge, J., & Zhou, C. (2005). Computer-aided detection of breast masses on full field digital mammograms. *Medical Physics, 32*(9), 2827–2838. Available from https://doi.org/10.1118/1.1997327, https://www.ncbi.nlm.nih.gov/pmc/articles/PMC2742215/.

Wilkinson, L., Thomas, V., & Sharma, N. (n.d). Microcalcification on mammography: Approaches to interpretation and biopsy. *The British Journal of Radiology, 90*(1069), 20160594. https://www.ncbi.nlm.nih.gov/pmc/articles/PMC5605030/. Available from https://doi.org/10.1259/bjr.20160594.

Wong, A. K. C., & Sahoo, P. K. (1989). A gray-level threshold selection method based on maximum entropy principle. *IEEE Transactions on Systems, Man, and Cybernetics, 19*(4), 866–871. Available from https://doi.org/10.1109/21.35351.

Yin, F. F., Giger, M. L., Doi, K., Vyborny, C. J., & Schmidt, R. A. (1994). Computerized detection of masses in digital mammograms: Automated alignment of breast images and its effect on bilateral-subtraction technique. *Medical Physics, 21*(3), 445–452. Available from https://doi.org/10.1118/1.597307.

Yu, S. N., Li, K. Y., & Huang, Y. K. (2006). Detection of microcalcifications in digital mammograms using wavelet filter and Markov random field model. *Computerized Medical Imaging and Graphics, 30*(3), 163–173. Available from https://doi.org/10.1016/j.compmedimag.2006.03.002, https://linkinghub.elsevier.com/retrieve/pii/S0895611106000309.

Zhang, X., Homma, N., Goto, S., Kawasumi, Y., Ishibashi, T., Abe, M., Sugita, N., & Yoshizawa, M. (2013). A hybrid image filtering method for computer-aided detection of microcalcification clusters in mammograms. *Journal of Medical Engineering, 2013*615254. Available from https://doi.org/10.1155/2013/615254, https://www.ncbi.nlm.nih.gov/pmc/articles/PMC4782620/.

Zhang, Y., Ji, T. Y., Li, M. S., & Wu, Q. H. (2016). Identification of power disturbances using generalized morphological open-closing and close-opening undecimated wavelet. *IEEE Transactions on Industrial Electronics, 63*(4), 2330–2339. Available from https://doi.org/10.1109/TIE.2015.2499728.

Zhang, L., Qian, W., Sankar, R., Song, D., & Clark, R. (2001). A new false positive reduction method for MCCs detection in digital mammography. In 2001 IEEE International Conference on Acoustics, Speech, and Signal Processing. *Proceedings (Cat. No.01CH37221)*. (Vol. 2, pp. 1033–1036). doi: 10.1109/ICASSP.2001.941095.

Zhou, C., Chan, H. P., Petrick, N., Helvie, M. A., Goodsitt, M. M., Sahiner, B., & Hadjiiski, L. M. (2001). Computerized image analysis: Estimation of breast density on mammograms. *Medical Physics, 28*(6), 1056–1069. Available from https://doi.org/10.1118/1.1376640, http://doi.wiley.com/10.1118/1.1376640.

Zhu, Y., & Huang, C. (2012). An improved median filtering algorithm for image noise reduction. *Physics Procedia, 25*, 609–616. Available from https://doi.org/10.1016/j.phpro.2012.03.133, https://www.sciencedirect.com/science/article/pii/S1875389212005494.

CHAPTER

Mammographic breast density segmentation

Bashar Rajoub[1], Hani Qusa[2], Hussein Abdul-Rahman[1] and Heba Mohamed[3]

[1]*Faculty of Engineering Technology and Science, Higher Colleges of Technology, Dubai, United Arab Emirates*

[2]*Faculty of Computer Information Science, Higher Colleges of Technology, Dubai, United Arab Emirates*

[3]*Faculty of Health Sciences, Higher Colleges of Technology, Dubai, United Arab Emirates*

6.1 Introduction

6.1.1 Breast cancer

Breast cancer is the second most common cancer in women, with approximately 8%−12% of all women worldwide developing the disease in their lifetime (DeSantis et al., 2011; Ferlay et al., 2013; What is breast cancer?, 2018). Breast cancer development is linked to genetic abnormalities, which can arise due to the natural aging process (90% of cancers) or because of an abnormality inherited from parents (5%−10% of cancers). Genetic mutations can "turn on or off" specific genes in a cell, leading to uncontrolled cell division and replication. Over time, more cells are produced resulting in a formation of tumor, which can be benign or malignant. Benign tumors grow slowly and remain localized, while malignant tumors can invade nearby healthy tissue and make their way into the underarm lymph nodes. This provides a pathway into other parts of the body, enabling the spread of cancer (What is breast cancer?, 2018).

Breast cancer usually begins in the milk-producing glands (the lobules) or in the passages that drain milk from the lobules to the nipple (the ducts). Breast cancer can also begin in the fatty and fibrous connective tissues of the breast (the normal tissues). Symptoms of breast cancer include: painless lump in the breast; change of size, shape, or color of the breast; pain in the breast region; nipple is pulled or retracted; rash or itch around the nipples; nipple bleeding; skin over the breast is thickened; and peeling, scaling, or flaking of the skin on the breast or nipple (What are the symptoms of breast cancer? | CDC, 2021).

Breast cancer can be graded into clinical stages to show the severity and progress of the disease. The five year survival rates for stage I breast cancers are close to 100%, 93% for stage 2, 72% for stage 3, and 22% for stage 4. Descriptions of the main stages are listed below (Breast cancer, 2019; Breast cancer stages, 2018; What Are the Stages of Breast Cancer?, 2021):

Artificial Intelligence and Image Processing in Medical Imaging. DOI: https://doi.org/10.1016/B978-0-323-95462-4.00006-6

Stage 0: Known as ductal carcinoma in situ (DCIS), the cells are limited to within the ducts and have not invaded surrounding tissues.

Stage 1: At this stage, the tumor measures up to 2 cm across. It has not affected any lymph nodes, or there are small groups of cancer cells in the lymph nodes.

Stage 2: The tumor is 2 cm across, and it has started to spread to nearby nodes, or it is 2–5 cm across and has not spread to the lymph nodes.

Stage 3: The tumor is up to 5 cm across, and it has spread to several lymph nodes, or the tumor is larger than 5 cm and has spread to a few lymph nodes.

Stage 4: The cancer has spread to distant organs, most often the bones, liver, brain, or lungs.

It is estimated that 65% of all cancer cases diagnosed/detected in an advanced stage of the disease (Othman et al., 2015). Early detection strategies help increase the proportion of breast cancers detected at an early stage of the disease. As such, increasing the chances of full curative treatment and improving survival rates of patients (Ng & Muttarak, n.d.). Early detection strategies involve promoting awareness of early signs and symptoms of breast cancer and adopting a systematic screening program that tests asymptomatic population of certain risk groups to identify cases of breast cancer (Yip et al., 2008).

6.1.2 Mammographic screening

Early detection of signs of breast cancer can improve the chances of survival. Mammographic screening is one of the most reliable and effective methods for early detection of cancer, even before physical symptoms appear. Mammography is a specific type of imaging that uses a low-dose X-ray system (around 30 kVp) to examine the breast tissue. Mammograms are two-dimensional (2D) grayscale images that correspond to the superposition of three-dimensional breast tissue (black showing background and low density tissue, while white indicates dense areas).

Mammography offers high-quality images of breast tissue and is considered among the most reliable and effective methods for early detection of cancer (DeSantis et al., 2011; Greif, 2010; Tabar et al., 2003). Full-field digital mammography (FFDM) offers wide dynamic range, low noise, enhanced image quality, and lower X-ray dose (Lai et al., 2008). Proper breast positioning is crucial for successful diagnosis and reducing imaging artifacts. The breast is compressed by a special mammography unit in order to prevent motion blur and improve image quality. This also allows us to reduce the required radiation dose.

During the mammography procedure, each breast is imaged separately with two views: the mediolateral oblique (MLO) view and the craniocaudal (CC) view. Each view shows different appearances and details of breast tissue. The CC view provides a top view information, whilc the MLO view is used to observe the side view information.

Annual mammographic screening is recommended for women at normal risk, beginning at age 40 (NCCN National Comprehensive Cancer Network – Home, 2021). Wang et al. showed that mammography is less effective for subjects less than 40 years of age with dense breasts and for small lesions or tumors as they are difficult to see with the naked eye (Wang et al., 2017). In the UK, women aged between 50 and 70 years are invited for breast screening every three years (Breast Screening I Breast Cancer I Cancer Research UK, 2021). It is recommended that screening of women over the age of 50 is undertaken regularly to detect changes in the breast parenchyma that could lead to the development of the disease (Trop & Deck, 2006).

Mammographic screening is a challenging task. It is difficult to classify abnormalities whether malignant or benign, even for expert clinicians. Mammographic appearance of the breast tissue varies due to differences in the tissue composition, variations in shape, size, density, and other morphological features across the screened population. This increases the difficulty for a radiologist to correctly analyze the patterns and may result in a wrong diagnosis.

Different experts might have substantial variability in interpreting the same mammograms (Nishikawa, 2007). Moreover, human factors and limitations, such as fatigue, may result in inaccurate risk assessments, benign lesions misdiagnosed as cancers (false positives), and missed cancerous lesions (false negatives).

6.1.3 Mammography datasets

Nowadays, expert labeled mammography imaging data is rapidly becoming more accessible to researchers. Learning from this data can be used to create models that can diagnose and detect disease, thus improving diseases treatment. The Breast Cancer Digital Repository (BCDR) (Guevara Lopez et al., 2012), the Mammographic Image Analysis Society (MIAS) database (The Mammographic Image Analysis Society Digital Mammogram Database – ScienceOpen, 2021), the Digital Database for Screening Mammography (DDSM) (Heath et al., 1998) and the "Curated Breast Imaging Subset of DDSM" (CBIS-DDSM) dataset (Clark et al., 2013) are considered the most commonly used public databases in the scientific community.

The MIAS database comprises 322 mediolateral-oblique view (MLO) mammography images stored in portable network graphics format, with a resolution of 1024×1024 pixels. The images are annotated, indicating left and right breast, pathology, type, coordinates, and sizes of lesions present in the image.

SureMaPP (Bruno, 2021) is a recently published dataset of 343 hand labeled mammograms containing suspicious abnormalities (both benignant and malignant) and calcifications. Mammograms are available with two different spatial resolutions: 3584×2816 pixels and 5928×4728 pixels. The dataset can be downloaded from: https://mega.nz/folder/Ly5g0agB#-QL9uBEvoP8rNig8JBuYfw.

The Nijmegen database contains 40 mammogram images taken from 21 patients. The mammograms show clustered microcalcifications. The database is developed by N. Karssemeijer from the National Expert and Training Center for Breast Cancer Screening and the Department of Radiology at the University of Nijmegen, the Netherlands. All images were screen film mammography images recorded using various types of equipment. The images were corrected for the inhomogeneity of the light source and stored as 2048 × 2048 pixels (Karssemeijer et al., 1991; Nishikawa, 1998; Strickland & Hahn, 1996).

The "mini Mammography Image Analysis Society" (mini-MIAS) dataset consists of 322 images. The images size is 1024 × 1024 pixels and are stored in gray-scale portable gray map (PGM) format (Suckling et al., 2015). The dataset contains three different types of mammograms: dense glandular, fatty, and fatty glandular. The categories are further divided into normal, benign, and malignant cases.

The data includes annotations showing the background tissue type, class, and severity of the abnormality, the coordinates of the center of irregularities, and the approximate radius of a circle enclosing the abnormal region in pixels. The class distributions reveal that the MIAS dataset is imbalanced as the distribution is not uniform (207 normal, 64 benign, and 51 malignant). The min-MIAS dataset is already very small, as such undersampling might be a poor strategy as it may discard useful features from the dropped images.

The DDSM is a collection of mammograms from the following sources: Massachusetts General Hospital, Wake Forest University School of Medicine, Sacred Heart Hospital, and Washington University of St Louis School of Medicine (Lee et al., 2017). The DDSM database holds 2620 scanned film mammography studies distributed in 43 volumes. The dataset contains 695 normal, 870 benign, 141 benign without callback, and 914 malignant cases with verified pathology information. Each study includes two images of each breast, acquired in craniocaudal (CC) and mediolateral views that have been scanned from the film-based sources by four different scanners with a resolution between 50 and 42 microns, 16 bits, and an average size of 3000 × 4800 pixels for a total of 9916 radiographs. These images were stored using the Lossless Joint Pictures Expert Group standard. Each case may have one or more associated pathological lesion segmentations in MLO/CC/both image views of the same breast.

The DDSM contains clinical information, and labels provided in additional ACSII files include: the radiologists' observations and characterizations of the shape and margins of the lesions/masses or calcifications if present; pathology: benign, benign without call-back, or malignant, the subtlety of the lesion (between 1 and 5); the distribution of calcifications; and the ground truth of breast imaging reporting and data system (BI-RADS) tissue type classification from 0 to 5 (Heath et al., 1998). The dataset can be downloaded from http://www.eng.usf.edu/cvprg/Mammography/Database.html.

The region-of-interest (ROI) annotations for the abnormalities in the DDSM show a general position of lesions. Therefore, segmentation algorithms are needed for accurate feature extraction. The CBIS-DDSM dataset is an updated and standardized version of the older DDSM dataset. The dataset is curated by trained mammographers and provides higher quality imagery offered by the digital imaging and communications in medicine (DICOM) standard. The dataset includes decompressed images; updated ROI segmentation, bounding boxes and pathologic diagnosis (e.g., type of mass, grade of tumor, and cancer stage) for the training data by trained mammographers. The scans contain mammograms obtained using the bilateral CC and mediolateral oblique (MLO) views.

The "Curated Breast Imaging Subset of DDSM" (CBIS-DDSM) dataset is available online from The Cancer Imaging Archive (Clark et al., 2013). The dataset consists of 753 calcification cases (small flecks of calcium usually clustered together) and 891 mass cases (e.g., cysts or lumps). The CBIS-DDSM dataset is split into stratified training/testing sets. This makes it possible to compare the performance of different models. The dataset contains 10,239 images gathered from 1566 patients across 6775 subjects. The images are larger than 3000×5000 pixels, stored in DICOM format covering 163.6 Gb of disk space.

The BCDR database is supplied by the Faculty of Medicine "Centro Hospitalar São João" at the University of Porto, Portugal (Guevara Lopez et al., 2012). The BCDR dataset contains cases of 1010 patients with mammography images, clinical history, lesion segmentation, and selected precomputed image-based descriptors. Biopsy-proven annotations showing the contour of the lesions are also shown in either or both of the MLO and CC views. Mammograms in the dataset are also labeled according to the density of the breast using the BI-RADS standard (Breast cancer digital repository, 2021).

The BancoWeb LAPIMO (acronym of "*Labor*átorio de Análise e Processamento de Imagens Médicas e Odontológicas") database (Matheus & Schiabel, 2011) is a recent public databases, which contains 320 cases, 1473 images (MLO and CC views) divided in normal, benign, and malignant cases. The images are labeled with background patient information and few ROI annotations are available for some images. Textual description of BI-RADS categories is available for all images.

The INbreast database contains 410 images (MLO and CC) from 115 patients acquired at the Breast Center in CHSJ, Porto. The images files are FFDM with a solid contrast resolution of 14 bits acquired by a MammoNovation Siemens scanner. Ground truth annotations were based on the lesions contour made by radiologists (Moreira et al., 2012).

The Image Retrieval in Medical Applications (IRMA) project combines four mammographic databases: DDSM, MIAS, the Lawrence Livermore National Laboratory (LLNL), and routine images from the Rheinisch-Westfälische Technische Hochschule (RWTH) Aachen. Standardized coding of tissue type, tumor staging, ROI annotations, and lesion description according to the ACR and BI-RADS tissue codes. The database contains 10,509 images normal cases (12 volumes), cancer cases (15 volumes), and benign cases (14 volumes). Each case has at least a pathological lesion (PL) segmentation in the MLO and CC images of the same breast (de Oliveira et al., 2010; Oliveira et al., n.d.).

6.2 Breast imaging reporting and data system categories of density and tissue patterns

Breast imaging reporting and data system (BI-RADS) standards have been used in the literature to reduce complexity in breast imaging evaluation. The American College of Radiology (ACR) developed the BI-RADS framework for risk assessment and quality assurance through standardized breast imaging terminology, report organization, assessment structure, and a classification system for mammography, ultrasound, and MRI of the breast (Breast imaging reporting & data system, 2021). BI-RADS also provides guidance for clinicians on how to detect/classify pathological lesions. BI-RADS can also provide performance data, such as the positive predictive value and the percentage negative cancers obtained from follow up and outcome monitoring. Thus providing quality assurance metrics that screening programs can use to help improve patient care (Breast imaging reporting & data system, 2021).

BI-RADS has been widely adopted in clinical practice and screening programs throughout the world (Bihrmann et al., 2008; Dilhuydy, 2007; Taplin et al., 2002). Several studies have shown that the use of BI-RADS improves cancer detection rate and helps in planning better postexamination choices (Orel et al., 1999). However, other studies questioned the effectiveness of BI-RADS due to the subjective nature of reporting mammographic findings. For example, (Obenauer et al., 2005) reported that BI-RADS lexicon did not reduce the anatomical variability in interpreting mammograms before and after the use of BI-RADS. Geller et al. showed that BI-RADS 0 and 3 categories had the highest variability and recommended additional education about the use of these categories in clinical assessments (Geller et al., 2002).

BI-RADS offers two reporting schemes, the first reports classification of breast structures into seven BI-RADS categories (Breast imaging reporting & data system, 2021; Understanding your diagnosis, 2021), while the second scheme is concerned with assessment of breast density into four BI-RADS categories. ACR guidelines (Burnside et al., 2009) for BIRADS density reporting addresses the fact that the sensitivity of mammography decreases as the breast density increases. BIRADS offers radiologists a density classification system using four density categories (low to high density) that are defined as follows:

BIRADS I: The breast is almost entirely fatty with the fibrous and glandular tissue, making up less than 25% of the breast.

BIRADS II: There is some scattered fibroglandular tissue (low-density) with fibrous and glandular tissue occupying between 25% and 50% of the breast.

BIRADS III: The breast is heterogeneously dense (Isodense). Fibrous and glandular tissue occupy between 51% and 75% of the breast.

BIRADS IV: The breast is extremely dense. The fibrous and glandular tissue occupy over 75% of the breast.

Observing small masses, cysts or tumors can be challenging at BIRADS III and IV levels, which can lead to missing some cancers. According to the breast imaging lexicon described in the BI-RADS atlas, mammographic abnormalities such as masses, calcifications and microcalcifications, and architectural distortion are among the targets, which shall lead to patient recall for further assessment (Burnside et al., 2009).

6.3 Breast imaging reporting and data system categories of mammogram findings

Studies of breast imaging are divided into seven categories, numbered 0 through 6. The categories are as follows (Burnside et al., 2009; Understanding mammogram reports | mammogram results, 2021).

BI-RADS 0 (assessment incomplete)—Need further assistance. This means a possible abnormality may not be clearly seen or defined. Additional imaging evaluation and/or comparison to prior mammograms is needed to see if there have been changes in the area over time.

BI-RADS 1 (normal or negative)—No evidence of lesion. The breasts look symmetrical with no masses (lumps), distorted structures, or suspicious calcifications and there is no significant abnormality to report.

BI-RADS 2 benign (noncancerous) finding. This describes findings on a mammogram, which is benign, such as benign calcifications, calcified lesion with high density, lymph nodes in the breast, or calcified fibroadenomas. This helps help when comparing to future mammograms and ensures that the mammogram will not be misinterpreted.

BI-RADS 3 probably benign finding–Follow-up in a short time frame is suggested. The findings in this category are 98% benign (not cancer, e.g., noncalcified circumscribed mass/obscured mass). The benign findings are not expected to change over time but must be examined to see if an area of concern changes over time.

BI-RADS 4 (suspicious abnormality). Biopsy should be considered. Finding do not definitely look like cancer but is suspicious enough to recommend a biopsy (e.g., microlubulated mass). This category is often divided further: 4A: Finding with a low likelihood of being cancer; 4B: Finding with a moderate likelihood of being cancer, and 4C: Finding with a high likelihood of being cancer but not as high as Category 5.

BI-RADS 5: Highly suggestive of malignancy–Appropriate action should be taken. The findings have a high chance (at least 95%) of being cancer and biopsy is needed (e.g., the presence of indistinct and speculated mass.)

BI-RADS 6: Known biopsy-proven malignancy–Appropriate action should be taken. This category is only used for findings on a mammogram that have already been shown to be cancer by a previous biopsy. Mammograms may be used in this way to see how well the cancer is responding to treatment.

Tahocest et al. (1995) used three groups of features to classify mammograms into the four patterns described by Wolfe (1976) (N1, P1, P2, and Dy). Small square ROIs and large irregular ROIs were manually selected for feature extraction. Discriminant analysis to select among different feature extraction techniques, including Fourier transform and local-contrast analysis (derived from the square ROIs), and gray-level distribution and quantification (derived from the large irregular ROIs). The results show differences in agreement among radiologists and computer classification, depending on the Wolfe pattern. The scheme performed well for classifying pattern Dy but failed in classifying the P1 pattern.

In (Boumaraf et al., 2020), a method to classify mammographic masses into four assessment categories according to BI-RADS is proposed. Mass regions are first enhanced using histogram equalization and then segmented using region growing techniques. A total of 130 handcrafted BI-RADS features are then extracted from the shape, margin, and density of each mass, together with the mass size and the patient's age, as mentioned in BI-RADS mammography. The most clinically significant BI-RADS features are selected using genetic algorithm optimization. The model evaluated on the DDSM dataset achieved a classification accuracy of 84.5%. The authors consider this performance as state of art for BI-RADS classification of breast masses.

6.4 Breast density segmentation

Normal anatomy of the breast comprises of nodular densities, which appear as small bright blobs that normally correspond to terminal ductal lobular units (TDLUs); linear densities correspond to either ducts, fibrous strands, or blood vessels; homogeneous densities correspond to hyperplastic breast changes and fibrous tissue, which could obscure the underlying normal TDLU and ducts; and radiolucent areas represent adipose tissue and appear as dark regions.

Tabár et al. proposed five mammographic parenchymal patterns based on the composition of the four mammographic structures. Patterns I to III have a lower risk of developing breast cancer compared with Patterns IV and V, which are considered high risk. The five patterns are defined in terms of the relative compositions of the four mammographic structures as follows (He et al., 2015; Tabar et al., 2011):

Pattern I: [25%, 15%, 35%, 25%], the mammogram shows normal fibroglandular tissue with partial fatty replacement. Pathological changes are clearly perceived even if the breast is radiologically "dense."

Pattern II: [2%, 14%, 2%, 82%], the mammogram is mostly radiolucent fatty tissue, which makes it easy for detecting pathological lesions.

Pattern III: similar to Pattern II except that the retroareolar prominent ducts are often associated with periductal fibrosis.

Pattern IV: [49%, 19%, 15%, 17%], mammogram is dominated by prominent nodular, linear densities, which make it harder to detect pathological lesions on the mammogram.

Pattern V: [2%, 2%, 89%, 7%], mammogram is dominated by extensive homogeneous structureless fibrous tissue, which makes it very hard to detect small pathological lesions.

Breast density is a consistent and strong risk factor for breast cancer in several populations and across age (Boyd et al., 1995, 2007; Vachon et al., 2007). Density segmentation can be used for estimating breast cancer risk. This requires processing the images by analyzing regions and extracting features that can characterize the various patterns in the image. The goal of breast density segmentation is to map multiple subregions of the breast into different density groupings, which subsequently can be used to estimate breast cancer risk.

Density segmentation methods partition the breast region into different regions, each with a specific range of densities, providing us with a quantitative analysis of breast density. For example, we can estimate the overall density of the breast and also obtain a model of breast tissue composition in terms of percentage or relative proportions of dense/medium and fatty (low density) tissue. Breast density segmentation can be also used as a feature generation step for machine learning classifiers to estimate breast cancer risk and detect mammographic abnormalities.

6.4.1 Thresholding techniques

In (Keller et al., 2012), the breast region within the mammogram is segmented via an automated thresholding scheme. A straight line Hough transform is used to extract the pectoral muscle region. The algorithm then applies adaptive FCM clustering and trains a support vector machine (SVM) classifier to identify, which clusters within the breast tissue are likely fibroglandular. The predictions are aggregated into final dense tissue segmentation where percent density (PD) can be computed and incorporated into risk assessment models.

Density changes can occur due to different physiological processes, such as aging and hormone replacement therapy (Raundahl et al., 2006). Ursin et al. (1998) obtained mammograms from 19 women to quantify the changes in mammographic density over 12 months following treatment. Dense tissue regions were determined by interactive thresholding and mammographic density is estimated using the percentage of dense tissue in the breast region. The resulting PD estimates were compared with the expert outlining method of Wolfe et al. (1987) and showed highly correlated results ($r > 0.85$). Raundahl et al. quantified changes in mammographic appearance using the Hessian eigenvalues at different scales as features (Raundahl et al., 2008). The mammogram is subsequently divided into four structurally different areas by clustering the Hessian features and estimating a density score based on the relative size of the segmented areas.

Percent breast density has been suggested to be a useful marker for breast cancer risk (Byng et al., 1994). PD measures the percentage of fibroglandular tissue in the breast by dividing the area of dense tissue by the area of the entire breast region. Byng et al. (1994) used interactive thresholding and a histogram segmentation method to estimate mammographic PD density. The estimated PD was found to be well correlated with a six category classification of density obtained by expert radiologists (In this scheme, the categories were defined by none, $<10\%$, $1\%-25\%$, $25\%-50\%$,$50\%-75\%$ and $>75\%$ of the projected area of the breast appearing as mammographic density).

Histogram density segmentation methods (HSMs) rely on pixel intensity values for modeling the underlying tissue characteristics. However, determining the optimal threshold values for segmentation remains a complex task, often necessitating the involvement of expert radiologists, as such, they are considered labor intensive. Lu et al. (2007) modified the interactive thresholding-based method of Byng et al. (1994) by using a statistical model-based approach for computing breast density. Mathematical models were constructed to find optimal threshold values, therefore removing the expert observer-based segmentation from the loop. This was done using a multiple regression model to capture the interactions between the instrument image acquisition parameters (e.g., breast compression thickness, radiological thickness, radiation dose, compression force, etc.) and the image pixel intensity statistics of the imaged breasts. The model showed that the acquisition parameters were strong predictors of the threshold values from experts. Automatic threshold calculations obtained using the regression model were used to estimate breast density, thus removing the HSM step altogether.

Heine and Velthuizen (2000) proposed a statistical model to discriminate fibroglandular and fatty tissue based on a chi-squared probability analysis. The variance of the image was computed over a small moving window. A global threshold variance was determined to label each pixel as fatty or nonfatty based on the fact that the fatty tissue has a smaller variance compared with the nonfatty tissue.

6.4.2 Texture analysis

Texture analysis has been shown to provide excellent features for classification of breast region into glandular and fatty tissue (Miller & Astley, 1992). Texture features can be obtained from gray level opening operations and filtering of mammogram images with a texture filter bank. A texture filter bank consists of several kernels representing various textures, such as edges, lines, spots, or ripples.

He et al. (2008) and Miller & Astley (1992) developed a texture-based segmentation technique and investigated its potential for mammographic risk assessment. The method segments mammograms into four regions corresponding to four mammographic building blocks; nodular, linear, homogeneous, and radiolucent, as described by Tabár's tissue model. Here, a filter bank consisting of Gaussian, Laplacian of Gaussian and second-order derivatives of Gaussian at four scales were used to generate textons. Discriminating textons for different tissue types are selected based on texton ranking, outlier detection, and visual assessment. The method showed realistic segmentation results with respect to Tabár building blocks (albeit the over representation of linear structure tissue and the under representation of nodular and radiolucent tissue). The relative proportion of the Tabár building blocks within mammographic images can then be used to assign a risk score.

Li et al. (2020) proposed texture feature descriptor for classifying mammographic breast density based on local binary patterns (LBP), its variant method, local quinary patterns (LQP) and an extended LQP with rotation invariance (RILQP). The results on the INBreast and MIAS datasets show classification accuracies of 82.50% and 80.30%.

In (Gong et al., 2006), texture-based statistical modeling was used to segment mammographic images into different density regions. Textons were learned by clustering image patch vectors extracted from a square neighborhood surrounding each pixel. A hidden Markov random field (HMRF) algorithm is used to generate the final segmentation by utilizing the spatial coherence of texture to propagate information acquired at each pixel to its surroundings.

In (Oliver et al., 2010), fatty and dense tissue were modeled based on statistical analysis of the local image patches surrounding each pixel within the breast region. The Karhunen-Loeve transform (PCA) and linear discriminant analysis were then applied to train a model to segment each pixel into fatty or dense tissue.

6.4.3 Classification and clustering-based segmentation

Breast density classification aims at classifying mammographic images according to breast density categories. Bueno et al. (2011) classified breast parenchymal density according to the BIRADS density categories. Statistical features were used to characterize the texture of the whole breast, obtained from gray level histograms and co-occurrence matrices. A hierarchical classification procedure was used based on the kNN classifier and a combination of linear Bayes normal, SVM, and principal component analysis. The results showed up to 84% of samples were correctly classified.

Ferrari et al. (2004) proposed a breast density model for the segmentation of the fibroglandular disk using a mixture of Gaussians to model different density categories. Keller et al. (2011) applied a multiclass fuzzy c-means clustering algorithm for density-based cancer risk estimation. The fibroglandular tissue is segmented based on an optimal number of clusters which were derived by the tissue properties (encoded by the number of modes in the gray level histogram derived of the specific mammogram). The final segmentation of the dense and fatty structures was achieved through cluster agglomeration using linear discriminant analysis.

A strong correlation was observed between the output of the computerized algorithm and the radiologist-provided ground truth, demonstrating its efficacy with $r = 0.83$ ($P < .001$).

Saha et al. (2001) used scale-based fuzzy connectivity methods to segment dense tissue regions from fat within breasts. The breast region is first segmented and used to derive parameters and create a fuzzy connectivity of dense regions. The connectivity scene is segmented using an automatic threshold selection method. The segmented dense tissue regions were then used to characterize mammographic patterns using a set of density and area related parameters such as the sum of intensity values of pixels in the dense region, the sum of intensity values of pixels in the fatty region, the ratio between them, the area of the dense region and the ratio between the areas of the dense and fatty regions. The method has the potential to evaluating risk and remove the subjectivity inherent in interactive threshold selection techniques.

In (Zwiggelaar et al., 2010) the distribution of local gray level appearances is used to model the local texture information into a gray level appearance (LGA) histogram. Subsequently, a unique LGA number that represents the local texture is extracted from the histogram. Class conditional LGA histograms were generated for the images from the same BIRADS class. This is then used to segment mammograms into four BIRADS density regions where each pixel is labeled with the corresponding BIRADS class.

Adel et al. (2007) used Bayesian theory and a maximum a posteriori method to train a Markov random field (MRF) model. The model is then used to segment regions of interest in the mammographic image in terms of density levels fatty/fibroglandular/nonfatty. Lee and Nishikawa (2018) presented a deep learning model for breast density segmentation. A fully convolutional network is used to segment the breast and the dense fibroglandular areas. The breast PD is then computed for the left and the right breast in the MLO and CC views of the same women. PD estimates were used to classify the BI-RADS density by applying the proposed deep learning model. The results are compared against radiologists' BI-RADS density assessments and against state-of-the-art algorithm called the laboratory for individualized breast radiodensity assessment (LIBRA). The results showed the model agrees with BI-RADS estimates and has excellent ability to separate each sub-BI-RADS breast density class while LIBRA failed.

Boehm et al. (2008) used discriminant analysis to classify breast parenchymal density into three categories: fibrosis, involution atrophy, and normal. Features based on the intensity information and topologic analysis were used. For intensity based features, the 20th percentile, median, and mean values were calculated from gray level histograms. For topologic analysis, Minkowski functionals (Thompson, 1996) were used to extract the area and perimeter of a binary image pattern as well as the number of connected components.

Oliver et al. (2005) used the k-means and decision tree algorithms to classify breast parenchymal density into fatty, dense, and glandular tissue. The leave-one-woman-out strategy for training and testing was adopted. Gray-level information is used to initialize the segmentation step of the breast profile. The k-Means algorithm is then used to cluster the different tissue types of the mammograms. Texture and gray level information were extracted using morphological features and texture features based on co-occurrence matrices. The k-nearest neighbors (kNN) classifier and the ID3 decision tree classifier were used to classify breast density as dense, fatty, or glandular.

Oliver et al. (2006) reviewed different strategies for extracting features for breast tissue density segmentation. Features were extracted from the whole breast region and from sub-regions within the breast area. Segmentation of the breast according to the distance to the skin-line (Karssemeijer, 1998), and segmentation of the breast according to internal breast tissue, were addressed. Segmentation based on the Fractal approach results in a pixelated segmentation, while the statistical approach obtains larger and clearly separated regions. Fuzzy C-Means was also used and resulted in a slightly improved performance. Texture features derived from co-occurrence matrices and a set of morphological features were calculated (Oliver et al., 2005). A Bayesian classifier combining the kNN and the C4.5 decision tree algorithms was used for classifying breast density into the four BIRADS categories. The results show that segmentation with focus on the internal breast tissue, got 82% of correct classification based on a leave-one-woman-out methodology.

Oliver et al. (2008) tested the individual performance of the kNN classifier, the C4.5 decision tree classifier, and the combination of the two classifiers using the Bayesian approach on two public databases. The methodology can be summarized as follows: (1) the breast tissue is segmented into fatty and dense categories; (2) morphological and texture features are extracted from the segmented breast areas; (3) sequential forward selection algorithm is used to select the most discriminant features; and (4) a Bayesian combination of several classifiers is used to make the final prediction. The results showed density assessments using the automatic and expert-based BI-RADS are strongly correlated.

Subashini et al. (2010) extracted 14 statistical features from the breast region and trained the SVM classifier using different combinations of these features. Nine features, including mean, standard deviation, smoothness, skewness, uniformity, kurtosis, average histogram, modified standard deviation, and modified skewness, were selected for breast tissue classification with three density classes. The classifier accuracy obtained was 95.44%.

Tzikopoulos et al. (2011) used density estimation and detection of asymmetry for breast density classification and segmentation. The breast boundary and the pectoral muscle were segmented. Features for breast density were extracted from the segmented breast area. The features include statistical features extracted from a set of regions within the breast and features related to the fractal dimension; computed based on the power spectrum. The differences of these features from the two images of each pair of mammograms are used to detect breast asymmetry using a one-class SVM classifier. Also, classification of mammograms according to the three density classes (fatty, glandular, and dense) was tested using classification and regression trees (CARTs), kNN and SVM classifiers.

In (Kallenberg et al., 2011) a number of features were extracted from the breast region such as location, intensity, texture and global context. The sequential floating forward selection algorithm produced an optimal feature subset which is used to train a breast density segmentation model based on an ensemble of five neural networks.

Liasis et al. (2012) combined different texture features for mammographic breast density classification. Texture descriptors based on scale invariant feature transforms, LBPs, and texton histograms. Breast density classification based on the individual histograms of the texture features and their combination is evaluated on the MIAS database. The results showed that the combination of the statistical distributions of all texture features produced the highest classification accuracy of 93% using SVM classifier.

6.5 Breast cancer risk assessment

Mammographic density and mammographic parenchymal patterns are both strong predictive markers of breast cancer risk (Boyd et al., 1995, 2007; Tabar et al., 2011; Vachon et al., 2007; Wolfe, 1976). BI-RADS assigns four categories describing the breast density. Breast tissue descriptors enable clinicians to describe and communicate mammographic features. They have shown to be good predictors of breast cancer risk (increased breast density is correlated with increased risk of breast cancer) (Mandelson et al., 2000). Women with the highest quartile of mammographic density have four to six times increased risk of breast cancer compared with women with little or no dense tissue (McCormack & dos Santos Silva, 2006).

Stone et al. (2009) showed that percent dense area is negatively correlated with body mass index (percent dense area is calculated as a ratio of dense area and the total area of the breast thus, largely affected by the area of nondense tissue in the breast). However, dense area, defined as the number of pixels in the dense region, is independent of other breast cancer risk factors. Stone et al. (2009) used multivariate linear regression to evaluate associations between the two measures of mammographic density and breast cancer risk factors. Dense area was found to be a simpler, independent of other breast cancer risk factors and a reliable biomarker for risk prediction.

Manduca et al. (2009) evaluated the association of many image texture features with the risk of breast cancer. Five families of texture features derived from Markovian co-occurrence matrices were used to model the appearance of pixels; run-length features (computes runs of similar gray level values in an image); law features (a set of filters that can capture different structures in an image); wavelet features (multiresolution analysis that can enhance textures); and Fourier features (to describe the coarseness of textures). A bootstrap approach capable of handling a large number of correlated variables was used to identify the strongest features associated with breast cancer risk. The strongest features include PD, Markovian, run length, Laws, wavelet, and Fourier features. The results show that breast cancer cases have stronger textural properties at coarse scales in their screening mammograms than controls.

Zheng et al. (2015) used texture features for breast cancer risk assessment. Here, instead of extracting mammographic texture within selected regions of interest (ROIs), the authors adopted an extended texture model that can capture more of the entire parenchymal texture structure in the breast region. A lattice-based strategy that relies on a regular grid virtually overlaid on the mammographic image. A local window centered on each lattice point to extract gray-level histogram statistics, co-occurrence, run-length and structural features (e.g., edge-enhancing, LBP, and fractal dimension). This generates a full texture map that characterizes the parenchymal patterns over the entire breast. The best AUC value generated was 0.85 based on 106 mammograms from cases with 318 age-matched controls.

Zwiggelaar et al. (2004) presented methods for detecting linear structures in mammograms, and for classifying them into anatomical types (vessels, spicules, ducts, etc.). The line operator, orientated bins, Gaussian derivatives, and ridge detector were used to estimate line-strength, orientation and scale at each pixel in the mammogram. The results showed the importance of linear structures in detecting and classifying mammographic abnormalities. Classification results were compared with expert annotations using ROC analysis demonstrated useful discrimination between the anatomical classes.

Petroudi et al. (2003) modeled mammographic parenchymal appearance using texture models. The training set consists of a representative set of mammograms with their associated BI-RADS density classification. This stage consists of three steps: breast area segmentation, filtering, and clustering. A statistical distribution is derived from the responses of a rotationally invariant filter bank. The k-means algorithm was employed to compute the cluster centers, representing the texton dictionary for all training sets. With 4 BI-RADS categories and 10 textons per class, the dictionary holds 40 textons. Mammogram images are then labeled by assigning a texton to each breast tissue pixel according to the nearest model and the chi^2 distribution comparison. A texton histogram belonging to the same class was computed for the image constituted the models of parenchymal patterns for that class. The set of texton histograms defines the breast parenchymal density models. Note that each histogram represents a model for the density with which it is associated—there are as many models as training images.

Castella et al. (2007) proposed a density-based method for breast cancer risk assessment of digital mammograms. Four 256×256 pixel ROIs per mammogram were manually selected and used to extract a set of 36 statistical features. The features were derived from gray level histograms, gray level co-occurrence matrices, neighborhood gray tone difference matrix as well as from the primitive matrix and the fractal dimension. Bayesian classifier based on the measure of Mahalanobis distance, a naïve Bayesian classifier, and linear discriminant analysis were trained to predict the density class attributed to the test ROI. The classifiers obtained excellent performances when tested in the two-class problem reduction.

Li et al. (2008) applied power law spectral analysis for breast cancer risk assessment. An ROI was first manually selected from the central breast region behind the nipple. The discrete Fourier transform was then performed on the ROI using the Hanning window. The corresponding power spectrum was calculated and used to estimate the exponent of the power law. The results demonstrated the usefulness of the exponent of power law as a feature for breast cancer risk assessment.

Huo et al. (2000) extracted 14 features representing parenchymal patterns from the central breast region to identify useful mammographic features that are associated with predictors of breast cancer risk. The feature set comprised features based on the absolute gray level values; features based on gray level histogram analysis (balance and skewness); features based on the spatial relationship among gray levels within the ROI (contrast, coarseness, and features derived from co-occurrence matrices); and features based on Fourier analysis. The results suggest that women who are at high risk tend to have dense breasts and their mammographic patterns tend to be coarse and low in contrast.

Li et al. (2005) extracted a set of ROI-based texture features to assess the mammographic parenchymal patterns: (1) features based on absolute values of gray levels, (2) features based on gray-level histogram analysis, (3) features based on spatial relationships among gray levels, (4) features based on fractal analysis, (5) features based on edge frequency, and (6) features based on Fourier transform analysis. The fractal dimension of the image is estimated using the conventional box-counting and the Minkowski methods. The aim of the study was to classify women from a high-risk gene-mutation carrier group and a low-risk group. The results showed that gene-mutation carriers and women at low risk of developing breast cancer have different mammographic patterns: gene-mutation carriers tend to have denser breast tissue, coarser texture, and lower contrast mammographic images. In terms of features, the high-risk group has a negative value in skewness measure, a higher coarseness measure, a lower fractal dimension measure, and a smaller edge gradient measure (Li et al., 2005). The robustness of this method was later tested on a large clinical dataset of full-field digital mammograms with 456 cases, including 53 women with BRCA1/2 gene mutations, 75 women with unilateral cancer, and 328 low-risk women (Li et al., 2012). The results confirmed the findings of their previous study (Li et al., 2005).

In (He et al., 2010, 2011), mammographic image segmentation based on Tabár tissue modeling. Such approaches can be useful for mammographic risk assessment as well as for quantitative evaluation of the variation in characteristic mixture of breast tissue over time. Tabár texture signatures were generated by extracting texture features from annotated image patches. The features are composed of 2D cumulative histograms that describe (1) the gray level distribution of pixels at different radial distances from the central pixel within local windows and (2) the gray level distribution of pixels at different angles with respect to the central pixel within local windows. The third part of the texture signature is extracted from a modified co-occurrence matrix with $d = 1$ and $\theta = 0$, representing the frequency of some gray-level configuration and the magnitude and variance between two adjacent pixels. The extracted features were clustered and used to learn mammographic models for Tabár building blocks. Breast tissue is then segmented by labeling each pixel according to the Tabár model outputs. The MIAS database was used in a quantitative and qualitative evaluation with respect to mammographic risk assessment based on both Tabár and BIRADS risk categories. The method showed good consistency with expert radiologist's annotations. Classification accuracies of 86% and 87% were observed, corresponding to the low and high-risk categories for Tabar and BIRADS.

The sensitivity of mammography can be significantly reduced by increased breast density of the fibroglandular tissue, causing masked tumors. He et al. (2012) proposed a mammographic density segmentation approach using three categories: dense, semi-dense, and nondense tissue. Gray level histogram based features were derived from multiresolution local windows and used to build models of Tabár mammographic building blocks. A Bayesian classifier was trained based on a novel binary model matching pattern to achieve mammographic segmentation. Results showed that the probability based classifier is robust in dealing with inter and intraclass variation. Also, substantial agreements in mammographic risk classification between the classification results and ground truth provided by expert screening radiologists. The total classification accuracies were 93% and 85% in Tabár's categories and the corresponding low- and high- risk category; 88% and 78% accuracies in BIRADS categories and the corresponding low- and high-risk category.

6.6 Conclusion

In conclusion, research has shown that mammographic density and parenchymal patterns are strong predictors of breast cancer risk. Several studies have explored different methods for evaluating these features, including texture analysis and image segmentation. The negative correlation between percent dense area and breast cancer risk, as shown by Stone et al. (2009) and the use of texture features for risk assessment by Zheng et al. (2015) and Huo et al. (2000) suggest the potential for these features to be used in clinical settings for early detection and risk assessment. Furthermore, the proposed density-based method by Castella et al. (2007) and the ROI-based texture features by Li et al. have the potential to improve the accuracy of breast cancer diagnosis. Breast density estimation is still challenging due to low contrast and significant natural variability of density in appearance and texture in mammographic images. Breast density segmentation can be viewed as a density classification problem where dense tissues in the mammographic images can be separated correctly. Methods for breast density estimation must overcome the low signal-to-noise ratio in examined images and segment the images automatically.

Overall, these studies demonstrate the importance of mammographic density and parenchymal patterns in breast cancer detection and the potential for advanced imaging techniques to enhance breast cancer screening and diagnosis.

References

Adel, M., Rasigni, M., Bourennane, S., & Juhan, V. (2007). Statistical segmentation of regions of interest on a mammographic image. *EURASIP Journal on Advances in Signal Processing*, *2007*(1), 049482. Available from https://doi.org/10.1155/2007/49482, https://asp-eurasipjournals.springeropen.com/articles/10.1155/2007/49482.

Bihrmann, K., Jensen, A., Olsen, A. H., Njor, S., Schwartz, W., Vejborg, I., & Lynge, E. (2008). Performance of systematic and non-systematic ('opportunistic') screening mammography: A comparative study from Denmark. *Journal of Medical Screening*, *15*(1), 23–26. Available from https://doi.org/10.1258/jms.2008.007055.

Boehm, H. F., Schneider, T., Buhmann-Kirchhoff, S. M., Schlossbauer, T., Rjosk-Dendorfer, D., Britsch, S., & Reiser, M. (2008). Automated classification of breast parenchymal density: Topologic analysis of X-ray attenuation patterns depicted with digital mammography. *American Journal of Roentgenology*, *191*(6), W275–W282. Available from https://doi.org/10.2214/AJR.07.3588, https://www.ajronline.org/doi/10.2214/AJR.07.3588.

Boumaraf, S., Liu, X., Ferkous, C., & Ma, X. (2020). A new computer-aided diagnosis system with modified genetic feature selection for BI-RADS classification of breast masses in mammograms. *BioMed Research International*, *2020*, e7695207. Available from https://doi.org/10.1155/2020/7695207, https://www.hindawi.com/journals/bmri/2020/7695207/.

Boyd, N. F., Byng, J. W., Jong, R. A., Fishell, E. K., Little, L. E., Miller, A. B., Lockwood, G. A., Tritchler, D. L., & Yaffe, M. J. (1995). Quantitative classification of mammographic densities and breast cancer risk: Results from the Canadian National Breast Screening Study. *Journal of the National Cancer Institute*, *87*(9), 670–675. Available from https://doi.org/10.1093/jnci/87.9.670.

Boyd, N. F., Guo, H., Martin, L. J., Sun, L., Stone, J., Fishell, E., Jong, R. A., Hislop, G., Chiarelli, A., Minkin, S., & Yaffe, M. J. (2007). Mammographic density and the risk and detection of breast cancer. *The New England Journal of Medicine*, *356*(3), 227–236. Available from https://doi.org/10.1056/NEJMoa062790.

Breast cancer. (2019). Breast cancer survival rates are rising as screening and treatment improve. However, breast cancer is still the most invasive cancer in women. Read on to learn more. Breast cancer: Symptoms, causes, and treatment Breast cancer. https://www.medicalnewstoday.com/articles/37136.

Breast cancer digital repository. (2021). https://bcdr.eu/information/about.

Breast cancer stages. (2018). Breast cancer stages: 0 Through IV & more. Breast cancer stages. https://www.cancercenter.com/cancer-types/breast-cancer/stages.

Breast imaging reporting & data system. (2021). https://www.acr.org/Clinical-Resources/Reporting-and-Data-Systems/Bi-Rads.

Breast screening | Breast cancer | Cancer Research UK. (2021). https://www.cancerresearchuk.org/about-cancer/breast-cancer/getting-diagnosed/screening/breast-screening.

Bruno, A. (2021). alessandrobruno10/suremapp. https://github.com/alessandrobruno10/suremapp.

Bueno, G., Vállez, N., Déniz, O., Esteve, P., Rienda, M. A., Arias, M., & Pastor, C. (2011). Automatic breast parenchymal density classification integrated into a CADe system. *International Journal of Computer Assisted Radiology and Surgery*, *6*(3), 309–318. Available from https://doi.org/10.1007/s11548-010-0510-z, http://link.springer.com/10.1007/s11548-010-0510-z.

Burnside, E. S., Sickles, E. A., Bassett, L. W., Rubin, D. L., Lee, C. H., Ikeda, D. M., Mendelson, E. B., Wilcox, P. A., Butler, P. F., & D'Orsi, C. J. (2009). The ACR BI-RADS® experience: Learning from history. *Journal of the American College of Radiology*, *6*(12), 851–860. Available from https://doi.org/10.1016/j.jacr.2009.07.023, https://www.jacr.org/article/S1546-1440(09)00390-1/abstract.

Byng, J. W., Boyd, N. F., Fishell, E., Jong, R. A., & Yaffe, M. J. (1994). The quantitative analysis of mammographic densities. *Physics in Medicine and Biology*, *39*(10), 1629–1638. Available from https://doi.org/10.1088/0031-9155/39/10/008, https://iopscience.iop.org/article/10.1088/0031-9155/39/10/008.

Castella, C., Kinkel, K., Eckstein, M. P., Sottas, P.-E., Verdun, F. R., & Bochud, F. O. (2007). Semiautomatic mammographic parenchymal patterns classification using multiple statistical features. *Academic Radiology*, *14*(12), 1486–1499. Available from https://doi.org/10.1016/j.acra.2007.07.014, https://www.sciencedirect.com/science/article/pii/S1076633207004059.

Clark, K., Vendt, B., Smith, K., Freymann, J., Kirby, J., Koppel, P., Moore, S., Phillips, S., Maffitt, D., Pringle, M., Tarbox, L., & Prior, F. (2013). The cancer imaging archive (TCIA): Maintaining and operating a public information repository. *Journal of Digital Imaging*, *26*(6), 1045–1057. Available from https://doi.org/10.1007/s10278-013-9622-7, http://link.springer.com/10.1007/s10278-013-9622-7.

de Oliveira, J. E. E., Machado, A. M. C., Chavez, G. C., Lopes, A. P. B., Deserno, T. M., & Araújo, A. d. A. (2010). MammoSys: A content-based image retrieval system using breast density patterns. *Computer Methods and Programs in Biomedicine*, *99*(3), 289–297. Available from https://doi.org/10.1016/j.cmpb.2010.01.005, https://www.sciencedirect.com/science/article/pii/S0169260710000180.

DeSantis, C., Siegel, R., & Jemal, A. (2011). Breast cancer facts & figures 2011–2012. 36.

Dilhuydy, M. H. (2007). Breast imaging reporting and data system (BI-RADS) or French "classification ACR" What tool for what use? A point of view. *European Journal of Radiology*, *61*(2), 187–191. Available from https://doi.org/10.1016/j.ejrad.2006.08.032.

Ferlay, J., Steliarova-Foucher, E., Lortet-Tieulent, J., Rosso, S., Coebergh, J. W. W., Comber, H., Forman, D., & Bray, F. (2013). Cancer incidence and mortality patterns in Europe: Estimates for 40 countries in 2012. *European Journal of Cancer (Oxford, England: 1990)*, *49*(6), 1374–1403. Available from https://doi.org/10.1016/j.ejca.2012.12.027.

Ferrari, R. J., Rangayyan, R. M., Borges, R. A., & Frère, A. F. (2004). Segmentation of the fibro-glandular disc in mammogrms using Gaussian mixture modelling. *Medical & Biological Engineering & Computing*, *42*(3), 378–387. Available from https://doi.org/10.1007/BF02344714, http://link.springer.com/10.1007/BF02344714.

Geller, B. M., Barlow, W. E., Ballard-Barbash, R., Ernster, V. L., Yankaskas, B. C., Sickles, E. A., Carney, P. A., Dignan, M. B., Rosenberg, R. D., Urban, N., Zheng, Y., & Taplin, S. H. (2002). Use of the American College of Radiology BI-RADS to report on the mammographic evaluation of women with signs and symptoms of breast disease. *Radiology*, *222*(2), 536–542. Available from https://doi.org/10.1148/radiol.2222010620, https://pubs.rsna.org/doi/abs/10.1148/radiol.2222010620.

Gong, Y. C., Brady, M., Petroudi, S., Astley, S. M., Brady, M., Rose, C., & Zwiggelaar, R. (2006). Texture based mammogram classification and segmentation (pp. 616–625). Available from https://doi.org/10.1007/11783237_83.

Greif, J. M. (2010). Mammographic screening for breast cancer: An invited review of the benefits and costs. *Breast (Edinburgh, Scotland)*, *19*(4), 268–272. Available from https://doi.org/10.1016/j.breast.2010.03.017.

Guevara Lopez, M. A., Posada, N., Moura, D., Pollán, R., Franco-Valiente, J., Ortega, C., Del Solar, M., Díaz-Herrero, G., Ramos, I., Loureiro, J., Fernandes, T., & Araújo, B. (2012). BCDR: A Breast Cancer Digital Repository. *BCDR*, 1065–1066.

He, W., Denton, E. R. E., Stafford, K., & Zwiggelaar, R. (2011). Mammographic image segmentation and risk classification based on mammographic parenchymal patterns and geometric moments. *Biomedical Signal Processing and Control*, *6*(3), 321–329. Available from https://doi.org/10.1016/j.bspc.2011.03.008, https://linkinghub.elsevier.com/retrieve/pii/S174680941100036X.

He, W., Denton, E. R. E., Zwiggelaar, R., Maidment, A. D. A., Bakic, P. R., & Gavenonis, S. (2012). *Mammographic segmentation and risk classification using a novel binary model based bayes classifier breast imaging* (pp. 40–47). Berlin, Heidelberg: Springer Berlin Heidelberg. Available from http://link.springer.com/10.1007/978-3-642-31271-7_6.

He, W., Denton, E. R. E., Zwiggelaar, R., Martí, J., Oliver, A., Freixenet, J., & Martí, R. (2010). *Mammographic image segmentation and risk classification using a novel texture signature based methodology digital mammography* (pp. 526–533). Berlin, Heidelberg: Springer Berlin Heidelberg. Available from http://link.springer.com/10.1007/978-3-642-13666-5_71.

He, W., Juette, A., Denton, E. R. E., Oliver, A., Martí, R., & Zwiggelaar, R. (2015). A review on automatic mammographic density and parenchymal segmentation. *International Journal of Breast Cancer*, *2015*, 1–31. Available from https://doi.org/10.1155/2015/276217, http://www.hindawi.com/journals/ijbc/2015/276217/.

He, W., Muhimmah, I., Denton, E. R. E., Zwiggelaar, R., & Krupinski, E. A. (2008). *Mammographic segmentation based on texture modelling of tabár mammographic building blocks digital mammography* (pp. 17–24). Berlin, Heidelberg: Springer Berlin Heidelberg. Available from http://link.springer.com/10.1007/978-3-540-70538-3_3.

Heath, M., Bowyer, K., Kopans, D., Kegelmeyer, P., Moore, R., Chang, K., Munishkumaran, S., Karssemeijer, N., Thijssen, M., Hendriks, J., & van Erning, L. (1998). *Current status of the digital database for screening mammography* Digital Mammography: Nijmegen, 1998 Dordrecht: Springer Netherlands. Available from https://doi.org/10.1007/978-94-011-5318-8_75.

Heine, J. J., & Velthuizen, R. P. (2000). A statistical methodology for mammographic density detection. *Medical Physics*, *27*(12), 2644–2651. Available from https://doi.org/10.1118/1.1323981.

Huo, Z., Giger, M. L., Wolverton, D. E., Zhong, W., Cumming, S., & Olopade, O. I. (2000). Computerized analysis of mammographic parenchymal patterns for breast cancer risk assessment: Feature selection. *Medical Physics*, *27*(1), 4–12. Available from https://doi.org/10.1118/1.598851, https://onlinelibrary.wiley.com/doi/abs/10.1118/1.598851.

Kallenberg, M. G. J., Lokate, M., van Gils, C. H., & Karssemeijer, N. (2011). Automatic breast density segmentation: An integration of different approaches. *Physics in Medicine and Biology*, *56*(9), 2715–2729. Available from https://doi.org/10.1088/0031-9155/56/9/005.

Karssemeijer, N. (1998). Automated classification of parenchymal patterns in mammograms. *Physics in Medicine and Biology*, *43*(2), 365–378. Available from https://doi.org/10.1088/0031-9155/43/2/011, https://iopscience.iop.org/article/10.1088/0031-9155/43/2/011.

Karssemeijer, N., Colchester Alan, C. F., & Hawkes David, J. (1991). A stochastic model for automated detection of calcifications in digital mammograms. *Information Processing in Medical Imaging*, 227–238. Available from https://doi.org/10.1007/BFb0033756.

Keller, B. M., Nathan, D. L., Wang, Y., Zheng, Y., Gee, J. C., Conant, E. F., & Kontos, D. (2012). Estimation of breast percent density in raw and processed full field digital mammography images via adaptive fuzzy c-means clustering and support vector machine segmentation. *Medical Physics*, *39*(8), 4903–4917. Available from https://doi.org/10.1118/1.4736530, https://www.ncbi.nlm.nih.gov/pmc/articles/PMC3416877/.

Keller, B., Nathan, D., Wang, Y., Zheng, Y., Gee, J., Conant, E., Kontos, D., Fichtinger, G., Martel, A., & Peters, T. (2011). *Adaptive multi-cluster fuzzy C-means segmentation of breast parenchymal tissue in digital mammography medical image computing and computer-assisted intervention – MICCAI 2011* (pp. 562–569). Berlin, Heidelberg: Springer Berlin Heidelberg. Available from http://link.springer.com/10.1007/978-3-642-23626-6_69.

Lai, C.-J., Shaw, C. C., Geiser, W., Chen, L., Arribas, E., Stephens, T., Davis, P. L., Ayyar, G. P., Dogan, B. E., Nguyen, V. A., Whitman, G. J., & Yang, W. T. (2008). Comparison of slot scanning digital mammography system with full-field digital mammography system. *Medical Physics*, *35*(6Part1), 2339–2346. Available from https://doi.org/10.1118/1.2919768, https://aapm.onlinelibrary.wiley.com/doi/abs/10.1118/1.2919768.

Lee, R. S., Gimenez, F., Hoogi, A., Miyake, K. K., Gorovoy, M., & Rubin, D. L. (2017). A curated mammography data set for use in computer-aided detection and diagnosis research. *Scientific Data*, *4*(1), 170177. Available from https://doi.org/10.1038/sdata.2017.177, http://www.nature.com/articles/sdata2017177.

Lee, J., & Nishikawa, R. M. (2018). Automated mammographic breast density estimation using a fully convolutional network. *Medical Physics*, *45*(3), 1178–1190. Available from https://doi.org/10.1002/mp.12763, https://onlinelibrary.wiley.com/doi/abs/10.1002/mp.12763.

Li, H., Giger, M. L., Lan, L., Bancroft Brown, J., MacMahon, A., Mussman, M., Olopade, O. I., & Sennett, C. (2012). Computerized analysis of mammographic parenchymal patterns on a large clinical dataset of full-field digital mammograms: Robustness study with two high-risk datasets. *Journal of Digital Imaging*, *25*(5), 591–598. Available from https://doi.org/10.1007/s10278-012-9452-z, https://www.ncbi.nlm.nih.gov/pmc/articles/PMC3447101/.

Li, H., Giger, M. L., Olopade, O. I., & Chinander, M. R. (2008). Power spectral analysis of mammographic parenchymal patterns for breast cancer risk assessment. *Journal of*

Digital Imaging, *21*(2), 145–152. Available from https://doi.org/10.1007/s10278-007-9093-9, https://www.ncbi.nlm.nih.gov/pmc/articles/PMC3043857/.

Li, H., Giger, M. L., Olopade, O. I., Margolis, A., Lan, L., & Chinander, M. R. (2005). Computerized texture analysis of mammographic parenchymal patterns of digitized mammograms1. *Academic Radiology*, *12*(7), 863–873. Available from https://doi.org/10.1016/j.acra.2005.03.069, https://www.sciencedirect.com/science/article/pii/S1076633205003478.

Li, H., Mukundan, R., & Boyd, S. (2020). Robust texture features for breast density classification in mammograms. In: *16th International Conference on Control, Automation, Robotics and Vision* (ICARCV) (pp. 454–459). Available from https://doi.org/10.1109/ICARCV50220.2020.9305431.

Liasis, G., Pattichis, C., & Petroudi, S. (2012). Combination of different texture features for mammographic breast density classification (pp. 732–737). Available from https://doi.org/10.1109/BIBE.2012.6399758.

Lu, L.-J. W., Nishino, T. K., Khamapirad, T., Grady, J. J., Leonard, M. H., & Brunder, D. G. (2007). Computing mammographic density from a multiple regression model constructed with image-acquisition parameters from a full-field digital mammographic unit. *Physics in Medicine and Biology*, *52*(16), 4905–4921. Available from https://doi.org/10.1088/0031-9155/52/16/013, https://www.ncbi.nlm.nih.gov/pmc/articles/PMC2691417/.

Mandelson, M. T., Oestreicher, N., Porter, P. L., White, D., Finder, C. A., Taplin, S. H., & White, E. (2000). Breast density as a predictor of mammographic detection: Comparison of interval- and screen-detected cancers. *Journal of the National Cancer Institute*, *92*(13), 1081–1087. Available from https://doi.org/10.1093/jnci/92.13.1081.

Manduca, A., Carston, M. J., Heine, J. J., Scott, C. G., Pankratz, V. S., Brandt, K. R., Sellers, T. A., Vachon, C. M., & Cerhan, J. R. (2009). Texture features from mammographic images and risk of breast cancer. *Cancer Epidemiology, Biomarkers & Prevention*, *18*(3), 837–845. Available from https://doi.org/10.1158/1055-9965.EPI-08-0631, https://doi.org/10.1158/1055-9965.EPI-08-0631.

Matheus, B. R. N., & Schiabel, H. (2011). Online mammographic images database for development and comparison of CAD schemes. *Journal of Digital Imaging*, *24*(3), 500–506. Available from https://doi.org/10.1007/s10278-010-9297-2, https://www.ncbi.nlm.nih.gov/pmc/articles/PMC3092049/.

McCormack, V. A., & dos Santos Silva, I. (2006). Breast density and parenchymal patterns as markers of breast cancer risk: A meta-analysis. *Cancer Epidemiology, Biomarkers & Prevention*, *15*(6), 1159–1169. Available from https://doi.org/10.1158/1055-9965.EPI-06-0034, https://doi.org/10.1158/1055-9965.EPI-06-0034.

Miller, P., & Astley, S. (1992). Classification of breast tissue by texture analysis. *Image and Vision Computing*, *10*(5), 277–282. Available from https://doi.org/10.1016/0262-8856(92)90042-2, https://www.sciencedirect.com/science/article/pii/0262885692900422.

Moreira, I. C., Amaral, I., Domingues, I., Cardoso, A., Cardoso, M. J., & Cardoso, J. S. (2012). INbreast: Toward a full-field digital mammographic database. *Academic Radiology*, *19*(2), 236–248. Available from https://doi.org/10.1016/j.acra.2011.09.014.

NCCN National Comprehensive Cancer Network – Home. (2021). https://www.nccn.org.

Ng, K. H., & Muttarak, M. (n.d). Advances in mammography have improved early detection of breast cancer, 6.

Nishikawa, R. M. (1998). Mammographic databases. *Breast Disease*, *10*(3–4), 137–150. Available from https://doi.org/10.3233/bd-1998-103-414.

Nishikawa, R. M. (2007). Current status and future directions of computer-aided diagnosis in mammography. *Computerized Medical Imaging and Graphics*, *31*(4–5), 224–235. Available from https://doi.org/10.1016/j.compmedimag.2007.02.009, https://linkinghub.elsevier.com/retrieve/pii/S0895611107000171.

Obenauer, S., Hermann, K. P., & Grabbe, E. (2005). Applications and literature review of the BI-RADS classification. *European Radiology*, *15*(5), 1027–1036. Available from https://doi.org/10.1007/s00330-004-2593-9, http://link.springer.com/10.1007/s00330-004-2593-9.

Oliveira, J. E. E., Gueld, M. O., Araújo, A. D. A., Ott, B., & Deserno, T. M. (n.d). Towards a standard reference database for computer-aided mammography.

Oliver, A., Freixenet, J., Bosch, A., Raba, D., Zwiggelaar, R., Marques, J. S., Pérez de la Blanca, N., & Pina, P. (2005). *Automatic classification of breast tissue pattern recognition and image analysis* (pp. 431–438). Berlin, Heidelberg: Springer Berlin Heidelberg. Available from http://link.springer.com/10.1007/11492542_53.

Oliver, A., Freixenet, J., Martí, J., Pérez, E., Pont, J., Denton, E. R. E., & Zwiggelaar, R. (2010). A review of automatic mass detection and segmentation in mammographic images. *Medical Image Analysis*, *14*(2), 87–110. Available from https://doi.org/10.1016/j.media.2009.12.005, https://linkinghub.elsevier.com/retrieve/pii/S1361841509001492.

Oliver, A., Freixenet, J., Martí, R., Pont, J., Pérez, E., Denton, E. R. E., & Zwiggelaar, R. (2008). A novel breast tissue density classification methodology. *IEEE Transactions on Information Technology in Biomedicine*, *12*(1), 55–65. Available from https://doi.org/10.1109/TITB.2007.903514.

Oliver, A., Freixenet, J., Martí, R., Zwiggelaar, R., Larsen, R., Nielsen, M., & Sporring, J. (2006). *A Comparison of breast tissue classification techniques medical image computing and computer-assisted intervention – MICCAI 2006* (pp. 872–879). Berlin, Heidelberg: Springer Berlin Heidelberg. Available from http://link.springer.com/10.1007/11866763_107.

Orel, S. G., Kay, N., Reynolds, C., & Sullivan, D. C. (1999). BI-RADS categorization as a predictor of malignancy. *Radiology*, *211*(3), 845–850. Available from https://doi.org/10.1148/radiology.211.3.r99jn31845.

Othman, A., Ahram, M., Al-Tarawneh, M. R., & Shahrouri, M. (2015). Knowledge, attitudes and practices of breast cancer screening among women in Jordan. *Health Care for Women International*, *36*(5), 578–592. Available from https://doi.org/10.1080/07399332.2014.926900.

Petroudi, S., Kadir, T., & Brady M. (2003). Automatic classification of mammographic parenchymal patterns, 1, 798–801. Available from https://doi.org/10.1109/IEMBS.2003.1279885.

Raundahl, J., Loog, M., & Nielsen, M. (2006). SPIE Mammographic density measured as changes in tissue structure caused by HRT. In: *Medical Imaging 2006: Image Processing 6144* (pp. 141–148). Available from https://www.spiedigitallibrary.org/conference-proceedings-of-spie/6144/61440G/Mammographic-density-measured-as-changes-in-tissue-structure-caused-by/, https://doi.org/10.1117/12.653021.full 6144.

Raundahl, J., Loog, M., Pettersen, P., Tanko, L. B., & Nielsen, M. (2008). Automated effect-specific mammographic pattern measures. *IEEE Transactions on Medical Imaging*, *27*(8), 1054–1060. Available from https://doi.org/10.1109/TMI.2008.917245.

Saha, P. K., Udupa, J. K., Conant, E. F., Chakraborty, D. P., & Sullivan, D. (2001). Breast tissue density quantification via digitized mammograms. *IEEE Transactions on Medical Imaging*, *20*(8), 792–803. Available from https://doi.org/10.1109/42.938247.

Stone, J., Warren, R. M. L., Pinney, E., Warwick, J., & Cuzick, J. (2009). Determinants of percentage and area measures of mammographic density. *American Journal of Epidemiology*, *170*(12), 1571–1578. Available from https://doi.org/10.1093/aje/kwp313, https://academic.oup.com/aje/article-lookup/doi/10.1093/aje/kwp313.

Strickland, R. N., & Hahn, H. I. (1996). Wavelet transforms for detecting microcalcifications in mammograms. *IEEE Transactions on Medical Imaging*, *15*(2), 218–229. Available from https://doi.org/10.1109/42.491423.

Subashini, T. S., Ramalingam, V., & Palanivel, S. (2010). Automated assessment of breast tissue density in digital mammograms. *Computer Vision and Image Understanding*, *114*(1), 33–43. Available from https://doi.org/10.1016/j.cviu.2009.09.009, https://www.sciencedirect.com/science/article/pii/S1077314209001593.

Suckling, J., Parker, J., Dance, D., Astley, S., Hutt, I., Boggis, C., Ricketts, I., Stamatakis, E., Cerneaz, N., Kok, S., Taylor, P., Betal, D., & Savage, J. (2015). Mammographic image analysis society (MIAS) database v1.21. Available from https://www.repository.cam.ac.uk/handle/1810/250394. https://doi.org/10.1810/250394.

Tabar, L., Tot, T., & Dean, P. B. (2011). *Breast cancer - The art and science of early detection with mammography: Perception, interpretation, histopathologic correlation* (1st edition, p. 1231)Thieme.

Tabar, L., Yen, M.-F., Vitak, B., Chen, H.-H. T., Smith, R. A., & Duffy, S. W. (2003). Mammography service screening and mortality in breast cancer patients: 20-year follow-up before and after introduction of screening. *Lancet (London, England)*, *361* (9367), 1405–1410. Available from https://doi.org/10.1016/S0140-6736(03)13143-1.

Tahocest, P. G., Correal, J., Soutot, M., & Vidalt, J. J. (1995). Computer-assisteddiagnosis: The classification of mammographic breast parenchymal patterns. *Physics in Medicine & Biology*, *40*(1), 16.

Taplin, S. H., Ichikawa, L. E., Kerlikowske, K., Ernster, V. L., Rosenberg, R. D., Yankaskas, B. C., Carney, P. A., Geller, B. M., Urban, N., Dignan, M. B., Barlow, W. E., Ballard-Barbash, R., & Sickles, E. A. (2002). Concordance of breast imaging reporting and data system assessments and management recommendations in screening mammography. *Radiology*, *222*(2), 529–535. Available from https://doi.org/10.1148/radiol.2222010647.

The mammographic image analysis society digital mammogram database – ScienceOpen. (2021). https://www.scienceopen.com/document?vid = 119a594e-fa40-4bf8-9eac-950ceddb7ed1.

Thompson, A. C. (1996). Cambridge core minkowski geometry. https://www.cambridge.org/core/books/minkowski-geometry/BEB8FE99553CABD2BECD623887C879B8.

Trop, I., & Deck, W. (2006). Should women 40 to 49 years of age be offered mammographic screening? *Canadian family physician Médecin de famille canadien*, *52*, 1050–1052.

Tzikopoulos, S. D., Mavroforakis, M. E., Georgiou, H. V., Dimitropoulos, N., & Theodoridis, S. (2011). A fully automated scheme for mammographic segmentation and classification based on breast density and asymmetry. *Computer Methods and Programs in Biomedicine*, *102*(1), 47–63. Available from https://doi.org/10.1016/j.cmpb.2010.11.016, https://www.sciencedirect.com/science/article/pii/S0169260710002944.

Understanding mammogram reports | mammogram results. (2021). https://www.cancer.org/cancer/breast-cancer/screening-tests-and-early-detection/mammograms/understanding-your-mammogram-report.html.

Understanding your diagnosis. (2021). https://www.cancer.org/treatment/understanding-your-diagnosis.html.

Ursin, G., Astrahan, M. A., Salane, M., Parisky, Y. R., Pearce, J. G., Daniels, J. R., Pike, M. C., & Spicer, D. V. (1998). The detection of changes in mammographic densities. *Cancer Epidemiology, Biomarkers & Prevention*, *7*(1), 43–47.

Vachon, C. M., van Gils, C. H., Sellers, T. A., Ghosh, K., Pruthi, S., Brandt, K. R., & Pankratz, V. S. (2007). Mammographic density, breast cancer risk and risk prediction. *Breast Cancer Research*, *9*(6), 217. Available from https://doi.org/10.1186/bcr1829, http://breast-cancer-research.biomedcentral.com/articles/10.1186/bcr1829.

Wang, J., Nishikawa, R. M., & Yang, Y. (2017). Global detection approach for clustered microcalcifications in mammograms using a deep learning network. *Journal of Medical Imaging*, *4*(2), 024501. Available from https://doi.org/10.1117/1.JMI.4.2.024501, http://medicalimaging.spiedigitallibrary.org/article.aspx?doi = 10.1117/1.JMI.4.2.024501.

What are the stages of breast cancer? (2021). https://www.webmd.com/breast-cancer/stages-grades-breast-cancer.

What are the symptoms of breast cancer? | CDC. (2021). https://www.cdc.gov/cancer/breast/basic_info/symptoms.htm.

What is breast cancer? (2018). https://www.breastcancer.org/symptoms/understand_bc/what_is_bc.

Wolfe, J., Saftlas, A. F., & Salane, M. (1987). Mammographic parenchymal patterns and quantitative evaluation of mammographic densities: A case-control study. *American Journal of Roentgenology*, *148*(6), 1087–1092. Available from https://doi.org/10.2214/ajr.148.6.1087, https://www.ajronline.org/doi/10.2214/ajr.148.6.1087.

Wolfe, J. N. (1976). Breast patterns as an index of risk for developing breast cancer. *American Journal of Roentgenology*, *126*(6), 1130–1137. Available from https://doi.org/10.2214/ajr.126.6.1130, https://www.ajronline.org/doi/10.2214/ajr.126.6.1130.

Yip, C.-H., Smith, R. A., Anderson, B. O., Miller, A. B., Thomas, D. B., Ang, E.-S., Caffarella, R. S., Corbex, M., Kreps, G. L., & McTiernan, A. (2008). Breast health global initiative early detection panel, guideline implementation for breast healthcare in low- and middle-income countries: Early detection resource allocation. *Cancer*, *113* (8 Suppl), 2244–2256. Available from https://doi.org/10.1002/cncr.23842.

Zheng, Y., Keller, B. M., Ray, S., Wang, Y., Conant, E. F., Gee, J. C., & Kontos, D. (2015). Parenchymal texture analysis in digital mammography: A fully automated pipeline for breast cancer risk assessment. *Medical Physics*, *42*(7), 4149–4160. Available from https://doi.org/10.1118/1.4921996, https://onlinelibrary.wiley.com/doi/abs/10.1118/1.4921996.

Zwiggelaar, R., Astley, S. M., Boggis, C. R. M., & Taylor, C. J. (2004). Linear structures in mammographic images: Detection and classification. *IEEE Transactions on Medical Imaging*, *23*(9), 1077–1086. Available from https://doi.org/10.1109/TMI.2004.828675.

Zwiggelaar, R., Martí, J., Oliver, A., Freixenet, J., & Martí, R. (2010). Local greylevel appearance histogram based texture segmentation. *Digital Mammography*, 175–182. Available from https://doi.org/10.1007/978-3-642-13666-5_24.

CHAPTER

A mathematical resolution in selecting suitable magnetic field-based breast cancer imaging modality: a comparative study on seven diagnostic techniques

7

Ilker Ozsahin[1,2], Natacha Usanase[1,3], Berna Uzun[1,4], Dilber Uzun Ozsahin[1,5,6] and Mubarak Taiwo Mustapha[1,3]

[1]*Operational Research Centre in Healthcare, Near East University, Nicosia/TRNC, Mersin-10, Turkey*

[2]*Brain Health Imaging Institute, Department of Radiology, Weill Cornell Medicine, New York, NY, United States*

[3]*Department of Biomedical Engineering, Near East University, Nicosia/TRNC, Mersin-10, Turkey*

[4]*Department of Mathematics, Near East University, Nicosia/TRNC, Mersin-10, Turkey*

[5]*Department of Medical Diagnostic Imaging, College of Health Science, University of Sharjah, Sharjah, United Arab Emirates*

[6]*Research Institute for Medical and Health Sciences, University of Sharjah, Sharjah, United Arab Emirates*

7.1 Introduction

Breast cancer is one of the main sources of cancer fatality in women and the most prevalent form of malignancy within this population. It is defined as an uncontrolled growth of normal cells that occurs in the milk ducts or milk-supplying lobules of the breast (Ng & Sudharsan, 2004). These malignant cells can invade and destroy nearby normal tissue and spread through the bloodstream or lymphatic system. Several factors can increase the risk of breast cancer: age, inherited gene mutations (such as BRCA1 or BRCA2), hormone therapies, being overweight, alcohol consumption, lack of exercise, and ionizing radiation exposure (Chui et al., 2012; Crystal et al., 2003). Above two million new cases of malignancy were estimated in 2020, and above six hundred thousand cases were estimated as fatalities (Cancer Today, 2022). Furthermore, by the extent a lump becomes evident, cancer has advanced to at least stage 2 cancer (Devi & Anandhamala, 2018). Therefore it is beneficial to do early diagnosis testing to help detect cancer early, when it is most treatable. The early

Artificial Intelligence and Image Processing in Medical Imaging. DOI: https://doi.org/10.1016/B978-0-323-95462-4.00007-8

diagnostic techniques involve breast self-examinations, regular mammograms, and other yearly clinical tests (Friebel et al., 2014; Zhou et al., 2015).

Mammography imaging, which employs low-energy 20−30 keV X-rays, is currently the standard diagnosing approach (Iranmakani et al., 2020). Additionally, magnetic resonance imaging (MRI) is the second technique, although this method has a high sensitivity for cancer detection, it may also pick up false positives; its specificity (true negative) is therefore low. The high cost and prolonged scanning process of MRI make it unsuitable for use as a common breast imaging technique (Iranmakani et al., 2020). However, breast MRI uses low-energy frequency waves and a magnetic field to provide detailed images of the breast's internal tissues in a noninvasive and nonionizing manner (Bhushan et al., 2021). In women who have already received a breast cancer diagnosis, an MRI can be performed to assess the extent of the malignancy and search for spread malignancies. Applying MRI, tumors smaller than or equal to 2 cm in size have been precisely diagnosed and measured (Bhushan et al., 2021). The American College of Radiology has suggested using MRI for several purposes, including high-risk population screening (Niell et al., 2014). As a result, suspicious breast cancers that frequently go undetected by clinical, mammographic, and ultrasound examinations can be detected by MRI (Torrisi et al., 2019).

By adjusting the concepts of truth and falsehood in their most basic forms, fuzzy logic is a computer technique that investigates logical frameworks (Brans & de Smet, 2016; Mareschal, 2005). It is especially well suited to the development of knowledge-based techniques, weighing the effects of products about to be released into the population, and preparing and processing various types of equipment, such as diagnostic tools. Integer, geometric, dynamic, and linear programming are a few of the modeling techniques used in fuzzy logic (Brans & de Smet, 2016; Mareschal, 2005). Moreover, fuzzy logic is combined with various techniques, including preference ranking for organization method for enrichment evaluation (PROMETHEE); fuzzy PROMETHEE; a multicriteria decision-making model, to discover the optimal solution in challenging situations where multiple criteria are to be considered and a critical decision is to be made. Fuzzy PROMETHEE has been used in medicine and is currently being used in different sectors to improve decision-making and develop creative solutions because of its applicability and effectiveness in assessing diverse alternatives with several competing criteria (Ozsahin, Uzun Ozsahin, et al., 2021).

In recent years, several new magnetic field-based breast cancer imaging techniques have been developed, including magnetic resonance spectroscopy (MRS), dynamic contrast-enhanced MRI (DCE-MRI), diffusion-weighted imaging, MR elastography (MRE), ultrafast MRI, and abbreviated MRI. While these techniques show promise, their relative effectiveness and potential clinical utility compared to established imaging modalities such as mammography and MRI are not well understood. In addition, breast cancer is a complex disease, and there is still a significant need for more effective diagnostic techniques. The diagnosis is accomplished by these comprehensive methodologies that largely entail numerous criteria, which can be tricky and risky if done inappropriately (Bezerra et al., 2013). Furthermore, since people with smaller tumors at the moment of diagnosis are at a high mortality risk if misdiagnosed and also have a minimal survival rate if not treated early, early detection of the disease is essential for a good prognosis and efficient treatment (Hamajima et al., 2012). Some of the research gaps that this study is addressing are:

1. Lack of comparative studies: There is a need for more comparative studies that evaluate the efficacy of both the old and the newly developed imaging modalities as well as those improved, especially techniques using magnetic fields.
2. Continuous need for mathematical modeling incorporation in disease diagnosis: Mathematical algorithms are becoming increasingly applied in medical research, as they can provide insights into the underlying mechanisms of diseases and guide the development of new treatments. Therefore there is a need for a study that uses mathematical models to evaluate breast cancer imaging modalities that could help in identifying the strengths and weaknesses of different techniques and provide guidance for the development of new imaging technologies by new inventions or improvement of the existing ones.
3. Understanding the impact of diverse imaging parameters on the final clinical outcome: The efficacy of imaging modalities can be affected by various factors such as sensitivity, specificity, spatial resolution, and imaging time. A study elaborating on the effects that such variables can have in deciding the type of imaging technique to use, mostly in cancer conditions, could help identify the most effective imaging protocols for diagnosis.
4. Patient-specific imaging: Breast cancer is a heterogeneous disease, and different patients may require different imaging procedures depending on the stage of cancer. Therefore a study that evaluates imaging modalities could help identify patient-specific imaging protocols that maximize diagnostic accuracy.

As a result, this study aims to compare seven imaging methods that use a magnetic field to screen tumors, namely, abbreviated MRI, MRS, DW-MRI, DCE-MRI, ultrafast MRI, MRE, and conventional MRI, depending on several criteria using fuzzy PROMETHEE. The novelty of the study lies in its attempt to use a mathematical model to select the most appropriate technique for breast cancer imaging from the new and old techniques that use the power of a magnetic field. Moreover, this decision-making technique can be valuable in easing the decision-making process when identifying the ultimate effective modality for breast cancer diagnosis and could lead to a more accurate and efficient diagnosis.

The remaining part of the study is structured as follows: part two entails the imaging modalities being compared and the proposed mathematical approach. Part three shows and discusses the results from the application of fuzzy PROMETHEE in breast cancer diagnosis. Finally, part four concludes the study.

7.2 Materials and methods

7.2.1 Magnetic field imaging modalities

Magnetic field-based breast cancer imaging techniques are methods used to help in the diagnosis and monitoring of breast cancer, as well as to assess the effectiveness of other treatments. These therapeutic approaches rely on their magnetic properties to screen and detect tumors. Several different techniques are going to be explained in this section.

7.2.1.1 Magnetic resonance imaging

MRI is a diagnostic approach that creates precise images of bones, soft tissues, and organs using a strong magnetic field and radiofrequency pulses (Devi & Anandhamala, 2018). The American Cancer Society (ACS) recommends using an MRI to diagnose breast cancer since it prevents high-risk women from dangerous radiation. However, this type of diagnosis is more costly than other imaging technologies, increases the likelihood of false positive results (Maxwell et al., 2017), lacks selectivity, and only when combined with other screening methods can it identify a probable lesion with sufficient specificity (Devi & Anandhamala, 2018). Additionally, this technique relies on the magnetic properties of certain atomic nuclei to produce high-resolution images of the body. Conventional MRI machines use a large, cylinder-shaped magnet to generate a strong, static magnetic field. The patient is placed inside the magnet and exposed to a series of radiofrequency pulses, which cause the atomic nuclei within the body to emit a faint radio signal (Arrigoni et al., 2018). This signal is detected by sensors that are placed around the area being imaged. The data collected by the sensors is then processed by a computer to generate a well-detailed image of the body. MRI is commonly used to diagnose and monitor a wide spectrum of diseases such as breast cancer, an example is illustrated in Fig. 7.1, and brain as well as spinal cord injuries, cardiovascular disease, cancer, and joint and muscle disorders (Arrigoni et al., 2018). Though it has the ability to detect small lesions, it also has pitfalls such as long acquisition time, high cost, and high rate of false positives.

7.2.1.2 Abbreviated magnetic resonance imaging

Abbreviated MRI is a diagnostic methodology with a short imaging duration than a traditional MRI scan (Gao & Heller, 2020). It is typically used to assess a specific area of the body and may be recommended if a full-body MRI scan is not necessary (Sheth & Abe, 2017). Abbreviated MRI is generally considered to be safe and well-tolerated by most patients. When compared to standard mammography and conventional MRI, abbreviated MRI has a faster acquisition time, lower cost, and lower false positive rate. However, it has lower sensitivity. Also, as with any medical procedure, there are some risks, such as allergic reactions to the contrast agent if it is used and other adverse effects associated with this method (Sheth & Abe, 2017). Furthermore, a general standardized abbreviated MRI guideline is yet to be established. An example of abbreviated MRI protocol is shown in Fig. 7.2.

7.2.1.3 Ultrafast magnetic resonance imaging

Ultrafast MRI refers to a type of MRI scan that can capture images of the body more quickly than traditional MRI scanners. This is achieved through the use of advanced imaging techniques and specialized hardware that allows for faster data acquisition and processing (Gao & Heller, 2020). Ultrafast MRI has several

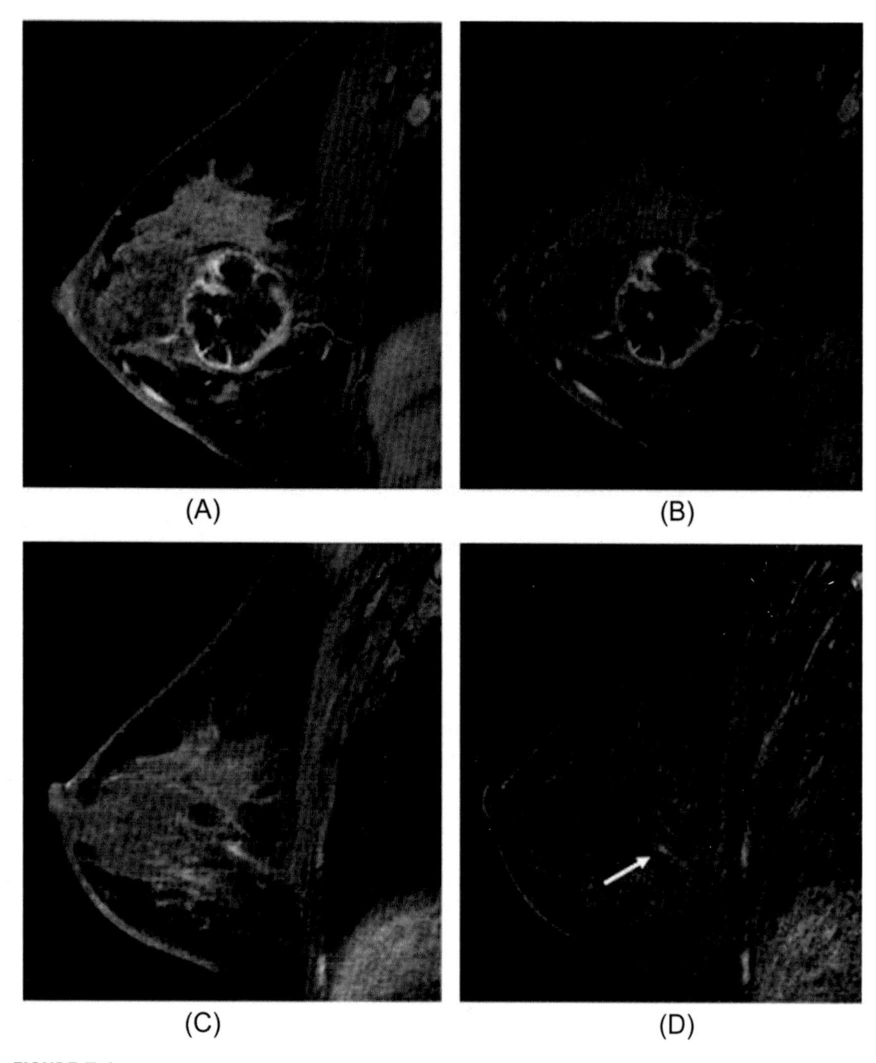

FIGURE 7.1

A pre (A and B) and post (C and D) therapy magnetic resonance imaging scan of a female patient with invasive ductal carcinoma.

From Mann, R. M., Cho, N., & Moy, L. (2019). Breast MRI: State of the art. Radiology, 292(3), 520–536. https://doi.org/10.1148/radiol.2019182947.

advantages over traditional MRI. For one, it allows for faster scans, which can be particularly beneficial for patients who are anxious or claustrophobic or for those who need to undergo multiple scans in a short period. Ultrafast MRI also produces high-quality images, like the one seen in Fig. 7.3, compared to those

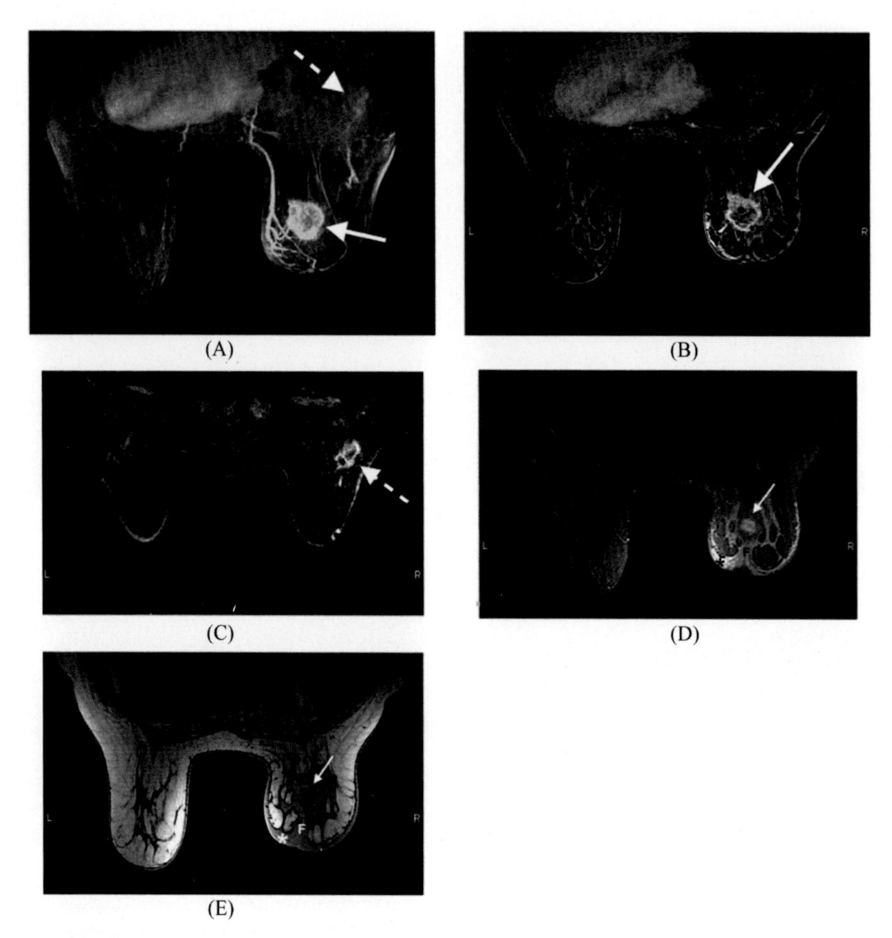

FIGURE 7.2

A pictorial protocol sample of abbreviated magnetic resonance imaging. (A) Maximum intensity projection image illustrates a 3.8 × 3.2 × 4.6 cm rim enhancing centrally necrotic mass in the right breast (arrow). (B) Contrast-enhanced axial T1-weighted fat-suppressed image demonstrates the same rim-enhancing mass in the right breast (arrow). (C) Contrast-enhanced axial T1-weighted fat-suppressed image demonstrates an irregular 2.8 × 2.0 cm spiculated mass, compatible with metastatic lymph nodes (dashed arrow). (D) Contrast-enhanced axial T2-weighted fat-suppressed image. (E) Non-contrast-enhanced T1-weighted axial image without fat suppression.

From Houser, M., Barreto, D., Mehta, A., & Brem, R. F. (2021). Current and future directions of breast MRI. Journal of Clinical Medicine, 10(23). https://doi.org/10.3390/jcm10235668.

produced by traditional MRI scanners (Onishi et al., 2020). Additionally, ultrafast MRI is often more comfortable for patients, as it produces less noise and vibration during the scan. Ultrafast MRI is used to diagnose a wide range of medical

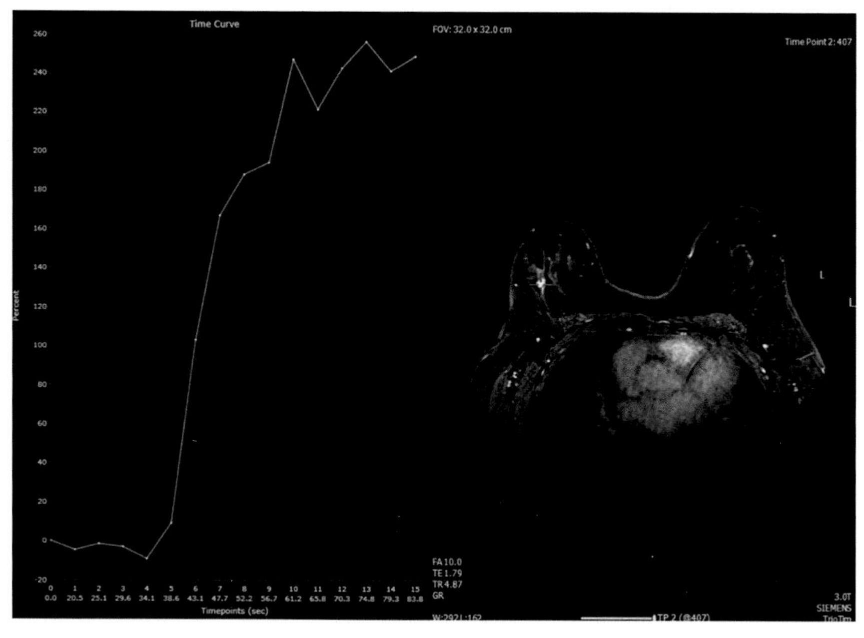

FIGURE 7.3

An ultrafast magnetic resonance imaging axial postcontrast T1-weighted image of invasive ductal carcinoma for a 54-year-old female patient.

From Gao, Y. & Heller, S. L. (2020). Abbreviated and ultrafast breast MRI in clinical practice. Radiographics, 40(6), 1507–1527. https://doi.org/10.1148/rg0.2020200006.

conditions, including injuries, diseases, and abnormalities of organs, bones, and tissues. It is particularly useful for imaging the breast, the brain, the heart, and other organs that require rapid motion or high-resolution images. Ultrafast MRI is also used in research to study the function and structure of the body's tissues and organs in greater detail (Gao & Heller, 2020).

7.2.1.4 Dynamic contrast-enhanced magnetic resonance imaging

Dynamic contrast agent-enhanced breast MRI is an MRI test that involves contrast agents to enhance the visibility of certain tissues or structures within the body (Rahbar & Partridge, 2016). This agent is typically administered intravenously and consists of a paramagnetic substance that can be detected by the MRI scanner (Rahbar & Partridge, 2016). DCE-MRI is commonly used in medical imaging to visualize the vasculature and perfusion in various organs and tissues. It can also be used to evaluate the effectiveness of diverse disease detection, monitoring, and treatments, including cancer, cardiovascular disease, and neurological disorders (Trimboli et al., 2017). DCE-MRI exams are now included in the standard acquisition protocol for the evaluation of breast cancer where physicians can learn the shape and vascularization of the tumor (Trimboli et al., 2017). Contrast kinetics, curve features, and tumor characterization are three important variables to consider while using DCE-MRI in breast cancer as seen in Fig. 7.4. However, this technique has limited specificity and can produce false positives due to benign breast lesions.

7.2.1.5 Magnetic resonance elastography

Information on the mechanical characteristics of tissue in vivo can be obtained using MRE (Ehman et al., 2012). Breast MRE measures the stiffness or elasticity of breast tissues after the application of external stress (Patel et al., 2021). Practically, the stiffness of malignant tumors is mostly above that of benign tumors and normal breast tissues, see Fig. 7.5. Additionally, to ascertain the biomechanical characteristics of each organ's tissue and any changes brought on by disease, MRE is performed (Iranmakani et al., 2020). One of the key components for assessing MRE is adequate shear wave transmission in the target organ. Although the manual mass examination makes the operation easier for the breast, it can be challenging for deeper organs like the pancreas and heart (Mortezazadeh et al., 2020).

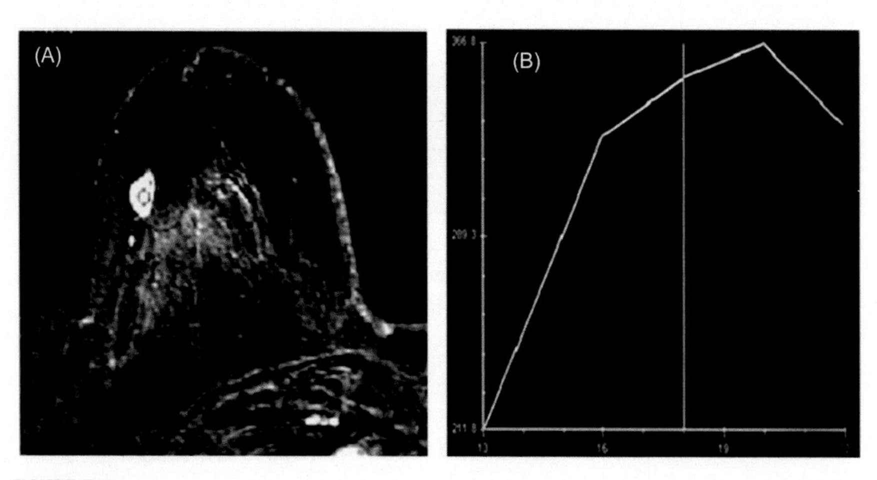

FIGURE 7.4

Dynamic contrast-enhanced magnetic resonance (DCE MR) imaging image of a patient with a determined malignant breast tumor. The image shows homogenously enhanced and hyperintense breast lesions and a plot of the signal intensity vs the number of frames. (A) Transverse DCE MR image of a patient with histopathologically proven malignant breast tumor. (B) Plot of the signal intensity-versus-frame number obtained in the enhanced lesion area.

From Sharma, U., Sah, R. G., & Jagannathan, N. R. (2008). Magnetic Resonance Imaging (MRI) and Spectroscopy (MRS) in Breast Cancer. Magnetic Resonance Insights, 2, MRI.S991. https://doi.org/10.4137/mri.s991.

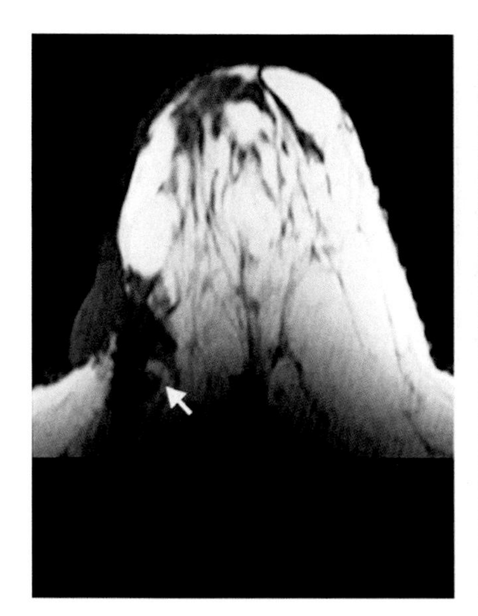

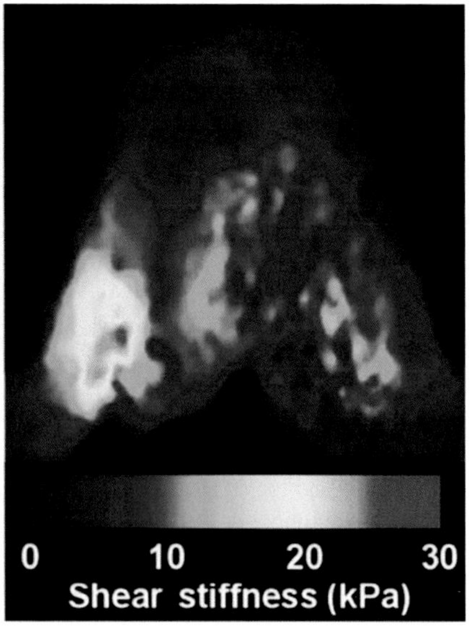

FIGURE 7.5

A magnetic resonance elastography lateral breast T1-weighted image of invasive ductal carcinoma for a 43-year-old female patient.

From Kwee, T. C., Basu, S., Saboury, B., Alavi, A., & Torigian, D. A. (2012). Functional oncoimaging techniques with potential clinical applications. Frontiers in Bioscience - Elite, 4(3), 1081–1096. http://www.bioscience.org/fbs/getfile.php?FileName= /2012/v4e/af/443/443.pdf.

7.2.1.6 Diffusion-weighted imaging

This is a type of unenhanced MRI that addresses some of the limitations of conventional breast MRI by assessing the movement of water molecules within tissues (Fusco et al., 2021). It is often used to evaluate the brain and spine, as well as other organs and tissues (Bhushan et al., 2021). Perfection in differentiating benign from malignant breast lesions as well as the evaluation and forecasting of therapy efficacy are possible advantages of diffusion-weighted imaging (DWI) techniques (Durur-Subasi, 2019). Breast cancer can now be detected using DWI, especially in thick breasts. DWI is quick (scan duration: 120–180 seconds), less expensive than other contrast-enhanced MRI methods, and does not necessitate intravenous gadolinium injection. Depending on Brownian molecular movements, DWI is an extremely sensitive approach, even in micrometer-sized materials. The image is created by the water in the biological tissue (Iranmakani et al., 2020). Because of its increased cellular density, breast cancer typically has restricted water diffusion, which is measured by DWI-MRI as having a low apparent diffusion coefficient (ADC) value, see Fig. 7.6. The distinguishing of breast cancer has been accomplished using ADC values (Fusco et al., 2021). DWI-MRI is a

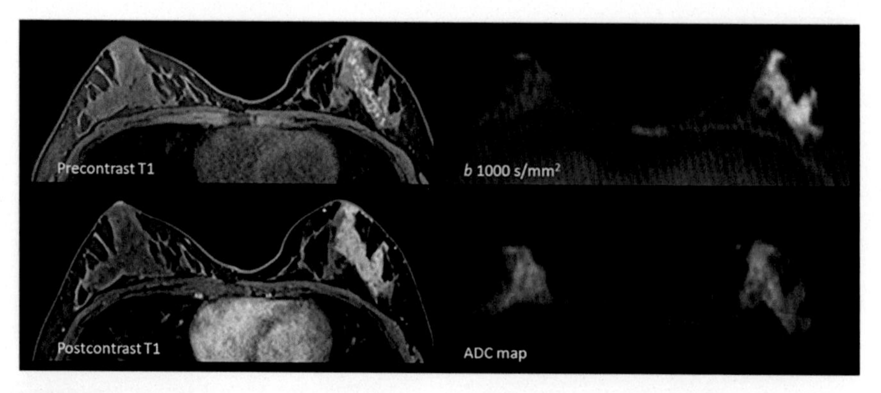

FIGURE 7.6

A diffusion-weighted magnetic resonance imaging image of invasive ductal carcinoma for a 23-year-old patient.

From Durur-Subasi, I. (2019). DW-MRI of the breast: A pictorial review. Insights into Imaging, 10(1). https://doi.org/10.1186/s13244–019–0745–3.

safe and well-tolerated medical imaging technique. It does not expose the patient to ionizing radiation and has few risks or side effects, some of which are claustrophobia where people may feel anxious while in the MRI machine, allergic reactions due to the used contrast agent, and noise produced by the machine during the exam.

7.2.1.7 Magnetic resonance spectroscopy

Body fluids, cell extracts, and tissue samples can be subjected to MRS, a diagnostic approach that measures the chemical makeup of tissues or organs in the body (Bhushan et al., 2021). MRS is often used to identify changes in the levels of certain chemical components such as neurotransmitters or amino acids, which can be indicators of certain medical conditions. It is safe, has no ionizing radiation, and is well-tolerated by many patients. It can detect changes in metabolites associated with cancer such as total choline and lipids, an early therapeutic response marker, and thus can be used for monitoring treatment response as illustrated in Fig. 7.7. Although breast MRS has made significant development over the past 10 years, there are still several issues to be resolved before using this method, such as the optimization of analysis techniques and the complexity of acquisition processes (Bhushan et al., 2021). Combining MRS and MRI can be used to learn more about the chemical composition of breast lesions. Uses for the data generated by this technique include tracking the effectiveness of cancer treatments and enhancing lesion diagnosis precision (Iranmakani et al., 2020).

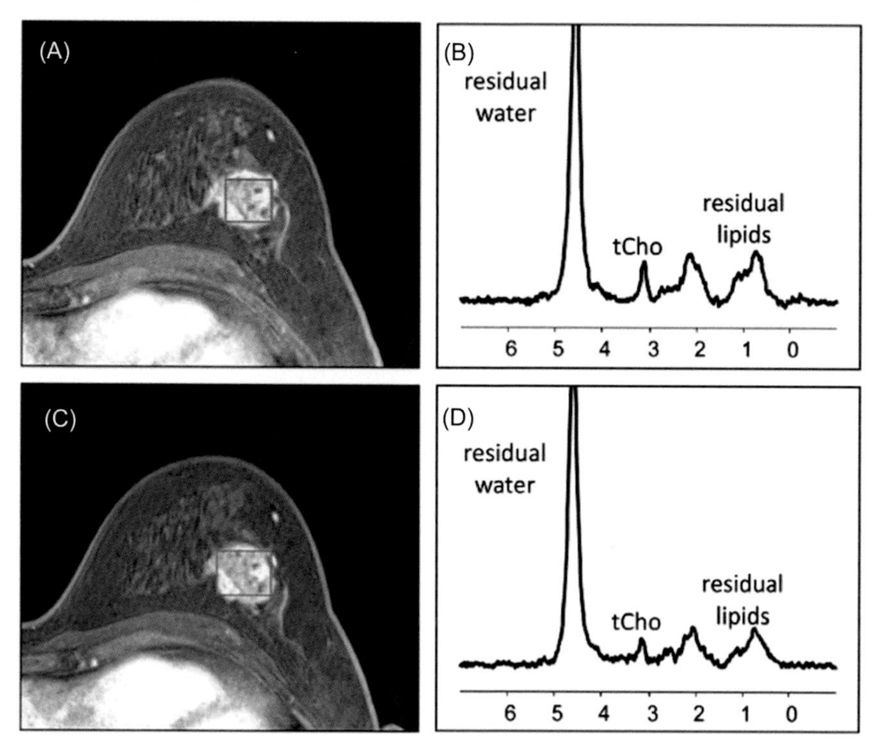

FIGURE 7.7

An example of breast magnetic resonance spectroscopy for pre and postmonitoring of treatment response. (A) Pretreatment postcontrast gradient echo (repetition time [TR]/echo time [TE] 5 4.4/1.5 milliseconds) 1.5 T MR image of a 51-year-old woman with invasive ductal carcinoma. (B) Corresponding single-voxel water- and lipid-suppressed spectrum showing residual water and lipids and a tCho peak at 3.2 ppm, acquired with PRESS (Point-Resolved Spectroscopy), 5.1 mL, TR/TE 5 3000/125 milliseconds, 128 averages, CHESS (Chemical Shift Selective) water suppression, and MEGA/BASING (Mescher-Garwood technique/band-selective inversion with gradient dephasing) lipid suppression. Absolute quantification of tCho using T2-corrected water as internal reference gave [tCho] 5 4.1 mmol/kg. (C) Posttreatment scan on day 1 showing voxel placement. (D) Posttreatment spectrum, acquired with same parameters except voxel size 5.4 mL. Although no anatomic changes are evident, MR spectroscopy quantification showed a decrease in [tCho] to 1.5 mmol/kg. The subject demonstrated pathologic complete response at surgery.

From Fardanesh, R., Marino, M. A., Avendano, D., Leithner, D., Pinker, K., & Thakur, S. B. (2019). Proton MR spectroscopy in the breast: Technical innovations and clinical applications. Journal of Magnetic Resonance Imaging, 50(4), 1033–1046. https://doi.org/10.1002/jmri.26700.s.

7.2.2 Fuzzy preference ranking for organization method for enrichment evaluation

Multi criteria decision making is a field of study that focuses on methods and tools for making decisions that involve multiple conflicting criteria. These criteria may be objective, such as cost or efficiency, or subjective, such as personal preferences or values (Ozsahin, Uzun Ozsahin, et al., 2021). Fuzzy logic is a mathematical system that allows for the representation and manipulation of imprecise or vague information (Brans & de Smet, 2016; Mareschal, 2005). It is often used in combination with traditional Boolean logic to create more flexible and sophisticated decision-making systems (Brans & de Smet, 2016; Mareschal, 2005). It can also be used to make approximate or probabilistic inferences based on imprecise or incomplete data and to handle complex, real-world situations where the boundaries between different categories are not clear-cut (Brans & de Smet, 2016; Mareschal, 2005).

PROMETHEE is a method that can be used in both individual and group decision-making situations. It is based on the principle of outranking, which means that an alternative is considered superior to another alternative if it is better on at least one and/or many criteria. PROMETHEE has applications in a wide range of sectors, including medicine, technology, ecology, energy use, water management, agribusiness, the educational system, and finance (Balcioglu et al., 2023; Behzadian et al., 2010; Chen et al., 2011; Usanase et al., 2022; Uzun et al., n.d.). Fuzzy PROMETHEE is a variant of the PROMETHEE method that incorporates the use of fuzzy logic. Fuzzy PROMETHEE is often used in situations where it is difficult to quantify the importance of each criterion or where the criteria are imprecise values. It allows for the incorporation of vague preferences and uncertainty in the decision-making process.

7.2.3 Application of fuzzy preference ranking for organization method for enrichment evaluation in imaging diagnosis

When utilizing PROMETHEE, the decision-maker requires two forms of information: firstly, the weights, which are the importance values of the criteria. Weights are assigned to criteria to reflect their relative importance in the decision-making process. This is to give more weight to criteria that are deemed more critical or significant in achieving the desired outcome. Secondly, the determination of the preference function, which considers the positive and negative differences between the performance of two alternatives on each criterion and aggregates these differences using a weighting function that reflects the relative importance of each criterion (Brans & Vincke, 1985). Based on specific conditions, the preference function assigns scores to the ranks of two possibilities and calculates a preference degree from 0 to 1. The preference functions to be employed, the rankings, and the techniques to choose the optimal function for a specific scenario were presented by Brans and Vincke (1985). The main objective of the fuzzy

PROMETHEE method is to create an evaluation of various fuzzy sets. In this study, two scenarios were considered: the first scenario represents the preference of hospital management, and the second scenario represents the patient's preference. Furthermore, a fuzzy scale is defuzzified using the Yager index (Abdullah et al., 2018), and a linguistic scale rule is followed as indicated in Table 7.1.

In this study, the cost, sensitivity, specificity, spatial resolution of the device, the requirement of diffusion of water molecules, the requirement of a gadolinium injection, portability, duration of the treatment, overall market availability, skin rash effects, and cause of headache are all the chosen criteria to be used in the evaluation of magnetic field-based imaging techniques (abbreviated MRI, MRS, DW-MRI, DCE-MRI, ultrafast MRI, MRE, and conventional MRI) in the diagnosis of breast cancer. Following data collection, the Gaussian preference function was used for each of the criteria as shown in Table 7.2 to provide preference for the alternatives depending on the standard deviation of the criteria. The decision lab program was therefore used afterwards.

7.3 Results

Table 7.3 illustrates a complete ranking of magnetic field-based breast cancer imaging modalities by fuzzy PROMETHEE based on the hospital's preference from which abbreviated MRI ranked the first alternative with a net flow of 0.2766 followed by MRS, DW-MRI, DCE-MRI, ultrafast MRI, and MRE having a net flow of 0.0356, −0.0003, −0.0059, −0.0725, −0.0794, respectively. Whereas, conventional MRI ranked seventh with a net flow of −0.1542.

Table 7.1 The linguistic importance fuzzy scale of chosen criteria.

Evaluation scale	Triangular fuzzy scale	Important ratings of criteria for the first scenario	Important ratings of criteria for the second scenario
VH	(0.75,1,1)	Sensitivity, specificity, spatial resolution	Sensitivity, specificity, duration
H	(0.50,0.75,1)	Cost of the device, duration, gadolinium injection	Skin rash, headache, gadolinium injection
M	(0.25,0.50,0.75)	Skin rash, headache	Diffusion of water molecules
L	(0,0.25,0.50)	Portability, availability	Portability, availability, spatial resolution,
VL	(0,0,0.25)	Diffusion of water molecules	Cost of the device

Table 7.2 PROMETHEE application for the magnetic field-based breast cancer imaging modalities.

Criteria	Cost of device	Sensitivity	Specificity	Spatial resolution	Diffusion of the water molecule	Requires injection	Portable	Duration	Availability	Skin rash	Headache
Units	($)	(%)	(%)	(μm)	–	–	–	Minutes	–	–	–
Preference											
(min/max)	Min	Max	Max	Min	Max	Min	Max	Min	Max	Min	Min
Preference Function	Gaussian	Gaussian	Gaussian	Gaussian	Gaussian	Gaussian	Gaussian	Gaussian	Gaussian	Gaussian	Gaussian
Evaluations											
MRI	150,000	75–100	83–98.4	25–100	No	Yes	No	30–60	Highly available	Yes	Yes
MRS	35,000–150,000	93	70	25	No	Yes	Yes	45–60	Available	No	No
DCE-MRI	32,000	89–99	37–86	25–100	No	Yes	Yes	10–20	Available	Yes	Yes
DWI	100,000	83	84	25–100	Yes	No	No	2–3	Highly available	No	No
MRE	30,000–125,000	90–100	37–80	25–100	No	Yes	Yes	21–38	Available	No	No
Ultrafast MRI	110,000	92.4	78.2	50–200	No	Yes	No	2	Less available	Yes	Yes
Abbreviated MRI	500	91.4	76.2	35–75	No	Yes	No	<10	Available	No	No

Table 7.3 The complete ranking depends on the hospital's preference.

Rank	Imaging modalities	Net flow	Positive outranking flow	Negative outranking flow
1st	Abbreviated MRI	0.2766	0.4184	0.1418
2nd	MRS	0.0356	0.3086	0.273
3rd	DW-MRI	−0.0003	0.2884	0.2887
4th	DCE-MRI	−0.0059	0.2459	0.2517
5th	Ultrafast MRI	−0.0725	0.237	0.3095
6th	MRE	−0.0794	0.2182	0.2976
7th	Conventional MRI	−0.1542	0.2081	0.3623

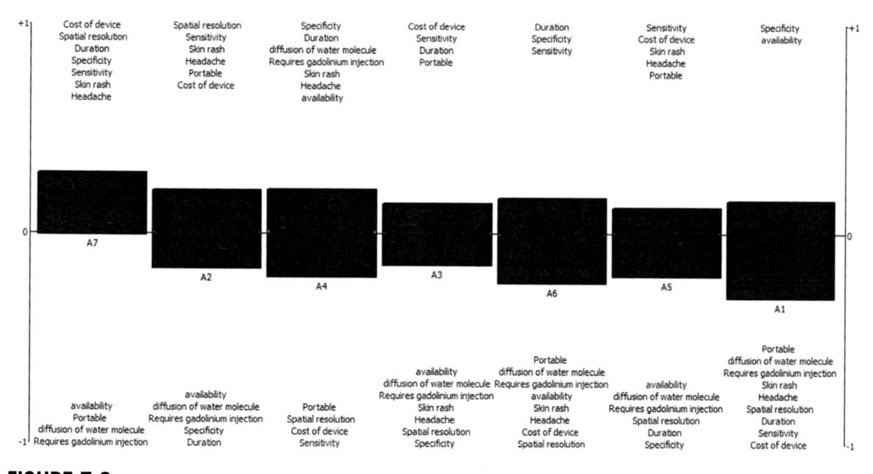

FIGURE 7.8

A rainbow representation for fuzzy PROMETHEE evaluation results of the first scenario (A1: conventional MRI, A2: MRS, A3: DCE-MRI, A4: DW-MRI, A5: MRE, A6: ultrafast, A7: abbreviated MRI).

MRS, magnetic resonance spectroscopy; DW-MRI, diffusion-weighted imaging MRI; DCE-MRI, dynamic contrast-enhanced MRI; ultrafast MRI; MRE, magnetic resonance elastography.

Fig. 7.8 shows the positive and negative sides of each alternative that resulted in the above complete ranking.

Table 7.4 shows a complete ranking of magnetic field-based breast cancer imaging modalities by fuzzy PROMETHEE based on the patient's preference from which ultrafast MRI ranked the first alternative with a net flow of 0.1223 followed by abbreviated MRI, DW-MRI, DCE-MRI, conventional MRI, and MRE having a net

Table 7.4 The complete ranking based on the patient's preference.

Rank	Imaging modalities	Net flow	Positive outranking flow	Negative outranking flow
1st	Ultrafast MRI	0.1223	0.2436	0.1212
2nd	Abbreviated MRI	0.1159	0.2490	0.1331
3rd	DW-MRI	0.0840	0.2663	0.1823
4th	DCE-MRI	−0.0552	0.1590	0.2143
5th	Conventional MRI	−0.0618	0.1922	0.2540
6th	MRE	−0.1018	0.1449	0.2467
7th	MRS	−0.1034	0.1493	0.2526

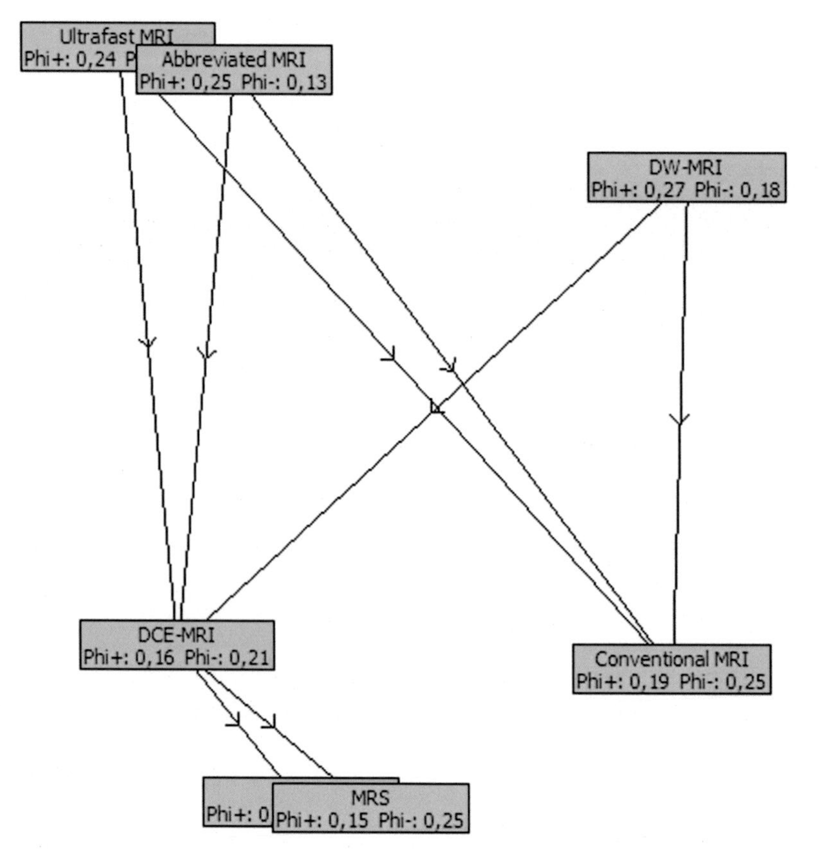

FIGURE 7.9

A network view representation for fuzzy PROMETHEE evaluation results of the second scenario.

flow of 0.1159, 0.0840, −0.0552, −0.0618, −0.1018, respectively. Whereas, MRS ranked seventh with a net flow of −0.1034. A graphical representation of this complete outranking flow among the competing alternatives is found in Fig. 7.9.

7.4 Sensitivity analysis

After obtaining the complete ranking based on the patient's preference, a sensitivity analysis was performed to assess the robustness of the fuzzy PROMETHEE to changes in the criteria weights. The study evaluates the impact of varying the criteria weights using different scenarios and shows that the ranking of the diagnostic techniques varies across different weight configurations, indicating that the results are robust and highly sensitive to changes in the criteria weights. For this analysis, in the current study, the availability criterion in the first scenario was newly assigned a 0.75 wt, which differs from the 0.25 wt that was assigned to it before. Due to the sensitiveness of the fuzzy PROMETHEE, as new weight is assigned, the ranking changes, whereby after running the PROMETHEE program, as shown in Table 7.5, only five rankings have not changed: 1st, 2nd, 3rd, 4th, and 7th ranks. The fifth and sixth rankings were altered. This analysis demonstrates that the fuzzy PROMETHEE method is very sensitive to changes in the weights assigned to the criteria. This means that even small changes in the weightings can significantly alter the rankings of the options. However, the analysis also shows that the applied method is robust and can provide reliable results under different parameter configurations. Therefore, the method can be used with confidence in practical applications involving cancer conditions, where the weighting of criteria may vary depending on the specific context. Furthermore, results show that the availability of magnetic field-based breast cancer imaging devices has a huge impact on the decision to be taken depending on the preference of hospital management since this may affect the cost of the device and the provision of reliable service.

Table 7.5 The complete ranking of the first scenario.

Rank	Alternatives	Phi	Phi +	Phi −
1st	Abbreviated MRI	0.249	0.377	0.128
2nd	MRS	0.024	0.274	0.250
3rd	DW-MRI	0.009	0.265	0.257
4th	DCE-MRI	0.000	0.225	0.225
5th	MRE	−0.069	0.199	0.268
6th	Ultrafast MRI	−0.072	0.214	0.286
7th	Conventional MRI	−0.141	0.189	0.329

7.5 Discussion

Breast cancer is still the leading fatal health condition in females' gender worldwide. From the results of this study, it is highly recommended to use novel MRI techniques because of their huge contribution and fewer side effects on breast cancer patients who are known to have a very weak immune system, which is one of the main reasons why some of the adverse effects of techniques were considered such as headache, skin rash. The availability, the cost of the devices, the duration of the test, the sensitivity and specificity of the imaging devices, the ability of the machine to detect the tumors even in assessing water molecules as well as the requirement of injection, which is one of the main reasons of skin rash and diverse allergies because some people react to these reagents, have all been considered in the current study for a better analysis.

From the demonstration in Fig. 7.8, the results of the current study are in alignment with that of McGraw (2022), which has shown that abbreviated breast MRI is not restricted to dense breast tissue and can depict abnormalities on a mammogram, not only this but also in their research, patients felt comfortable with the whole procedure and the test was cost-effective when compared to traditional techniques. Though MRS is still evolving in the diagnostic field of breast cancer, due to its high sensitivity, minimal cost, and fewer side effects, it is a highly recommended diagnostic test as proven in the study of Katz-Brull et al. (2002), where 83% sensitivity and 85% specificity resulted when MRS was used to diagnose breast cancer. Moreover, recently different stakeholders such as the radiology department, oncology department, surgeons, and artificial intelligence experts are collaborating to enhance the diagnosis of cancer and come up with a better treatment and/or management plan that can lower the continuous elevation of death rates caused by different malignancies (Molla et al., 2021; Mustapha et al., 2022; Ozsahin, Sheshakli, et al., 2021; Precious Onakpojeruo et al., 2022; Tuzkaya et al., 2019; Uzun Ozsahin et al., 2023). This rapid improvement in diagnostic procedures nowadays has reached the imaging sector and tremendous advancements have been found that show better performance of the novel magnetic-based imaging techniques when compared to conventional MRI (Chhetri et al., 2020; Hu & Liu, 2020). As of this study, though MRI can still be considered a better option when contrasted with other traditional imaging modalities, the conventional MRI ranked the least preferable compared to other magnetic-based imaging modalities in the hospital management's preference scenario, first scenario, which is an encouraging result of the continuous development being made in imaging field that is leading to increased specificity, positive predictive value, and distinction of tumor subtypes and can enable more useful applications for breast cancer screening and therapy.

This is the first study to analyze and compare diverse magnetic field-based breast cancer imaging modalities using a mathematical algorithm, fuzzy PROMETHEE. For further research, more criteria and alternatives can be added, this highly depends on how the applied algorithm is sensitive to any small changes. Therefore whether there is an addition or deletion of diverse criteria and alternatives of magnetic field-based breast cancer imaging modalities as well as the weights assigned to the criteria, there is a probability that the findings of this study can change. However, there are also some limitations to this study. The current study is limited by the availability of data since the applied data is not originally from a hospital facility therefore more research incorporating real-life data is encouraged. Finally, the study was not able to cover individual patient factors that may influence the choice of imaging modality, since cancer cases differ in different patients considering the age, the type of cancer, and the severity of cancer as well.

7.6 Conclusion

This study aimed at analyzing and comparing seven mostly used magnetic field-based imaging modalities in the diagnosis of breast cancer using fuzzy PROMETHEE considering two perspectives one on hospital management and the other one on the patient. Many adjustments and improvements have been made over the years to the conventional MRI from which different additional modalities were developed, for correcting the imaging results as well as providing highly sensitive and specific techniques to support the early diagnosis of breast cancer. Of the seven evaluated approaches, abbreviated MRI ranked as the first alternative when the preference of the hospital management was considered while ultrafast MRI ranked as the first alternative when the preference of the patient was considered. Fuzzy PROMETHEE has proven significant potency in evaluating the seven imaging modalities depending on multiple criteria. This mathematical model can serve as a decision tool in breast cancer diagnosis by supporting physicians in giving the right imaging diagnostic test.

References

Abdullah, L., Chan, W., & Afshari A. (2018). Application of PROMETHEE method for green supplier selection: A comparative result based on preference functions. https://doi.org/10.1007/s40092-018-0289-z.

Arrigoni F., Calloni S., Huisman T.A.G.M., Chiapparini L. (2018). Conventional MRI. Handbook of clinical neurology; 154, 219–234. https://doi.org/10.1016/B978-0-444-63956-1.00013-8.

Balcioglu, O., Balcioglu1, O., Usanase2, N., Uzun3, B., Ozsahin4, I., & Ozsahin5, U. (2023). A comparative analysis of DOACs Vs warfarin for venous thromboembolism treatment in renal insufficiency. *Turkish Journal of Vascular Surgery*, *32*, 42–50. Available from https://doi.org/10.9739/tjvs.2022.09.018.

Behzadian, M., Kazemzadeh, R. B., Albadvi, A., & Aghdasi, M. (2010). PROMETHEE: A comprehensive literature review on methodologies and applications. *European Journal of Operational Research, 200*, 198–215. Available from https://doi.org/10.1016/J.EJOR.2009.01.021.

Bezerra, L. A., Oliveira, M. M., Araújo, M. C., Viana, M. J. A., Santos, L. C., Santos, F. G. S., et al. (2013). Infrared imaging for breast cancer detection with proper selection of properties: From acquisition protocol to numerical simulation. *Multimodality Breast Imaging: Diagnosis and Treatment*, 285–332. Available from https://doi.org/10.1117/3.1000499.CH11.

Bhushan, A., Gonsalves, A., & Menon, J. U. (2021). Current state of breast cancer diagnosis, treatment, and theranostics. *Pharmaceutics*, 13. Available from https://doi.org/10.3390/pharmaceutics13050723.

Brans, J. P., & de Smet, Y. (2016). PROMETHEE methods. *International Series in Operations Research and Management Science, 233*, 187–219. Available from https://doi.org/10.1007/978-1-4939-3094-4_6.

Brans, J. P., & Vincke, Ph (1985). Note—A preference ranking organisation method. *Management Science, 31*, 647–656. Available from https://doi.org/10.1287/mnsc.31.6.647.

Cancer Today. (2022). https://gco.iarc.fr/today/online-analysis-table? v = 2020&mode = cancer &mode_population = continents &population = 900 &populations = 900 & key = asr&sex = 0&cancer = 39&type = 1&statistic = 5&prevalence = 0&population_group = 0&ages_group%5B%5D = 0 &ages_group % 5B%5D = 17 &group_cancer = 1&include_nmsc = 0 &include_nmsc_other = 1. Accessed 22.12.22.

Chen, Y. H., Wang, T. C., & Wu, C. Y. (2011). Strategic decisions using the fuzzy PROMETHEE for IS outsourcing. *Expert Systems with Applications, 38*, 13216–13222. Available from https://doi.org/10.1016/j.eswa.2011.04.137.

Chhetri, A., Li, X., & Rispoli, J. V. (2020). Current and emerging magnetic resonance-based techniques for breast cancer. *Frontiers in Medicine (Lausanne), 7*, 175. Available from https://doi.org/10.3389/FMED.2020.00175/BIBTEX.

Chui, J. H., Pokrajac, D. D., Maidment, A. D. A., & Bakic, P. R. (2012). Towards breast anatomy simulation using GPUs. *Lecture Notes in Computer Science (Including Subseries Lecture Notes in Artificial Intelligence and Lecture Notes in Bioinformatics), 7361 LNCS*, 506–513. Available from https://doi.org/10.1007/978-3-642-31271-7_65/COVER.

Crystal, P., Strano, S. D., Shcharynski, S., & Koretz, M. J. (2003). Using sonography to screen women with mammographically dense breasts. *AJR. American Journal of Roentgenology, 181*, 177–182. Available from https://doi.org/10.2214/AJR.181.1.1810177.

Devi, R. R., & Anandhamala, G. S. (2018). Recent trends in medical imaging modalities and challenges for diagnosing breast cancer. *Biomedical and Pharmacology Journal, 11*, 1649–1658. Available from https://doi.org/10.13005/bpj/1533.

Durur-Subasi, I. (2019). DW-MRI of the breast: A pictorial review. *Insights Imaging, 10*, 1–9. Available from https://doi.org/10.1186/S13244-019-0745-3/FIGURES/12.

Ehman, R. L., Glaser, K. J., & Manduca, A. (2012). Review of MR elastography applications and recent developments. *Journal of Magnetic Resonance Imaging, 36*, 757–774. Available from https://doi.org/10.1002/JMRI.23597.

Friebel, T. M., Domchek, S. M., & Rebbeck, T. R. (2014). Modifiers of cancer risk in BRCA1 and BRCA2 mutation carriers: Systematic review and meta-analysis. *Journal of the National Cancer Institute, 106*. Available from https://doi.org/10.1093/JNCI/DJU091.

Fusco, R., Granata, V., Raso, M. M., Vallone, P., de Rosa, A. P., Siani, C., et al. (2021). Blood oxygenation level dependent magnetic resonance imaging (MRI), dynamic contrast enhanced mri and diffusion weighted mri for benign and malignant breast cancer discrimination: A preliminary experience. *Cancers (Basel)*, *13*. Available from https://doi.org/10.3390/cancers13102421.

Gao, Y., & Heller, S. L. (2020). Abbreviated and ultrafast breast MRI in clinical practice. *Radiographics: A Review Publication of the Radiological Society of North America, Inc*, *40*, 1507–1527. Available from https://doi.org/10.1148/RG.2020200006/ASSET/IMAGES/LARGE/RG.2020200006.FIG21A.JPEG.

Hamajima, N., Hirose, K., Tajima, K., Rohan, T., Friedenreich, C. M., Calle, E. E., et al. (2012). Menarche, menopause, and breast cancer risk: Individual participant meta-analysis, including 118 964 women with breast cancer from 117 epidemiological studies. *The Lancet Oncology*, *13*, 1141–1151. Available from https://doi.org/10.1016/S1470-2045(12)70425-4.

Hu, Q., & Liu, S. (2020). Progresses of functional magnetic resonance imaging diagnosis in breast cancer. *Yangtze Medicine*, *04*, 85–96. Available from https://doi.org/10.4236/YM.2020.42009.

Iranmakani, S., Mortezazadeh, T., Sajadian, F., Ghaziani, M. F., Ghafari, A., Khezerloo, D., et al. (2020). A review of various modalities in breast imaging: Technical aspects and clinical outcomes. *Egyptian Journal of Radiology and Nuclear Medicine*, *51*, 1–22. Available from https://doi.org/10.1186/S43055-020-00175-5/FIGURES/26.

Katz-Brull, R., Lavin, P. T., & Lenkinski, R. E. (2002). Clinical utility of proton magnetic resonance spectroscopy in characterizing breast lesions. *JNCI: Journal of the National Cancer Institute*, *94*, 1197–1203. Available from https://doi.org/10.1093/JNCI/94.16.1197.

Mareschal B. (2005). Chapter 5: PROMETHEE methods outranking & decision open journal view project PROMETHEE MCDA methods view project.

Maxwell, A. J., Lim, Y. Y., Hurley, E., Evans, D. G., Howell, A., & Gadde, S. (2017). False-negative MRI breast screening in high-risk women. *Clinical Radiology*, *72*, 207–216. Available from https://doi.org/10.1016/J.CRAD.2016.10.020.

McGraw, M. (2022). Abbreviated breast MRI for breast cancer screening. *Oncology Times*, *44*. Available from https://doi.org/10.1097/01.COT.0000818692.76592.A7, 21–21.

Molla, M. U., Giri, B. C., & Biswas, P. (2021). Extended PROMETHEE method with pythagorean fuzzy sets for medical diagnosis problems. *Soft Computing*, *25*, 4503–4512. Available from https://doi.org/10.1007/S00500-020-05458-7.

Mortezazadeh, T., Gholibegloo, E., Riyahi, A. N., Haghgoo, S., Musa, A. E., & Khoobi, M. (2020). Glucosamine Conjugated gadolinium (III) oxide nanoparticles as a novel targeted contrast agent for cancer diagnosis in MRI. *Journal of Biomedical Physics and Engineering*, *10*, 25–38. Available from https://doi.org/10.31661/JBPE.V0I0.1018.

Mustapha, M. T., Ozsahin, D. U., Ozsahin, I., & Uzun, B. (2022). Breast cancer screening based on supervised learning and multi-criteria decision-making. *Diagnostics*, *12*, 1326. Available from https://doi.org/10.3390/DIAGNOSTICS12061326.

Ng, E. Y. K., & Sudharsan, N. M. (2004). Computer simulation in conjunction with medical thermography as an adjunct tool for early detection of breast cancer. *BMC Cancer*, *4*, 1–6. Available from https://doi.org/10.1186/1471-2407-4-17/FIGURES/4.

Niell, B. L., Gavenonis, S. C., Motazedi, T., Chubiz, J. C. ott, Halpern, E. P., Rafferty, E. A., et al. (2014). Auditing a breast MRI practice: Performance measures for screening and diagnostic breast MRI. *Journal of the American College of Radiology*, *11*, 883–889. Available from https://doi.org/10.1016/J.JACR.2014.02.003.

Onishi, N., Sadinski, M., Hughes, M. C., Ko, E. S., Gibbs, P., Gallagher, K. M., et al. (2020). Ultrafast dynamic contrast-enhanced breast MRI may generate prognostic imaging markers of breast cancer. *Breast Cancer Research*, *22*, 1–13. Available from https://doi.org/10.1186/S13058-020-01292-9/FIGURES/8.

Ozsahin, D. U., Sheshakli, S., Kibarer, A. G., Denker, A., & Duwa, B. B. (2021). Analysis of early stage breast cancer treatment techniques. *Applications of Multi-Criteria Decision-Making Theories in Healthcare and Biomedical Engineering*, 71–80. Available from https://doi.org/10.1016/B978-0-12-824086-1.00005-0.

Ozsahin I., Uzun Ozsahin D., Uzun B. (2021). Applications of multi-criteria decision-making theories in healthcare and biomedical engineering.

Patel, B. K., Samreen, N., Zhou, Y., Chen, J., Brandt, K., Ehman, R., et al. (2021). MR elastography of the breast: Evolution of technique, case examples, and future directions. *Clinical Breast Cancer*, *21*, e102–e111. Available from https://doi.org/10.1016/J.CLBC.2020.08.005.

Precious Onakpojeruo E., Uzun B., Uzun Ozsahin D. (2022). Hydrogel-based drug delivery nanoparticles with conventional treatment approaches for cancer tumors; A comparative study using MCDM technique. Available from https://doi.org/10.21203/rs.3.rs-2116197/v1.

Rahbar, H., & Partridge, S. C. (2016). Multiparametric MR imaging of breast cancer. *Magnetic Resonance Imaging Clinics of North America*, *24*, 223–238. Available from https://doi.org/10.1016/J.MRIC.2015.08.012.

Sheth, D., & Abe, H. (2017). Abbreviated MRI and accelerated MRI for screening and diagnosis of breast cancer. *Topics in Magnetic Resonance Imaging: TMRI*, *26*, 183–189. Available from https://doi.org/10.1097/RMR.0000000000000140.

Torrisi, L., Restuccia, N., & Torrisi, A. (2019). Study of gold nanoparticles for mammography diagnostic and radiotherapy improvements. *Reports of Practical Oncology & Radiotherapy*, *24*, 450–457. Available from https://doi.org/10.1016/J.RPOR.2019.07.005.

Trimboli, R. M., Codari, M., Khouri Chalouhi, K., Ioan, I., lo Bue, G., Ottini, A., et al. (2017). Correlation between voxel-wise enhancement parameters on DCE-MRI and pathological prognostic factors in invasive breast cancers. *La Radiologia Medica*, *123*, 91–97. Available from https://doi.org/10.1007/S11547-017-0809-8.

Tuzkaya, G., Sennaroglu, B., Kalender, Z. T., & Mutlu, M. (2019). Hospital service quality evaluation with IVIF-PROMETHEE and a case study. *Socio-economic Planning Sciences*, *68*, 100705. Available from https://doi.org/10.1016/J.SEPS.2019.04.002.

Usanase N., Uzun B., Ozsahin U. (2022). The preference ranking of gold nanoparticle synthesis methods using a multi-criteria decision-making model.

Uzun, O., Hüseyin, G., Berna, U., & James. L. (Eds.) (n.d). D. Professional practice in earth sciences application of multi-criteria decision analysis in environmental and civil engineering.

Uzun Ozsahin, D., Ikechukwu Emegano, D., Uzun, B., & Ozsahin, I. (2023). The systematic review of artificial intelligence applications in breast cancer diagnosis. *Diagnostics*, *13*, 45. Available from https://doi.org/10.3390/DIAGNOSTICS13010045.

Zhou, Y., Chen, J., Li, Q., Huang, W., Lan, H., & Jiang, H. (2015). Association between breastfeeding and breast cancer risk: Evidence from a meta-analysis. *Breastfeeding Medicine: The Official Journal of the Academy of Breastfeeding Medicine*, *10*, 175–182. Available from https://doi.org/10.1089/BFM.2014.0141.

CHAPTER

BI-RADS-based classification of breast cancer mammogram dataset using six stand-alone machine learning algorithms

Ilker Ozsahin[1,2], Berna Uzun[1,3], Mubarak Taiwo Mustapha[1,4], Natacha Usanese[1,4], Melize Yuvali[1,5] and Dilber Uzun Ozsahin[1,6,7]

[1] *Operational Research Centre in Healthcare, Near East University, Nicosia/TRNC, Mersin-10, Turkey*

[2] *Brain Health Imaging Institute, Department of Radiology, Weill Cornell Medicine, New York, NY, United States*

[3] *Department of Mathematics, Near East University, Nicosia/TRNC, Mersin-10, Turkey*

[4] *Department of Biomedical Engineering, Near East University, Nicosia/TRNC, Mersin-10, Turkey*

[5] *Department of Biostatistics, Near East University, Nicosia/TRNC, Mersin-10, Turkey*

[6] *Department of Medical Diagnostic Imaging, College of Health Science, University of Sharjah, Sharjah, United Arab Emirates*

[7] *Research Institute for Medical and Health Sciences, University of Sharjah, Sharjah, United Arab Emirates*

8.1 Introduction

Breast cancer is the most frequent noncutaneous malignancy type in the female gender; one in every eight women in the United States is affected by this type of malignancy. With 287,850 expected new cases, among which 43,250 were expected to die in 2022, breast cancer is still in the leading places among the most lethal female carcinomas worldwide (Abba et al., 2020). This type of malignancy describes a condition where cells develop out of control in the breast organ, both in the female and male gender. In addition, breast tumors are the aberrant growth of cells that take the appearance of lumps, microclassifications, and artificial distortions (Abiodun et al., 2018). More than two million females are diagnosed with this type of malignancy each year, making it the most prevalent noncutaneous cancer in women on a global scale (Applications of Support Vector Machine SVM, 2018). As a result, an early cancer diagnosis is essential for effective treatment and to boost the survival probability (Abiodun et al., 2018).

Artificial Intelligence and Image Processing in Medical Imaging. DOI: https://doi.org/10.1016/B978-0-323-95462-4.00008-X

Different methodologies for medical imaging are frequently performed to diagnose breast cancer and track the development of tumors (Abiodun et al., 2018). Mammography, ultrasound (US), magnetic resonance imaging (MRI), computed tomography (CT), positron emission tomography, and microwave imaging are some of the used techniques (Ara et al., 2021). The most popular and effective diagnostic method for breast cancer is mammography. Breast US is rarely used to diagnose breast cancer because it cannot identify early signs such as microcalcifications (Shaikh et al., 2021). Observational studies indicate a 40% decrease in fatality (Bansal et al., 2022; Boateng & Abaye, 2019) following mammography tests. Regardless of the benefits of imaging systems in breast cancer diagnosis, various radiological and laboratory procedures are still used for final confirmation to make an accurate diagnosis (Abiodun et al., 2018).

Although mammograms remain advantageous in detecting breast cancer in its early stages using low-dose X-rays, as with other imaging systems, there are still drawbacks to mammography, whereby a needle biopsy is ordered for 15 out of every 1000 women who have had a mammogram, and 10–13 of those biopsies will reveal false positives (Boateng et al., 2020). Therefore it is essential to incorporate additional classification techniques into cancer-diagnostic techniques. To do so, medical professionals, in corporation with radiologists, have started diagnosing breast cancer in a very detailed pattern in which two views are recorded for every breast throughout mammography: the mediolateral oblique, which is a side view, and the craniocaudal, which is a top view (Breast Imaging Reporting & Data System and American College of Radiology, n.d.), depending on the breast imaging reporting and data system (BI-RADS), which was developed in 1986 as a system for breast anomalies classification (Broeders et al., 2012). The BI-RADS provides mammography reports that include classifications for the characterization of breast carcinoma stages, enabling uniform breast imaging output (Cancer of the Breast Female—Cancer Stat Facts, 2023). The classes are labeled from 0 to 6, although the seventh group was recently introduced for the approved malignant state (Cheng & Li, 2012).

To provide improved image processing, such as segmentation and identification for diagnosis, picture refining, and feature extraction, an adjustment in the radiology department procedures, mainly the final stage of reporting results, was needed from early ages (Connelly, 2016). As a result, artificial intelligence (AI) approaches, both deep learning (DL) and machine learning (ML) subsets, have been employed in disease diagnosis and treatment for both classification and image processing purposes (Debelee et al., 2020; Elter et al., 2007). This is due to the success of various AI applications in medicine (Favati et al., 2022; Ghali et al., 2020; Ghosh, 2019) and other sectors (Giri & Saravanakumar, 2017; Haruna et al., 2021), which have shown a high impact on the accuracy and high precision of results. Computer-aided diagnosis (CAD) methods have been developed to use automated feature extraction and classification to help doctors to diagnose and spot breast tumors (Hassanipour et al., 2019). Its objectives are effective cancer diagnosis, accurate interpretation, error reduction, and cost-effectiveness. According to a recent study by Trister et al. (Heer et al., 2020), new methods founded on convolutional neural networks (CNNs) enhanced clinical diagnostic performance and boosted breast imaging productivity.

The application of AI technologies makes breast cancer diagnosis more efficient by decreasing mortality rates by 30%–70%, which aids radiologists in their ability to evaluate images and spot anomalies in the earliest phases (Huang et al., 2020). Though AI models are beneficial in the classification of malignancies, there is still a long road in handling vast and complex data and a tremendous decrease in the mortality rates of breast cancer, mainly in the female gender, where it is most prevalent. Moreover, radiologists frequently use mammography tests to search for the most typical indicators of breast tumors, usually recording them into BI-RADS categories. Such assessment mainly aims to standardize and improve the clarity and consistency of the mammographic terminology used by radiologists and referring doctors (Islam & Poly, 2019). They do so by considering all important details such as the mass shape (round, oval, lobular, irregular), margin (circumscribed, microlobulated, obscured, ill-defined, spiculated), the density of the tumor (high, iso, low, fat-containing), as well as the age of the patient, all leading to the severity and category (benign or malignant) of breast cancer. This form of information is reported using BI-RADS. As a result, this study aims at classifying breast cancer depending on the BI-RADS categories and various breast tumor features using six ML algorithms. These algorithms are k-nearest neighbors (KNN), support vector machine (SVM), random forest (RF), artificial neural network (ANN), Naïve Bayes (NB), and logistic regression (LR). The key contributions of this study include but are not limited to the following:

1. Providing a comprehensive comparison of the performance of six ML models in the classification of breast cancer mammogram dataset.
2. Employing the BI-RADS classification system, which is widely used in clinical practice, to evaluate the performance of the applied models.
3. A helpful guide to improving CAD systems for mammography and improving the accuracy of breast cancer diagnosis, thus supporting radiologists and physicians in the early diagnosis of breast cancer and aiding the medical and educational domains.

8.1.1 Related work

8.1.1.1 Application of machine learning algorithms generally in medicine

A comparative analysis of several popular ML algorithms, including KNN, genetic algorithm, SVM, decision tree, and long short-term memory, was conducted in a study by Bansal et al. (2022). The authors evaluated the performance of these algorithms using several metrics, including accuracy, precision, recall, F1-score, and area under the receiver operating characteristic (ROC) curve. They concluded that the ML algorithms' performance varies depending on the dataset and the specific metrics used to evaluate them (Bansal et al., 2022). Additionally, in their study, Shafaf and Malek (2019) provided an insightful overview of the evolution of the application of ML techniques in emergency medicine. The paper highlights the ability of ML methods to enhance the accuracy of diagnosis, prognosis, and treatment decisions in emergency medicine. They discussed various ML algorithms that have been applied in emergency medicine, including decision trees, LR, ANN, and SVM, by providing detailed information on each algorithm, its strengths and limitations, and its specific applications in emergency medicine by highlighting several studies that have used these models to improve the accuracy of diagnosis and prediction of various conditions, such as acute myocardial infarction, sepsis, and acute kidney injury, and also the use of ML approaches in predicting patient outcomes, such as mortality and readmission rates (Bansal et al., 2022). The authors also discussed the challenges and limitations of ML applications in emergency medicine, such as the need for large amounts of high-quality data, the interpretability of the models, and the potential for bias in the data. Thus recommending further research, such as the need for more extensive studies, the development of more interpretable models, and the use of ML algorithms in real-time clinical decision-making (Khalid & Usman, 2021).

In their study, Hassanipour et al. (2019) aimed to contrast the performance of ANN models with LR models for health-related outcomes in trauma patients. Their observation and discussion concluded that ANN models might be a better alternative to LR models for predicting outcomes in trauma patients, especially for predicting nonmortality outcomes in both blunt and penetrating trauma patients (Hassanipour et al., 2019). In addition, Abiodun et al. (2018) covered various aspects of ANN applications in their study, including the types of ANN models, the training algorithms, the optimization techniques, and the applications in different domains, such as image processing, speech recognition, and natural language processing. They also highlighted the challenges and limitations of ANN applications, such as the need for large amounts of training data, the overfitting problem, and the interpretability of the models, as well as the potential future directions of ANN research, such as the development of more efficient algorithms, the incorporation of multidisciplinary domain knowledge into the models, and the integration of ANN with other technologies such as DL and big data (Abiodun et al., 2018). Moreover, in his study, Tyagi (2019) presents a comprehensive review of various image classification techniques used to diagnose neurological disorders of the brain using MRI images. The advantage of early detection and diagnosis of neurological disorders and the role of MRI in the diagnosis process and also a detailed overview of various imaging classifiers such as ANN, SVM, RF, KNN, CNN, and deep belief networks were all covered in this review (Tyagi, 2019).

8.1.1.2 Application of machine learning models in cancer conditions

In their study, Huang et al. (2020) highlighted the potential of AI models in revolutionizing cancer diagnosis and prognosis, particularly in improving accuracy, speed, and efficiency. They discussed different AI approaches that can be used for cancer diagnosis and prognosis and the challenges and limitations that must be addressed for AI to be effectively implemented in clinical practice. These include data quality and quantity, regulatory issues, ethical considerations, and the need for collaboration between different disciplines (Lahoura et al., 2021). Furthermore, Venmathi et al. (2015) review recent research on computer-aided mammography for breast cancer diagnosis. Their study covered various aspects of medical image classification, including the types of features used for classification, the evaluation methodologies, the performance metrics, and the ROC used to evaluate the effectiveness of the classification algorithms. The authors also discussed the challenges and limitations of using computer-aided mammography for breast cancer diagnoses, such as the need for large datasets, the sensitivity and specificity of the classification algorithms, and the interpretation of the results (Langarizadeh & Moghbeli, 2016).

Moreover, Kourou et al. (2015) discussed the potential use of ML algorithms in predicting cancer prognosis and improving patient outcomes. The authors describe the challenges in cancer prognosis, including the heterogeneity of cancer cells and the limitations of traditional statistical models. They argue that ML algorithms can be effective in analyzing large datasets and identifying patterns that can be used to guide cancer prognosis and treatment. Their study provided an overview of various ML algorithms, ANN, NB, SVM, decision tree, and different types of data, including genomics, imaging, and clinical data and their use in cancer conditions (Kourou et al., 2015).

The rest of this work is organized as follows: Part two defines the dataset used, its parameters and originality, detailed information on the applied models and how they have been applied, and part three illustrates the findings and discussion. Lastly, the study is concluded in part four.

8.2 Materials and methods

8.2.1 Description of dataset

Researchers increasingly need datasets to create diagnostic technologies and design, evaluate, and improve computational methods (Mehmood et al., 2021; Menezes et al., 2018). The dataset used in the current study was originally obtained from the Fraunhofer Institute for Integrated Circuits in the image processing and medical engineering department (BMT) (Mustapha et al., 2022). The dataset of features of breast cancer in mammograms was categorized using BI-RADS, including 961 instances, of which 516 are benign cases and 445 are malignant cases. The dataset is composed of five independent variables, which are: the age of the patient ranging from 18 to 96 (mean: 55.78, standard deviation [SDV]: 14.67), the mass shape (round = 1, oval = 2, lobular = 3, irregular = 4), the margin (circumscribed = 1, microlobulated = 2, obscured = 3, ill-defined = 4, speculated = 5), the density of the tumor (high = 1, iso = 2, low = 3, fat-containing = 4), as well as the severity (benign = 0, malignant = 1) of breast cancer, and one dependent variable and the BI-RADS-based assessment as the dependent variable. More on these five attributes are stated in Table 8.1.

Table 8.1 Dependent and independent variables with their significant role in breast cancer classification.

Variables	Significance
Patient's age (Nickson et al., 2012)	Age elevates the probability of breast cancer. The majority of diagnosed breast cancer cases are over the age of 50.
Mass shape (Obaid et al., n.d.)	An irregular shape indicates a higher possibility of malignant tumors, whereas round and oval tumors with defined margins are typically benign.
Margins (Ozsahin et al., 2023)	Focusing on how a lesion's margin interacts with the tissues around it is critical to distinguish cancerous and benign masses.
Breast density (Frequently Asked Questions about Mammography and the USPSTF Recommendations: A Guide for Practitioners and Semantic Scholar, n.d.)	The proportion of fibroglandular tissue to fat in a breast (the denser the breast, the higher the cancer risk).
Severity	The severity of breast cancer highly depends on the features mentioned above; margins, density, and shape. They all contribute to breast cancer categorization into either benign or malignant.
BI-RADS (Broeders et al., 2012)	This classification system simplifies mammographic assessments and standardized breast imaging presentation.
Class 1	Negative
Class 2	Benign
Class 3	Probably benign
Class 4	Malignancy suspicion
Class 5	Highly malignancy suspicion

BI-RADS, *breast imaging reporting and data system.*

Based on the above-mentioned descriptions, the retrieved dataset will classify the cases as either BI-RADS 1, BI-RADS 2, BI-RADS 3, BI-RADS 4, or BI-RADS 5. This classification is what radiologists use in interpreting if there is the presence or absence of breast carcinoma.

8.2.2 Proposed methodology

It is crucial to use different models due to the limitations in understanding how a particular algorithm is more effective in classifying an extensive dataset than the others (Salmi and Rustam, 2019). As a result, selecting a model for a specific dataset may be challenging. This study applied six ML models through the

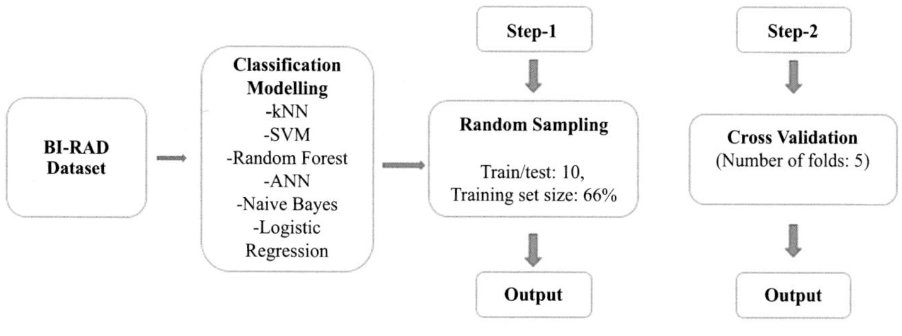

FIGURE 8.1

Classification workflow.

Orange (Version 3–3.34.0) program to classify breast cancer. These chosen algorithms are KNN, SVM, RF, ANN, NB, and LR. Their performance in finding cancerous tissue was tested on a BI-RADS dataset by dividing the available data into six categories.

With the help of the Orange program, the classification process, as shown in Fig. 8.1, is divided into two phases: in the first stage, the data are randomly categorized as training and test data (repeat: 10, training test size: 66%), and the success of the algorithms is determined. In the second stage, the ML algorithm's performance in classification was examined with cross-validation (Number of folds: 5).

8.2.2.1 K-nearest neighbors

A clear and simple way to implement, supervised ML technique known as KNN is applied to perform classification and regression tasks (Elter et al., 2007). The KNN algorithm believes that related variables are in close proximity. As a result, in KNN, the classification of a new data point is based on the k-nearest data points in the training set. The value of k is a hyperparameter that needs to be specified by the user. Therefore the algorithm calculates the distance between the new data point and every data point in the training set and then selects the k data points with the smallest distances. The new data point is then assigned the class label that is most common among its KNN (Elter et al., 2007). This means new data can be quickly and accurately categorized into specified classes by the KNN method. The number of closest neighbors in the consensus phase is indicated by the "K" factor. Parameter tuning, which involves selecting an acceptable value for K, is essential for increased accuracy (Elter et al., 2007).

In their study, Khorshid and Abdulazeez (2021) discussed in-depth use of the KNN model in breast cancer diagnosis by highlighting the application of different feature selection and extraction methods, the impact of choosing a distance metric, and the influence of the value of K in providing the outcome. They also emphasized the benefits and drawbacks of using KNN for breast cancer diagnosis by comparing its performance with other classifiers (Seyer Cagatan et al., 2022). The key advantage of KNN lies in its simplicity and interpretability. Although, its performance is highly dependent on the choice of k and the distance metric used to calculate the distances between data points. KNN also requires a lot of storage space to store the entire training set, and its prediction time can be slow for large datasets. Despite its limitations, KNN is still widely used in practice, particularly in cases where the number of features is relatively small and the dataset size is manageable (Elter et al., 2007).

8.2.2.2 Support vector machine

In 1995, Vapnik developed the SVM concept (Shafaf & Malek, 2019). Regression, pattern recognition, classification, and prediction problems were the focus of the proposed methodology. The idea behind SVM is comparable to earlier ML methods that compelled the construction of data-driven approaches (Shafaf & Malek, 2019). Structural risk optimization and statistical learning concepts are the two fundamental purposes of SVM. It offers unique characteristics that separate it from other ML algorithms like ANN, including the capacity to reduce data complexity, inaccuracy, and redundancy while enhancing performance (Shahid et al., 2019). It is generally utilized in classification rather than regression, which divides the data into two or multiple categories (Shahid et al., 2019). SVM is a supervised learning algorithm that finds the best possible boundary or hyperplane that separates the data points into different classes. This algorithm is popular because it handles nonlinear data using a kernel trick. This technique allows it to convert the original data into a higher dimensional space where it can be easily separated. The basic idea of SVM is to get the maximum margin hyperplane (i.e., the line that separates the classes with the largest possible margin) in a high-dimensional feature space. This hyperplane can then classify new data points (Shahid et al., 2019).

There are several types of SVM models, including linear SVM, which finds the hyperplane that best splits the data points, nonlinear SVM, which applies a kernel function to convert the data into a higher dimensional space where it can be more easily separated, support vector regression which is used for regression analysis, where the algorithm finds a hyperplane that best fits the data points, one-class SVM, which is used for anomaly detection, where the algorithm learns the characteristics of normal data points and can then identify outliers. SVM models have several advantages over other ML algorithms. They effectively handle high-dimensional data, can handle both linear and nonlinear data, and are relatively insensitive to outliers. However, SVM models can be sensitive to the choice of kernel function and can be computationally expensive for large datasets (Shahid et al., 2019). In their study, Ghali et al. (2020) covered the basics of SVM, such as the mathematical concepts and the training process, as well as the use of SVM in various aspects of cancer genomics, including cancer diagnosis, prognosis, and drug response prediction by highlighting the advantages of SVM in cancer genomics, such as the capacity to handle high-dimensional data, its ability to handle noise and outliers, and its ability to deal with nonlinear relationships between variables. Additionally, the paper discusses some of the challenges and limitations of SVM in cancer genomics, such as the need for large datasets and the potential for overfitting (Sharma & Mishra, 2022).

8.2.2.3 Random forest

This model is an independent supervised learning technique that builds several decision trees in an ensemble. This model is applied in both classification and regression cases. The RF algorithm classifies every decision tree, and the class with the most votes determines the prediction of the model (see Fig. 8.2) (Elter et al., 2007). In addition, utilizing randomly chosen subsets of the observational data and randomly chosen groups of the predictor variables, the RF model creates several trees. The voting procedure for a classification tree is then used to tally the predictions from the tree ensemble (Elter et al., 2007). It is to be considered that each decision tree in an RF model is trained on a randomly sampled subset of the training data and also uses a randomly selected subset of features. Therefore the output of the RF is the average prediction of all the decision trees (Sheppard et al., 2021).

Furthermore, RF is less prone to overfitting than single decision tree models, especially when the number of trees in the ensemble is large. RF can handle missing data and noisy data without significant loss of performance. It can also provide feature importance measures, which can be useful for feature selection and

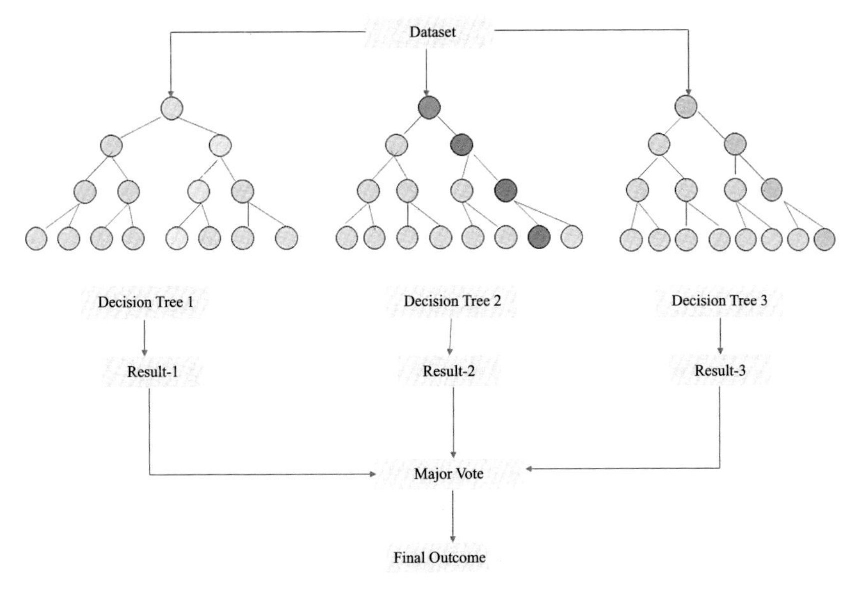

FIGURE 8.2

A pictorial representation of random forest workflow.

understanding the underlying patterns in the data. Overall, RF is a versatile and robust algorithm applied to a wide range of ML applications, specifically in clinical applications such as diagnosis and treatment planning (Sheppard et al., 2021).

8.2.2.4 Artificial neural network

This computer-based method simulates how biological neurons function in the human brain using computers to create predictive models for desired parameters (Shipe et al., 2019). They comprise various processing units, or neurons, that are organized into layers coupled by adjustable weights and biases. Every neuron collects input from other neurons and applies an activation function to produce an output propagated to the next layer. An input, hidden, and output layer comprise ANNs (see Fig. 8.3), which can be single or multisystems (Tan et al., 2021). Furthermore, training ANNs involve adjusting the weights of the connections between neurons to reduce the error between the predicted outputs and the actual outputs, which is typically through backpropagation. This process involves calculating the gradient of the error concerning the weights and adjusting the weights accordingly using an optimization technique such as stochastic gradient descent (Tayyebi et al., 2019). Thus, the uniqueness of ANNs comes from their exceptional data processing characteristics, mostly connected to nonlinearity, fault and noise tolerance, learning, and generalized capacities (Tayyebi et al., 2019).

There are several types of ANNs, namely feedforward neural networks (FNNs), recurrent neural networks (RNNs), CNNs, and deep neural networks (DNNs). Each type has its characteristics and applications (Bassett & Conner, 2003). The simplest ANNs are FNNs, employed for tasks like classification and regression. They consist of an input layer, one or more hidden layers, and an output layer, and the information flows only in one direction (Tayyebi et al., 2019). RNNs are used in applications that involve sequences, such as natural language

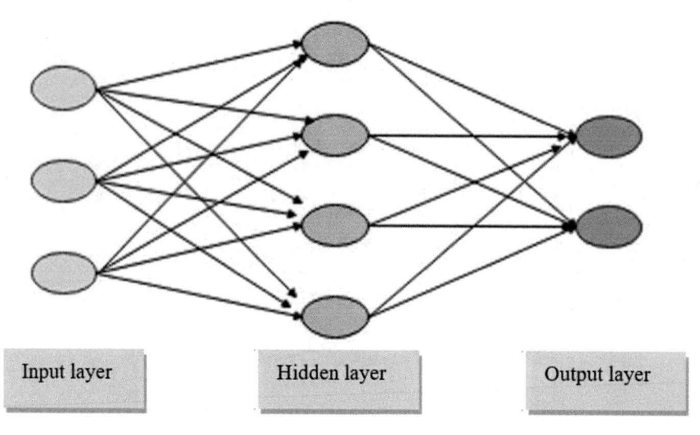

FIGURE 8.3

A pictorial representation of the workflow of artificial neural network.

processing and audio recognition. They have connections that allow the output of a neuron to be fed back as input to the same neuron or another neuron in the same or a previous layer. CNN is employed in image and video processing tasks (Tayyebi et al., 2019). They have convolutional layers that apply filters to the input image or video to extract features, followed by pooling layers that downsample the features. DNNs are ANNs with multiple layers, and they have been very successful in many projects, including face and voice recognition, natural language processing, and game-playing (Tayyebi et al., 2019).

8.2.2.5 Naïve Bayes

This is a probabilistic algorithm used for classification and prediction tasks. The fundamental idea behind this classifier is the Bayes theorem that estimates the likelihood of an occurrence happening provided certain past information; this model's theorem is based on the "naive" hypothesis that each set of features will operate independently of one another given the value of the class variable (Elter et al., 2007). Using previously known potential task factors, it calculates the conditional likelihood of an event occurring (The Nature of Statistical Learning Theory and SpringerLink, n.d.). Furthermore, there are different types of NB models, including multinomial NB, which is used for text classification tasks where the input features are the frequency of words or terms in a document; Bernoulli NB, which is similar to multinomial NB but considers the presence or absence of words or terms in a document rather than their frequency, and lastly, Gaussian NB that is used for continuous data, where the input features are assumed to follow a Gaussian distribution (The Nature of Statistical Learning Theory and SpringerLink, n.d.).

In general, the NB model is easy to implement, requires very little training data, can handle continuous and discrete data, is highly scalable, and is unaffected by redundant features. It is also computationally efficient, making it a popular choice for real-world applications such as classifying news, spam filtering, weather forecasting, diagnosing illnesses, and recognizing faces (The Radiology Assistant: Bi-RADS for Mammography and Ultrasound, 2013). In their study, Langarizadeh and Moghbeli (2016) sought to present a thorough review of the use of the NB approach in disease prediction. They systematically reviewed the existing literature on the application of NB. Also, they identified several limitations of NB in disease prediction, such as the need for large data, the potential for overfitting, and the difficulty of incorporating temporal and contextual information. The authors concluded that NB algorithms have the potential to be a valuable tool for disease prediction, but further research was needed to address the limitations and validate the performance of the NB model in real-world settings (Trister et al., 2017).

8.2.2.6 Logistic regression

Binary classification uses supervised learning techniques such as LR (Elter et al., 2007). This model can determine whether cancer is benign or malignant. As in linear regression, a "logistic function" in the shape of an "S" is used to fit the line to the data rather than a straight line (Elter et al., 2007). LR can produce probabilities and categorize fresh samples using continuous and discrete data, making it a powerful ML technique. The idea of "maximum likelihood," which selects the best fit of a distribution to the data, is the foundation for its operation. Depending on a set of independent attributes, LR determines the likelihood of a specific outcome, whereby mostly the value of the dependent variable ranges between 0 and 1, resulting in a probability (Tyagi, 2019). It is easier to manipulate data when it is fitted with a distribution pattern, either normal, exponential, or gamma (Tyagi, 2019). Predictive modeling and classification typically employ this kind of computational approach.

LR models are widely used in various fields, including finance, marketing, and healthcare (Tyagi, 2019; Usman et al., 2022). Some of the key features and applications of LR models include binary classification, where the output variable is either 0 or 1 and linear separability, whereby the LR model assumes that the input data is linearly separable, which means that it can be divided into two classes using a linear decision boundary (Uzun Ozsahin, Balcioglu et al., 2022). Furthermore, unlike other classification models such as decision trees or RF, LR models provide a probability score for each prediction, allowing for a more nuanced evaluation of the model's performance. LR model is relatively easy to interpret, as the model's coefficients represent the impact of every input attribute on the target. This model can also be regularized to prevent overfitting and improve generalization performance (Uzun Ozsahin, Balcioglu et al., 2022).

8.3 Results

8.3.1 Statistical analysis of the dataset

Descriptive analysis is a statistical method commonly used in scientific research to summarize and describe the main characteristics of a dataset. It involves using different tools and techniques to organize, summarize, and present data meaningfully so that patterns and relationships within the data can be easily identified and interpreted. This form of analysis is an important step in research because it helps researchers better understand their data, which in turn can lead to more informed decisions about subsequent analysis and interpretation. In this study, the need for a descriptive analysis was basically to identify the characteristics of the applied dataset and understand the distribution of the data by providing statistical measures such as mean, mode, SDV, sample variance, range, and other measures that help in understanding the central tendency, variability, and spread of the data as seen in Table 8.2. Descriptive analysis also helps identify any outliers, missing values, and data imbalances that can affect the effectiveness of the ML approaches.

Furthermore, correlation analysis is a statistical method to inspect the relationship between two or more factors in a scientific study. The analysis allows

Table 8.2 Descriptive analysis of the used dataset.

	Mean	Mode	SDV	SV	Kurtosis	Skewness	Range	Min	Max
Age	55.78	67.00	14.67	215.26	−0.31	−0.22	78.00	18.00	96.00
Shape	2.78	4.00	1.24	1.54	−1.57	−0.30	3.00	1.00	4.00
Margin	2.81	1.00	1.57	2.46	−1.63	−0.05	4.00	1.00	5.00
Density	2.92	3.00	0.35	0.12	12.90	−3.07	3.00	1.00	4.00
Severity	0.49	0.00	0.50	0.25	−2.00	0.06	1.00	0.00	1.00

SDV, *standard deviation;* SV, *simple variance.*

researchers to determine whether two variables are related and how strongly they are related. The first step in correlation analysis is to identify the variables to be analyzed. These variables can be measured in different ways such as numerical, categorical, or ordinal scales. Once the variables are identified, mathematical formulas are applied to calculate the correlation coefficient, which measures the strength and direction of the relationship between the two parameters. When conducting correlation analysis, it is important to consider the sample size and significance level. As a result, a larger sample size generally provides more accurate results, while a smaller sample size may lead to unreliable results. Results were analyzed once the correlation coefficient was computed. Whether the positive or negative correlation coefficient will affect how this interpretation is made. A positive value means that the two variables are strongly related, which means that as one variable rises, the other rises. A negative value means that the two parameters are inversely correlated, which means that as one variable increases, the other variable falls. If the coefficient is 0, there is no correlation between the two elements.

Because AI and other computational methods incorporate additional parameters to fine-tune models, they are now seen as more robust and more complicated. Therefore the uncertainty of the results obtained from the model variance increases with increasing complexity. As mentioned above, a statistical evaluation has been conducted through the correlation method to analyze the correlation between the independent and dependent attributes considered in this study. From the results of the correlation analysis shown in Table 8.3, all five independent variables (patient age, shape, margin, density, severity) have a weak positive or direct correlation to the dependent variable (BI-RADS); however, only the correlation relationship between margin and shape was higher/stronger with a correlation coefficient of 0.738 compared to the correlation coefficient between severity and shape, as well as severity and margin, which was moderate (0.565, 0.574, respectively).

Table 8.3 The correlation analysis between independent and dependent variables.

	BI-RADS category	Age	Shape	Margin	Density	Severity
BI-RADS category	1					
Patient age	0.095	1				
Shape	0.180	0.380	1			
Margin	0.158	0.421	0.738	1		
Density	0.028	0.052	0.074	0.124	1	
Severity	0.223	0.455	0.565	0.574	0.069	1

BI-RADS, *breast imaging reporting and data system.*

8.3.2 The classification results of proposed methodology

Table 8.4 shows the effectiveness of this study's six different ML approaches. According to the area under the curve (AUC), LR surpassed other ML models with an AUC of 84.6%; the findings have also shown the high performance of ANN (84.1%) and Naïve Bayes (83.9%) while SVM (82.7%), RF (82.1%), and KNN (80.7%) models showed lower performance. The classification accuracy (CA) results show that SVM (80.2%) had the highest accuracy rate compared to the other five algorithms. ANN, LR, Naïve Bayes, RF, and KNN showed less accuracy in breast cancer classification as the CA results showed 79.9%, 79.8%, 77.1%, 75.8%, and 75.4%, respectively.

Considering the F1 results, defined as the harmonic mean of sensitivity and recall, the proposed ML models achieved high success rates close to each other, falling in a range of 0.73 to 0.78. Furthermore, in Table 8.5 below, we classified six different ML algorithms using the data at hand with cross-validation without any commands. According to the results of this application, LR (85.3%), NB (84.1%), and ANN (84.6%) were successful.

Table 8.4 Random sampling results (average over classes).

Applied ML models	AUC	CA	F1	Precision	Recall
KNN	0.807	0.754	0.735	0.718	0.754
SVM	0.827	0.802	0.781	0.764	0.802
RF	0.821	0.758	0.743	0.729	0.758
ANN	0.841	0.799	0.778	0.763	0.799
NB	0.839	0.771	0.757	0.752	0.771
LR	0.846	0.798	0.776	0.759	0.798

ANN, *artificial neural network;* AUC, *area under the curve;* KNN, *k-nearest neighbor;* LR, *logistic regression;* ML, *machine learning;* NB, *Naïve Bayes;* RF, *random forest;* SVM, *support vector machine.*

Table 8.5 Cross-validation results (number of folds: 5, average over classes).

Applied ML models	AUC	CA	F1	Precision	Recall
KNN	0.798	0.751	0.730	0.711	0.751
SVM	0.837	0.808	0.786	0.768	0.808
RF	0.818	0.760	0.745	0.733	0.760
ANN	0.846	0.805	0.783	0.765	0.805
NB	0.841	0.784	0.768	0.758	0.784
LR	0.853	0.798	0.776	0.756	0.798

ANN, *artificial neural network;* AUC, *area under the curve;* KNN, *k-nearest neighbor;* LR, *logistic regression;* ML, *machine learning;* NB, *Naïve Bayes;* RF, *random forest;* SVM, *support vector machine.*

8.4 Discussion

Breast cancer is a serious problem in the healthcare sector, and early detection is crucial for successful treatment. From the statistical analysis, the correlation analysis between the five considered independent variables and the dependent variable shows that all variables have a weak correlation with BI-RADS, with the highest correlation being between margin and shape. This implies that the classification of breast cancer based on these parameters is complex and requires a more sophisticated approach, such as ML. Over the years, ML algorithms have shown promise in improving disease diagnosis and treatment accuracy. This study applied six stand-alone ML models to classify breast cancer in a mammogram dataset based on BI-RADS criteria. The classification results demonstrate that the LR algorithm outperformed the other models regarding AUC. LR had an AUC from 84.6% after random sampling to 85.3% after cross-validation, indicating its superior performance in correctly identifying malignant and benign breast tumors. This result is in line with other research (Uzun Ozsahin, Ikechukwu Emegano, et al., 2022; Venmathi et al., 2015) that demonstrated the efficiency and superior efficacy of LR in detecting breast cancer. The ANN and NB models also showed high performance in breast cancer classification, with AUCs of 84.1% and 83.9% after random sampling changing to 84.6% and 84.1% after cross-validation, respectively.

On the other hand, SVM, RF, and KNN models showed relatively lower performance with AUC values of 82.7%, 82.1%, and 80.7%, respectively. Regarding CA, the SVM model had the highest accuracy rate at 80.2% compared to other algorithms. However, its performance in terms of AUC was relatively lower than LR, ANN, and Naïve Bayes. This finding suggests that SVM may be better suited for situations where CA is more important than overall predictive power. On the other hand, RF and KNN models showed lower performance in both AUC and CA. Therefore these results suggest that applying ML models to improve the accuracy of breast cancer classification is crucial in ensuring timely and appropriate medical interventions.

As seen in this work, success in applying ML algorithms in breast cancer classification was observed. It is practically observed that ML algorithms have high performance, which renders their broad applicability in classifying different diseases (What Are the Risk Factors for Breast Cancer? | CDC, 2023) and many other clinical programs. Additionally, in comparison to a similar study conducted in 2022 (What Does It Mean to Have Dense Breasts? | CDC, 2023) in which classification was made with four different algorithms, and the ML algorithm that gave the best results was RF (AUC: 59%, sensitivity: 56%, specificity: 61%, acc: 58%) (What Does It Mean to Have Dense Breasts? | CDC, 2023), the RF algorithm in the current study showed higher performance with AUC of 82%. In 2019, a classification study was conducted on breast tissue using images with ML approaches, and high success was achieved; it was observed that ML algorithms intervened by radiologists gave more successful results (Wickramasinghe & Kalutarage, 2021). Furthermore, in 2021, Wu et al. applied four different ML algorithms (KNN, NB, decision tree [DT], SVM) to classify breast tissue. The used dataset was composed of 1222 samples (it is essential to use an extensive dataset because ML algorithms learn by an experience, like the human brain). Among the applied algorithms, the SVM algorithm showed superiority in giving highly accurate results (Wu & Hicks, 2021). In addition, Ara et al. (2021) applied ML algorithms for categorizing malignant and benign breast tumors. They aimed to create a structure with classifiers and to develop an automatic diagnosis system for breast prediagnosis. According to their analysis, RF and SVM outperformed LR, KNN, DT, and NB methods (Yuvalı et al., 2022).

Regarding the above-reported comparative literature, the final outcome in this study shows the high performance of the applied models in classifying breast cancer into BI-RADS categories based on the mass shape, margin, density, age of the patient, and severity of cancer. However, this study is limited due to using only six ML models. Therefore further studies are needed to explore the use of larger datasets to validate the performance of these algorithms and apply more ML algorithms or other AI techniques such as DL, to improve breast cancer CA for a better and higher level boosting early breast cancer diagnosis leading to an early treatment plan. Moreover, further research is required to validate the findings of this study and optimize the performance of these applied algorithms in real-world clinical settings.

8.5 Conclusion

It is seen that ML algorithms are still being tested for many vital diseases in the field of health, considering that health problems with fatal consequences such as cancer are vast and have a complex structure; therefore they should be examined together with more than one factor. Today, many programs have been developed in which ML algorithms are applied, and day by day, it is seen that even the small sample in the dataset does not reduce the success of the ML algorithm. As a result, the success seen in the current study of ML algorithms in breast cancer classification based on BI-RADS is promising for the future of medicine. ML algorithms are preferred because they are neither limited to the size of the dataset nor the data type. The dataset can be numerical or image and does not prevent the models from classification.

In conclusion, the study's findings demonstrate the potential of ML algorithms in accurately classifying breast malignancies depending on the BI-RADS parameters. The findings confirm that LR, ANN, and Naïve Bayes algorithms had the best performance, while SVM, RF, and kNN algorithms had a relatively lower performance.

References

Abba, S. I., Usman, A. G., & IŞIK, S. (2020). Simulation for response surface in the HPLC optimization method development using artificial intelligence models: A data-driven approach. *Chemometrics and Intelligent Laboratory Systems*, *201*, 104007. Available from https://doi.org/10.1016/J.CHEMOLAB.2020.104007.

Abiodun, O. I., Jantan, A., Omolara, A. E., Dada, K. V., Mohamed, N. A., & Arshad, H. (2018). State-of-the-art in artificial neural network applications: A survey. *Heliyon*, *4* (11), e00938. Available from https://doi.org/10.1016/j.heliyon.2018.e00938.

Applications of Support Vector Machine (SVM) Learning in cancer genomics. (2018). Cancer Genomics & Proteomics, 15(1). https://doi.org/10.21873/cgp.20063.

Ara, S., Das, A., & Dey, A. (2021). Malignant and benign breast cancer classification using machine learning algorithms. *2021 International Conference on Artificial Intelligence, ICAI, 2021*, 97–101. Available from https://doi.org/10.1109/ICAI52203.2021.9445249.

Bansal, M., Goyal, A., & Choudhary, A. (2022). A comparative analysis of K-nearest neighbor, genetic, support vector machine, decision tree, and long short term memory algorithms in machine learning. *Decision Analytics Journal*, *3*, 100071. Available from https://doi.org/10.1016/j.dajour.2022.100071.

Bassett, L. W., & Conner, K. M. S. (2003). The abnormal mammogram. In R. C. Bast, D. W. Kufe, R. E. Pollock, R. R. Weichselbaum, J. F. Holland, & E. Frei (Eds.), *Holland-Frei cancer medicine* (6th edn). Hamilton (ON): BC Decker. Available from https://www.ncbi.nlm.nih.gov/books/NBK12642/.

Boateng, E. Y., & Abaye, D. A. (2019). A review of the logistic regression model with emphasis on medical research. *Journal of Data Analysis and Information Processing*, *07*(04), 190–207. Available from https://doi.org/10.4236/jdaip.2019.74012.

Boateng, E. Y., Otoo, J., & Abaye, D. A. (2020). Basic tenets of classification algorithms K-nearest-neighbor, support vector machine, random forest and neural network: A review. *Journal of Data Analysis and Information Processing*, *08*(04), 341–357. Available from https://doi.org/10.4236/jdaip.2020.84020.

Breast Imaging Reporting & Data System | American College of Radiology. (n.d.). Retrieved November 21, 2022, from https://www.acr.org/Clinical-Resources/Reporting-and-Data-Systems/Bi-Rads.

Broeders, M., Moss, S., Nystrom, L., Njor, S., Jonsson, H., Paap, E., Massat, N., Duffy, S., Lynge, E., & Paci, E. (2012). The impact of mammographic screening on breast cancer mortality in Europe: A review of observational studies. *Journal of Medical Screening*, *19* (SUPPL. 1), 14–25. Available from https://doi.org/10.1258/JMS.2012.012078/ASSET/IMAGES/LARGE/10.1258_JMS.2012.012078-FIG2.JPEG.

Cancer of the Breast (Female)—Cancer Stat Facts. (2023). SEER. Retrieved April 11, 2023, from https://seer.cancer.gov/statfacts/html/breast.html.

Cheng, L., & Li, X. (2012). Breast imaging reporting and datasystem (BI-RADS) of magnetics resonance imaging: Breast mass. *Gland Surgery*, *1*(1), 624–674. Available from https://doi.org/10.3978/J.ISSN.2227-684X.2012.05.01.

Connelly, R., Gayle, V., & Lambert, P. S. (2016). Statistical modeling of key variables in social survey data analysis. *Methodological Innovations, 9*. Available from https://doi.org/10.1177/2059799116638002.

Debelee, T. G., Schwenker, F., Ibenthal, A., & Yohannes, D. (2020). Survey of deep learning in breast cancer image analysis. *Evolving Systems*, *11*(1), 143–163. Available from https://doi.org/10.1007/S12530-019-09297-2/FIGURES/1.

Elter, M., Schulz-Wendtland, R., & Wittenberg, T. (2007). The prediction of breast cancer biopsy outcomes using two CAD approaches that both emphasize an intelligible decision process: Prediction of breast biopsy outcomes using CAD approaches. *Medical Physics*, *34*(11), 4164–4172. Available from https://doi.org/10.1118/1.2786864.

Favati, B., Borgheresi, R., Giannelli, M., Marini, C., Vani, V., Marfisi, D., Linsalata, S., Moretti, M., Mazzotta, D., & Neri, E. (2022). Radiomic applications on digital breast tomosynthesis of BI-RADS category 4 calcifications sent for vacuum-assisted breast biopsy. *Diagnostics*, *12*(4). Available from https://doi.org/10.3390/DIAGNOSTICS12040771.

Frequently Asked Questions about Mammography and the USPSTF Recommendations: A Guide for Practitioners | Semantic Scholar. (n.d.). Retrieved November 21, 2022, from https://www.semanticscholar.org/paper/Frequently-Asked-Questions-about-Mammography-and-%3A-Berg-Hendrick/38c7972f647f32fd9499dae4a62acda03f951cfe.

Ghali, U. M., Usman, A. G., Chellube, Z. M., Degm, M. A. A., Hoti, K., Umar, H., & Abba, S. I. (2020). Advanced chromatographic technique for performance simulation of anti-Alzheimer agent: An ensemble machine learning approach. *SN Applied Sciences*, *2*(11). Available from https://doi.org/10.1007/s42452-020-03690-2.

Ghosh, A. (2019). Artificial intelligence using open source BI-RADS data exemplifying potential future use. *Journal of the American College of Radiology*, *16*(1), 64–72. Available from https://doi.org/10.1016/J.JACR.2018.09.040.

Giri, P., & Saravanakumar, K. (2017). Breast cancer detection using image processing techniques. *Oriental Journal of Computer Science and Technology*, *10*(2), 391–399. Available from https://doi.org/10.13005/ojcst/10.02.19.

Haruna, S. I., Malami, S. I., Adamu, M., Usman, A. G., Farouk, A., Ali, S. I. A., & Abba, S. I. (2021). Compressive strength of self-compacting concrete modified with rice husk ash and calcium carbide waste modeling: A feasibility of emerging emotional intelligent model (EANN) versus traditional FFNN. *Arabian Journal for Science and Engineering*, *46*(11), 11207–11222. Available from https://doi.org/10.1007/S13369-021-05715-3.

Hassanipour, S., Ghaem, H., Arab-Zozani, M., Seif, M., Fararouei, M., Abdzadeh, E., Sabetian, G., & Paydar, S. (2019). Comparison of artificial neural network and logistic regression models for prediction of outcomes in trauma patients: A systematic review and meta-analysis. *Injury*, *50*(2), 244–250. Available from https://doi.org/10.1016/j.injury.2019.01.007.

Heer, E., Harper, A., Escandor, N., Sung, H., McCormack, V., & Fidler-Benaoudia, M. M. (2020). Global burden and trends in premenopausal and postmenopausal breast cancer: A population-based study. *The Lancet Global Health*, *8*(8), e1027–e1037. Available from https://doi.org/10.1016/S2214-109X(20)30215-1.

Huang, S., Yang, J., Fong, S., & Zhao, Q. (2020). Artificial intelligence in cancer diagnosis and prognosis: Opportunities and challenges. *Cancer Letters*, *471*, 61–71. Available from https://doi.org/10.1016/j.canlet.2019.12.007.

Islam, Md. M., & Poly, T. N. (2019). Machine learning models of breast cancer risk prediction. *[Preprint]. Bioinformatics*. Available from https://doi.org/10.1101/723304.

Khalid, G. M., & Usman, A. G. (2021). Application of data-intelligence algorithms for modeling the compaction performance of new pharmaceutical excipients. *Future Journal of Pharmaceutical Sciences*, *7*(1), 1–11. Available from https://doi.org/10.1186/S43094-021-00183-W.

Khorshid, S. F., & Abdulazeez, A. M. (2021). *Breast cancer diagnosis based on K-nearest neighbors: A review*.

Kourou, K., Exarchos, T. P., Exarchos, K. P., Karamouzis, M. V., & Fotiadis, D. I. (2015). Machine learning applications in cancer prognosis and prediction. *Computational and Structural Biotechnology Journal*, *13*, 8–17. Available from https://doi.org/10.1016/j.csbj.2014.11.005.

Lahoura, V., Singh, H., Aggarwal, A., Sharma, B., Mohammed, M. A., Damaševičius, R., Kadry, S., & Cengiz, K. (2021). Cloud computing-based framework for breast cancer diagnosis using extreme learning machine. *Diagnostics*, *11*(2), 241. Available from https://doi.org/10.3390/DIAGNOSTICS11020241.

Langarizadeh, M., & Moghbeli, F. (2016). Applying naive bayesian networks to disease prediction: A systematic review. *Acta Informatica Medica*, *24*(5), 364. Available from https://doi.org/10.5455/aim.2016.24.364-369.

Mehmood, M., Ayub, E., Ahmad, F., Alruwaili, M., Alrowaili, Z. A., Alanazi, S., Humayun, M., Rizwan, M., Naseem, S., & Alyas, T. (2021). Machine learning enabled early detection of breast cancer by structural analysis of mammograms. *Computers, Materials and Continua*, *67*(1), 641–657. Available from https://doi.org/10.32604/cmc.2021.013774.

Menezes, G. L., Winter-Warnars, G. A., Koekenbier, E. L., Groen, E. J., Verkooijen, H. M., & Pijnappel, R. M. (2018). Simplifying breast imaging reporting and data system classification of mammograms with pure suspicious calcifications. *Journal of Medical Screening*, *25*(2), 82–87. Available from https://doi.org/10.1177/0969141317715281.

Mustapha, M. T., Ozsahin, D. U., Ozsahin, I., & Uzun, B. (2022). Breast cancer screening based on supervised learning and multi-criteria decision-making. *Diagnostics*, *12*(6), 1326. Available from https://doi.org/10.3390/DIAGNOSTICS12061326.

Nickson, C., Mason, K. E., English, D. R., & Kavanagh, A. M. (2012). Mammographic screening and breast cancer mortality: A case–control study and meta-analysis. *Cancer Epidemiology, Biomarkers & Prevention*, *21*(9), 1479–1488. Available from https://doi.org/10.1158/1055-9965.EPI-12-0468.

Obaid, O. I., Mohammed, M. A., Ghani, M. K. A., Mostafa, S. A., & Taha, F. (n.d.). Evaluating the performance of machine learning techniques in the classification of wisconsin breast cancer. International Journal of Engineering.

Ozsahin, D. U., Mustapha, M. T., Uzun, B., Duwa, B., & Ozsahin, I. (2023). Computer-aided detection and classification of monkeypox and chickenpox lesion in human subjects using deep learning framework. *Diagnostics*, *13*(2), 292. Available from https://doi.org/10.3390/DIAGNOSTICS13020292.

Salmi, N., & Rustam, Z. (2019). Naïve bayes classifier models for predicting the colon cancer. IOP Conference Series: *Materials Science and Engineering, 546*(5), 052068. Available from https://doi.org/10.1088/1757-899x/546/5/052068.

Seyer Cagatan, A., Taiwo Mustapha, M., Bagkur, C., Sanlidag, T., & Ozsahin, D. U. (2022). An alternative diagnostic method for C. neoformans: Preliminary results of deep-learning based detection model. *Diagnostics*, *13*(1), 81. Available from https://doi.org/10.3390/DIAGNOSTICS13010081.

Shafaf, N., & Malek, H. (2019). Applications of machine learning approaches in emergency medicine; A review article. *Archives of Academic Emergency Medicine*, *7*(1), 34. PMID:31555764; pmc6732202.

Shahid, N., Rappon, T., & Berta, W. (2019). Applications of artificial neural networks in health care organizational decision-making: A scoping review. *PLoS One*, *14*(2), e0212356. Available from https://doi.org/10.1371/journal.pone.0212356.

Shaikh, K., Krishnan, S., & Thanki, R. (2021). Artificial Intelligence in Breast Cancer Early Detection and Diagnosis. Available from https://doi.org/10.1007/978-3-030-59208-0.

Sharma, A., & Mishra, P. K. (2022). Performance analysis of machine learning based optimized feature selection approaches for breast cancer diagnosis. *International Journal of Information Technology*, *14*(4), 1949–1960. Available from https://doi.org/10.1007/s41870-021-00671-5.

Sheppard, V. B., Sutton, A. L., Hurtado-De-Mendoza, A., He, J., Dahman, B., Edmonds, M. C., Hackney, M. H., & Tadesse, M. G. (2021). Race and patient-reported symptoms in adherence to adjuvant endocrine therapy: A report from the women's hormonal initiation and persistence study. *Cancer Epidemiology, Biomarkers & Prevention : A Publication of the American Association for Cancer Research, Cosponsored by the American Society of Preventive Oncology*, *30*(4), 699–709. Available from https://doi.org/10.1158/1055-9965.EPI-20-0604.

Shipe, M. E., Deppen, S. A., Farjah, F., & Grogan, E. L. (2019). Developing prediction models for clinical use using logistic regression: An overview. *Journal of Thoracic Disease, 11(S4)*, S574–S584. Available from https://doi.org/10.21037/jtd.2019.01.25.

Tan, M., Al-Shabi, M., Chan, W. Y., Thomas, L., Rahmat, K., & Ng, K. H. (2021). Comparison of two-dimensional synthesized mammograms versus original digital mammograms: A quantitative assessment. *Medical and Biological Engineering and Computing*, *59*(2), 355–367. Available from https://doi.org/10.1007/S11517-021-02313-1/TABLES/5.

Tayyebi, S., Hajjar, Z., & Soltanali, S. (2019). A novel modified training of radial basis network: Prediction of conversion and selectivity in 1-hexene dimerization process. *Chemometrics and Intelligent Laboratory Systems*, *190*, 1–9. Available from https://doi.org/10.1016/J.CHEMOLAB.2019.05.005.

The Nature of Statistical Learning Theory | SpringerLink. (n.d.). Retrieved February 9, 2023, from https://link.springer.com/book/10.1007/978-1-4757-2440-0.

The Radiology Assistant: Bi-RADS for Mammography and Ultrasound. (2013). Retrieved November 21, 2022, from https://radiologyassistant.nl/breast/bi-rads/bi-radsfor-mammography-and-ultrasound-2013.

Trister, A. D., Buist, D. S. M., & Lee, C. I. (2017). Will machine learning tip the balance in breast cancer screening? *JAMA Oncology*, *3*(11), 1463–1464. Available from https://doi.org/10.1001/JAMAONCOL.2017.0473.

Tyagi, V. (2019). A review on image classification techniques to classify neurological disorders of brain MRI. *2019 International Conference on Issues and Challenges in Intelligent Computing Techniques (ICICT)*, 1–4. Available from https://doi.org/10.1109/ICICT46931.2019.8977658.

Usman, A. G., Ghali, U. M., Degm, M. A. A., Muhammad, S. M., Hincal, E., Kurya, A. U., Işik, S., Hoti, Q., & Abba, S. I. (2022). Simulation of liver function enzymes as determinants of thyroidism: A novel ensemble machine learning approach. *Bulletin of the National Research Centre*, *46*(1), 1–10. Available from https://doi.org/10.1186/S42269-022-00756-6.

Uzun Ozsahin, D., Balcioglu, O., Usman, A. G., Ikechukwu Emegano, D., Uzun, B., Abba, S. I., Ozsahin, I., Yagdi, T., & Engin, C. (2022). Clinical modelling of RVHF using pre-operative variables: A direct and inverse feature extraction technique. *Diagnostics*, *12*(12), 3061. Available from https://doi.org/10.3390/DIAGNOSTICS12123061.

Uzun Ozsahin, D., Ikechukwu Emegano, D., Uzun, B., & Ozsahin, I. (2022). The systematic review of artificial intelligence applications in breast cancer diagnosis. *Diagnostics*, *13*(1), 45. Available from https://doi.org/10.3390/DIAGNOSTICS13010045.

Venmathi, A. R., Dr, E. N., Ganesh., & Kumaratharan, N. (2015). A review of medical image classification and evaluation methodology for breast cancer diagnosis with computer aided mammography. *International Journal of Applied Engineering Research*, *10*, 30045–30054.

What Are the Risk Factors for Breast Cancer? | CDC. (2023). Retrieved February 6, 2023, from https://www.cdc.gov/cancer/breast/basic_info/risk_factors.htm.

What Does It Mean to Have Dense Breasts? | CDC. (2023). Retrieved November 21, 2022, from https://www.cdc.gov/cancer/breast/basic_info/dense-breasts.htm.

Wickramasinghe, I., & Kalutarage, H. (2021). Naive bayes: Applications, variations and vulnerabilities: A review of literature with code snippets for implementation. *Soft Computing*, *25*(3), 2277–2293. Available from https://doi.org/10.1007/s00500-020-05297-6.

Wu, J., & Hicks, C. (2021). Breast cancer type classification using machine learning. *Journal of Personalized Medicine*, *11*(2), 61. Available from https://doi.org/10.3390/JPM11020061.

Yuvalı, M., Yaman, B., & Tosun, Ö. (2022). Classification comparison of machine learning algorithms using two independent CAD datasets. *Mathematics*, *10*(3), 311. Available from https://doi.org/10.3390/MATH10030311.

CHAPTER

9 Artificial intelligence in cardiovascular imaging: advances and challenges

Mohanad Alkhodari[1,2], Mostafa Moussa[1] and Salam Dhou[3]

[1]*Healthcare Engineering Innovation Center (HEIC), Department of Biomedical Engineering, Khalifa University, Abu Dhabi, United Arab Emirates*

[2]*Cardiovascular Clinical Research Facility (CCRF), Radcliffe Department of Medicine, University of Oxford, Oxford, United Kingdom*

[3]*Department of Computer Science and Engineering, American University of Sharjah, Sharjah, United Arab Emirates*

9.1 Introduction to cardiovascular imaging research

With the rapid evolution of interventional cardiology and cardiovascular treatment procedures, including transcatheter therapies, over the last two decades (Kerneis et al., 2019), the need for noninvasive cardiovascular imaging has grown significantly to ensure an accurate initial assessment of cardiovascular diseases prior to surgical intervention. Globally, cardiovascular diseases are considered the leading cause of death with an estimated 32% of all deaths. Every year, it is estimated that cardiovascular diseases take nearly 17.9 million lives most of which are due to heart attacks and strokes (World Health Organization, 2021). Therefore, cardiovascular imaging could play a pivotal role in the early detection of these diseases ensuring proper and timely cardiac management, counseling, and medication. Moreover, analyzing cardiac images taken using multiple modalities, patient clinical profiles, and information about the pathology could pave the way toward new discoveries in cardiovascular diagnostics, especially when integrated within a framework of the most recent advances in computerized algorithms. However, it remains challenging to interpret thousands of patient data from different imaging modalities given the number of available experts, that is, cardiologists, and the existence of other limitations associated with costs, time management, and screening analysis.

The emergence of artificial intelligence (AI) tools in the field of cardiovascular imaging could aid in analyzing this large amount of patient data in a timely manner with less dependency on medical expertise at a very early stage in the assessment of disease (Dey et al., 2019). In addition, AI could facilitate image-guided interventions and clinical procedures owing to its potential in learning from big data and in discovering new characteristics or disease-related anomalies

Artificial Intelligence and Image Processing in Medical Imaging. DOI: https://doi.org/10.1016/B978-0-323-95462-4.00009-1

that may not yet be well-known to medical doctors (Seetharam et al., 2021). In cardiovascular imaging, AI could assist in streamline image interpretation, reconstruction, and quantification during patients' diagnosis. Moreover, it has the capacity to assist in early decision-making regarding diseases, especially in many ambiguous cases with high capabilities in handling quality control and workflow of information between doctors and patients (Sermesant et al., 2021).

Motivated by the aforementioned, this chapter presents the most recent advances of AI in the field of cardiovascular imaging with emphasis on cardiac computed tomography (CT), cardiac magnetic resonance imaging (MRI), and cardiac ultrasound (echocardiography) research. The role of AI within these imaging modalities lies in providing a better understanding of the pathogenesis of many cardiovascular diseases. It can be used in many tasks, including image clustering, pattern recognition, disease identification, risk prognosis, intervention guidance, and other medical information estimation/quantification. This chapter also highlights the most critical challenges facing AI in cardiovascular imaging research nowadays with take-home messages serving as a guidance for future research.

9.2 Artificial intelligence in cardiac computed tomography

Cardiac CT has emerged as a prominent imaging modality in recent years for its high spatial and temporal resolutions in inspecting the cardiovascular system (Nicol et al., 2019). It allows clinicians to visualize blood vessel stenosis, especially in the coronary arteries, that results from the accumulation of plaques blocking the flow of blood. Combined with the CT technology, angiography-based techniques that relies on the injection of an iodine-containing contrast material allow for a better examination of these plagues in a three-dimensional (3D) manner (Abdelrahman et al., 2020). Therefore cardiac CT helps in guiding the assessment of coronary calcification through scoring the level of calcium due to plagues in arteries (Jin et al., 2021). Alongside imaging the vascular system, cardiac CT aids in myocardial perfusion analysis, left ventricular function evaluation, and heart rhythm and cardiac arrhythmias diagnosis (Oikonomou et al., 2019). The integration of AI in cardiac CT could expand the frontiers of this imaging modality by allowing the automation and expedition of many processes, which could lead to faster, yet accurate, clinical decision making in all of the aforementioned applications (van den Oever et al., 2020). A short review of the most recent advances of AI in cardiac CT is provided in the following subsections and in Table 9.1.

9.2.1 Assessment of coronary calcification

The most abundant mineral in the human body is calcium, and approximately 1% of it is found dissolved in the bloodstream. An increased accumulation of calcium

Table 9.1 Recent advances of artificial intelligence in cardiac computed tomography.

Study	Year	Application	Patients	AI Algorithm	Findings/ performance
Assessment of coronary calcification					
Al'Aref et al. (2020)	2019	Calcium scoring and prediction of obstructive coronary artery disease (CAD)	35,281	Boosted ensemble (XGBoost)	Area under the curve (AUC): >0.88 CAD prediction AUC: 0.87
Wang et al. (2020)	2019	Quantification of coronary artery calcium score	530	3D ResNet neural network	Agreement of 0.77 (CI: 0.73–0.81) 13% reclassification rate
van Velzen et al. (2020)	2020	Automated coronary calcium scoring	7,240	Two CNN blocks	Risk prediction agreement of 0.90 Intraclass correlation coefficients (ICCs): 0.79–0.97
Wolterink et al. (2015)	2020	Automatic calcium scoring	530	Decision trees and CNN	ICCs: >0.95 Accuracy: >0.90
Wang et al. (2021)	2021	Enhancing CT images for a better calcium scoring	96	Convolutional neural network (CNN)	Significant reduction in the noise (*P*-value <.001) Reduced Agatston, volume, and mass scores
Eng et al. (2021)	2021	Automated coronary calcium scoring	866	Encoder-decoder CNN (ImageNet)	Agreement of 0.89 for gated CT Sensitivity: >0.80 Positive predictive value (PPV): >0.87
Myocardial perfusion analysis					
Xiong et al. (2015)	2015	Prediction of stenosis using CTP image features during resting states	140	Random forest, Ada-boost, and Naïve Bayes	AUC: 0.73 Accuracy: 0.70 Sensitivity: 0.79 Specificity: 0.64
Han et al. (2018)	2016	Prediction of ischemia from rest computed tomography perfusion (CTP)	252	Gradient-boosting	AUC: 0.75 [CI: 0.69–0.81] Reclassification improvement of 0.52 (*P*-value < .001)

(*Continued*)

Table 9.1 Recent advances of artificial intelligence in cardiac computed tomography. *Continued*

Study	Year	Application	Patients	AI Algorithm	Findings/ performance
Coronary CT angiography					
Motwani et al. (2017)	2017	Predicting mortality from features extracted using cCTA images	10,030	Logit-boosting	AUC: 0.79 Predicting 5-year all-cause mortality significantly better than existing cCTA metrics alone
Coenen et al. (2018)	2018	Evaluating the diagnostic accuracy of blood fractional flow reserve (FFR)	351	Neural networks	AUC: 0.84 Accuracy: 0.85 Pearson $R = 0.997$ with computational fluid dynamics (CFD)-FFR
Lossau née Elss et al. (2019)	2019	Estimate and correct motion in cCTA	19	CNN	Motion accuracy: 13.37 ± 1.21 mm Magnitude accuracy: 0.77 ± 0.09 mm
Tatsugami et al. (2019)	2019	Enhancing coronary CT angiography (cCTA) images	30	Deep learning-based image restoration and CNN	Lower noise level using DLR with 18.5 ± 2.8 Hounsfield units (HU) Higher image quality score of 3.85
Zreik et al. (2019)	2019	Characterizing plaques and corresponding stenosis	163	Recurrent CNN (RCNN)	Plaques detection accuracy: 0.77 Stenosis detection accuracy: 0.80
Carolina (2020)	2020	Acute chest pain assessment from CT-FFR derived from triple-rule-out CT angiography	159	Neural networks	CT-FFR of ≤ 0.80 served as a better predictor for any MACE than stenosis of $\geq 50\%$ at CT angiography

within coronary arteries is universally common in coronary artery disease (CAD) patients causing tissue stiffening, plaque rupture, and heart failure (Williams et al., 2019). Coronary artery calcification (CAC) is known to be highly affecting men more than women, especially in patients with history of dyslipidemia, chronic kidney disease, diabetes, hypertension, and increased C-reactive protein levels (Mori et al., 2018; Nakao et al., 2018). The most commonly used CAC screening technique is the electron beam CT (EBCT) through the Agatston method (Agatston et al., 1990) owing to its ability of reducing image artifacts caused by heart and respiration motion during breath-holding. In addition, a multidetector CT (MDCT) technology has shown high correlation in the measurements of calcium scores in electrocardiography (ECG)-gated and nongated cardiac CT imaging (Azour et al., 2017). However, even though recent advances in imaging technologies have established a better understanding of CAC, AI, and big data techniques could potentially help in extending this knowledge by providing a more comprehensive map for the function and evaluation of this condition, which could eventually lead to better diagnostic protocols (Rogers et al., 2018).

Machine and deep learning-based techniques for the purpose of CAC reconstruction, detection, quantification, and labeling were presented in many recent studies (Fig. 9.1) (Al'Aref et al., 2020; Eng et al., 2021; van Velzen et al., 2020; Wang

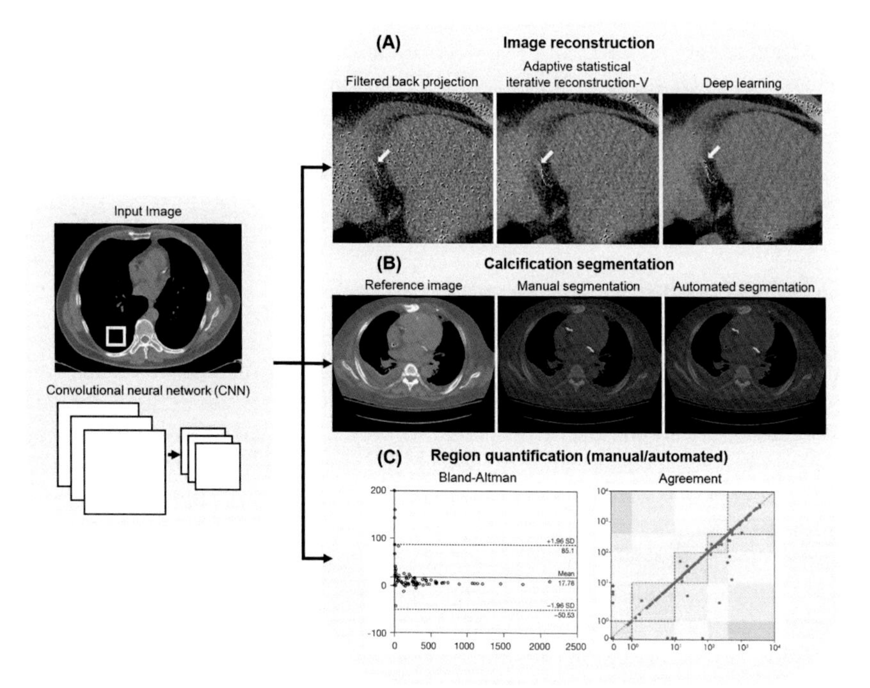

FIGURE 9.1 Deep learning in coronary artery calcification.

(A) Reconstruction of computed tomography images relative to regular filter back projection and adaptive statistical iterative reconstruction-V techniques, (B) calcification segmentation compared to expert's manual extraction, and (C) quantification of calcified regions and agreement with ground truth calculations.

et al., 2020; Wang et al., 2021; Wolterink et al., 2015; Wolterink et al., 2016). Wang et al. (2021) have found a significant reduction in the noise (P-value $< .001$) of the reconstructed images using deep learning, which have led to a better CAC assessment in terms of Agatston, mass, and volume scores of calcified regions. Furthermore, Eng et al. (2021) and van Velzen et al. (2020) have showed a strong agreement (Cohen's kappa (κ) > 0.89) between deep learning quantification of segmented regions and the ground truth obtained through manual segmentation of experts. In addition, their high CAC scores, that is, CAC >1, were successfully detected through deep learning with a sensitivity and positive predictive value of >0.8 and >0.87, respectively. They have concluded that deep learning could be a promising robust technique that have showed high levels of correlations of up to 0.98 with currently used techniques. Al'Aref et al. (2020) have developed gradient-boosting machine learning (ML) trained models for the prediction of patient with obstructive CAD (area under the curve [AUC] >0.88). Their work included a more comprehensive modeling technique that relies on various clinical variables alongside CT-based variables, including age, sex, symptoms, and other cardiac risk factors. Moreover, Wang et al. (2020) have showed no significant difference in calcium scoring between deep learning and manually derived values. However, they have observed a 13% reclassification rate of risk stratification for CAC from a stage to the following one. Other studies by Wolterink et al. (2015) have utilized ML (decision tree-based) and convolutional neural networks (CNNs) (Wolterink et al., 2016) to obtain additional features such as size, shape, and location of the calcified regions. They have reached overall sensitivity values of 0.87 and 0.71, respectively, in detecting regions of higher risk for developing severe CAC relative to manual experts' annotations.

9.2.2 Myocardial perfusion analysis

Myocardial CT perfusion (CTP) is a newly emerged imaging technique used in the diagnosis of myocardial ischemia. An ischemia is a medical term that refers to the lack of oxygenated blood reaching an organ such as the brain or the heart. In the case of heart muscles (myocardium), any partial or complete blockage of the blood by plagues causes the tissues to start dying and in worst cases develop heart attacks if any raptures took place (Monti et al., 2020). The followed protocol for CTP imaging, as in other imaging modalities, involves evaluating heart muscles under stress (hyperemia) and at rest (baseline) (Danad et al., n.d.). In stress CTP, an iodinated contrast agent is first injected to attenuate X-rays directly targeting the agent in tissues during its first pass through the myocardium. Therefore, any perfusion or anatomy defects can be detected and visualized with a high resolution (Seitun et al., 2016). On the other hand, the resting CTP does not require the injection of contrast materials during the imaging procedure. It was shown that a reduced rest hyperemic blood flow in stenotic lesions took place in rest-only CTP (Han et al., 2018). Many investigations have been carried out to suggest AI as a promising tool to advance the diagnostics of coronary ischemia in CTP imaging.

Very few studies (Fig. 9.2) have investigated the use of AI in CTP. Han et al. (2018) have trained a gradient-boosting ML model on a dataset of 252 patients for the prediction of ischemia from at rest CTP images. The authors have shown a significant improvement in utilizing CTP over regular CT stenosis imaging with an increase in the AUC from 0.68 to 0.75. In addition, Xiong et al. (2015) evaluated three different ML algorithms, namely random forest, Ada-boost, and Naïve Bayes, on automatically segmented and delineated left ventricular regions using CTP image features during resting states. They utilized features such as the normalized perfusion intensity, transmural perfusion ratio, and myocardial wall thickness and achieved the highest accuracy with the Ada-boost algorithm (AUC = 0.73) in discriminating patients with significant coronary stenosis related to a ground truth standard obtained from quantitative coronary angiography.

9.2.3 Coronary computed tomography angiography

Coronary CT angiography (cCTA) is a medical imaging technique that relies on the injection of an iodine-containing contrast agent to enhance CT scans when examining coronary arteries (Maroules et al., 2019). It is highly recommended by the current guidelines of the European society of cardiology (ESC) (Montalescot et al., 2013) for the assessment of patients with stable chest pain and obstructive CAD. In the assessment of coronary artery stenosis, cCTA allows for noninvasively calculating the fractional flow reserve (FFR), which is highly regarded as the gold-standard in hemodynamic assessment (Fihn et al., 2015). FFR represents the pressure difference across regions with stenosis and helps in determining the impact of arteries narrowing on the supply of oxygenated blood to the heart

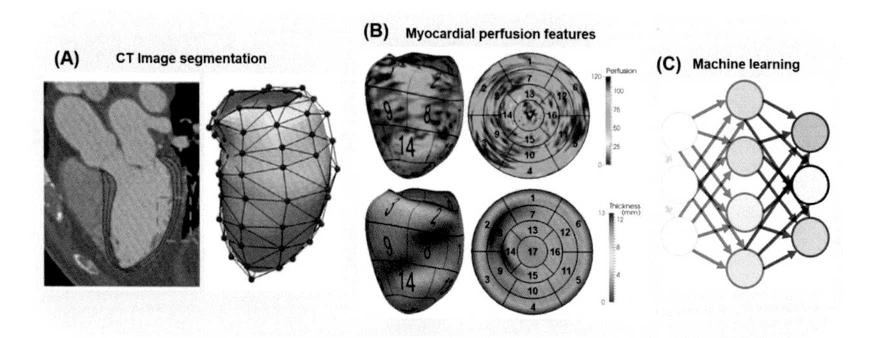

FIGURE 9.2 Using machine learning for the prediction of myocardial ischemia (coronary stenosis) from myocardial CT perfusion images.

(A) CT image acquisition and segmentation in two-dimensional and three-dimensional representations, (B) myocardial perfusion features extraction including perfusion strength and thickness of coronary arteries, and (C) training machine learning models to predict the existences of any ischemia or high-risk stenosis.

(Nørgaard et al., 2019). For example, a 0.80 FFR shows a 20% drop in blood pressure because of a stenosis, which is a representation of the maximum amount of blood flow through stenotic arteries relative to normal ones. Calculations of FFR from cCTA images have significantly improved the diagnostic procedures with less reliance on invasive cardiac catherization techniques (Coenen et al., 2018). In addition, cCTA has demonstrated an effective prognostic value in predicting additional major adverse cardiac events (MACE), including total death, hospitalization due to heart failure, stroke, and revascularization (coronary interventions) alongside the regular cardiac risk factors (Cho et al., 2018). Therefore, the information acquired after plaque detection and quantification in cCTA acts as a rich material for AI algorithms (Fig. 9.3), including characteristics of plaques (calcified or noncalcified), plaques burden, and coronary artery blood flow.

Several studies in literature have investigated the use of AI algorithms in FFR assessment (Carolina, 2020; Coenen et al., 2018; Gohmann et al., 2021; Ihdayhid & Zekry, 2020; Tesche et al., 2018). For example, Coenen et al. (Coenen et al., 2018) have compared ML-based CT-FFR and computational fluid dynamics (CFD)-FFR with invasive FFR in the diagnostic accuracy of blood flow

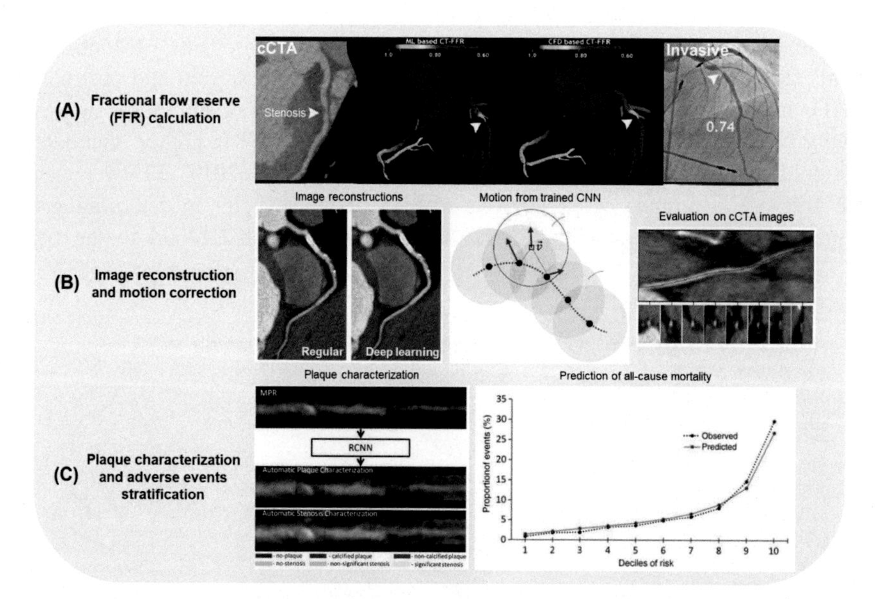

FIGURE 9.3 Artificial intelligence integration within coronary computed tomography angiography (cCTA) assessment of coronary artery disease.

(A) Estimation of fractional flow reserve levels and modeling them on three-dimensional (3D) coronary models, (B) using machine learning in cCTA image enhancements (noise reduction) and motion correction and detection, and (C) characterizing plaques and predicting the occurrences of stenosis alongside stratifying all-cause mortality rates and adverse cardiac risks.

estimations in a pregenerated 3D coronary model. Their study has shown that neural network-based estimation of FFR exhibited an excellent correlation with the computationally extensive CFD-FFR (Pearson $R = 0.997$) with almost similar AUC values of 0.84 compared to 0.69 using the reference method. In addition, Martin et al. (Carolina, 2020) evaluated ML-based CT-FFR derived from triple-rule-out CT angiography for acute chest pain (ACP) assessment. Their neural network algorithm provided higher agreement on the decision regarding the presence of a disease compared to regular CT angiography. Moreover, ML-based CT-FFR of ≤ 0.8 served as a better predictor for any MACE than stenosis of $\geq 50\%$ at CT angiography. It further improved the specificity of ACP detection at emergency department settings. Besides FFR assessment, multiple studies (Lossau (née Elss) et al., 2019; Tatsugami et al., 2019) have utilized AI for image reconstruction and motion correction to enhance the diagnostic protocols in cCTA. Tatsugami et al. (2019) have compared a deep learning-based image restoration (DLR) algorithm with the current hybrid iterative reconstruction (IR) approach and assessed their qualities in enhancing cCTA images. Their approach, which consisted of 10-layer deep neural network, showed a lower noise level using DLR with 18.5 ± 2.8 (vs 23.0 ± 4.6 in IR) Hounsfield units (HU) with a higher image quality score of 3.85 (2.96 in IR). Moreover, Lossau et al. (Lossau née Elss et al., 2019) have developed an ML algorithm to estimate and correct motion in cCTA. Authors have utilized CNNs to estimate underlying 2D motion vectors from 2.5D image patches with an average accuracy of 13.37 degrees $\pm$ 1.21 degrees mm and 0.77 degrees $\pm$ 0.09 degrees mm in predicting the motion direction and magnitude, respectively. Lastly, AI was used in cCTA for plaque characterization (van Assen et al., 2019; Zreik et al., 2019) and risk stratification (Motwani et al., 2017; van Rosendael et al., 2018). Zreik et al. (Zreik et al., 2019) have developed recurrent CNN (RCNN) on reconstructed multiplanar reformatted cCTA images to characterize plaques and corresponding stenosis. They have achieved overall accuracy levels of 0.77 and 0.80 in the detection of coronary plaques and stenosis, respectively. In addition, Motwani et al. (Motwani et al., 2017) have developed a logit-boosting ML algorithm for predicting mortality from features extracted using cCTA images such as segment stenosis score and modified Duke index. Their algorithm achieved an overall AUC of 0.79 in predicting all-cause mortality compared with 0.61 using the regular Framingham risk score.

9.3 Artificial intelligence in cardiac magnetic resonance imaging

MRI is an imaging technique that utilizes magnetic resonance (field strength of 1.5 T or 3 T, the former more commonly) and radiowave signals that get absorbed and emitted by protons in the body to map anatomical features. Cardiac MRI (CMR) is simply the application of MRI to image the heart (Assomull et al.,

2007). The limitations associated with the implementation of MRI are not many and are mainly limited to its cost, poor temporal resolution, and inability to use in imaging patients with metallic prostheses or implants (Grover et al., 2015). Despite there being several types of imaging sequences, we consider the two main types: gradient-echo and spin-echo (Baird & Warach, 1998). They are both useful in visualizing heart function quantitatively and qualitatively, depending on the application and the required level of characterization. Velocity mapping is based on gradient-echo, wherein blood and fat appear white, making it better suited for flow analysis, T1-weighted spin-echo differs in that blood appears black, and T2-weighted spin-echo can highlight myocardial inflammation or edema (Grover et al., 2015).

CMR is used for cardiac image construction, reconstruction, and segmentation to assess angiogenesis, regional perfusion, cardiac myocyte formation, myocardial or heart viability in general (Rickers et al., 2005; Saeed et al., 2015), and ventricular function via ejection fraction (Assomull et al., 2007; Saeed et al., 2015; Syed et al., 2010) when suspecting symptoms that could entail various conditions. These conditions include atherosclerosis, cardiac amyloidosis (Karamitsos et al., 2009), cardiomyopathy (hypertrophic, Peterzan et al., 2016, nonischemic Shehata et al., 2008), heart failure (Shan et al., 2004; Benza et al., 2008), CAD (Nandalur et al., 2007), aneurysms (Assomull et al., 2007), myocardial ischemia (Saeed et al., 2015), heart diseases like aortic stenosis and other congenital heart diseases (Alfakih et al., 2004; Bohbot et al., 2020), pulmonary arterial hypertension (Caruthers et al., 2003), and cardiac tumors among others (Assomull et al., 2007; Saeed et al., 2015). It is abundantly clear that CMR is versatile in its utility, but considering the time it takes to produce usable images in the first place, diagnosing these conditions would certainly benefit from a more automated approach, as opposed to expert inspection.

AI, or more particularly ML, has rapidly become a ubiquitous tool in medical imaging applications, and MRI is no different. ML and deep learning techniques serve to aid in better and faster CMR imaging acquisition, analysis, and reporting (Edalati et al., 2022; Fotaki et al., 2021), and diagnosis of some of the above-mentioned conditions in an automated fashion to both support physicians and offset the imaging modality's temporal resolution. Furthermore, explainable ML can provide better quality control, as well as provide users with insights into the ML models and their decisions or diagnoses, allowing better interpretability and reproducibility of these models (Fotaki et al., 2021; Kart et al., 2023). U-Net, a CNN developed for biomedical image segmentation (Ronneberger et al., 2015), is frequently used in works involving MRI and AI or ML. Though many recent works use ML in MRI, we focus on those that make use of AI for cardiac segmentation or image reconstruction for ventricular function assessment in order to evaluate the effects of CVD risk factors or to indirectly identify cardiomyopathies (S. Li, Zhang, et al., 2021; Suinesiaputra et al., 2022; Yang et al., 2020), congenital heart disease (CHD) (Diller et al., 2020; Karimi-Bidhendi et al., 2020), and directly classify cardiomyopathies (Li, Chen, et al., 2021; Puyol-Antón et al., 2018; You et al., 2021). A brief summary of the recent advances of AI in CMR is provided in the following subsection and in Table 9.2.

9.3.1 Segmentation and ventricular function assessment

Segmentation in medical imaging involves extracting the region of interests (ROIs), such as the ventricles, myocardium, and atria of the heart, and clearly labeling them in order to identify the required areas for studies involving cardiovascular diseases. Consequently, cardiac MRI segmentation (Fig. 9.4) is the first part of any such study. Some works, such as Li, Zhang, et al. (2021) and Yang et al. (2020), perform cardiac segmentation solely for the purpose of training and evaluating ML methods to do just that, whereas others perform segmentation for the assessment of heart function via ventricular ejection fraction and indexed mass and volume among other metrics, like Suinesiaputra et al. work (Suinesiaputra et al., 2022).

Li, Zhang, et al. (2021) aimed to perform LV, RV, and myocardium segmentation by making use of autoencoders with U-Net as the encoder and a conditional generative adversarial network (GAN) as the decoder with 350 subjects' CMR images from the MICCAI 2020 multicenter, multivendor, and multidisease cardiac segmentation (M&Ms) challenge (Campello et al., 2021). They obtain an average dice similarity coefficient (DSC) for all three segmentation objectives of 0.86 and 0.85 at end-diastole (ED) and end-systole (ES), respectively, and an average Hausdorff distance (HD) of 10.6 and 11.2 for ED and ES, respectively. Similarly, Yang et al. (Yang et al., 2020) used a dilated block adversarial network (DBAN) with 150 subjects' short-axis MRI images from the MICCAI 2017 Automated Cardiac Diagnosis Challenge dataset (Bernard et al., 2018) and the Right Ventricle Segmentation Challenge dataset (Petitjean et al., 2015) and sunnybrook cardiac data datasets (Radau et al., 2009) for comparison. The DBAN provided better performance than all other tested methods with an average DSC of 0.92 and 0.89 for LV, RV, and myocardium segmentation and an average HD of 8.7 and 10.8 at ED and ES, respectively, as well as demonstrating comparable performance with the other two datasets and to similar works. Suinesiaputra et al. (2022), however, begin with cardiac MRI segmentation for ventricular function evaluation. They use CMR data of 5098 subjects with VGGNet to detect landmarks in various MRI views, U-Net for segmentation, and a finite element LV model for atlas construction with comparison against manual analysis via logistic regression (LR). They obtained an average error distance of less than 2.5 mm for landmark detection, an average DSC of approximately 0.9 at ED and ES for myocardium and LV, and a difference between automated and manual analyses of less than 1.0 ± 2.6 mL/m^2 for LV indexed volume, 3.0 ± 6.4 g/m^2 for indexed mass, and $0.6\% \pm 3.3\%$ for ejection fraction.

Table 9.2 Recent advances of artificial intelligence in cardiac magnetic resonance imaging.

Study	Year	Application	Patients	AI Algorithm	Findings/ performance
Segmentation and ventricular function assessment					
Yang et al. (2020)	2018	Left ventricle, right ventricle, & myocardium segmentation in short-axis MRI.	150	Dilated Block Adversarial Network + U-Net	Average Dice similarity coefficient = 0.92 in end-diastole and 0.89 in end-systole Average Hausdorff distance (HD) = 8.7 in ED and 10.8 in ES
Li, Zhang, et al. (2021)	2021	Left ventricle, right ventricle, & myocardium segmentation	350	U-Net + Generative Adversarial Network (GAN)	Average Dice = 0.86 in end-diastole and 0.85 in end-systole Average HD = 10.6 in end-diastole and 11.2 in end-systole
Suinesiaputra et al. (2022)	2022	Landmark detection, segmentation, and atlas validation to evaluate the use of the proposed methodology for cardiovascular disease risk factors	5098	VGGNet, U-Net, Finite Element Analysis, Logistic Regression	Maximum average error distance: short-axis = 2.07 ± 1.11 with intraclass correlation coefficient = 0.995 Average segmentation Dice $\approx$ 0.9 at end-diastole and in end-systole for myocardium and left ventricle, small differences in left ventricle functions between automated

(Continued)

Table 9.2 Recent advances of artificial intelligence in cardiac magnetic resonance imaging. *Continued*

Study	Year	Application	Patients	AI Algorithm	Findings/ performance
					model and manual. P-value between area under curve (AUC) of all risk factors > 0.01 -> no statistical difference; the automated model is good.
Congenital heart disease analysis					
Karimi-Bidhendi et al. (2020)	2020	Augmentation of cardiac MRI datasets and comparison of neural network performance with manual analysis for congenital heart diseases	64	GAN + Fully convolutional network Synthetically Augmented Dataset	Average Dice = 0.91 in end-diastole and 0.87 in end-systole for LV Average Dice = 0.85 in end-diastole and 0.81 in end-systole for RV
Diller et al. (2020)	2020	Augmentation of cardiac MRI datasets of congenital heart disease for use in training models to detect them	303	Progressive-GAN + U-Net	Long axis left ventricle Average Dice = 0.978 Long axis right ventricle Average Dice = 0.981 Long axis right atrium Average Dice = 0.986 Short axis left ventricle Average Dice = 0.987 Short axis right ventricle Average Dice = 0.965

(*Continued*)

Table 9.2 Recent advances of artificial intelligence in cardiac magnetic resonance imaging. *Continued*

Study	Year	Application	Patients	AI Algorithm	Findings/ performance
Assessment of cardiomyopathy					
Puyol-Antón et al. (2018)	2018	Automated left ventricle atlas formation and classification of dilated cardiomyopathy using ultrasound data with a model trained with MRI and ultrasound data	69	Linear Discriminant Analysis and Laplacian Support Vector Machine	Accuracy = 94.32% Sensitivity = 93.00% Specificity = 96.57%
Li, Chen, et al. (2021)	2021	Automated segmentation and classification of dilated cardiomyopathy using MRI data	134	K-Nearest Neighbor, Support Vector Machine, AdaBoost, and Random Forest	Accuracy = 95.50% Sensitivity = 91.46% Specificity = 94.45%
You et al. (2021)	2021	Diagnosis of hypertrophic obstructive cardiomyopathy using MRI data and deep learning	104	Double branch CNN + Long short-term memory network	Accuracy = 96.79% Sensitivity = 95.24% Precision = 97.42% F1-Score = 94.57%

9.3.2 Congenital heart disease analysis

CHDs are among the most common birth defects and require regular imaging follow-ups, due to their fatal nature if left unchecked. MRI is often used for cardiac imaging in infants with CHDs due to the high spatial resolution, noninvasiveness, and due to the fact that it does not make use of ionizing radiation. ML and deep learning techniques can be used in the detection of CHDs with CMR images directly from the image or indirectly by measuring blood flow or heart function (Bratt et al., 2019). However, their utility is not limited to diagnosis; they can also be used to synthetically enhance the generally small CHD CMR datasets to provide better training sets for ML, or more particularly, deep learning algorithms (Fig. 9.5), such as Karimi-Bidhendi et al. and Diller et al. works (Diller et al., 2020; Karimi-Bidhendi et al., 2020).

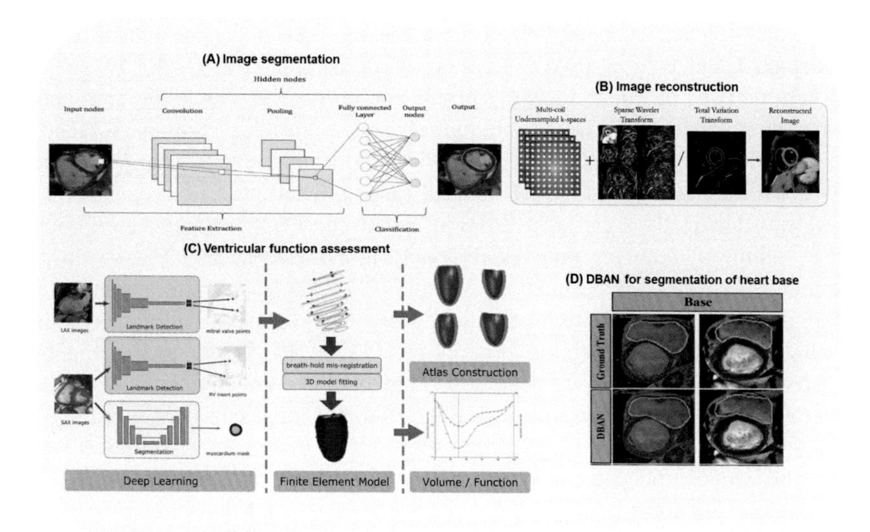

FIGURE 9.4 Deep learning for cardiac magnetic resonance image segmentation and ventricular function estimation.

(A) Image segmentation, (B) image reconstruction, (C) landmark detection and ventricular function assessment, and (D) dilated block adversarial network segmentation of heart base compared to manual segmentation.

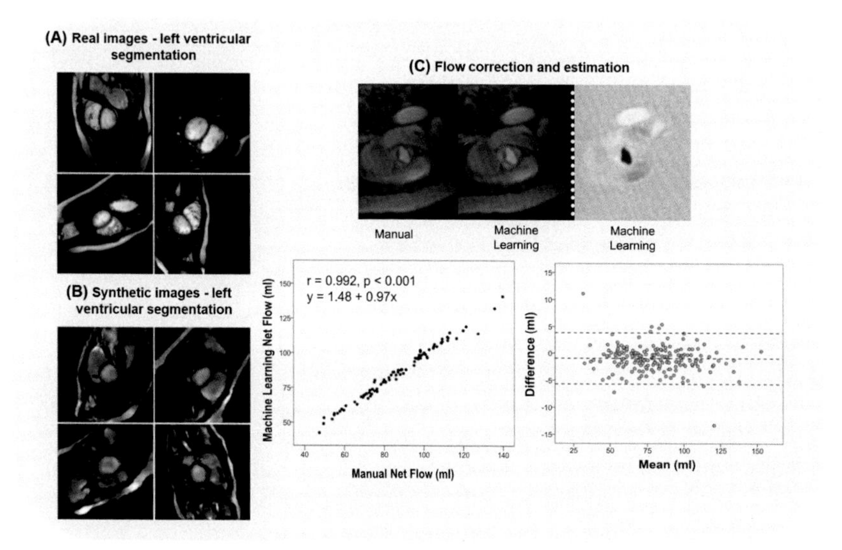

FIGURE 9.5 Analysis of congenital heart diseases using deep learning

left ventricular segmentation in (A) real cardiac magnetic resonance (CMR) images and (B) synthetic CMR images. (C) Manual and automated segmentation in a patient with trileaflet aortic valve, correlation of net flow, and Bland-Altman plot of flow estimation.

Karimi-Bidhendi et al. (2020) proposed using GAN to create a synthetically augmented CHD dataset (SAD) from a set of 64 pediatric CMRs of subjects between the ages of 2 and 18 with various conditions, such as aortic arch anomaly, cardiomyopathy, CAD, double outlet right ventricle, pulmonary stenosis, transposition of great arteries, tetralogy of Fallot, and truncus arteriosus, and a fully convolutional network (FCN) is used for segmentation. The proposed model shows no statistically significant difference between FCN-SAD and ground truth in ES volume of both LV and RV, but only in ES volume of LV, as well as a minimum ICC of 0.967. Diller et al. (2020) similarly seek to augment 303 Tetralogy-of-Fallot-afflicted individuals' CMR image dataset to allow better deep learning application by training progressive GANs (PG-GAN) then using both datasets to train U-Net for segmentation. They obtained an average DSC of 0.978 for long-axis LV, 0.981 for long-axis RV, 0.986 for long-axis right atrium (RA), 0.987 for short-axis LV, and 0.965 for short-axis RV images. In addition to the high DSC values, the percent variation was also small, at a maximum of 0.035 for short-axis RV images.

9.3.3 Assessment of cardiomyopathy

Cardiomyopathy is a general term that refers to disease or weakening of the heart muscle, causing difficulty in pumping blood to the body, and possibly leading to heart failure. The main types of cardiomyopathy include dilated cardiomyopathy (DCM), and hypertrophic cardiomyopathy (HCM), all of which share symptoms like arrhythmias, or lower extremity swelling but have different causes. DCM is characterized by the ventricles dilating and thinning, whereas HCM is characterized by thickening or hypertrophy of the heart muscle, both making it more difficult to pump blood. Both types, however, may not cause symptoms in most cases, but in cases in which they do, cardiomyopathy can be life-threatening. Taking this into consideration, CMR has been crucial in detecting cardiomyopathies, and robust automated methods (Fig. 9.6) of cardiomyopathy detection from CMR would surely save physicians time and effort.

Puyol-Antón et al. (2018) sought to train a multimodal system with a database of 69 subjects' MRI and ultrasound (US)-derived cardiac motion atlases from the STACOM'11 database (Tobon-Gomez et al., 2013) and local hospital datasets. The regional relationships between MRI and US after segmentation are used to train multiview linear discriminant analysis (MLDA) and multiview Laplacian SVM (MvLapSVM) with eightfold repeated stratified cross-validation with this data after canonical correlation analysis (CCA) and Kernel CCA for multiview dimensionality reduction, with a global and regional approach. MvLapSVM with the regional approach yields the best performance with an accuracy of 94.32%, a sensitivity of 93.00%, and a specificity of 96.57%, also showing a statistically significantly better performance than without the multimodal atlas. In their work, Li, Chen, et al. (2021) also sought to develop an automated method for the classification of DCM but use one

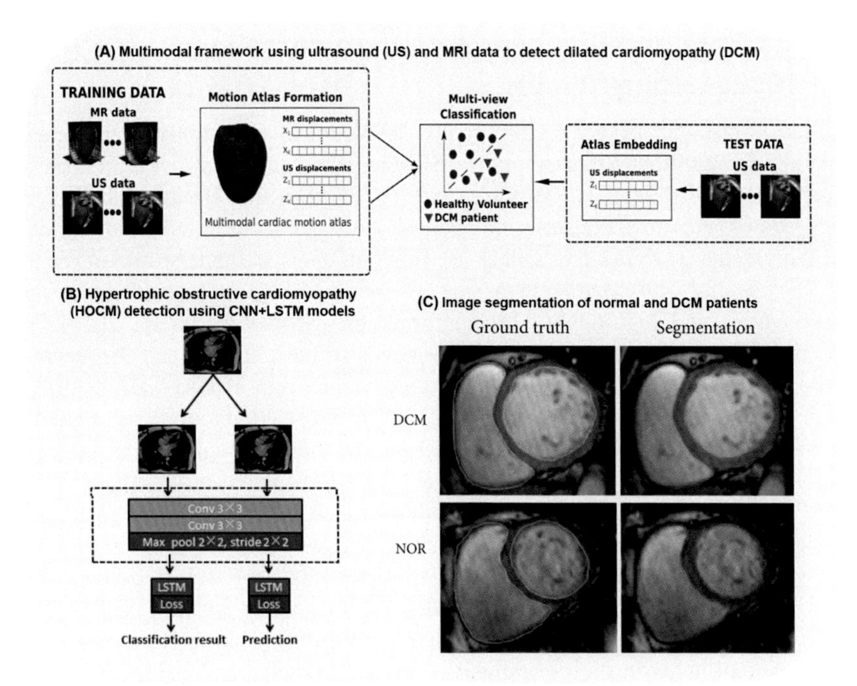

FIGURE 9.6 The use of deep learning for the assessment of cardiomyopathy.

(A) Multimodal framework of detecting dilated cardiomyopathy (DCM) using ultrasound (US) and magnetic resonance imaging data, (B) cardiac magnetic resonance and convolutional neural network and long short-term memory layers used for hypertrophic obstructive cardiomyopathy detection, and (C) image segmentation of a patient with DCM and a normal patient.

modality. Cardiac Cine-MRI data from 134 subjects are obtained, LV and RV are segmented (LV with a DSC of 0.87), ED and ES phases are highlighted, the images are processed and features extracted, and finally, data is fed into k-nearest neighbor, SVM, boosted decision trees with adaptive boosting algorithm (AdaBoost), and random forest (RF), of which the latter gave the best performance with an accuracy of 95.50%, a sensitivity of 91.46%, and a specificity of 94.45%. The work presented by You et al. (2021) focuses on the diagnosis of hypertrophic obstructive cardiomyopathy on CMR data obtained from 104 subjects to train and test CNNs and long short-term memory networks, as shown in Fig. 9.6 to form a double-branched, multitask neural network. Their proposed network yielded an accuracy of 96.79%, a sensitivity of 95.24%, a precision of 97.42, and an F1-score of 94.57%, far exceeding the performance of ResNet50 and DenseNet with a 17.76% and 10.54% increase in F1-score, respectively.

9.4 Artificial intelligence in cardiac ultrasound (echocardiography)

Cardiac ultrasound, or echocardiography, is the most commonly used imaging modality for cardiovascular imaging (Siegersma et al., 2019). It relies on the transmission and reflection of sound waves (>20 kHz) that are generated by piezoelectric crystals to reconstruct detailed hemodynamic information of the heart and the vascular system. It is widely used by clinicians as the first line of diagnosis for cardiac diseases owing to its noninvasiveness, portability, cost-effectiveness, and high-quality image formation (Seetharam et al., 2019). The type of echocardiography depends on the needs of the clinician and the accuracy of the required image. It could include transthoracic, transesophageal, doppler, or stress echocardiography (Aly et al., 2021). The most commonly used types are the transthoracic echocardiography (TTE) and transesophageal echocardiography (TEE). In the standard TTE, the clinician spreads a gel over the chest to ease sound waves transmission and presses a transducer against the skins to transmit and record echoes from the heart. On the other hand, TEE requires the insertion of a gastroprobe through the esophagus to record the heart echoes, thus, it provides more-detailed images that were difficult to acquire in the standard chest echocardiogram (Shillcutt & Bick, 2013). Combined with these two types, doppler ultrasound helps in the assessment of blood flow changes inside and outside the heart, while stress ultrasound provides better visualization of the coronary arteries abnormalities that only appear during physical activities (Fuad Jan & Jamil Tajik, 2017). Although echocardiography produces high-quality images, it still heavily dependent on the clinicians' experience and their subjective interpretation of images leading in some cases to an inaccurate assessment (Alsharqi, Woodward et al., 2018). Therefore, AI could play a significant role in reducing human error associated with image analysis, which would eventually lead to better treatment plans. A brief review of the use of AI in echocardiography is provided in the following subsections and in Table 9.3.

9.4.1 Left ventricular systolic and diastolic assessment

The left side of the heart includes the left ventricle, which is mainly responsible for pumping the oxygenated blood to the rest of the body. Any dysfunctionality in the left ventricle leads to developing symptoms such as fatigue and dyspnea (Fletcher et al., 2021). The severe abnormalities in the function of the blood filling and ejection mechanisms by the left ventricle could eventually lead to heart failure. In worst cases, the progression of heart failure indicates the occurrence of CAD and arterial hypertension (Alkhodari & Jelinek, Karlas et al., 2021). To assess the systolic and diastolic function of the heart, echocardiography is a primary imaging modality that is routinely used in clinical practice. It allows for observing the left ventricular ejection fraction (LVEF) metric, which is the ratio

Table 9.3 Recent advances of artificial intelligence in cardiac ultrasound (echocardiography).

Study	Year	Application	Patients	AI Algorithm	Findings/ performance
Left ventricular systolic and diastolic assessment					
Choi et al. (n.d.)	2020	Prediction of systolic heart failure	1295	Classification and regression trees	Accuracy: 99.6% Concordance rate: 98%
Ghorbani et al. (n.d.)	2020	Estimating left ventricle function and cardiac risk	6106	Convolutional neural network (CNN)	R^2: up to 0.74 AUC: up to 0.89
Zhuang et al. (2021)	2021	Left ventricle segmentation	496	YOLOv3 CNN	Mean absolute distance: 2.57 Hausdorff distance: 6.68
Duffy et al. (2022)	2022	Detection of left ventricle hypertrophy	23,745	DeepLabv3 spatiotemporal CNN	Cardiac amyloidosis AUC: 0.79 Hypertrophy AUC: 0.89
Applications in stress echocardiography					
Omar et al. (2018)	2018	Identifying wall motion abnormality	61	VGG-16 CNN	Accuracy: 75%
Kusunose et al. (2020)	2020	Identifying regional wall motion abnormality	400	CNN	AUC: 0.99
Bennasar et al. (2020)	2020	Predicting CAD	529	Support vector machine (SVM) and random forest	Accuracy: 70.32% Sensitivity-specificity: 70.24%
Upton et al. (2022)	2021	Predicting CAD	23,000	SVM, random forest, logistic regression	AUC: up to 0.93 Sensitivity: 90.5% Specificity: 92.7%.

(*Continued*)

Table 9.3 Recent advances of artificial intelligence in cardiac ultrasound (echocardiography). *Continued*

Study	Year	Application	Patients	AI Algorithm	Findings/ performance
Analysis of valvular heart diseases					
Andreassen et al. (2020)	2019	Segmentation of mitral valve leaflets	111	U-Net CNN	Mean error: 2 ± 1.96 mm Mean Dice coefficients: up to 0.8
Corinzia et al. (2020)	2020	Segmentation of mitral valve	85	Unsupervised multilayer perceptron and neural networks	Dice coefficients: 0.48 Benchmark performance: 0.45
Vafaeezadeh et al. (2021)	2021	Detection of normal and prosthetic mitral valves	2044	CNN with transfer learning	Accuracy: 98% AUC: 0.99
Zhang et al. (2021)	2021	Classification of mild, moderate, and severe mitral regurgitation	1427	CNN	Accuracy: up to 91% Macro F1: 0.91 Micro F1: 0.92

of blood pumped at each contraction to the total amount of blood within the left ventricle (Nagueh et al., 2016). Although functional assessment was made possible through echocardiograms, it is still complex to accurately analyze due to the different structural and functional changes that take place during the cardiac cycle. Therefore AI could potentially aid in the assessment by screening for the presence of a systolic or diastolic anomalies and quantification of the grade of such abnormalities. Among the applications of AI in left ventricle assessment, several studies have demonstrated the efficacy of machine/deep learning methods in estimating diastolic and systolic function, predicting left ventricle hypertrophy, and segmenting the left ventricle for better quantifications (Fig. 9.7).

In the area of heart failure analysis, Choi et al. (n.d.) developed a classification and regression trees to diagnose the occurrence of a preserved ejection fraction, that is, LVEF greater than 55%, from echocardiographic variables. Their model that reached up to 99% compared to expert evaluation relied on variables such as the index mass, septal velocity, and tricuspid regurgitation velocity. Ghorbani et al. (n.d.) provided deep learning models for estimating cardiac diastolic/systolic function and for predicting cardiovascular risk associated with systemic phenotypes. Their CNN model was capable of estimating the left ventricle systolic and diastolic volumes alongside LVEF with an R^2 accuracy up to 0.74. In

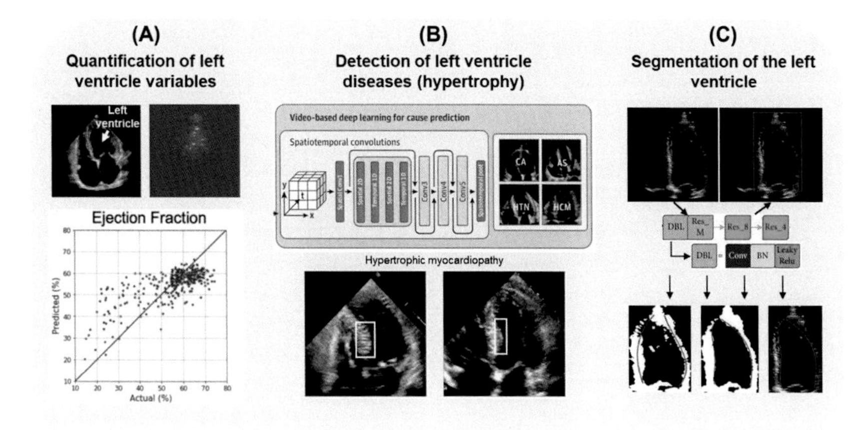

FIGURE 9.7 The role of artificial intelligence in echocardiography for the assessment of the systolic and diastolic cardiac phases.

(A) Quantification of left ventricular variables, including left ventricle ejection fraction, (B) detection of left ventricle abnormalities such as hypertrophy and cardiac amyloidosis, and (C) segmentation of the left ventricle for accurate assessment of echocardiographic measures.

addition, the prediction of left ventricle diseases was commonly used as an application of AI within echocardiography. Duffy et al. (2022) utilized deep learning to detect and characterize left ventricle hypertrophy. Their algorithm that had DeepLabv3 spatiotemporal convolutional architecture reached an overall AUC of 0.79 and 0.89 in detecting cardiac amyloidosis and hypertrophic cardiomyopathy. On another note, echocardiogram segmentation was used to analyze the left ventricle and the myocardium associated with it. Zhuang et al. (2021) developed an automated CNN algorithm for the localization of the apex and bottom of the left ventricle from echocardiography images. Their segmentation approach has reached a mean absolute distance score of 2.57 and HD of 6.68.

9.4.2 Applications in stress echocardiography

Stress echocardiography is the most commonly used test for CAD assessment with around four million tests performed yearly in the United States (Ladapo et al., 2014). It concerns imaging the heart during or after applying a stressor such as exercising on a treadmill or being exposed to drugs such as dobutamine (Augustine et al., 2014). The clinician will keep monitoring heart rates until it reaches the peak level where he will perform an ultrasound test. This test allows the clinician to get enough information about the functionality of heart muscles and filling/pumping mechanism (Alsharqi, Upton, et al., 2018). Although stress echocardiography is widely used as a functional cardiac imaging modality, its

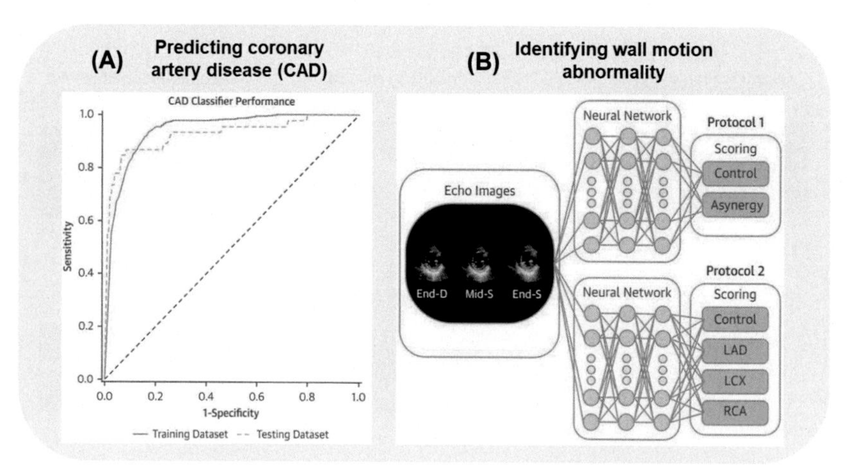

FIGURE 9.8 Artificial intelligence applications in stress echocardiography

(A) predicting coronary artery disease from features extracted from echocardiograms under stress and (B) identification of wall motion abnormality using deep learning.

interpretation is considered one of the most challenging activities for clinicians (Patricia A. Pellikka, 2022). It is heavily affected by the rapid heart rates, respiratory artifacts, and the relaxation of the heart during recovery. There have been many attempts to counter such difficulties through comparisons between stress and at-rest images and enhancing agents (Pellikka et al., 2020), however, accurate assessment is still lacking. Therefore, AI tools could be potential solutions for a better interpretability of stress echocardiogram in many applications, including the detection of CAD and identification of wall motion abnormality (Fig. 9.8).

In the area of CAD diagnosis, Upton et al. (2022) have developed a ML-based approach (using SVM, random forest, and LR) to identify patients with severe CAD. Their best-performing model selected up to 31 features, including wall motion abnormality and endocardial velocity and achieved an AUC value of up to 0.93, sensitivity of 90.5%, and specificity of 92.7%. In addition, Bennasar et al. (2020) utilized a minimal number of clinical and demographical features such as prior history of CAD, sex, and other prescribed medications to predict the presence of a significant CAD. Their SVM classifier that was trained on 529 patient data achieved an overall accuracy up to 70.32% with sensitivity and specificity measures of 70.24%. For wall motion abnormality detection, Omar et al. (Omar et al., 2018) trained a VGG-16 CNN with 3D Dobutamine stress echocardiography and its corresponding Bull's-Eye plot. Their model achieved the highest levels of performance in identifying abnormal wall motion of the cardiac muscle with an accuracy of 75%. Moreover, Kusunose et al. (2020) investigated the use of CNN to improve the detection of regional wall motion abnormalities (RWMAs). Their algorithm identified the presence of abnormalities with an AUC of 0.99 in 400 patient data (1200 echocardiograms).

9.4.3 Analysis of valvular heart diseases

Valvular heart diseases (VHD) are a major cause of cardiovascular diseases and are causing high mortality rate worldwide (Draper et al., 2019). Any damage that exists in the four heart valves, that is, the mitral, tricuspid, aortic, and pulmonary valves, could cause potential damage in the flow of blood through cardiac arteries. A damage may occur as a result of stenosis (narrowing of arteries), plaques, or hardened valves, thus, force, direction, and time affecting blood flow and the closing and opening of the valves are interrupted (Alkhodari & Fraiwan, 2021; Zeng et al., 2016). The current gold standard for evaluating VHD is the transthoracic echocardiography, and it is always advised to provide early diagnosis to allow optimum and timely medication (Nedadur et al., 2022). However, the interpretation of millions of echocardiograms of patients at-risk of VHD is considered burdensome. In addition, it requires clinical expertise and proper training due to the difficulty in detecting valves in echocardiography images (Writing Committee Members et al., 2021). The development of AI techniques could potentially transform the assessment of VHD, thus, preventing many cases from reaching severe conditions. There have been many studies that investigate the use of AI in VHD analysis for several applications including automated segmentation of valves images and detection of VHD (Fig. 9.9).

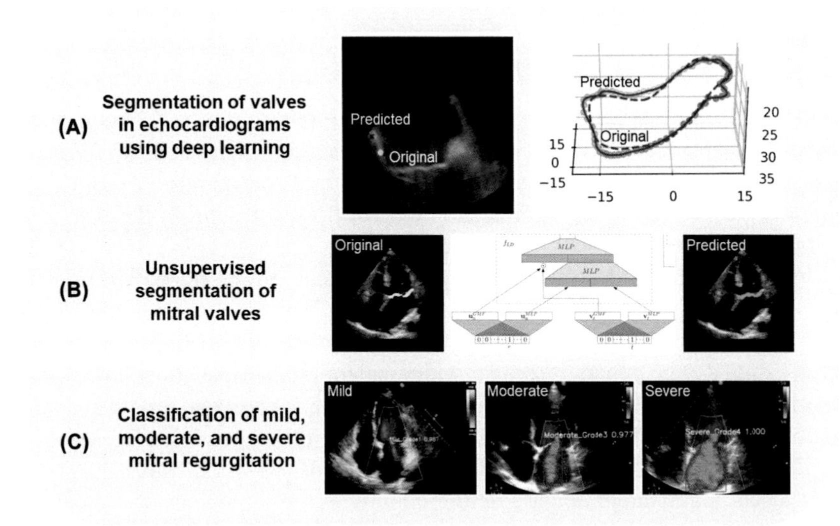

FIGURE 9.9 The use of artificial intelligence in valves analysis in echocardiography.

(A) Segmentation of valves in echocardiograms using trained deep learning models, (B) unsupervised segmentation of mitral valves using multilayer perceptron and neural networks, and (C) automatic classification of mild, moderate, and severe mitral regurgitation.

In the segmentation of valves in echocardiograms, Andreassen et al. (2020) utilized 111 multiframe 3D TEE echocardiography images to detect mitral valve leaflets. They have developed a deep learning approach based on CNN that was trained on detecting mitral annulus coordinates. Their approach had a mean error of only 2 ± 1.9 mm after enforcing continuity around the predicted annulus points. Moreover, Corinzia et al. (2020) developed an unsupervised approach based on low-dimensional embedding of 3D echocardiograms using neural networks to segment the mitral valve. In their work, they have used multilayer perceptron and neural networks to detect mitral valve regions in 85 TTE echocardiograms. They have achieved a dice coefficient of 0.48 with overall benchmark performance of 0.45. Vafaeezadeh et al. (2021) proposed a CNN approach based on transfer learning to identify patients with normal mitral valves from patients with prosthetic mitral valves in TTE echocardiograms. Their algorithm utilized a total of 2044 images and achieved an overall accuracy measure of 98% with an AUC up to 0.99. In the area of the detection of abnormal valves in echocardiograms, Zhang et al. (2021) trained CNN models on classifying mild, moderate, and severe mitral regurgitation (MR) from 1427 doppler echocardiography cases. Their algorithm had an overall accuracy of up to 91% with a macro F1 and micro F1 coefficients of 0.91 and 0.92, respectively.

9.5 Challenges and future directions

Although AI could potentiality facilitate new protocols to ease the assessment of many cardiovascular diseases in cardiac imaging, it is still at a nascent stage in this field and many challenges should be carefully addressed. Therefore proper future directions should be followed to enrich the use of AI in clinical practice.

Most machine and deep learning algorithms could suffer from data bias that takes place during training (Petersen et al., 2019). For example, training models on specific patient cohort, although returns promising performance levels, may lack the ability to be generalized over wider patient populations. In addition, studies carried out in wealthy populations only, that have enough equipment and computation tools for AI, could be biased in terms of evaluating diseases as it discards information about patients in poorer parts of the world. Thus, AI should be tested on large patient cohorts before coming out with conclusions about cardiac diseases.

Black-box mechanism in current AI tools still raises uncertainty concerns to clinicians on how decisions were made and reached by the developed algorithm (Rudin, 2019). The lack of interpretability creates a major confusion when diagnosing cardiovascular diseases, because the cause of such abnormalities is not defined properly. It is always advised to develop the upcoming AI tools with an explainability mechanism to guide medical doctors' decisions and allow them to understand the roots of the problem, especially with the ability of deep learning to be trained on thousands of data. In cardiac imaging, images are a rich material for such approaches as it could be easier to point out regions within the image that derives the decisions of the trained model regarding diseases or any abnormal behaviors in the cardiovascular system.

Sharing patient information to be used for AI development is another challenge in cardiovascular imaging (Nundy et al., 2019). Although a participant signs a consent form agree to on the contribution on any further studies, one could not simply assume that patient and the public are happy for healthcare data exchange, especially when severe diseases take place. The nonreversibility concern should be carefully handled when a patient decides to withdraw his consent from the study, as pretrained machine and deep learning models do not allow the removal of single patient data from trained parameters. Moreover, another concern arises when sharing patient data with commercial partners when building AI algorithms. Therefore, a trust should be built between patients and developers to maintain practical and acceptable solutions.

To ensure a bright AI future in clinical practice within cardiology, several key messages to the researchers and medical practitioners should be carefully considered in future. Joint development from both sides is key to success. AI experts should be able to understand the challenges doctors face in dealing with patients in clinical situations. On the other hand, clinicians should be trained carefully on becoming AI and data experts. In addition, concerns related to privacy, ethics, and commercial partnerships should be heavily studied to ensure a balance between data access and patient satisfaction. Lastly, it has become clear that a successful integration of AI in cardiology will not be possible without a proper collaboration between clinicians, researchers, computer scientists, and patients to establish new protocols in handling diseases and thus, enhancing the quality of life for many patients worldwide.

9.6 Conclusion

This chapter covered the most recent advances of AI in cardiovascular imaging which have showed a strong potential in guiding diagnostic and treatment protocols by ensuring accurate and timely assessment of many diseases from trained machine perspective. There is still room for research to enhance the implementation of AI in clinical settings and thus, leveraging next-generation healthcare that focuses on personalized medicine according to patient-specific conditions owning to the fact that AI can handle and learn from big patient data and draw conclusion that aid clinicians in providing accurate and timely decision making.

References

Abdelrahman, K. M., Chen, M. Y., Dey, A. K., Virmani, R., Finn, A. V., Khamis, R. Y., Choi, A. D., Min, J. K., Williams, M. C., Buckler, A. J., Taylor, C. A., Rogers, C., Samady, H., Antoniades, C., Shaw, L. J., Budoff, M. J., Hoffmann, U., Blankstein, R., Narula, J., & Mehta, N. N. (2020). Coronary computed tomography angiography from clinical uses to emerging technologies: JACC STATE-OF-THE-ART REview. *Journal of the American College of Cardiology*, *10*, 1226–1243. Available from https://doi.org/10.1016/J.JACC.2020.06.076.

Agatston, A. S., Janowitz, W. R., Hildner, F. J., Zusmer, N. R., Viamonte, M., & Detrano, R. (1990). Quantification of coronary artery calcium using ultrafast computed tomography. *Journal of the American College of Cardiology*, *4*, 827–832. Available from https://doi.org/10.1016/0735-1097(90)90282-T.

Al'Aref, S. J., Maliakal, G., Singh, G., van Rosendael, A. R., Ma, X., Xu, Z., Al Hussein Alawamlh, O., Lee, B., Pandey, M., Achenbach, S., Al-Mallah, M. H., Andreini, D., Bax, J. J., Berman, D. S., Budoff, M. J., Cademartiri, F., Callister, T. Q., Chang, H. J., Chinnaiyan, K., … Shaw, L. J. (2020). Machine learning of clinical variables and coronary artery calcium scoring for the prediction of obstructive coronary artery disease on coronary computed tomography angiography: Analysis from the CONFIRM registry. *European Heart Journal*, *3*, 359–367. Available from https://doi.org/10.1093/EURHEARTJ/EHZ565, https://pubmed.ncbi.nlm.nih.gov/31513271/.

Alfakih, K., Reid, S., Jones, T., & Sivananthan, M. (2004). Assessment of ventricular function and mass by cardiac magnetic resonance. *European Radiology*, 1813–1822.

Alkhodari, M., & Fraiwan, L. (2021). Convolutional and recurrent neural networks for the detection of valvular heart diseases in phonocardiogram recordings. *Computer Methods and Programs in Biomedicine*, 105940. Available from https://doi.org/10.1016/J.CMPB.2021.105940.

Alkhodari, M., Jelinek, H. F., Karlas, A., Soulaidopoulos, S., Arsenos, P., Doundoulakis, I., Gatzoulis, K. A., Tsioufis, K., Hadjileontiadis, L. J., & Khandoker, A. H. (2021). Deep learning predicts heart failure with preserved, mid-range, and reduced left ventricular ejection fraction from patient clinical profiles. *Frontiers in Cardiovascular Medicine*, 1604. Available from https://doi.org/10.3389/FCVM.2021.755968/BIBTEX.

Alsharqi, M., Woodward, W. J., Mumith, J. A., Markham, D. C., Upton, R., & Leeson, P. (2018). Artificial intelligence and echocardiography. *Echo Research and Practice*, *4*, R115. Available from https://doi.org/10.1530/ERP-18-0056, https://www.ncbi.nlm.nih.gov/pmc/articles/PMC6280250/.

Alsharqi, M., Upton, R., Mumith, A., & Leeson, P. (2018). *Artificial intelligence: A new clinical support tool for stress echocardiography*, *8*, 513–515. Available from https://doi.org/10.1080/17434440.2018.1497482, https://www.tandfonline.com/doi/abs/10.1080/17434440.2018.1497482.

Aly, I., Rizvi, A., Roberts, W., Khalid, S., Kassem, M. W., Salandy, S., du Plessis, M., Tubbs, R. S., & Loukas, M. (2021). Cardiac ultrasound: An anatomical and clinical review. *Translational Research in Anatomy*, 100083. Available from https://doi.org/10.1016/J.TRIA.2020.100083.

Andreassen, B. S., Veronesi, F., Gerard, O., Solberg, A. H. S., & Samset, E. (2020). Mitral annulus segmentation using deep learning in 3-D transesophageal echocardiography.

IEEE Journal of Biomedical and Health Informatics, *4*, 994–1003. Available from https://doi.org/10.1109/JBHI.2019.2959430.

Assomull, R. G., Pennell, D. J., & Prasad, S. K. (2007). Cardiovascular magnetic resonance in the evaluation of heart failure. *Heart (British Cardiac Society)*, 985–992.

Augustine, D., Ayers, L. V., Lima, E., Newton, L., Lewandowski, A. J., Davis, E. F., Ferry, B., & Leeson, P. (2014). Dynamic release and clearance of circulating microparticles during cardiac stress. *Circulation Research*, *1*, 109–113. Available from https://doi.org/10.1161/CIRCRESAHA.114.301904, https://pubmed.ncbi.nlm.nih.gov/24141170/.

Azour, L., Kadoch, M. A., Ward, T. J., Eber, C. D., & Jacobi, A. H. (2017). Estimation of cardiovascular risk on routine chest CT: Ordinal coronary artery calcium scoring as an accurate predictor of Agatston score ranges. *Journal of Cardiovascular Computed Tomography*, *1*, 8–15. Available from https://doi.org/10.1016/J.JCCT.2016.10.001, https://www.journalofcardiovascularct.com/article/S1934-5925(16)30238-6/abstract.

Baird, A. E., & Warach, S. (1998). Magnetic resonance imaging of acute stroke. *Journal of Cerebral Blood Flow & Metabolism*, 583–609.

Bennasar, M., Banks, D., Price, B. A., & Kardos, A. (2020). Minimal patient clinical variables to accurately predict stress echocardiography outcome: Validation study using machine learning techniques. *JMIR Cardio*, *4*(1). Available from https://doi.org/10.2196/16975, https://pubmed.ncbi.nlm.nih.gov/32469316/.

Benza, R., Biederman, R., Murali, S., & Gupta, H. (2008). Role of cardiac magnetic resonance imaging in the management of patients. *Journal of the American College of Cardiology*, 1683–1692.

Bernard, O., Lalande, A., Zotti, C., Cervenansky, F., Yang, X., Heng, P.-A., Cetin, I., Lekadir, K., Camara, O., & Ballester, M. A. G. Others. (2018). Deep learning techniques for automatic MRI cardiac multi-structures. *IEEE Transactions on Medical Imaging*, 2514–2525.

Bohbot, Y., Renard, C., Manrique, A., Levy, F., Maréchaux, S., Gerber, B. L., & Tribouilloy, C. (2020). Usefulness of cardiac magnetic resonance imaging in aortic stenosis. *Circulation: Cardiovascular Imaging*, e010356.

Bratt, A., Kim, J., Pollie, M., Beecy, A. N., Tehrani, N. H., Codella, N., Perez-Johnston, Rocio, Palumbo, M. C., Alakbarli, J., & Colizza, W.Others. (2019). Machine learning derived segmentation of phase velocity encoded. *Journal of Cardiovascular Magnetic Resonance*, 1–11.

Campello, V. M., Gkontra, P., Izquierdo, C., Martín-Isla, C., Sojoudi, A., Full, P. M., Maier-Hein, K., Zhang, Y., He, Z., & Ma, J.Others. (2021). Multi-centre, multi-vendor and multi-disease cardiac segmentation: *IEEE Transactions on Medical Imaging*, 3543–3554.

Carolina, S. S. M. (2020). Value of machine learning-based coronary CT fractional flow reserve applied to triple-rule-out CT angiography in acute chest pain • content codes. *Radiology: Cardiothoracic Imaging*, *3*, e190137. Available from https://doi.org/10.1148/ryct.2020190137.

Caruthers, S. D., Lin, S. J., Brown, P., Watkins, M. P., Williams, T. A., Lehr, K. A., & Wickline, S. A. (2003). Practical value of cardiac magnetic resonance imaging for clinical. *Circulation*, 2236–2243.

Cho, I., Al'aref, S. J., Berger, A., Ó Hartaigh, B., Gransar, H., Valenti, V., Lin, F. Y., Achenbach, S., Berman, D. S., Budoff, M. J., Callister, T. Q., Al-Mallah, M. H.,

Cademartiri, F., Chinnaiyan, K., Chow, B. J. W., Delago, A., Villines, T. C., Hadamitzky, M., Hausleiter, J., ... Min, J. K. (2018). Prognostic value of coronary computed tomographic angiography findings in asymptomatic individuals: A 6-year follow-up from the prospective multicentre international CONFIRM study. *European Heart Journal*, *11*, 934–941. Available from https://doi.org/10.1093/EURHEARTJ/EHX774, https://pubmed.ncbi.nlm.nih.gov/29365193/.

Choi, D.-J., Park, J. J., Ali, T., Lee, S., (n.d.) ARTICLE Artificial intelligence for the diagnosis of heart failure. Available from: https://doi.org/10.1038/s41746-020-0261-3.

Coenen, A., Kim, Y. H., Kruk, M., Tesche, C., De Geer, J., Kurata, A., Lubbers, M. L., Daemen, J., Itu, L., Rapaka, S., Sharma, P., Schwemmer, C., Persson, A., Schoepf, U. J., Kepka, C., Yang, D. H., & Nieman, K. (2018). Diagnostic accuracy of a machine-learning approach to coronary computed tomographic angiography–Based fractional flow reserve result from the MACHINE Consortium. *Circulation: Cardiovascular Imaging*, *6*, 7217. Available from https://doi.org/10.1161/CIRCIMAGING.117.007217, https://www.ahajournals.org/doi/abs/10.1161/CIRCIMAGING.117.007217.

Corinzia, L., Laumer, F., Candreva, A., Taramasso, M., Maisano, F., & Buhmann, J. M. (2020). Neural collaborative filtering for unsupervised mitral valve segmentation in echocardiography. *Artificial Intelligence in Medicine*, 101975. Available from https://doi.org/10.1016/J.ARTMED.2020.101975.

Danad, I., Szymonifka, J., Schulman-Marcus, J., Min, J. K. Static and dynamic assessment of myocardial perfusion by computed tomography. Available from: https://academic.oup.com/ehjcimaging/article/17/8/836/1748290. https://doi.org/10.1093/ehjci/jew044.

Dey, D., Slomka, P. J., Leeson, P., Comaniciu, D., Shrestha, S., Sengupta, P. P., & Marwick, T. H. (2019). Artificial intelligence in cardiovascular imaging: JACC state-of-the-art review. *Journal of the American College of Cardiology*, *11*, 1317–1335. Available from https://doi.org/10.1016/J.JACC.2018.12.054.

Diller, G.-P., Vahle, J., Radke, R., Vidal, M. L. B., Fischer, A. J., Bauer, U. M. M., Sarikouch, S., Berger, F., Beerbaum, P., & Baumgartner, H. (2020). Utility of deep learning networks for the generation of artificial cardiac, Others *BMC Medical Imaging*, 1–8.

Draper, J., Subbiah, S., Bailey, R., & Chambers, J. B. (2019). Murmur clinic: Validation of a new model for detecting heart valve disease. *Heart (British Cardiac Society)*, *1*, 56–59. Available from https://doi.org/10.1136/HEARTJNL-2018-313393, https://heart.bmj.com/content/105/1/56.abstract.

Duffy, G., Cheng, P. P., Yuan, N., He, B., Kwan, A. C., Shun-Shin, M. J., Alexander, K. M., Ebinger, J., Lungren, M. P., Rader, F., Liang, D. H., Schnittger, I., Ashley, E. A., Zou, J. Y., Patel, J., Witteles, R., Cheng, S., & Ouyang, D. (2022). High-throughput precision phenotyping of left ventricular hypertrophy with cardiovascular deep learning. *JAMA Cardiology.*. Available from https://doi.org/10.1001/JAMACARDIO.2021.6059, https://jamanetwork.com/journals/jamacardiology/fullarticle/2789370.

Edalati, M., Zheng, Y., Watkins, M. P., Chen, J., Liu, L., Zhang, S., Song, Y., Soleymani, S., Lenihan, D. J., & Lanza, G. M. (2022). Implementation and prospective clinical validation of AI-based planning. *Medical Physics*, 129–143.

Eng, D., Chute, C., Khandwala, N., Rajpurkar, P., Long, J., Shleifer, S., Khalaf, M. H., Sandhu, A. T., Rodriguez, F., Maron, D. J., Seyyedi, S., Marin, D., Golub, I., Budoff, M., Kitamura, F., Takahashi, M. S., Filice, R. W., Shah, R., Mongan, J., ... Patel, B. N. (2021). Automated coronary calcium scoring using deep learning with multicenter external validation. *NPJ Digital Medicine*, *1*. Available from https://doi.org/10.1038/S41746-021-00460-1, https://pubmed.ncbi.nlm.nih.gov/34075194/.

Fihn, S. D., Blankenship, J. C., Alexander, K. P., Bittl, J. A., Byrne, J. G., Fletcher, B. J., Fonarow, G. C., Lange, R. A., Levine, G. N., Maddox, T. M., Naidu, S. S., Ohman, E. M., Smith, P. K., Anderson, J. L., Halperin, J. L., Albert, N. M., Bozkurt, B., Brindis, R. G., Curtis, L. H., ... Shen, W. K. (2015). 2014 ACC/AHA/AATS/PCNA/SCAI/STS focused update of the guideline for the diagnosis and management of patients with stable ischemic heart disease: A report of the American College of Cardiology/American Heart Association Task Force on Practice Guidelines, and the American Association for Thoracic Surgery, Preventive Cardiovascular Nurses Association, Society for Cardiovascular Angiography and Interventions, and Society of Thoracic Surgeons. *Journal of Thoracic and Cardiovascular Surgery*, *3*, e5–e23. Available from https://doi.org/10.1016/J.JTCVS.2014.11.002/ATTACHMENT/E2893B99-97E5-422B-A7A2-7A51888C4E74/MMC1.PDF, https://www.jtcvs.org/article/S0022-5223(14)01776-0/abstract.

Fletcher, A. J., Lapidaire, W., & Leeson, P. (2021). Machine learning augmented echocardiography for diastolic function assessment. *Frontiers in Cardiovascular Medicine*, 879. Available from https://doi.org/10.3389/FCVM.2021.711611.

Fotaki, A., Puyol-Antón, E., Chiribiri, A., Botnar, R., Pushparajah, K., & Prieto, C. (2021). Artificial Intelligence in Cardiac MRI: Is clinical adoption forthcoming? *Frontiers in Cardiovascular Medicine*.

Fuad Jan, M., & Jamil Tajik, A. (2017). Modern imaging techniques in cardiomyopathies. *Circulation Research*, *7*, 874–891. Available from https://doi.org/10.1161/CIRCRESAHA.117.309600, https://pubmed.ncbi.nlm.nih.gov/28912188/.

Ghorbani, A., Ouyang, D., Abid, A., He, B., Chen, J. H., Harrington, R. A., Liang, D. H., Ashley, E. A., Zou, J. Y. Deep learning interpretation of echocardiograms. Available from: https://doi.org/10.1038/s41746-019-0216-8.

Gohmann, R. F., Pawelka, K., Seitz, P., Majunke, N., Heiser, L., Renatus, K., Desch, S., Lauten, P., Holzhey, D., Noack, T., Wilde, J., Kiefer, P., Krieghoff, C., Lücke, C., Gottschling, S., Ebel, S., Borger, M. A., Thiele, H., Panknin, C., ... Gutberlet, M. (2021). Combined coronary CT-angiography and TAVR planning for ruling out significant coronary artery disease: Added value of machine-learning–based CT-FFR. *JACC: Cardiovascular Imaging*. Available from https://doi.org/10.1016/J.JCMG.2021.09.013.

Grover, V. P. B., Tognarelli, J. M., Crossey, M. M. E., Cox, I. J., Taylor-Robinson, S. D., & McPhail, M. J. W. (2015). Magnetic resonance imaging: Principles and techniques: Lessons for. *Journal of Clinical and Experimental Hepatology*, 246–255.

Han, D., Lee, J. H., Rizvi, A., Gransar, H., Baskaran, L., Schulman-Marcus, J., ó Hartaigh, B., Lin, F. Y., & Min, J. K. (2018). Incremental role of resting myocardial computed tomography perfusion for predicting physiologically significant coronary artery disease: A machine learning approach. *Journal of Nuclear Cardiology: Official Publication of the American Society of Nuclear Cardiology*, *1*, 223–233. Available from https://doi.org/10.1007/S12350-017-0834-Y, https://pubmed.ncbi.nlm.nih.gov/28303473/.

Ihdayhid, A. R., & Zekry, S. B. (2020). Machine learning CT FFR: The evolving role of on-site techniques. *Radiology: Cardiothoracic Imaging*, *3*. Available from https://doi.org/10.1148/RYCT.2020200228/ASSET/IMAGES/LARGE/RYCT.2020200228.FIG2.JPEG, https://pubs.rsna.org/doi/abs/10.1148/ryct.2020200228.

Jin, H. Y., Weir-McCall, J. R., Leipsic, J. A., Son, J. W., Sellers, S. L., Shao, M., Blanke, P., Ahmadi, A., Hadamitzky, M., Kim, Y. J., Conte, E., Andreini, D., Pontone, G., Budoff, M. J., Gottlieb, I., Lee, B. K., Chun, E. J., Cademartiri, F., Maffei, E., ... Chang, H. J. (2021). The relationship between coronary calcification and the natural history of coronary artery disease. *JACC: Cardiovascular Imaging*, *1*, 233–242. Available from https://doi.org/10.1016/J.JCMG.2020.08.036.

Karamitsos, T. D., Francis, J. M., Myerson, S., Selvanayagam, J. B., & Neubauer, S. (2009). The role of cardiovascular magnetic resonance imaging in heart failure. *Journal of the American College of Cardiology*, 1407–1424.

Karimi-Bidhendi, S., Arafati, A., Cheng, A. L., Wu, Y., Kheradvar, A., & Jafarkhani, H. (2020). Springer Fully-automated deep-learning segmentation of pediatric cardiovascular. *Journal of Cardiovascular Magnetic Resonance*, 1–24, 22.

Kart, T., Alkhodari, M., Lapidaire, W., & Leeson, P. (2023). Modelling relations between blood pressure, cardiovascular phenotype and clinical factors using large scale imaging data. *European Heart Journal-Cardiovascular Imaging, jead, 161*. Available from https://doi.org/10.1093/ehjci/jead161.

Kerneis, M., Nafee, T., Yee, M. K., Kazmi, H. A., Datta, S., Zeitouni, M., Afzal, M. K., Jafarizade, M., Walia, S. S., Qamar, I., Pitliya, A., Kalayci, A., Al Khalfan, F., & Gibson, C. M. (2019). Most promising therapies in interventional cardiology. *Current Cardiology Reports*, *4*(4), 1–8. Available from https://doi.org/10.1007/S11886-019-1108-X, https://link.springer.com/article/10.1007/s11886-019-1108-x.

Kusunose, K., Abe, T., Haga, A., Fukuda, D., Yamada, H., Harada, M., & Sata, M. (2020). A deep learning approach for assessment of regional wall motion abnormality from echocardiographic images. *JACC. Cardiovascular Imaging*, *2*(Pt 1), 374–381. Available from https://doi.org/10.1016/J.JCMG.2019.02.024, https://pubmed.ncbi.nlm.nih.gov/31103590/.

Ladapo, J. A., Blecker, S., & Douglas, P. S. (2014). Physician decision making and trends in the use of cardiac stress testing in the United States: An analysis of repeated cross-sectional data. *Annals of Internal Medicine*, *7*, 482–490. Available from https://doi.org/10.7326/M14-0296, https://pubmed.ncbi.nlm.nih.gov/25285541/.

Li, M., Chen, Y., Mao, Y., Jiang, M., Liu, Y., Zhan, Y., Li, X., Su, C., Zhang, G., & Zhou, X. (2021). Diagnostic Classification of Patients with Dilated Cardiomyopathy Using. *Computational and Mathematical Methods in Medicine*.

Li, S., Zhang, Y., Yang, X. (2021). 2021 1402-1405 semi-supervised cardiac MRI segmentation based on generative adversarial.

Lossau (née Elss), T., Nickisch, H., Wissel, T., Bippus, R., Schmitt, H., Morlock, M., & Grass, M. (2019). Motion estimation and correction in cardiac CT angiography images using convolutional neural networks. *Computerized Medical Imaging and Graphics: The Official Journal of the Computerized Medical Imaging Society*. Available from https://doi.org/10.1016/J.COMPMEDIMAG.2019.06.001, https://pubmed.ncbi.nlm.nih.gov/31299452/.

Maroules, C. D., Rajiah, P., Bhasin, M., & Abbara, S. (2019). Current evidence in cardiothoracic imaging: Growing evidence for coronary computed tomography angiography as a first-line test in Stable Chest Pain. *Journal of Thoracic Imaging*, *1*, 4–11. Available from https://doi.org/10.1097/RTI.0000000000000357, https://journals.lww.com/thoracicimaging/Fulltext/2019/01000/Current_Evidence_in_Cardiothoracic_Imaging_0.3.aspx.

Montalescot, G., Sechtem, U., Achenbach, S., Andreotti, F., Arden, C., Budaj, A., Bugiardini, R., Crea, F., Cuisset, T., Di Mario, C., Ferreira, J. R., Gersh, B. J., Gitt, A. K., Hulot, J. S., Marx, N., Opie, L. H., Pfisterer, M., Prescott, E., Ruschitzka, F., ... Yildirir, Aylin (2013). 2013 ESC guidelines on the management of stable coronary artery disease: The task force on the management of stable coronary artery disease of the European Society of Cardiology. *European Heart Journal*, *38*, 2949–3003. Available from https://doi.org/10.1093/EURHEARTJ/EHT296, http://europepmc.org/article/MED/23996286.

Monti, C. B., Codari, M., van Assen, M., De Cecco, C. N., & Vliegenthart, R. (2020). Machine learning and deep neural networks applications in computed tomography for coronary artery disease and myocardial perfusion. *Journal of Thoracic Imaging*, S58–S65. Available from https://doi.org/10.1097/RTI.0000000000000490, https://pubmed.ncbi.nlm.nih.gov/32195886/.

Mori, H., Torii, S., Kutyna, M., Sakamoto, A., Finn, A. V., & Virmani, R. (2018). Coronary artery calcification and its progression: What does it really mean? *JACC: Cardiovascular Imaging*, *1*, 127–142. Available from https://doi.org/10.1016/J.JCMG.2017.10.012.

Motwani, M., Dey, D., Berman, D. S., Germano, G., Achenbach, S., Al-Mallah, M. H., Andreini, D., Budoff, M. J., Cademartiri, F., Callister, T. Q., Chang, H. J., Chinnaiyan, K., Chow, B. J. W., Cury, R. C., Delago, A., Gomez, M., Gransar, H., Hadamitzky, M., Hausleiter, J., ... Slomka, P. J. (2017). Machine learning for prediction of all-cause mortality in patients with suspected coronary artery disease: A 5-year multicentre prospective registry analysis. *European heart journal*, *7*, 500–507. Available from https://doi.org/10.1093/EURHEARTJ/EHW188, https://pubmed.ncbi.nlm.nih.gov/27252451/.

Nagueh, S. F., Smiseth, O. A., Appleton, C. P., Byrd, B. F., Dokainish, H., Edvardsen, T., Flachskampf, F. A., Gillebert, T. C., Klein, A. L., Lancellotti, P., Marino, P., Oh, J. K., Popescu, B. A., & Waggoner, A. D. (2016). Recommendations for the evaluation of left ventricular diastolic function by echocardiography: An update from the american society of echocardiography and the european association of cardiovascular imaging. *Journal of the American Society of Echocardiography: Official Publication of the American Society of Echocardiography*, *4*, 277–314. Available from https://doi.org/10.1016/J.ECHO.2016.01.011, https://pubmed.ncbi.nlm.nih.gov/27037982/.

Nakao, Y. M., Miyamoto, Y., Higashi, M., Noguchi, T., Ohishi, M., Kubota, I., Tsutsui, H., Kawasaki, T., Furukawa, Y., Yoshimura, M., Morita, H., Nishimura, K., Kada, A., Goto, Y., Okamura, T., Tei, C., Tomoike, H., Naito, H., & Yasuda, S. (2018). Sex differences in impact of coronary artery calcification to predict coronary artery disease. *Heart (British Cardiac Society)*, *13*, 1118–1124. Available from https://doi.org/10.1136/HEARTJNL-2017-312151, https://heart.bmj.com/content/104/13/1118.abstract.

Nandalur, K. R., Dwamena, B. A., Choudhri, A. F., Nandalur, M. R., & Carlos, R. C. (2007). Diagnostic performance of stress cardiac magnetic resonance imaging in the. *Journal of the American College of Cardiology*, 1343–1353.

Nedadur, R., Wang, B., Tsang, W., & Street, E. (2022). Artificial intelligence for the echocardiographic assessment of valvular heart disease. *Heart (British Cardiac Society)*, 1–8. Available from https://doi.org/10.1136/heartjnl-2021-319725, http://heart.bmj.com/.

Nicol, E. D., Norgaard, B. L., Blanke, P., Ahmadi, A., Weir-McCall, J., Horvat, P. M., Han, K., Bax, J. J., & Leipsic, J. (2019). The future of cardiovascular computed tomography: Advanced analytics and clinical insights. *JACC: Cardiovascular Imaging*, *6*, 1058–1072. Available from https://doi.org/10.1016/J.JCMG.2018.11.037.

Nundy, S., Montgomery, T., & Wachter, R. M. (2019). Promoting trust between patients and physicians in the era of artificial intelligence. *JAMA: The Journal of the American Medical Association*, *6*, 497–498. Available from https://doi.org/10.1001/JAMA.2018.20563, https://pubmed.ncbi.nlm.nih.gov/31305873/.

Nørgaard, B. L., Fairbairn, T. A., Safian, R. D., Rabbat, M. G., Ko, B., Jensen, J. M., Nieman, K., Chinnaiyan, K. M., Sand, N. P., Matsuo, H., Leipsic, J., & Raff, G. (2019). Coronary ct angiography-derived fractional flow reserve testing in patients with stable coronary artery disease: Recommendations on interpretation and reporting. *Radiology: Cardiothoracic Imaging*, *5*. Available from https://pubs.rsna.org/doi/abs/10.1148/ryct.2019190050, https://doi.org/10.1148/RYCT.2019190050/ASSET/IMAGES/LARGE/RYCT.2019190050.FIG4.JPEG.

Oikonomou, E. K., West, H. W., & Antoniades, C. (2019). Cardiac computed tomography. *Arteriosclerosis, Thrombosis, and Vascular Biology*, *11*, 2207–2219. Available from https://doi.org/10.1161/ATVBAHA.119.312899, https://www.ahajournals.org/doi/abs/10.1161/ATVBAHA.119.312899.

Omar, H. A., Domingos, J. S., Patra, A., Upton, R., Leeson, P., & Alison, J. (2018). Noble, Quantification of cardiac bull's-eye map based on principal strain analysis for myocardial wall motion assessment in stress echocardiography. *Proceedings - International Symposium on Biomedical Imaging*, 1195–1198. Available from https://doi.org/10.1109/ISBI.2018.8363785.

Pellikka, M. D. P. A. (2022). Artificially intelligent interpretation of stress echocardiography: The future is now*. *Cardiovascular Imaging*. Available from https://doi.org/10.1016/J.JCMG.2021.110.010, https://www.jacc.org/doi/10.1016/j.jcmg.2021.11.010.

Pellikka, P. A., Arruda-Olson, A., Chaudhry, F. A., Hui Chen, M., Marshall, J. E., Porter, T. R., & Sawada, S. G. (2020). New York, guidelines for performance, interpretation, and application of stress echocardiography in ischemic heart disease: From the American Society of echocardiography. *Journal of the American Society of Echocardiography*, 1–41.e8. Available from https://doi.org/10.1016/j.echo.2019.07.001.

Petersen, S. E., Abdulkareem, M., & Leiner, T. (2019). Artificial intelligence will transform cardiac imaging—Opportunities and challenges. *Frontiers in Cardiovascular Medicine*, 133. Available from https://doi.org/10.3389/FCVM.2019.00133/BIBTEX.

Peterzan, M. A., Rider, O. J., & Anderson, L. J. (2016). The role of cardiovascular magnetic resonance imaging in heart failure. *Cardiac Failure Review*, *115*.

Petitjean, C., Zuluaga, M. A., Bai, W., Dacher, J.-N., Grosgeorge, D., Caudron, J., Ruan, S., Ayed, I. B., Cardoso, M. J., & Chen, H.-C. (2015). Right ventricle segmentation from cardiac MRI: A collation study, Others *Medical Image Analysis*, 187–202.

Puyol-Antón E. Ruijsink B. Gerber B. Amzulescu M.S., Langet H., Craene M.D., Schnabel J.A. Piro P. King A.P. (2018). 2018 IEEE Transactions on Biomedical Engineering 956-966 IEEE Regional multi-view learning for cardiac motion analysis: Application to 66.

Radau, P., Lu, Y., Connelly, K., Paul, G., Dick, A., & Wright, G. (2009). Evaluation framework for algorithms segmenting short axis cardiac MRI. *The MIDAS Journal-Cardiac MR Left Ventricle Segmentation Challenge*.

Rickers, C., Wilke, N. M., Jerosch-Herold, M., Casey, S. A., Panse, P., Panse, N., Weil, J., Zenovich, A. G., & Maron, B. J. (2005). Utility of cardiac magnetic resonance imaging in the diagnosis of. *Circulation*, 855–861.

Rogers, M. A., & Aikawa, E. (2018). Cardiovascular calcification: Artificial intelligence and big data accelerate mechanistic discovery. *Nature Reviews Cardiology*, *16*(5),

261–274. Available from https://doi.org/10.1038/s41569-018-0123-8, https://www.nature.com/articles/s41569-018-0123-8.

Ronneberger O. Fischer P. Brox T. (2015) 2015 234-241 U-net: Convolutional networks for biomedical image segmentation.

Rudin, C. (2019). Stop explaining black box machine learning models for high stakes decisions and use interpretable models instead. *Nature Machine Intelligence 2019 1:5. (5)*, 206–215. Available from https://doi.org/10.1038/s42256-019-0048-x, https://www.nature.com/articles/s42256-019-0048-x.

Saeed, M., Van, T. A., Krug, R., Hetts, S. W., & Wilson, M. W. (2015). Cardiac MR imaging: Current status and future direction. *Cardiovascular Diagnosis and Therapy, 290.*

Seetharam, K., Shrestha, S., & Sengupta, P. P. (2019). Artificial intelligence in cardiac imaging. *US Cardiology Review*, *2*, 110–116. Available from https://doi.org/10.15420/USC.2019.19.2.

Seetharam, K., Shrestha, S., & Sengupta, P. P. (2021). Cardiovascular imaging and intervention through the lens of artificial intelligence. *Interventional Cardiology Review*. Available from https://doi.org/10.15420/ICR.2020.04.

Seitun, S., Castiglione Morelli, M., Budaj, I., Boccalini, S., Galletto Pregliasco, A., Valbusa, A., Cademartiri, F., & Ferro, C. (2016). Stress computed tomography myocardial perfusion imaging: A new topic in cardiology. *Revista espanola de cardiologia (English ed.)*, *2*, 188–200. Available from https://doi.org/10.1016/J.REC.2015.10.018, https://pubmed.ncbi.nlm.nih.gov/26774540/.

Sermesant, M., Delingette, H., Cochet, H., Jaïs, P., & Ayache, N. (2021). Applications of artificial intelligence in cardiovascular imaging. *Nature Reviews Cardiology*, *18*(8), 600–609. Available from https://doi.org/10.1038/s41569-021-00527-2, https://www.nature.com/articles/s41569-021-00527-2.

Shan, K., Constantine, G., Sivananthan, M., & Flamm, S. D. (2004). Role of cardiac magnetic resonance imaging in the assessment of myocardial. *Circulation*, 1328–1334.

Shehata, M. L., Turkbey, E. B., Vogel-Claussen, J., & Bluemke, D. A. (2008). Role of cardiac magnetic resonance imaging in assessment of nonischemic. *Topics in Magnetic Resonance Imaging*, 43–57.

Shillcutt, S. K., & Bick, J. S. (2013). Echo didactics: A comparison of basic transthoracic and transesophageal echocardiography views in the perioperative setting. *Anesthesia and Analgesia*, *6*, 1231–1236. Available from https://doi.org/10.1213/ANE.0B013E31828CBACA, https://pubmed.ncbi.nlm.nih.gov/23558842/.

Siegersma, K. R., Leiner, T., Chew, D. P., Appelman, Y., Hofstra, L., & Verjans, J. W. (2019). Artificial intelligence in cardiovascular imaging: State of the art and implications for the imaging cardiologist. *Netherlands Heart Journal*, *9*, 403. Available from https://doi.org/10.1007/S12471-019-01311-1, https://www.ncbi.nlm.nih.gov/pmc/articles/PMC6712136/.

Suinesiaputra, A., Mauger, C. A., Ambale-Venkatesh, B., Bluemke, D. A., Gade, J. D., Gilbert, K., Janse, M. H. A., Hald, L. S., Werkhoven, C., Wu, C. O., Lima, J. A. C., & Young, A. A. (2022). Deep learning analysis of cardiac MRI in legacy datasets: Multiethnic. *Frontiers in Cardiovascular Medicine*. Available from https://doi.org/10.3389/fcvm.2021.807728.

Syed, I. S., Glockner, J. F., Feng, D., Araoz, P. A., Martinez, M. W., Edwards, W. D., Gertz, M. A., Dispenzieri, A., Oh, J. K., & Bellavia, D. (2010). Role of cardiac magnetic resonance imaging in the detection of cardiac, Others *JACC: Cardiovascular Imaging*, 155–164.

Tatsugami, F., Higaki, T., Nakamura, Y., Yu, Z., Zhou, J., Lu, Y., Fujioka, C., Kitagawa, T., Kihara, Y., Iida, M., & Awai, K. (2019). Deep learning-based image restoration algorithm for coronary CT angiography. *European Radiology*, *10*, 5322–5329. Available from https://doi.org/10.1007/S00330-019-06183-Y, https://pubmed.ncbi.nlm.nih.gov/30963270/.

Tesche, C., De Cecco, C. N., Baumann, S., Renker, M., McLaurin, T. W., Duguay, T. M., Bayer, R. R., Steinberg, D. H., Grant, K. L., Canstein, C., Schwemmer, C., Schoebinger, M., Itu, L. M., Rapaka, S., Sharma, P., & Schoepf, U. J. (2018). Coronary CT angiography-derived fractional flow reserve: Machine learning algorithm versus computational fluid dynamics modeling. *Radiology*, *1*, 64–72. Available from https://doi.org/10.1148/RADIOL.2018171291, https://pubmed.ncbi.nlm.nih.gov/29634438/.

Tobon-Gomez, C., Craene, M. D., Mcleod, K., Tautz, L., Shi, W., Hennemuth, A., Prakosa, A., Wang, H., Carr-White, G., & Kapetanakis, S.Others. (2013). Benchmarking framework for myocardial tracking and deformation algorithms. *Medical image analysis*, *17*, 632–648, Elsevier.

Upton, R., Mumith, A., Beqiri, A., Parker, A., Hawkes, W., Gao, S., Porumb, M., Sarwar, R., Marques, P., Markham, D., Kenworthy, J., O'Driscoll, J. M., Hassanali, N., Groves, K., Dockerill, C., Woodward, W., Alsharqi, M., McCourt, A., Wilkes, E. H., ... Leeson, P. (2022). Automated echocardiographic detection of severe coronary artery disease using artificial intelligence. *JACC: Cardiovascular Imaging*. Available from https://doi.org/10.1016/J.JCMG.2021.10.013/SUPPL_FILE/MMC1.DOCX, https://www.jacc.org/doi/10.1016/j.jcmg.2021.10.013.

Vafaeezadeh, M., Behnam, H., Hosseinsabet, A., & Gifani, P. (2021). A deep learning approach for the automatic recognition of prosthetic mitral valve in echocardiographic images. *Computers in Biology and Medicine*, 104388. Available from https://doi.org/10.1016/J.COMPBIOMED.2021.104388.

van Assen, M., Varga-Szemes, A., Schoepf, U. J., Duguay, T. M., Hudson, H. T., Egorova, S., Johnson, K., Pierre, S. S., Zaki, B., Oudkerk, M., Vliegenthart, R., & Buckler, A. J. (2019). Automated plaque analysis for the prognostication of major adverse cardiac events. *European Journal of Radiology*, 76–83. Available from https://doi.org/10.1016/J.EJRAD.2019.04.013, https://pubmed.ncbi.nlm.nih.gov/31153577/.

van den Oever, L. B., Vonder, M., van Assen, M., van Ooijen, P. M. A., de Bock, G. H., Xie, X. Q., & Vliegenthart, R. (2020). Application of artificial intelligence in cardiac CT: From basics to clinical practice. *European Journal of Radiology*. Available from https://doi.org/10.1016/J.EJRAD.2020.108969, https://www.ejradiology.com/article/S0720-048X(20)30158-3/abstract.

van Rosendael, A. R., Maliakal, G., Kolli, K. K., Beecy, A., Al'Aref, S. J., Dwivedi, A., Singh, G., Panday, M., Kumar, A., Ma, X., Achenbach, S., Al-Mallah, M. H., Andreini, D., Bax, J. J., Berman, D. S., Budoff, M. J., Cademartiri, F., Callister, T. Q., Chang, H. J., ... Min, J. K. (2018). Maximization of the usage of coronary CTA derived plaque information using a machine learning based algorithm to improve risk stratification; insights from the CONFIRM registry. *Journal of Cardiovascular Computed Tomography*, *3*, 204–209. Available from https://doi.org/10.1016/J.JCCT.2018.04.011, https://moh-it.pure.elsevier.com/en/publications/maximization-of-the-usage-of-coronary-cta-derived-plaque-informat-2.

van Velzen, S. G. M., Lessmann, N., Velthuis, B. K., Bank, I. E. M., van den Bongard, D. H. J. G., Leiner, T., de Jong, P. A., Veldhuis, W. B., Correa, A., Terry, J. G., Carr, J. J., Viergever, M. A., Verkooijen, H. M., & Išgum, I. (2020). Deep learning for automatic calcium scoring in CT: Validation using multiple cardiac CT and chest CT protocols. *Radiology*, *1*, 66–79. Available from https://doi.org/10.1148/RADIOL.2020191621, https://pubmed.ncbi.nlm.nih.gov/32043947/.

Wang, W., Wang, H., Chen, Q., Zhou, Z., Wang, R., Zhang, N., Chen, Y., Sun, Z., & Xu, L. (2020). Coronary artery calcium score quantification using a deep-learning algorithm. *Clinical Radiology*, *3*, 237.e11–237.e16. Available from https://doi.org/10.1016/J.CRAD.2019.10.012, https://pubmed.ncbi.nlm.nih.gov/31718789/.

Wang, Y., Zhan, H., Hou, J., Ma, X., Wu, W., Liu, J., Gao, J., Guo, Y., & Zhang, Y. (2021). Influence of deep learning image reconstruction and adaptive statistical iterative reconstruction-V on coronary artery calcium quantification. *Annals of Translational Medicine*, *23*, 1726. Available from https://doi.org/10.21037/ATM-21-5548, https://www.ncbi.nlm.nih.gov/pmc/articles/PMC8743730/.

Williams, M. C., Moss, A. J., Dweck, M., Adamson, P. D., Alam, S., Hunter, A., Shah, A. S. V., Pawade, T., Weir-McCall, J. R., Roditi, G., van Beek, E. J. R., Newby, D. E., & Nicol, E. D. (2019). Coronary artery plaque characteristics associated with adverse outcomes in the SCOT-HEART study. *Journal of the American College of Cardiology*, *3*, 291–301. Available from https://doi.org/10.1016/J.JACC.2018.10.066.

Wolterink, J. M., Leiner, T., Takx, R. A. P., Viergever, M. A., & Išgum, I. (2015). Automatic coronary calcium scoring in non-contrast-enhanced ECG-triggered cardiac CT with ambiguity detection. *IEEE Transactions on Medical Imaging*, *9*, 1867–1878. Available from https://doi.org/10.1109/TMI.2015.2412651, https://pubmed.ncbi.nlm.nih.gov/25794387/.

World Health Organization. (2021). 2021 Cardiovascular diseases (CVDs) https://www.who.int/news-room/fact-sheets/detail/cardiovascular-diseases-(cvds).

Wolterink, J. M., Leiner, T., de Vos, B. D., van Hamersvelt, R. W., Viergever, M. A., & Išgum, I. (2016). Automatic coronary artery calcium scoring in cardiac CT angiography using paired convolutional neural networks. *Medical Image Analysis*, 123–136. Available from https://doi.org/10.1016/J.MEDIA.2016.04.004, https://pubmed.ncbi.nlm.nih.gov/27138584/.

Writing Committee MembersOtto, C. M., Nishimura, R. A., Bonow, R. O., Carabello, B. A., Erwin, J. P., Gentile, F., Jneid, H., Krieger, E. V., Mack, M., McLeod, C., O'Gara, P. T., Rigolin, V. H., Sundt, T. M., Thompson, A., & Toly, C. (2021). 2020 ACC/AHA guideline for the management of patients with valvular heart disease: Executive summary: A report of the American College of Cardiology/American Heart Association Joint Committee on clinical practice guidelines. *Journal of the American College of Cardiology*, *4*, 450–500. Available from https://doi.org/10.1016/j.jacc.2020.11.035, http://www.ncbi.nlm.nih.gov/pubmed/33342587.

Xiong, G., Kola, D., Heo, R., Elmore, K., Cho, I., & Min, J. K. (2015). Myocardial perfusion analysis in cardiac computed tomography angiographic images at rest. *Medical Image Analysis*, *1*, 77–89. Available from https://doi.org/10.1016/J.MEDIA.2015.05.010.

Yang, X., Zhang, Y., Lo, B., Wu, D., Liao, H., & Zhang, Y.-T. (2020). DBAN: Adversarial network with multi-scale features for cardiac MRI. *IEEE Journal of Biomedical and Health Informatics*, 2018–2028.

You, Y., Viktorovich, L. A., Qiu, J., Nikolaevich, K. A., & Vladimirovich, B. Y. (2021). Cardiac magnetic resonance image diagnosis of hypertrophic obstructive. *Computer Methods and Programs in Biomedicine*, 105889.

Zeng, Y., Sun, R., Li, X., Liu, M., Chen, S., & Zhang, P. (2016). Pathophysiology of valvular heart disease (Review). *Experimental and Therapeutic Medicine*, *4*, 1184–1188. Available from https://doi.org/10.3892/ETM.2016.3048/HTML, https://www.spandidos-publications.com/10.3892/etm.2016.3048.

Zhang, Q., Liu, Y., Mi, J., Wang, X., Liu, X., Zhao, F., Xie, C., Cui, P., Zhang, Q., & Zhu, X. (2021). Automatic assessment of mitral regurgitation severity using the mask R-CNN algorithm with color doppler echocardiography images. *Computational and Mathematical Methods in Medicine*. Available from https://doi.org/10.1155/2021/2602688.

Zhuang, Z., Jin, P., Noel, A., Raj, J., Yuan, Y., & Zhuang, S. (2021). *Automatic segmentation of left ventricle in echocardiography based on YOLOv3 model to achieve constraint and positioning*. Available from https://doi.org/10.1155/2021/3772129.

Zreik, M., Van Hamersvelt, R. W., Wolterink, J. M., Leiner, T., Viergever, M. A., & Išgum, I. (2019). A recurrent CNN for automatic detection and classification of coronary artery plaque and stenosis in coronary CT angiography. *IEEE Transactions on Medical Imaging*, *7*, 1588–1598. Available from https://doi.org/10.1109/TMI.2018.2883807, https://pubmed.ncbi.nlm.nih.gov/30507498/.

CHAPTER 10

Digital conversion and scaling of IgM and IgG antibody test results in COVID-19 diseases

Sayan Murat[1,2] and Sekeroglu Boran[3]

[1]*Faculty of Medicine, Clinical Laboratory, PCR Unit, Kocaeli University, Kocaeli, Turkey*
[2]*DESAM Research Institute, Near East University, Nicosia, Northern Cyprus, Mersin-10, Turkey*
[3]*Department of Software Engineering, Near East University, Nicosia, Northern Cyprus, Mersin-10, Turkey*

10.1 Introduction

The severe acute respiratory syndrome coronavirus-2 (SARS-CoV-2) pandemic was declared the coronavirus disease 2019 (COVID-19) pandemic on March 11, 2020. SARS-CoV-2 has affected 223 countries, more than 770 million confirmed COVID-19 cases, and almost six million deaths globally. The number of instances of COVID-19 infection continues to rise globally (World Health Organization, WHO Coronavirus Disease (COVID-19) Dashboard, 2023). Many countries were significantly affected and maintained lockdowns, physical distancing, quarantine, and travel restrictions to control the spread. These measures have greatly influenced the world economy. Therefore it has become crucial to develop diagnostic tests that provide reliable and rapid results on SARS-CoV-2 infection and host immune response to prevent future infections, increase treatment rates and follow-up, prevent deaths, and allow life to return to normal. Diagnostic manufacturers have increased their efforts to develop and produce these tests (Srivastava et al., 2020).

Several studies based on deep learning have been performed to detect COVID-19 using medical image scans (Dimililer & Sekeroglu, 2021; Sekeroglu & Ozsahin, 2020). Although more reliable datasets have been used in further studies (Tartaglione et al., 2020), biased, low-resolution, and inconsistent datasets considered in these studies have enabled PCR or rapid testing to be used as the primary tool for diagnosing COVID-19.

Thus, it is crucial to produce different diagnostic assays that provide reliable and rapid results regarding the SARS-CoV-2 infection and immune response in the host against the virus in a short timeframe to prevent future infections, enhance cure rate, prevent deaths, and normalize life (Yildirim et al., 2021). The COVID-19 pandemic is being fought by diagnostic test producers worldwide,

Artificial Intelligence and Image Processing in Medical Imaging. DOI: https://doi.org/10.1016/B978-0-323-95462-4.00010-8

who have worked to develop and produce quickly accessible test assays for identifying and treating COVID-19 infections. (The United States Food and Drug Administration FDA, U.S. Food and Drug Administration Kits: FDA, in Vitro Diagnostics, 2023).

Although detection of RNA gene targets (e.g., spike protein [S], an envelope protein [E], nucleocapsid protein (N), RNA-linked RNA polymerase [RdRp], open reading frame gene 1 [ORF 1a/b]) (Li et al., 2020; Sarıgül et al., 2020) by polymerase chain reaction (PCR) in nucleic acid amplification testing is susceptible and confirms the presence of COVID-19, only current infections can be detected. Additionally, patients who have recovered from the disease might be misdiagnosed using nucleic acid amplification testing (Ong et al., 2020; Sayan et al., 2020). Assay for anti-SARS-CoV-2 antibodies may help understand if someone has had a COVID-19 infection before and would play a significant role in identifying individuals who have cleared the virus and recovered from the disease. Moreover, the detection of antibodies may also help establish a measurable antibody response and estimate the duration of antibodies (immunoglobulins: IgM, IgG, IgA) responses by the human body, as well as providing a retrospective assessment of the size of the infected population for studies of epidemiological surveillance ("Centers for Disease Control and Interim Prevention Guidelines for Covid 19 Antibody Testing," 2020; Green et al., 2020).

Rapid and accurate laboratory assays are urgently desired to control and manage the COVID-19 pandemic (Vandenberg et al., 2021; West et al., 2021). However, the main challenge with novel SARS-CoV-2 antibody assays would be the sensitivity and specificity for detecting the antibodies test being selected (Péré et al., 2021). Different antibody assays' performance and effectiveness should be thoroughly evaluated for patient management.

In addition, the evaluation of the effectiveness of different blood samples, such as arterial (fingertip) and venous (serum and ethylene diamine tetra acetic acid [EDTA] blood) fractions, during the control and management of the pandemic is vital for the observation, measurement, and determination of the most informative sample for the level of IgM and IgG.

The numerical representation of the patient's antibody levels provides an objective analysis of the tests because of the differences between the results obtained from the medical doctors by visual inspection of kits. The numerical representation of the samples would produce consistent results for all models for each time; however, the levels acquired by the human experts may differ even for a single example for each analysis. The aim of the study is to assess the immunoglobulin IgM and IgG rapid immunochromatographic lateral flow assay for the rationalized use with digital conversion during the COVID-19 pandemic. Using quick test kits, this conversion will allow samples to be digitized and scaled consistently and identify the best blood samples for IgM and IgG level scaling. In addition to all the contributions mentioned above, noise and contamination that may occur on the antibody rapid tests and affect the visual observation of the tests are considered. A scale adjustment procedure is proposed using fundamental image processing techniques to monitor the antibodies' levels and vaccines' effects rapidly.

The rest of the chapter is organized as follows: Section 2 summarizes the considered dataset and proposed method in the Materials and Methods section. Section 3 presents performed experiments and the obtained results of the study. Discussions on the obtained results and the study's concluding remarks are presented in Sections 4 and 5, respectively.

10.2 Materials and methods

10.2.1 Dataset

Twenty-eight patients from Turkey (Ethical Approval Application No: YDU/2021/91–1295) were evaluated in this study. Eighteen of 28 patients were males, and the remaining 10 were females. All patients infected with COVID-19 were discharged after recovery from the clinic. One week after discharge, they were invited to the test.

Three types of blood samples (fingertip blood, EDTA blood, and serum) for each patient were collected and injected into qualitative rapid immunochromatographic LFA assay (RTA Laboratories Inc- MaxSure COVID-19 IgG/IgM Antibody Kit Istanbul-Turkey) to detect IgG and IgM antibodies to SARS-CoV-2, antibody tests. Rapid test results of each patient were obtained by injecting fingertip blood, EDTA blood, and serum samples into the kits, and a total of 168 fast test kit results were collected for IgM and IgG levels.

The images of the rapid test results were captured 15 minutes after the injection of the samples based on the suggestions of the quick test manufacturer, and a mobile phone camera was used. The camera angle and distance from the test kits were 180 degrees and 10 cm, respectively. Controlling natural illumination sources in the laboratory room provided constant illumination.

10.2.2 Proposed method

The result regions and the indicators of the rapid test kits were segmented from their frames manually. In this way, the losses of the information within the arrows showing the IgM and IgG values due to automatic segmentation errors and the lack of manual segmentation were eliminated. In addition, it facilitate medical doctors to repeatedly analyze the regions of the indicators independently in the results region produced by the kits. Fig. 10.1 shows the original and segmented images of patient 8 with the fingertip blood sample. "S," "C," "G," and "M" represent the sample input hole, rapid test control indicator, IgM-level indicator, and IgG-level indicator, respectively.

FIGURE 10.1

Original and manually segmented images for “patient 8” with the fingertip blood sample left to right: complete rapid COVID-19 rapid antibody test kit image, segmented result region image, segmented IgM indicator image, and segmented IgG indicator image.

Initially, all images were converted to grayscale to make images easier to process. Several grayscale conversion methods were proposed and investigated for several purposes (Güneş et al., 2016; Kanan et al., 2012). The ITU-R Recommendation BT.601 formula (International Telecommunication Union (ITU)- ITU R.- Recommendation Itu-r Bt.601–7., 2011) was considered because of giving the highest weight of red color, which is the region of interest, during the conversion. Eq. (10.1) shows the grayscale conversion formula.

$$G_L = 0.299R + 0.587G + 0.114B \tag{10.1}$$

R, G, and B values denote the red, green, and blue colors within the color image.

The darkest bar indicators in the tests provide the highest antibody production level. The darkest value in the grayscale-converted images (=0) was assigned to the top of the scale. Therefore the darkest points of the extracted IgM and IgG indicators were taken as reference values (Rp) and determined as the rapid antibody test results.

Since the darkest point determines the highest level, all reference values (Rp) were scaled between 0 and 100, and the initial scaling (L) was obtained for each test image.

On occasion, the backgrounds of the result regions of the kits smudge with red color tones. This smudge occurred as a similar or irrelevant color to the colors produced by the IgM and IgG indicators. Accordingly, it was aimed to reflect the contrast value between the background and the result indicators to the scale and provide a more sensitive antibody level. The Michelson contrast (Olsen et al., 2010; Wiebel et al., 2016) was used to calculate the ratio between the darkest and brightest points in the images for each result region. The formula of Michelson contrast is given in Eq. (10.2).

$$C_M z = \frac{\mathrm{Max}_{\mathrm{Pz}} - \mathrm{Min}_{\mathrm{Pz}}}{\mathrm{Max}_{\mathrm{Pz}} + \mathrm{Min}_{\mathrm{Pz}}} \tag{10.2}$$

$\mathrm{Max}_{\mathrm{P}}$ and $\mathrm{Min}_{\mathrm{P}}$ denote the maximum and minimum intensity values within the image z. Fig. 10.2 shows examples of the low and high-contrast-valued images obtained by Michelson contrast.

FIGURE 10.2

Example result region images for the contrast - left to right: an original image of "patient 1"; high contrast grayscale image of "patient 1" with the EDTA blood sample CM = 0.723; original image of "patient 16"; and low contrast grayscale image of "patient 16" with fingertip blood sample CM = 0.105.

Another problem observed was between the background of the result region and the indicators. Due to the formation of antibodies, the hands remain completely ground color with no result. This has made it crucial to determine the value of the information in the kits. It may cause a partially slipped representation of IgM or IgG values extracted by segmentation. Entropy (Singh & Kapoor, 2014; Ye et al., 2007) is the measure of the information within the image. It describes the amount of uncertainty or randomness there is in an image (Mello Román et al., 2019). The local entropy method (Hržić et al., 2019; Sekeroglu & Khashman, 2017; Yang et al., 2020) determines the randomness and degree of knowledge in the images and was used as a 5 × 5 segment. The most informative point was recorded as the entropy value of the picture. Eq. (10.3) shows the formula of image entropy.

$$E = -\sum_{i=0}^{255} p_i \log_2 p_i \tag{10.3}$$

where p_i is the probability of the gray level, and i is obtained from the normalized image histogram.

The direct use of the obtained entropy and contrast values in the scale-level adjustment will provide a linear increase for each test and indicator. Therefore a formula is proposed to minimize the effect of high-contrast images, increase the impact of low-contrast images on adjustment, and affect the adjustment as much as the contrast level of the region. The final scaling equation (AB_S) that considers all similarities and differences between the background and the result indicators is shown in Eq. (10.4).

$$AB_{sz} = L_{Tz} + (E_{max_z} . (1 - C_{Mz})) \tag{10.4}$$

where E_{max} is the maximum local entropy and L_T is the initial scale of the image z.

Even though IgM and IgG indicators did not occur in the rapid test kits, conversion of the indicators produces values because of the background color, noise, and shadow in the results region of the quick antibody test kits. Therefore after the adjustment procedure, the minimum antibody level was determined as $\Theta =$ 50%, where the lower values indicate no results from the kits or no antibody responses from the patients.

Fig. 10.3 demonstrates the detected information and scale adjustment on the rapid test kit results using the contrast and the local entropy. The highest

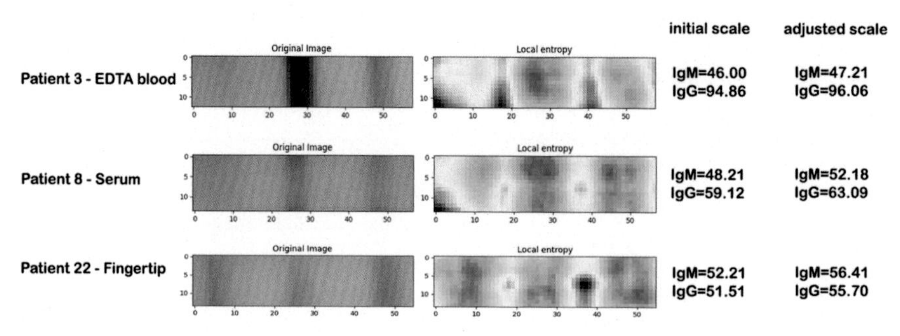

FIGURE 10.3

Effect of adjustment on rapid test results based on the contrast and local entropy.

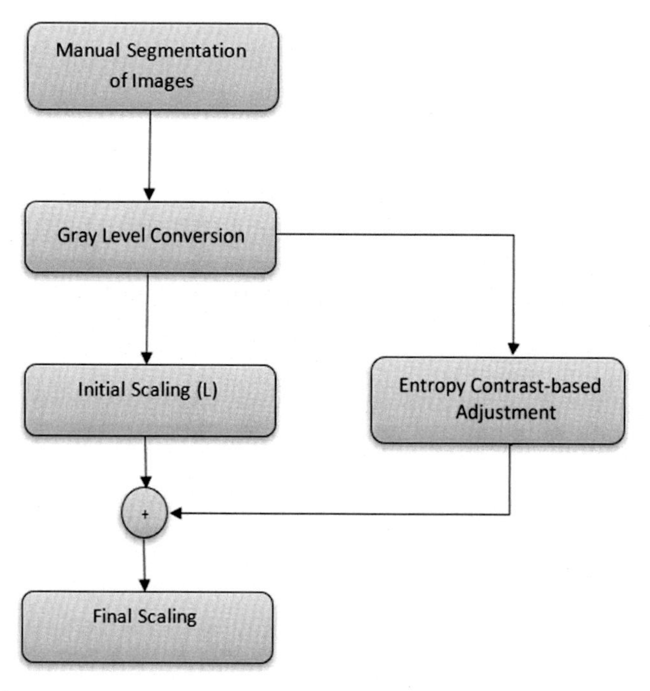

FIGURE 10.4

General block diagram of the proposed antibody scaling method.

difference was measured in "patient 3" in Fig. 10.3, and the effect of the local entropy was minimized. At the same time, the other results have a more significant impact because of indistinct indicators.

The general block diagram of the proposed method is shown in Fig. 10.4.

10.3 Experiments and results

Two varied experiments were performed to scale the IgM and IgG indicator results of the patients produced by each rapid COVID-19 antibody test to determine the highest antibody level on a patient basis.

10.3.1 Test kit-based scaling experiment

This experiment aimed to determine the most efficient blood sample for the rapid test kits, compare the scaling results obtained for each sample in antibody screening, and analyze the proposed method's ability.

IgM and IgG values for all kits of all patients were individually scaled and adjusted. Fig. 10.5 shows the scaled IgM and IgG values obtained for each blood sample and patient, respectively. It is clear to see from Fig. 10.5 that the IgM levels are close for all blood samples; however, the EDTA blood sample produced more stable results in IgG levels.

In IgM level scaling results, 21 of 28 patients had scaled as having antibody levels at least in one blood sample, and the remaining seven were negative in all tests. In nine of 21 patients whose antibodies were scaled, the highest IgM value was found in the serum test, while the other nine were in the fingertip test. The highest values of the remaining three patients were determined in the EDTA blood test. When the IgM-level results were considered independently for each test, 11, 17, and 13 patients were scaled below the limit by the fingertip, EDTA blood, and serum test, respectively.

The maximum and minimum standard deviation between the blood samples considered per patient was 7.16% and 1.31%, respectively. The average standard deviation of all test samples of IgM level was calculated as 4.29%.

In the IgG-level results, 24 of 28 patients had scaled as having antibody levels over the limit at least in a sample, and the remaining four were mounted below the limit in all tests. The numerical representation and the scaling of the tests showed that the EDTA blood test produced the highest IgG values for 19 of 24 patients.

When the IgG-level results were considered independently for each test, 10, four, and seven patients were scaled below the limit by the fingertip, EDTA blood, and serum test, respectively. The minimum standard deviation between the blood samples per patient was calculated as 2.88%, while the maximum and average standard deviations were calculated as 21.89% and 11.04%, respectively.

The adjustment on scaling was also analyzed using the indistinct test images and the tests that could not produce any antibody level. All test kits and blood samples were considered in the analysis.

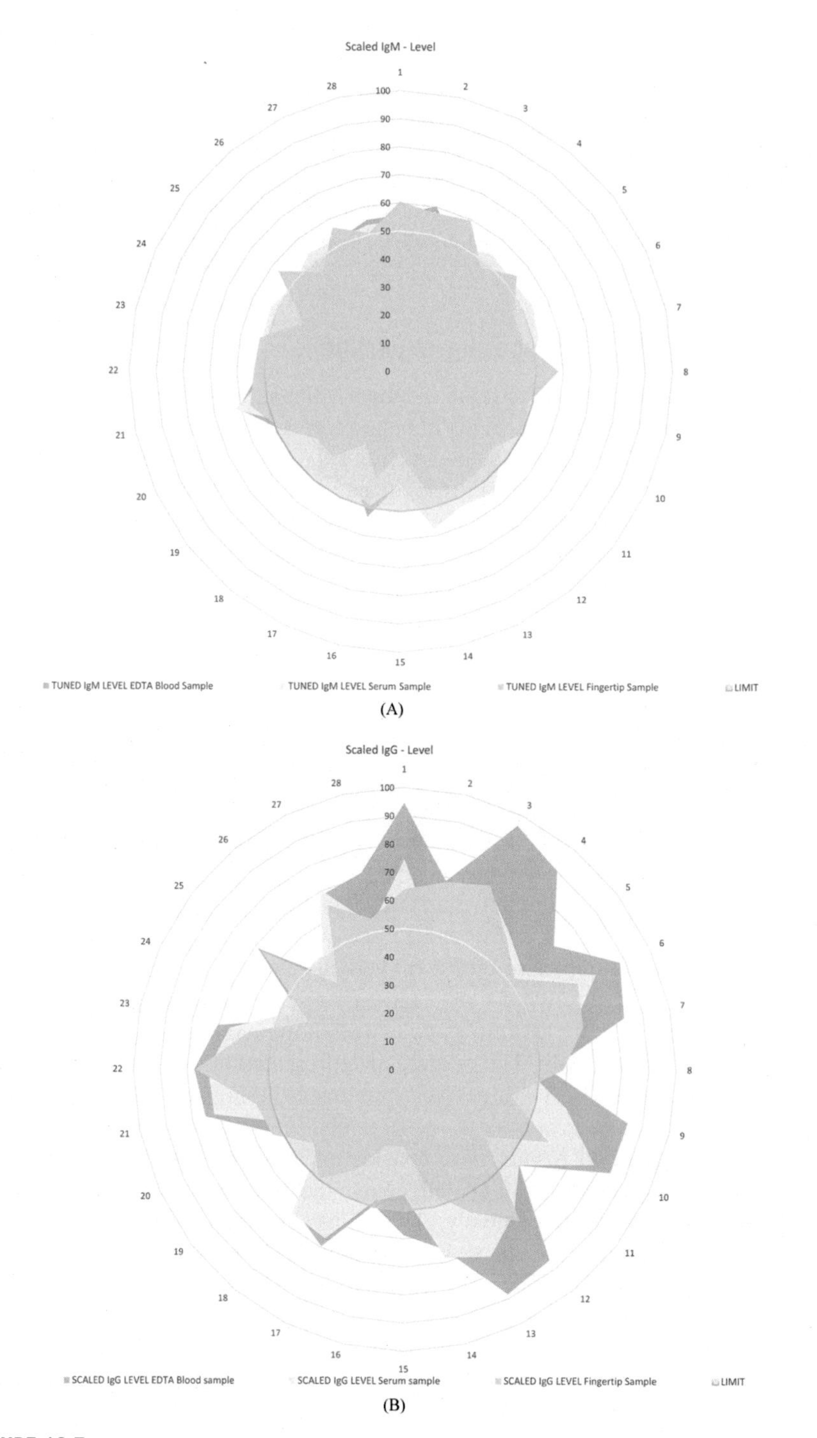

FIGURE 10.5

Scaled antibody levels for each test and corresponding patient. (A) IgM and (B) IgG.

Considering all the IgM values obtained, the highest adjustment made to the regular scaling was 4.36%, the lowest was 3.45%, and the mean adjustment level was 4.06% in the fingertip test. These rates were calculated as 4.31%, 1.20%, and 3.50% for the highest, lowest, and mean adjustments in the EDTA blood test, respectively. The serum test calculated the most elevated, softest, and mean adjustments as 4.42%, 3.00%, and 3.92% in the serum test.

10.3.2 Patient-based antibody scaling experiment

The minimum intensity values of IgM and IgG levels produced within the kits were also included in the first experiment to determine the optimal scale for each patient. The lowest values indicated the highest antibody level in the test kit images for each blood sample. They yielded optimized antibody levels of the patients by eliminating the false or low results produced by the different blood samples. Therefore the patients' highest antibody level was determined based on the indicators giving the best IgM and IgG scores among the three tests for each patient in terms of intensity. Table 10.1 shows the antibody levels of patients and the blood samples used to obtain the scale.

10.3.3 Evaluation of the proposed method

After obtaining the adjusted scale for IgM and IgG levels for all blood samples and patients, the results were compared to the original images, which the experts visually analyzed to provide an objective evaluation. The independent assessments were performed for the two experiments mentioned above.

The test kit-based scaling experiment used the accuracy metric for each blood sample and level (IgM and IgG) separately. In the fingertip blood sample, seven of 56 test results were measured below the defined limit even though they produced indistinct indicators in the result region. One result was measured above the limit. The accuracy of the fingertip blood sample was calculated as 85.71%. Three of the 56 results were assigned values below the limit in the EDTA blood sample, and two models were overlimited.

The accuracy of the proposed method in the EDTA blood sample was calculated as 91.07%. In the serum test, five and three of 56 pieces were assigned below and above the limits, and the accuracy was 85.71%.

The accuracy of IgM and IgG levels was also calculated separately. The accuracy was calculated as 80.95% when the IgM level was considered. A significant improvement was observed in the IgG-level results of the kit-based scaling experiment, especially in the EDTA blood test. The accuracy of the IgG level was measured as 92.85%.

The general accuracy of the proposed method without considering blood samples and levels was calculated as 86.90% (146/168).

In the evaluation of the patient-based antibody scaling experiment, the obtained scale of the patients' highest antibody level was used, and the accuracy

Table 10.1 Highest antibody levels for patients and corresponding test samples.

Patient no	Sample for IgM antibody test	Scaled IgM level, %	Sample for IgG antibody test	Scaled IgG level, %
1	Fingertip	60.36	EDTA blood	94.57
2	EDTA	60.36	EDTA blood	68.58
3	Fingertip	59.57	EDTA blood	96.07
4	Serum	53.60	EDTA blood	90.17
5	Fingertip	54.69	EDTA blood	70.52
6	Serum	53.98	EDTA blood	87.96
7	Serum	52.18	EDTA blood	82.94
8	Fingertip	58.30	Fingertip	59.65
9	–	–	EDTA blood	84.42
10	Fingertip	50.46	EDTA blood	84.19
11	–	–	EDTA blood	54.46
12	Serum	55.24	EDTA blood	86.29
13	Serum	53.01	EDTA blood	88.42
14	Serum	57.26	Serum	68.27
15	–	–	Fingertip	56.92
16	EDTA blood	53.15	–	–
17	–	–	Fingertip	69.62
18	–	–	Serum	65.28
19	–	–	–	–
20	–	–	Fingertip	54.32
21	Serum	61.97	EDTA blood	74.93
22	Fingertip	53.61	EDTA blood	77.32
23	Fingertip	52.80	EDTA blood	70.36
24	Serum	52.43	–	–
25	Fingertip	56.84	EDTA blood	68.74
26	Serum	53.35	–	–
27	Fingertip	56.61	Serum	70.50
28	EDTA blood	55.21	EDTA blood	71.38

EDTA, *ethylene diamine tetra acetic acid.*

of the proposed method was calculated. 23 patients' IgM levels were assigned to the correct group (82.14%). At the IgG level, the patient-based implementation of the proposed model set all patients to the right antibody level and achieved 100% accuracy. A total of 91.07% accuracy was achieved by the proposed model in the patient-based antibody scaling experiment. Tables 10.2 and 10.3 present the obtained kit-based and patient-based antibody scaling accuracy results of this study in detail.

Table 10.2 Kit-based antibody scaling results obtained in this study.

Kit-based antibody scaling						
Test	Sample	Accuracy	Test	Accuracy	Level	Accuracy
IgM level P	23/28	82.14%	EDTA blood	91.07%	G level	92.85%
IgG level P	25/28	89.28%				
IgM level E	23/28	82.14%	Fingertip	85.71%		
IgG level E	28/28	100%			M level	82.14%
IgM level S	23/28	82.14%			Serum	85.71%
IgG level S	25/28	89.28%				
Total	146/168	86.90%				

EDTA, *ethylene diamine tetra acetic acid.*

Table 10.3 Patient-based antibody scaling results obtained in this study.

Patient-based Antibody Scaling		
Level	Sample	Accuracy
IgG level	28/28	100%
IgM level	23/28	82.14%
Total	51/56	91.07%

10.4 Discussions

The obtained results were analyzed at several scalled antibody levels to perform an objective conclusion. The success of the numerical representation of the results occurred on the test kits, and the efficiencies of the kits for IgM and IgG levels were discussed independently.

The proposed method was capable of representing the blood samples on a numeric scale, with 86.90% and 91.07% in kit-based and patient-based experiments. These accuracy results were obtained by using the determined lower antibody limit $\Theta = 50$. The decrement of the limit value yields the correct scale of the test samples with an indistinct indicator on the result region and does not produce scale values above the limit. However, this would cause the scaling of the kit results above the limits without having the result indicators on the scale.

When the IgM and IgG level results were considered in the kit-based experiments, it was observed that more than 10% of decrement occurred for the success of the scaling of samples between IgG and IgM levels. In comparison, IgG and IgM level results were obtained as 92.85% and 82.14%, respectively (Table 10.3). This was caused by the indistinct, weak, and no results formation in the kits' IgM level regions. Therefore although they produced an indicator, nine of 84 IgM-level samples were not scaled above the limit. Their average scaling results were 45.85%.

A more accurate scale was obtained in determining IgG-level rankings because of more distinct result indicators. Only six of 84 samples were scaled below the limits, and all the samples that could not produce a result indicator scaled correctly. The average scaling results of these failed six IgG-level samples were 45.47%. The IgM and IgG-level results of the kit-based experiment showed that the samples below or above the limit failed with a low value. However, it should be noted that these samples were uncertain and indistinct because of the background colors in the result regions and could be concluded with long observation even by the experts.

To reach a general conclusion and interpret the results better, it was required to evaluate the test results separately. All tests produced the same accuracy result for the IgM-level test; however, the EDTA blood test achieved 100% accuracy in IgG-level, while fingertip and serum tests made 89.28% accuracy. These results showed that the EDTA blood test is the most effective test to observe and scale the IgG-level results of the patients.

When the patient-based experiments by scaling with the highest test results were considered, the IgM-level accuracy did not change; however, the IgG-level scaling achieved 100% accuracy. The number of test results showed that 70.83% of the IgG-level results were accurately scaled by the EDTA blood test, while the fingertip and serum results were used only in 16.66% and 12.50%. However, these numbers were distributed uncertainly in IgM-level experiments. The number of test results showed that 42.85%, 42.85%, and 14.28% of the IgM-level results were scaled by the fingertip, serum, and EDTA blood tests.

When all the obtained results were considered, it was observed that the EDTA blood test produced much more efficient and distinct results than other tests. This yields the EDTA blood test to be used accurately in vaccines and antibody level monitoring where IgG-level screening gains importance. In the IgM-level screening, it was observed that all tests produced the same results, but the serum and fingertip tests produced more significant results than the EDTA blood test. However, even though the EDTA blood test might reduce the significance of IgM-level indicators within the result regions, the results suggest that it can be effectively used for visual and scaled antibody-level monitoring.

Although the number of patients considered in the experiments is 28, the numerical representation of the antibody levels was performed independently. Therefore, using more samples would not change the initial scaling.

In addition, the initial and the following antibody-level records of each patient will also provide the monitoring of the antibody level in the body in a faster and more steady way. This would lead to the rapid analysis of the occurrence and disappearance of antibody levels in COVID-19 patients and vaccinated people.

10.5 Conclusion

This chapter proposed digital conversion and scaling of the rapid COVID-19 antibody test kits using fundamental image processing techniques.

Twenty-eight patients' fingertip, serum, and EDTA blood test kits were considered. The results showed that the proposed method could effectively scale the antibody levels and provide consistent results for the researchers.

The results may show that the proposed method accurately determines antibody levels for COVID-19 patients and vaccines. In addition, the analysis demonstrated that the EDTA blood sample produces more stable and efficient results in rapid test kits for IgM and IgG with 82.14% and 100% accuracy.

Monitoring the COVID-19 patients and vaccinated people's antibody levels in Turkey as a second stage of the research uses the proposed method to understand the antibody occurrence better and determine the vaccines' impact.

References

Centers for disease control and interim prevention guidelines for covid 19 antibody testing Centers for Disease Control (CDC). (2020).

Dimililer, K., & Sekeroglu, B. (2021). 8 25 2021/08/25 2021 International Conference on INnovations in Intelligent SysTems and Applications, INISTA 2021 - Proceedings 10.1109/INISTA52262.2021.9548637 9781665436038 Institute of Electrical and Electronics Engineers Inc. Cyprus The effect of discrete cosine transform on COVID-19 differentiation from chest X-Ray images: A preliminary study. Available from http://ieeexplore.ieee.org/xpl/mostRecentIssue.jsp?punumber = 9548323.

Green, K., Winter A., Dickinson, R. (2020). What test could potentially be used for the screening, diagnosis, and monitoring of covid-19, and what are their advantages and disadvantages?

Güneş, A., Kalkan, H., & Durmuş, E. (2016). Optimizing the color-to-grayscale conversion for image classification. *Signal, Image and Video Processing*, *5*, 853–860. Available from https://doi.org/10.1007/s11760-015-0828-7, http://www.springerlink.com/content/1863-1703.

Hržić, F., Štajduhar, I., Tschauner, S., Sorantin, E., & Lerga, J. (2019). Local-entropy based approach for X-Ray image segmentation and fracture detection. *Entropy*, *4*, 338. Available from https://doi.org/10.3390/e21040338.

International Telecommunication Union (ITU)- ITU R.- recommendation itu-r bt.601–7. (2011). Available from https://wwwituint/dms_pubrec/itu-r/rec/bt/R-REC-BT601-7-201103-I!!PDF-Epdf

Kanan, C., Cottrell, G. W., & Ben-Jacob, E. (2012). Color-to-grayscale: Does the method matter in image recognition? *PLoS One*, *1*, e29740. Available from https://doi.org/10.1371/journal.pone.0029740.

Li, N., Wang, P., & Geng, C. (2020). Molecular diagnosis of covid-19: Current situation and trend in china (review). *Experimental and Therapeutic Medicine*.

Mello Román, J., Vázquez Noguera, J., Legal-Ayala, H., Pinto-Roa, D., Gomez-Guerrero, S., & García Torres, M. (2019). Entropy and contrast enhancement of infrared thermal images using the multiscale top-hat transform. *Entropy*, *3*, 244. Available from https://doi.org/10.3390/e21030244.

Olsen, M.A., Hartung, D., Busch, C., & Larsen, R. (2010). 9 2010/09 Communications in Computer and Information Science. Available from https://doi.org/10.1007/978-3-642-14831-6_56 18650929 425–434. Denmark Contrast enhancement and metrics for biometric vein pattern recognition 93.

Ong, D. S. Y., de Man, S. J., Lindeboom, F. A., & Koeleman, J. G. M. (2020). Comparison of diagnostic accuracies of rapid serological tests and ELISA to molecular diagnostics in patients with suspected coronavirus disease 2019 presenting to the hospital. *Clinical Microbiology and Infection* (8). Available from https://doi.org/10.1016/j.cmi.2020.05.028, 1094-1094.e10, https://www.journals.elsevier.com/clinical-microbiology-and-infection.

Péré, H., Mboumba Bouassa, R.-S., Tonen-Wolyec, S., Podglajen, I., Veyer, D., & Bélec, L. (2021). Analytical performances of five SARS-CoV-2 whole-blood finger-stick IgG-IgM combined antibody rapid tests. *Journal of Virological Methods*, 114067. Available from https://doi.org/10.1016/j.jviromet.2021.114067.

Sarıgül, F., Doluca, O., Akhan, S., & Sayan, M. (2020). Investigation of compatibility of severe acute respiratory syndrome coronavirus 2 reverse transcriptase-PCR kits containing different gene targets during coronavirus disease 2019 pandemic. *Future Virology*, *8*, 515–524. Available from https://doi.org/10.2217/fvl-2020-0169, http://www.futuremedicine.com/loi/fvl.

Sayan, M., Sarigul Yildirim, F., Sanlidag, T., Uzun, B., Uzun Ozsahin, D., & Ozsahin, I. (2020). Capacity evaluation of diagnostic tests For COVID-19 using multicriteria decision-making techniques. *Computational and Mathematical Methods in Medicine*, 1–8. Available from https://doi.org/10.1155/2020/1560250.

Sekeroglu, B., & Khashman, A. (2017). 8 25 2017/08/25 ACM International Conference Proceeding Series 10.1145/3133264.3133272 9781450352956 96–102 Association for Computing Machinery Turkey Performance evaluation of binarization methods for document images. Available from http://portal.acm.org/ 131200.

Sekeroglu, B., & Ozsahin, I. (2020). Detection of COVID-19 from chest X-ray images using convolutional neural networks. *SLAS Technology.*, *6*, 553–565. Available from https://doi.org/10.1177/2472630320958376, http://journals.sagepub.com/toc/jlad/current.

Singh, K., & Kapoor, R. (2014). Image enhancement using exposure based sub image histogram equalization. *Pattern Recognition Letters*, *1*, 10–14. Available from https://doi.org/10.1016/j.patrec.2013.08.024.

Srivastava, N., Baxi, P., Ratho, R. K., & Saxena, S. K. (2020). Global trends in epidemiology of coronavirus disease 2019 (COVID-19). *Springer Science and Business Media LLC*, 9–21. Available from https://doi.org/10.1007/978-981-15-4814-7_2.

Tartaglione, E., Barbano, C. A., Berzovini, C., Calandri, M., & Grangetto, M. (2020). Unveiling COVID-19 from chest X-ray with deep learning: A hurdles race with small data. *International Journal of Environmental Research and Public Health*, *18*, 1–17. Available from https://doi.org/10.3390/ijerph17186933, https://www.mdpi.com/1660-4601/17/18/6933/pdf.

The United States Food and Drug Administration FDA, U.S. food and drug administration kits: FDA, in vitro diagnostics. (2023). Available from https://www.fda.gov/medical-devices/products-and-medical-procedures/in-vitro-diagnostics (Accessed September 2023).

Vandenberg, O., Martiny, D., Rochas, O., van Belkum, A., & Kozlakidis, Z. (2021). Considerations for diagnostic COVID-19 tests. *Nature Reviews. Microbiology*, *3*, 171–183. Available from https://doi.org/10.1038/s41579-020-00461-z, http://www.nature.com/nrmicro/index.html.

West, R., Kobokovich, A., Connell, N., & Gronvall, G. K. (2021). COVID-19 antibody tests: A valuable public health tool with limited relevance to individuals. *Trends in Microbiology*, *3*, 214–223. Available from https://doi.org/10.1016/j.tim.2020.11.002, http://www.elsevier.com/locate/tim.

Wiebel, C. B., Singh, M., & Maertens, M. (2016). Testing the role of Michelson contrast for the perception of surface lightness. *Journal of Vision*, *11*, 17. Available from https://doi.org/10.1167/16.11.17.

World Health Organization, WHO coronavirus disease (COVID-19) dashboard. (2023). Available from https://www.who.int/publications/m/item/covid-19-epidemiological-update—29-september-2023.

Yang, W., Cai, L., Wu, F., & Li, L. (2020). Image segmentation based on gray level and local relative entropy two dimensional histogram. *PLoS One*, *3*, e0229651. Available from https://doi.org/10.1371/journal.pone.0229651.

Ye, Z., Mohamadian, H., Pang, S.S., Iyengar, S. (2007). Proceedings of the 6th WSEAS International Conference on Information Security and Privacy 172–177 Image contrast enhancement and quantitative measuring of information flow.

Yildirim, F. S., Sayan, M., Sanlidag, T., Uzun, B., Ozsahin, D. U., Ozsahin, I., & Improta, G. (2021). Comparative evaluation of the treatment of COVID-19 with multicriteria decision-making techniques. *Journal of Healthcare Engineering*, 1–11. Available from https://doi.org/10.1155/2021/8864522.

CHAPTER

Artificial intelligence in dental research and practice

11

Snigdha Pattanaik[1], Shruti Singh[2], Debarchita Sarangi[3] and Emmanouil Evangelopoulos[4]

[1]*Department of Preventive and Restorative Dentistry, College of Dental Medicine, University of Sharjah, Sharjah, United Arab Emirates*

[2]*Department of Dentistry, All India Institute of Medical Sciences, Raebareli, Uttar Pradesh, India*

[3]*Department of Prosthetic Dentistry, Institute of Dental Sciences, Siksha O Anusandhan (Deemed to be University), Bhubaneswar, Odisha, India*

[4]*Department of Orthodontics, College of Dental Medicine, University of Sharjah, Sharjah, United Arab Emirates*

Artificial intelligence (AI) is the field in which machines are utilized to perform human tasks. Now the application of AI has extended to tasks that were performed only through human intelligence (HI). With digitalization, various conventional tasks that were earlier performed by HI are now being safely replaced. The application of AI further enhances economic productivity and efficiency of goods or services. This field is planned to be utilized in medicine, dentistry, and health care fields at different levels for diagnosis, treatment planning, treatment outcome prediction, and so on. This chapter will lay a basic foundation for understanding AI, the different components of AI, and its applications in the dental field. It also discusses various pros and cons pertaining to AI in the dental and medical fields.

11.1 Introduction

Artificial intelligence (AI) refers to performing intellectual tasks as performed by humans with the aid of machines and technology. AI is concerned with constructing an intelligent computer technology that exhibits features that can be associated with intelligence of the human mind and simulate human behaviors like learning, reasoning, problem solving, understanding language, and so on. This concept was proposed by Barr and Feigunnbaum (Ahmed et al., 2021; Barr et al., 1981).

AI has led to advancements in our day-to-day activities through facial recognition, self-driven cars, and so on. AI support has also been beneficial in medical

Artificial Intelligence and Image Processing in Medical Imaging. DOI: https://doi.org/10.1016/B978-0-323-95462-4.00011-X

fields, specifically in performing surgeries with the help of intelligent systems. The automated decision support systems help in disease diagnosis (Rahimy et al., 2013; Sun et al., 2020a; Yamada et al., 2019). The AI system has recently supported and developed personalized medicine, which helps in predicting disease-causing factors, diagnosing the disease, and selecting the best treatment for the concerned individual (Campbell et al., 2020; Cortés-Ciriano et al., 2020; Gálvez et al., 2020; Kalra et al., 2020; Ostaszewski et al., 2020).

AI is transforming dentistry also. Patient management tasks like booking and appointment scheduling require fewer staff members and can be done with fewer errors. AI also aids in assisting the complete treatment of a dental problem from its diagnosis to management (Chen et al., 2020). All of these tasks can be performed with better precision, specificity, sensitivity, and accuracy. In orthodontics, it can assist in identifying and classifying malocclusion. In oral medicine and radiology, it can assist in automatically detecting and classifying dental restorations on panoramic radiographs. In addition, it can also detect periodontal diseases, tooth abnormalities, root caries, facial defects, and bony lesions associated with a pathology or dental extraction (Cozzani et al., 2020). Automated detection of diseases based on the image and segmentation of image for detection of oral defects has been improved with the help of AI application in dentistry (Hatvani et al., 2018; Hung et al., 2019; Lee et al., 2018a; Tian et al., 2019; Xu et al., 2018). AI is also being utilized for robotics in dentistry (Grischke et al., 2020). Various fields in dentistry are being gradually explored through AI, with the emerging digital dentistry paradigm.

Data collection is the primary factor for the AI revolution wave in biomedicine. Powerful AI techniques like high-performance computing have sanctioned very precise and intuitive information retrieval from data collected. This part of AI involving data is termed a machine learning. It allows machines to learn a specific topic of interest from a certain accumulated dataset made available for information extraction. A machine is a set of algorithms executed in computer systems. Supervised learning techniques are used for information extraction, which helps in problem solving with a high success rate (Friedman et al., 2001). Supervised learning involves mapping a sample input to a desired output. This is based on input-output pairs of databases. With every new incoming sample, new prediction can be done depending on the function learned using training data (Friedman et al., 2001).

Deep learning (DL), natural language processing, cognitive computing, expert systems, robotics, and fuzzy logic (FL) are the various allied fields of AI. Machine learning forms an integral part of AI, which improves the ability of automated learning without being programmed. It aims to develop and allow automated learning without human intervention (Ahmed et al., 2021). DL refers to the process of running a basic dataset such as images or a part of an image with a carious lesion repetitively through the neural network (NN). The parameters of the model are adjusted to improve accuracy during training. In DL, multilayered NNs help in understanding hierarchical features in the data (Barr et al., 1981). A convolutional neural network (CNN) algorithm with a deep-learning basis has performed considerably well in various arenas like that of caries detection in periapical radiographs (Bouchahma et al., 2019), classification, and detection of impacted supernumerary teeth in patients on pan tomography, specifically in relation to maxillary anterior teeth (Kuwada et al., 2020). In terms of automated tooth segmentation and detection of periapical lesions in pan tomography, fully deep R-CNN models performed well (Ekert et al., 2019; Lee et al., 2020). The key aspects of AI are shown in Fig. 11.1.

Recent investigations have proved the application of artificial neural networks (ANNs) in the field of endodontics for reconfirming the location of the apical foramen on radiographs, thereby enhancing working length determination (Saghiri et al., 2012). In the field of restorative dentistry, they can be used in determining shade and Vickers hardness of bottom to top composites, which can further be used in debonding of composites (Arısu et al., 2018; Yamaguchi et al., 2019).

ANNs can be used as a light-curing unit in conservative treatments in shade selection of composites. They can also fulfill the requirement of 3D tooth preparation with more accuracy and precision (Otani et al., 2015; Wang et al., 2014). In the field of prosthodontics, the model can be used in designing prosthesis like removable partial dentures (RPD) by defining dental arches (Takahashi et al., 2021). AI in orthodontics can help analyze the treatment outcome on facial esthetics with its technology. The outcome of orthognathic treatment on facial esthetics and appearance of age can be analyzed with AI, with a new feature of objectively and reproducibly scoring facial attractiveness and apparent age (Patcas et al., 2019). Integrating the data of anterior teeth in the form of intraoral and facial images aids in the analysis of position and shape of maxillary anterior teeth for the clinician (Li et al., 2020).

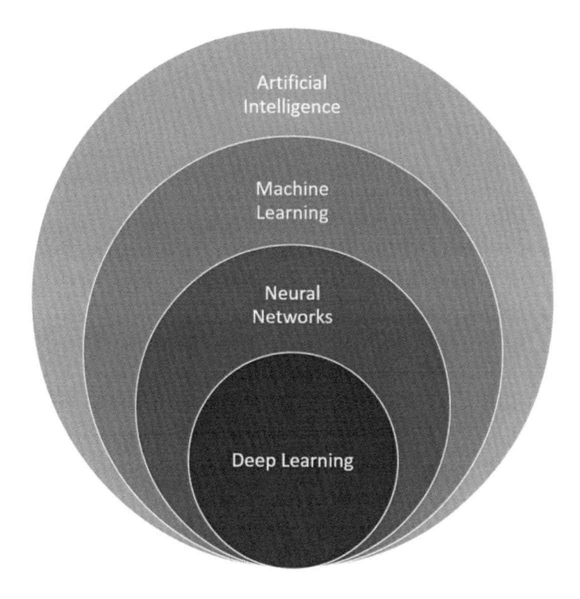

FIGURE 11.1

Key aspects of artificial intelligence.

Even though there has been a breakthrough in advancement of AI technology in the past decade, there is still uncertainty regarding information available to assist in a complete treatment plan for a dental disease including its diagnosis, planning, and management. This chapter provides a brief understanding of the application of AI in different fields of dentistry, and their outcomes are covered under each individual heading.

11.2 Difference between natural and computer intelligence

It was Alan Turing who theorized machines and digital computers long before their existence, and with the Turing machine, he invented the abstract conceptualization of a universal machine in 1950. many identify that as the predecessor of what we now know as the computer.

Alan Turing has contributed vast amounts of essential insights to the field of AI. He gave the Turing test for intelligence replacing the question "Can machines think?" by a question easier to test: "Can a machine simulate human thinking, in such a way that a human judge cannot distinguish that machine from a human?" (Arsiwala-Scheppach et al., 2023) The original test by Turing (1950) was described as the imitation game, as the premise was whether a machine could imitate a human (Arsiwala-Scheppach et al., 2023).

AI is the study and design performed with the help of various kinds of tools by the intelligent agent. These tools analyze the basic requirement of the environment and produce actions that help maximize success. Natural (human) intelligence is the defined quality of a human mind. It advances with the human's past experiences, situations, and unpredictable circumstances (Ha et al., 2018). Following are the differences between AI and HI:

- AI can work and provide service any hour of the day. It has a no-sleep formula, unlike human machines.
- HI may be driven by emotions, but AI has no prejudice. It handles all situations without bias.
- AI has a great speed.
- Human brains are great at multitasking. This puts HI at an upper hand. Humans require education, not "invention."
- Robotic brains cannot operate on complex movements; human brains can. Example: handling a four-year-old child.

Analyzing the above differences between human's and AI, we can say that neither of them can be ignored and each has its own importance in future. Ray Kurzweil rightly said, "We should not fear AI, rather it will act as a helping hand in enhancing us. In the way, as we get more intelligent, AI will make us smarter" (Sun et al., 2020b) (Figs. 11.2 and 11.3).

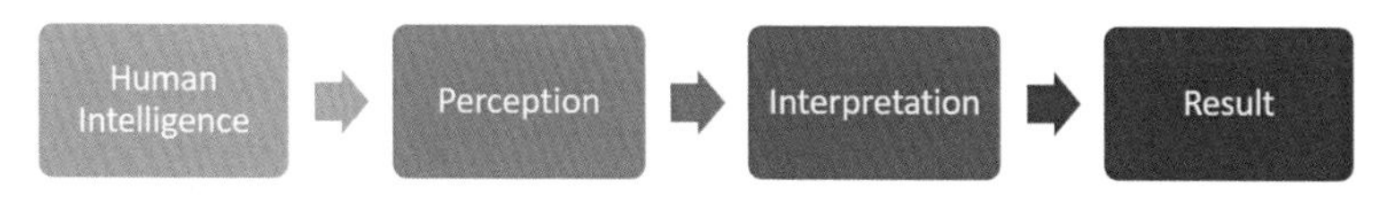

FIGURE 11.2

Representation of human intelligence.

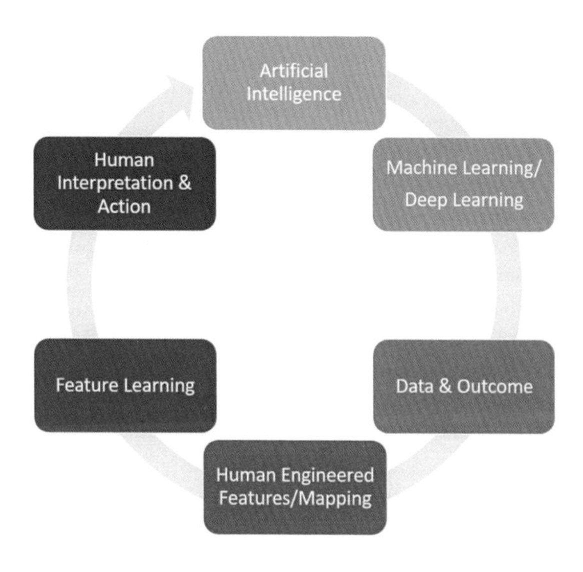

FIGURE 11.3

Schematic representation of artificial intelligence.

11.3 Deep learning and machine learning

Machine learning and manual algorithms are used for feature extraction by hand-crafted engineering (Chen et al., 2019). The application of deep machine learning in biosciences is a path-breaking milestone. Resurgence of NNs, incursion of Fuzzy Logic (FL) and labeling numerical data through linguistic terms are a few examples. The kernal method in the 1990s revolutionized machine learning. In the beginning of the 2000s, the method was the primary choice for recognizing the pattern of medium-sized datasets. Some other pattern recognition methods like Random Forest and XG Boost decision-tree–based methods were used to solve biomedical problems, including dentistry (Breiman, 2001; Chen et al., 2018; Cui et al., 2020; Dot et al., 2020; Du et al., 2018).

The FL offers its major advantages in dealing with uncertainty in data (Fig. 11.4) and has the capacity to provide solutions form of rule bases to experts (Alcalá-Fdez & Alonso, 2015; Zimmermann, 2011). FL has been applied

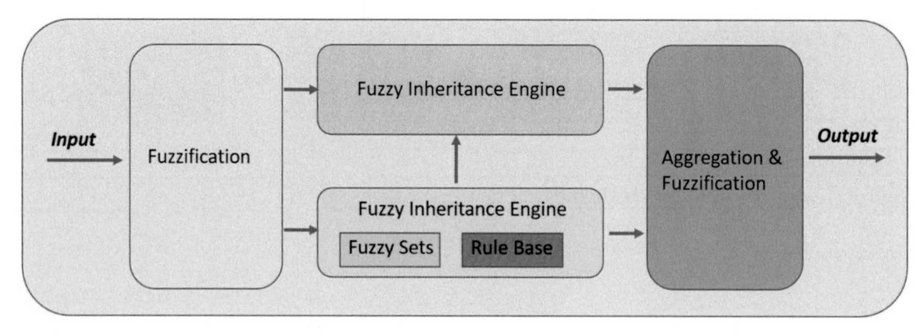

FIGURE 11.4

Fuzzy logic architecture.

specifically in esthetic dentistry for color designation (Chamorro-Martínez et al., 2016). Fuzzy colors allow the use of semantics in the automated operation and color description (Herrera et al., 2010) (Fig. 11.4).

As DL helps to build NN, which identifies patterns automatically for improving feature detection, it can be applied in fields in which situations and circumstances change periodically (stock market), or in fields with vast information in which human expertise may not be able to match technology (Chen et al., 2019).

DL utilizes NNs. The NNs consist of various layers that are assigned for different functions like input, max-pooling, convolution, and classification. DL approaches are classified as supervised, semisupervised, and unsupervised for supervision level. For deep supervised learning, labeled data are used, and ANNs, deep NNs, CNNs, and recurrent NNs are utilized. Partially labeled data are utilized by a semisupervised DL model, whereas an unsupervised model does not use any data or discover unknown data. Dimensionality reduction, clustering and generative techniques are used by unsupervised data (Chen et al., 2019).

ANNs used for supervised, unsupervised, and reinforcement learning problems are learning algorithms. These algorithms function on the basis of biological NNs. ANN structure constitutes operating neurons in the form of interconnected layers (Casalegno et al., 2019).

Neurons are the basic building blocks of ANN. They are grouped in layers such as input layer, output layer, and hidden layer (Schmidhuber, 2015). In the form of number of layers, these basic networks represent the building block for other DL models. One such model is CNNs which have shown to aid in various problem solving areas.

11.3.1 Convolutional neural networks

CNNs in their most basic form easily detect lines, textures, edges, and patterns in an image. They have the ability to automatically extract features from data. With application of learning algorithms to different layers, complex filters can be learned. These complex filters improve model performance compared with traditional methods and help detect complex shapes in images or signals. The complex filters can be learned in a straightforward manner in CNNs by applying the convolution operation, (Casalegno et al., 2019). CNNs are formed by different convolutional layers, pooling layers, and fully connected layers. By grouping convolutional operations, varied important features can be understood within the same layer. Convolutional layers remove information from the data and transform the data with a different representation of input values. The pooling layer decreases the data dimensions by amalgamating the outcomes of the clusters from one layer and then using it in the next layer as an individual neuron. The NN is formed by connecting every neuron in each layer, which in turn forms a series of dense connected layers. This operates over the convolutional layers in order to perform the operations (Schmidhuber, 2015).

The research of DL in dentistry is promising. ANN and specifically CNN have led to interesting results in diagnosis and prediction in fields like radiology and pathology. The major highlights are: identification of disease, segmentation of images, and application of generative adversarial NNs in correction of images (Casalegno et al., 2019).

11.4 General applications of artificial intelligence in the dental field

The field of dentistry has seen a significant increase in digitalization over the past 10 to 20 years. The shortage of medical and dental professionals in most developing countries particularly revives the need for special AI technology software that can aid in reducing cost, time, and medical errors (Machoy et al., 2020).

AI application in dentistry aids in making diagnoses more accurate and efficient. It requires great decision-making skills, practice, and knowledge to decide the best treatment modalities for a patient and to predict the prognosis of the treatment. Sometimes for some cases, a possibility may arise that a dentist might not have adequate information to make the appropriate clinical decision within a restricted time span for a specific case. In such cases, AI applications can direct the clinician toward the correct decision (Khanagar et al., 2021). Considering the AI applications in dentistry, as per investigation and studies, ANNs can be applied in multiple ways in clinical dentistry such as for locating the apex of the root canal on radiographs, thereby making the working length assessment more accurate, both for tooth preparation as well as orthodontic and prosthodontic treatment (Ahmed et al., 2021).

However, information in the form of literature in terms of the methods applied by AI in assisting in planning the whole treatment of a disease right from diagnosis to management remains unclear. Therefore, a detailed explanation of AI pertaining to individual field has been covered to understand the recent AI trends in different specialties of dentistry along with their outcome.

11.5 Dentistry and artificial intelligence: details of current applications

11.5.1 Artificial intelligence technologies in oral radiology and diagnostics

AI diagnostic models have been applied in the field of dental and maxillofacial radiology (DMFR) for locating root canal orifices, detecting vertical root fractures, and detecting interproximal caries along with other general findings. The studies were initially limited to preclinical findings data; but with the advancements in technology, further studies were taken up to convert these preclinical findings into clinical application. Various studies have been taken up with the aim of using DMFR methodologies for AI applications to identify areas of particular interest.

Since 2006, there has been a tremendous increase in the work by various authors to develop AI models for the maxillofacial region using clinical models. Majority of AI models developed mostly concentrated on clinical issues of jaws and teeth as DMFR imaging modalities use radiographs that focus on hard tissue conditions (Hung et al., 2020a). It was in 2009 that Flores et al. first used cone-beam computed tomography (CBCT) images for AI models to distinguish periapical cysts from granulomas as 2D images were previously used to build computer-assisted programs for the aid of clinical diagnosis. The 2D images included periapical, panoramic, and cephalometric radiographs (Hung et al., 2020a; Flores et al., 2009). It was after this that various studies used CBCT images to develop AI models for various clinical diagnosis. AI techniques application in DMFR focuses on automatic localization of cephalometric landmarks, osteoporosis, bone mineral density (BMD) diagnosis, differentiation of oral cysts and/or tumors, and identification of dental caries, periapical infections, and periodontitis.

11.5.1.1 Localizing cephalometric landmarks

AI models have improved the efficiency in treatment planning of orthodontic cases by automatically localizing cephalometric landmarks. Generally majority of orthodontists prefer digital tracing computer-aided software analysis over manual analysis as it is less cumbersome and saves time. But computer-aided tracing requires manual location of cephalometric landmarks. Therefore, even the computerized technique also holds chances of bias and error between different observers and is not accurate in locating the landmark. AI algorithm application has gradually tried to overcome these shortcomings and has proved a great help to specialists. Currently, CBCT images have replaced cephalometric radiographs in analysis owing to their better precision and clarity in 3D imaging. Hence in 2011, Cheng et al. used AI models in combination with CBCT images to localize cephalometric landmarks. As per results, AI models showed better accuracy, but the values did not meet the clinical requirements. Hence it was concluded that the AI models can be applied only for preliminary localization with manual correction before further analysis (Cheng et al., 2011; Flores et al., 2009; Hung et al., 2020a).

11.5.1.2 Bone density and osteoporosis

Considering the field of implant dentistry, alveolar bone density plays a very vital role in the success of implant dentistry. Patients with low BMD or with conditions like osteoporosis have high chance of resorption of marginal bone around implants, leading to compromised treatment. Patients with osteoporosis who are under bisphosphonates treatment develop high risk of osteonecrosis of the jaw following any minor oral surgery. Therefore AI models like DCNN and computer-assisted diagnosis (CAD) applications hold a good potential in early diagnosis of these bone conditions using panoramic radiographs. Considering the sensitivity, specificity, and accuracy of 95% as reported in various studies, these models hold a very high probability of applications in clinical practice (Khanagar et al., 2021).

11.5.1.3 Classifying and segmentation of oral cysts and/or tumors

Diagnosis of a cyst or tumor is always a challenge, and biopsy and histopathology are the gold standard in challenging and complicated cases as radiological findings do not give a concrete diagnosis. Application of AI models can provide an automated diagnosis of different jaw lesions and add value to the dental practice as general clinicians always find this array of diagnosis challenging.

There are four steps involved in AI models to classify cysts and tumors of the jaw. The first step involves lesion detection, which is done manually and depending on the data, the subsequent steps of segmentation and extraction of the data and features is done. After this, the final step of classifying the lesion into specific category is done. The AI model for cysts and tumors is not fully automated and does involve human aid for diagnosis. Studies have been carried out using different software for automated segmentation of radicular cysts, dentigerous cysts, and keratocytes. One such program using surgical navigation software like Feldkirchen Germany and iPlan, Brainlab AG was used to automatically segment keratocysts, yet further studies need to be conducted in developing AI models in this field to develop a more concrete completely automated program using 2D/3D images (Hung et al., 2020a).

11.5.1.4 Periodontitis/periapical disease

AI models have significantly contributed in differentiating periapical radiolucency due to a lesion or infection or due to periodontal factors like apical periodontitis and alveolar bone loss. Several authors proposed different models to identify degree of bone loss, along with the alveolar bone resorption. As per one study, DL CNN showed very good functioning in identifying periodontally compromised molars and premolars. Degree of alveolar bone loss was also predicted. There is a wide range of lesions that fall under periapical pathology, and studies have focused on classifying the extent of these lesions with the help of various AI models. To distinguish periapical cysts from granulomas, an AI model with application of CBCT images was proposed by Flores et al. The model showed promising results and was considered to add high value in clinical practice (Flores et al., 2009; Hung et al., 2020a; Khanagar et al., 2021).

11.5.1.5 Detection of dental caries

Dental caries is the most commonly prevalent oral disease in the population, globally leading to periapical infections if not treated and diagnosed in the right time. The diseases can very well be prevented with prophylactic measures, and its spread can be prevented with early detection and treatment. Caries detection models have been developed using nonclinical 2D images obtained from extracted teeth. Similar to the model of periapical diseases, a caries detection model based on DL algorithms was proposed by Lee et al. The model detected caries in maxillary molars and premolars. The model had high accuracy performance, but considering its use with 2D imaging, it had limitations of detecting only buccal and lingual decay. The model had a data input of teeth images of teeth without restorations. Hence, the efficiency of the model in detecting proximal decay and secondary caries was questionable. Another limitation was that its applicability on deciduous teeth remains unclear (Abdolali et al., 2016; Cheng et al., 2011; Flores et al., 2009; Hung et al., 2020a; Lee et al., 2018b; Rana et al., 2015).

Similar studies with the DL model were carried out using data of dental lesions in near-infrared transillumination. AI-based models using near-infrared-light transillumination (NILT) images showed satisfactory performance in diagnosing dental caries. Extending the area of diagnosis to root caries, proximal caries, and detecting periapical lesions, AI-based ANN models using bitewing radiographs were used. The results were encouraging (Khanagar et al., 2021).

Sjogren's syndrome and autoimmune conditions affecting the salivary glands leading to decreased salivation and cervical decay were studied using CNNs with CT images. Their performance in comparison to radiologists' diagnosis showed higher efficiency. Later, various studies were taken up to apply AI models of the CNN system to diagnose other serious conditions like lymph node metastasis and extranodal extensions with CT images. The results showed better sensitivity, specificity, and accuracy compared with specialists (Khanagar et al., 2021).

11.5.2 AI technologies in orthodontics

A successful orthodontic treatment involves a precise diagnosis, appropriate treatment planning, and correct prediction of prognosis (Khanagar et al., 2021). Removal of teeth for orthodontic treatment marks an important and difficult decision that varies depending on the clinician's experience. An AI expert system has been constructed for diagnosing extraction cases and to evaluate the efficacy of this model using NN. Four models of NN learning were developed using a back propagation algorithm and to evaluate cases of extraction. As a result of the system, 93% for diagnostic extraction and classification accuracy was seen with the models as compared with nonextraction diagnostics with detailing extraction pattern accuracy of 84%. This system had its own limitations considering that the data examined were limited to a small group of patients, yet the use of machine learning in NN applied to orthodontic systems can be a great help for general practitioners. As with general AI models, high performance can be achieved with quality and properly selected input data with systematic organization of models and generalization of the data to a large group of patient (Machoy et al., 2020) Fig. 11.5.

Several years before, a mathematical model was developed by researchers to optimize orthodontic treatment outcome. The model simulated whether or not to extract the teeth and formulated the morphologic traits, which would determine the tooth-extraction/nonextraction decisions. All successfully treated orthodontic patient records were collected, and the influence of dentofacial traits was determined. As per the coincidence rate, results between the actual treatments and the recommendation as per optimized model was found to be 90.4%. Artificial NN and hybrid genetic algorithm is currently being widely studied in the field of orthodontics with the objective to evaluate the size of premolars and canine teeth that did not erupt during tooth replacement. Measurements of a tooth

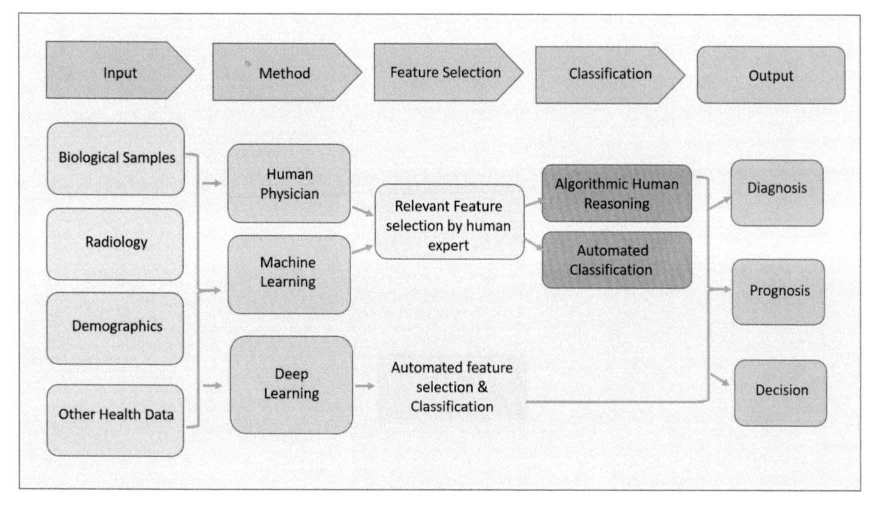

FIGURE 11.5

Summary of application of artificial intelligence.

model were used to collect the data (Machoy et al., 2020). Along with these networks, Bayesian network of AI models also proved very accurate in orthodontic treatment needs (Khanagar et al., 2021; Machoy et al., 2020).

Cephalometric landmarks identification in relation to AI technology applications has been widely studied. Studies have shown excellent accuracy in the computation of the landmarks (Khanagar et al., 2021; Park et al., 2019). Orthognathic surgery involves good treatment planning, and establishing accurate diagnosis marks the success of the treatment. Choi et al. studied lateral cephalometric radiographs with aid of a new AI model to decide the surgical or nonsurgical cases. This system showed effectiveness of a 96% success rate. He concluded the use of the model for orthognathic surgery case diagnosis owing to its promising results (Choi et al., 2019). AI algorithms have also been applied for understanding the progress in the growth of cervical vertebrae stages, with a mean accuracy of 77.02% when used besides cephalometric radiographs (Khanagar et al., 2021). Figs. 11.6 and 11.7 show AI application in the field of orthodontics (Bouletreau et al., 2019).

Another AI model of paraconsistent ANN (PANN) has been applied in the field of orthodontic diagnosis with the use of cephalometric variables. The PANN network used input data as patient cephalometric values. The values were compared with means drawn from normal individuals from the cephalometric point of view. The analysis aimed to establish cephalometric diagnosis by measuring skeletal and dental discrepancies. The outcome was expressed in degrees of anteroposterior, dental, and skeletal differences pertaining to incisors (upper and lower). The model also pointed out contradictions noted by orthodontists in the data, highlighting the contribution of such a system in orthodontics decision support (Machoy et al., 2020).

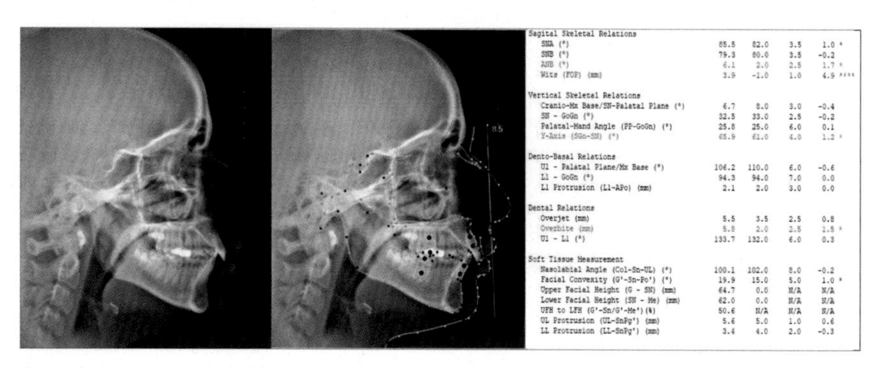

Sagital Skeletal Relations					
SNA (°)	85.5	82.0	3.5	1.0	*
SNB (°)	79.3	80.0	3.5	-0.2	
ANB (°)	6.1	2.0	2.5	1.7	*
Wits (FOP) (mm)	3.9	-1.0	1.0	4.9	****
Vertical Skeletal Relations					
Cranio-Mx Base/SN-Palatal Plane (°)	6.7	8.0	3.0	-0.4	
SN - GoGn (°)	32.5	33.0	2.5	-0.2	
Palatal-Mand Angle (PP-GoGn) (°)	25.8	25.0	6.0	0.1	
Y-Axis (SGn-SN) (°)	65.9	61.0	4.0	1.2	*
Dento-Basal Relations					
U1 - Palatal Plane/Mx Base (°)	106.2	110.0	6.0	-0.6	
L1 - GoGn (°)	94.3	94.0	7.0	0.0	
L1 Protrusion (L1-APo) (mm)	2.1	2.0	3.0	0.0	
Dental Relations					
Overjet (mm)	5.5	3.5	2.5	0.8	
Overbite (mm)	5.8	2.0	2.5	1.5	*
U1 - L1 (°)	133.7	132.0	6.0	0.3	
Soft Tissue Measurement					
Nasolabial Angle (Col-Sn-UL) (°)	100.1	102.0	8.0	-0.2	
Facial Convexity (G'-Sn-Po') (°)	19.9	15.0	5.0	1.0	*
Upper Facial Height (G - SN) (mm)	64.7	0.0	N/A	N/A	
Lower Facial Height (SN - Me) (mm)	62.0	0.0	N/A	N/A	
UFH to LFH (G'-Sn/G'-Me')(%)	50.6	N/A	N/A	N/A	
UL Protrusion (UL-SnPg') (mm)	5.6	5.0	1.0	0.6	
LL Protrusion (LL-SnPg') (mm)	3.4	4.0	2.0	-0.3	

FIGURE 11.6

Application of artificial intelligence in cephalometric tracing.

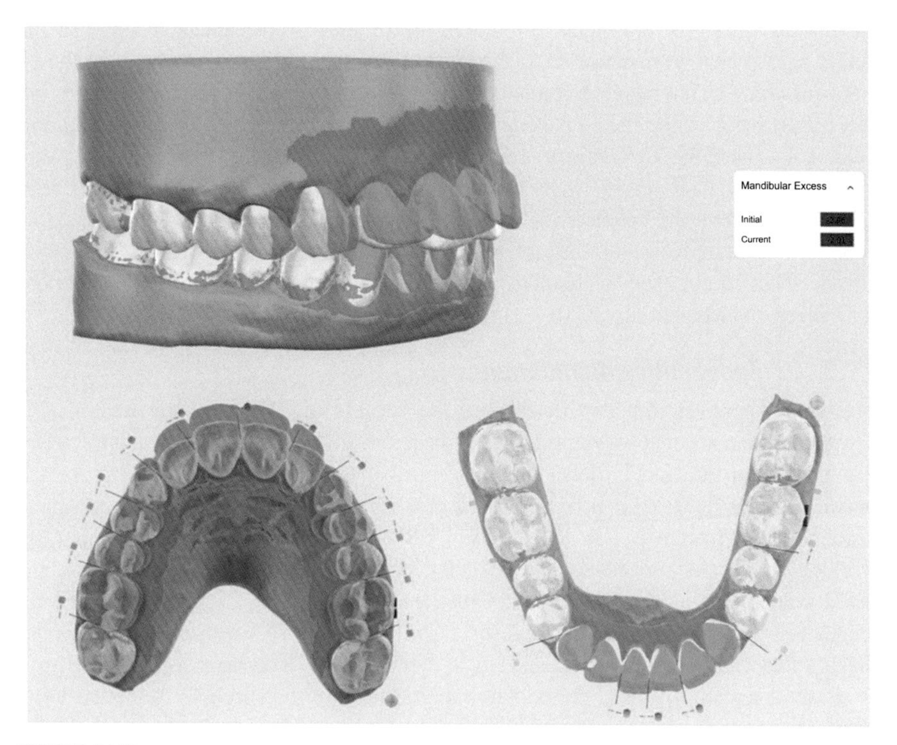

FIGURE 11.7

Application of artificial intelligence in orthodontic treatment planning.

11.5.3 Application of artificial intelligence technologies in conservative, endodontics, and prosthodontics

11.5.3.1 Conservative dentistry

Any dental restorations have a limited life span. Longevity of restoration relies on the type of material used and dental characteristic like type of cavity preparation. A study by Aliaga et al. used AI for performing analysis from a set of data from radiological information, graphs, and notes. His main objective was to find the most acceptable material for the cavity restorations and long-term surveillance of the reconstruction process. Cases were classified using an NN model with a multilayer perceptron, and the data were evaluated by a group of experts. He then concluded that this method could be used to access the type of reconstruction acceptable, thereby predicting feasibility and longevity of every procedure. Another advantage of the system database is that new acquired data can be incorporated and updated, which in turn can be used for variety of new cases in predicting the prognosis. Yet this technology has not been implemented to the clinical procedures (Aliaga et al., 2015). Along with color of the selected filling or prosthetic restoration, the filling or replacement material of the defect is of equal importance to the patient. In spite of the use of computer-based color-matching technologies, ceramic color matching with the color of the natural tooth still remains one of the most challenging tasks in esthetic dentistry. Even before computer color-matching in dentistry, back propagation neural networks (BPNNs) were introduced, but their results were not accurate and stable. GA was combined with BPNN to improve the precision of matching. The current improved BPNN has a low convergence rate. However, BPNN will have high practical application value with advancement in computer technology and will have better ability to provide services for patients (Machoy et al., 2020).

11.5.3.2 Endodontics application

The most important step that determines the success of an endodontic treatment is determining the accurate working length. It becomes important for a clinician to ensure that the instrumentation is done until the apical constriction as the prognosis of the treatment relies on it. AI can be applied over a wide area from detection of periapical lesions to root fractures. The ANN system has now been applied in determining the working length. As per studies, the accuracy of the system was exceptional as compared with professional endodontists. Other than showing 96% accuracy in working length determination, use of probabilistic neural networks (PNNs) in AI have shown diagnostic accuracy of 96.6% for of vertical root fractures (Khanagar et al., 2021).

Early detection of apical periodontitis might increase the success rate of the treatment. It is a common prevalent tooth infection leading to around 75% of radiolucent jaw lesions. Studies have reported that the use of the DL algorithm model CNNs have shown great efficiency and accuracy in the discriminatory ability to detect apical lesions on panoramic radiographs in comparison with highly experienced and trained dentists. AI in endodontics can be applied to fulfill various treatment needs. In terms of periapical lesions, CNN, DL, and ML can be applied. The CBCT scans and periapical radiographs of periapical lesion's detected by AI models can substantially improve the reliability and accuracy of diagnosis similar to or sometimes even better than highly trained and experienced specialists. ML ANN can be applied for detection of periapical pathology (could be a cyst or granuloma). The difference between the two periapical pathologies can be analyzed using gene expression in AI. ML, CNN, and PNN can be applied for detecting root fractures. The research in the area of root fracture is yet ongoing as limited work is reported, as mentioned earlier. In nonsurgical endodontic cases of root canal, a sound knowledge of root morphology and root canal modifications and variations becomes of prime importance to determine its success. Periapical radiographs and CBCT imaging have been frequently used for this purpose. Despite CBCT imaging being more accurate in assessing the root morphology and root canal configurations than periapical radiographs, radiation exposure forms a major limitation and issue of concern for patients. Root morphology along with its curvature and three-dimensional modification postinstrumentation can be determined using ML, CNN, and DL. Prediction of retreatment becomes a task for clinicians as success cannot be determined. AI methods using both case-based reasoning (CBR) and ML in combination proved very beneficial in predicting statistical probabilities for extraction. With the help of CBR based on previous problems, solutions can be created. Based on similar cases and problems, relevant information and knowledge can be integrated. The major advantage and strength of CBR is to be able to accurately predict retreatment outcome. The limitation of this system is that it functions completely dependent on the input data information. Any variability in the data system with difference of approach could create heterogeneity within the system. Currently with the aid of AI, cell viability can be predicted. A neuro-fuzzy inference system (NFIS) using ML and ANN could overcome challenges of microbe infection and follow protocols of regeneration procedures to predict cell viability (Aminoshariae et al., 2021).

11.5.3.3 Prosthodontics

In the field of prosthodontics, a Clinical Decision Support System (CDSS) model was applied specifically for RPDs. For prosthetic construction, it becomes important to understand the patient's oral conditions in relation to parts of denture components. For the purpose, an ontological paradigm was developed. It was a form of program showing similarity between input values and data of classical ontology cases and input patients. The values of the similarity metrics signified the efficiency of the model. Other than RPDs, CDSS were also applied in determining the teeth color after the bleaching procedure. Implementation of the regression model of CDSS was used and served as an intelligent part of the system. This model is composed of three parts: the input data of the patient, prebleaching color, and the prediction of outcome by post-treatment color. The system results were positive in predicting any color change using CDSS technique. They could predict the change in color using colorimetric value with an in-office whitening system. How far this system can be actually implemented in a clinical setup needs further research and studies with large input data in this field (Chen et al., 2016).

11.5.4 Artificial intelligence technologies in periodontics

Periodontal diseases are the most common oral diseases prevalent in humans. Its progression leads to early loss of teeth due to compromised alveolar bone support. In preliminary studies, various classified algorithms have been used to identify periodontal disease. The CAD system, a model of deep CNN algorithm, has been very effective in diagnosis and prediction of periodontal problems. It helps in diagnosing periodontal bone loss in dental pantographs and the teeth, which are periodontally compromised (Khanagar et al., 2021). Periodontal disease has been classified using various AI technology identification units like decision trees (DTs), SVM, and NN. To understand the accuracy of the system, patients were divided into two groups, and a total of six periodontal conditions were applied and checked using data on bone loss. Depending on the operation time and resolution, accuracy was checked. Among the identification unit, SVM and DTs achieved the best accuracy of 98% at classifying periodontal diseases, whereas ANN efficiency was estimated at 46%, which showed poor correlation between input and output variables. Therefore DT and SVM, though slightly complicated in reflecting all factors associated with the periodontal condition, were very clear and simple to be used as a decision-making tree, which would help in predicting periodontal diseases. Hence, this program proved as a very understandable and supportive diagnostic tool with high accuracy in not only diagnosing periodontal diseases but also opened a new area of study for identification of periodontal diseases (Machoy et al., 2020).

Aggressive periodontitis is a form of severe periodontitis that results from autoimmune conditions in patients. It is not a clear histopathological diagnosis, with no clear markers of microbiology or histopathology available. Even clinical markers have not been developed to date to differentiate patients between chronic periodontitis (CP) and aggressive periodontitis. *Actinobacillus actinomycetemcomitans* (Aa) and *Porphyromonas gingivalis* (Pg) are the main organisms involved in AgP. With the use of data of various blood and inflammatory markers, such as CD3, monocytes, neutrophils, eosinophils, lymphocytes, interleukin (IL) 1,2,4 levels, interferon gamma (INF-γ), and tumor necrosis factor α (TNF-α) from monocytes, along with CD4/CD8 ratio, ANNs DL mechanism was applied to differentiate and diagnose patients and treat AgP. Data on antibody levels of Aa and Pg were also used. NNs were developed to correctly classify patients into AgP or CP, and Akaike information criterion for matching probability was used. Cross entropy (CE) values were used in ANNs, depending on the reference of kernel density estimation (KDE). For all the above data sets, ANNs showed accuracy of 90% to 98% in classifying patients to AgP or CP. Among the data inputs, CE eosinophils, CD4/CD8 ratio, monocytes, and neutrophils proved to be the best predictors of AgP. Hence ANN in combination with CE values as per KDE are proven to be invaluable tools in classifying AgP and CP. The method was very convenient as compared with routine manual methods as the parameters used were easily obtained; for example, routine classification gave accurate diagnosis using simple and peripheral leukocyte count. This specific algorithm can even differentiate normal patients from periodontitis cases. This demands the use of specific therapeutic protocol's. Therefore, the technique is studied on limited samples, so in order to generalize it in the field, further studies and research need to be carried out on larger populations using a wider array of parameters with more advanced algorithmic models so this concept of personalized periodontitis treatment can be turned into reality (Papantonopoulos et al., 2014).

Along with periodontitis, mucosal diseases are widely prevalent in the oral cavity. These mucosal diseases are affected by salivary composition and microorganisms in the mouth. As per AI technologies, individual organisms can be classified from salivary microbes. Oral malodor called halitosis is one such problem, which can be identified using models that help classify compounds with volatile sulfur, methyl mercaptan, which cause yeast in the oral cavity. The models include the use of SVM, ANNs, and DTs. These models use the concentration of methyl mercaptan in the oral cavity air and terminal restriction fragment length polymorphisms (T-RFLPs) of the 16 S rRNA gene as data and code for the models of DL. Algorithm of SNN and ANN does not involve any organism or species, which produce inflammatory compounds in the oral environment, leading to halitosis. ANNs do not require even T-RFLP proportions (Nakano et al., 2014). Depending on the findings of ANN in mucosal diseases, many other models can be developed, one of which is NN, which can further aid in recognizing other systemic manifestations in oral mucosa, like recurrent rheumatoid arthritis ulcers. The input data need to be relevant, and the diagnosis involves the use of ANNs that use GA to optimize NN architecture. Using ANN models, a significant association between individuals, their specific environment and behavioral patterns, and their occurrence of recurrent aphthous ulcer has been found. This in turn holds great significance in screening patients of oral diseases and educating them (Dar-Odeh et al., 2010).

11.5.5 Artificial intelligence application in oral and maxillofacial surgery

The field of dentistry dealing with the diagnosis and treatment of oral cancer is maxillofacial surgery. As per WHO statistics, around 657,000 new cases of oral and pharyngeal cancers are detected every year, leading to 330,000 deaths (Khanagar et al., 2021). Any small diagnostic error may result in loss of a patient's life. AI technology has proved very vital in screening and diagnosing oral cancers. CNNs along with their improved programming have demonstrated high accuracy and ability in cancer detection with very promising outcome. CNNs along with confocal laser endomicroscopy images have shown great accuracy for oral squamous cell carcinoma diagnosis. This automated approach has lessened the clinician's burden (Khanday & Sofi, 2021). Some studies have used classifiers like NN, logistic regression, SVR, and adaptive NFIS to test the model with the functions. These classifiers include a data set on oral cancer prognosis. In addition, the data with genomic markers have shown better results in terms of prognosis of cancer (Chang et al., 2013).

AI technology can also predict facial swelling postextraction of teeth. An AI model based on ANN has been specifically developed for the extraction of impacted mandibular third molars, which majority of times leads to facial swelling postoperation. The models have demonstrated very promising results and can be very helpful for the clinicians for predicting the treatment prognosis (Fig. 11.8) (Khanagar et al., 2021).

11.5.6 Forensic odontology

AI technologies in the field of forensic odontology are relatively new. A forensic odontologist plays a vital role in identifying people suspected of criminal assaults for child abuse and sexual assault, which includes identification of lip marks and teeth bite trauma. Even in mass calamities, teeth become a vital structure in identification. Sometimes dental remains can prove extremely important to the families of the victim in providing justice when other relevant evidence has been removed. It becomes a moral duty of the dentists. There have been excellent outcomes of AI technology being applied in this field. CNN automated techniques

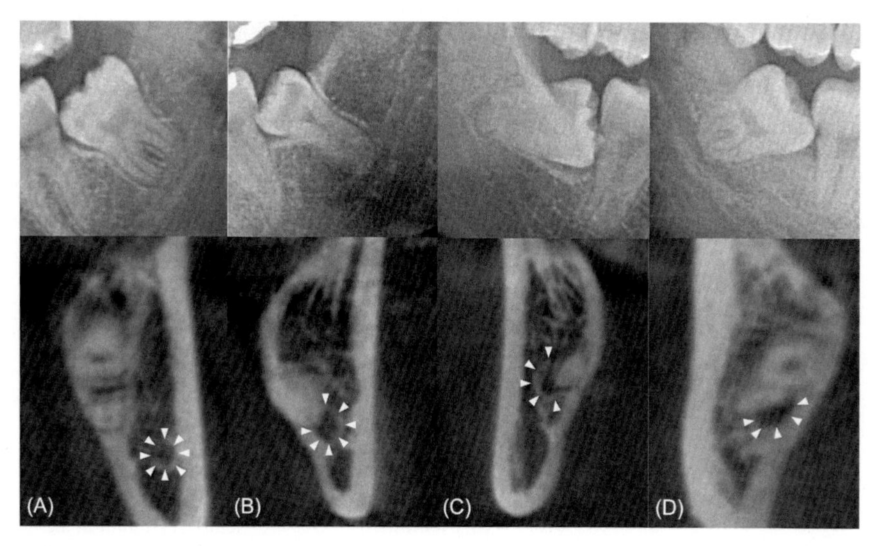

FIGURE 11.8

Artificial intelligence in positioning between mandibular third molar and inferior alveolar nerve on panoramic radiography.

From Choi, E., Lee, S., Jeong, E., Shin, S., Park, H., Youm, S., Son, Y., & Pang, K. M. (2022). Artificial intelligence in positioning between mandibular third molar and inferior alveolar nerve on panoramic radiography. Scientific Reports, 12(1). https://doi.org/10.1038/s41598-022-06483-2.

have been used in AI for determining lower third molar development stages, which helps in estimating the age of a person, and pantomographs can be used for gender determination and can predict the mandibular morphology. AI technique in forensic dentistry has shown very promising results (Khanagar et al., 2021).

11.6 Conclusion

The field of dentistry has seen a tremendous advancement and revolution due to the advent of AI in recent years. Various studies have shown that AI-powered automated systems have performed beyond expectations in different scenarios. They have been proved more accurate than dental specialists. Though they have their own limitations, they can be considered a valuable aid for clinicians in dental practice. They offer great help to clinicians in providing the best quality care by providing accurate diagnosis by predicting treatment and prognosis and provide better clinical decision-making ability. They help in reducing chair-side time, saving multiple steps, and by providing excellent infection control. AI systems can be used by dentists as a subordinate tool for improving the complete treatment outcome right from precise diagnosis to planning the right treatment. Machine learning systems also provide great aid to the nonspecialty dentists for good diagnostic support. These systems can serve as a second opinion for the dentists, thereby aiding in better diagnosis accuracy. AI models have proven very helpful in screening oral cancer patients and patients with osteoporosis, as a diagnostic tool in endodontics to locate root fractures which are difficult to locate manually, in forensic odontology, and in periodontology. Specifically in oral cancer patients, AI becomes a very valuable tool for diagnosing it in the early stages, thereby saving and improving the patient's quality of life. In spite of AI being widely applied in various specialties of dentistry, pedodontics and oral pathology still remain unexplored with its application and benefits. Considering the advancement AI has brought into various fields, further studies need to be taken up in the field of dentistry to lay a concrete foundation to utilize this technology for more efficient productivity for clinicians and improved diagnosis, treatment planning, and outcomes for patients.

11.7 Difficulties and challenges

Despite AI offering multiple aids to the clinical practice, the models of AI have their own limitations that might affect the dependability of the various models proposed.

1. The majority of AI models were studied and developed with the aid of a small number of databases and information retrieved from a specific institution over a specified period. The algorithms and projects presented in studies were validated only internally. Their effect on real clinical decision making is unclear. The data collected might lack heterogenicity. This would lead to lack of reliability and generalizability of the models, which in turn would lead to poor performance in clinical practice as the patient population along with imaging protocols and devices may differ (Hung et al., 2020b; Machoy et al., 2020).
2. In a medical dataset, small sample size becomes a common problem. Gathering effective and adequate samples is highly cost-effective, especially in oral cancer research, where the small sample size problem is more visible.
3. In terms of clinical decision support system (CDSS) validity, which is mostly applied in narrow domains under different conditions and technologies, it is questionable as it is not formally evaluated. Their development is also fragmented, and their current application is limited to very few clinics and hospital centers. There are many practical challenges that require attention. Successful implementation of CDSSs in dental practice can be determined only after overcoming the difficulties. Some of them are enlisted below:
   a. Majority of dentists have their own private practice, and a smaller population of dentists is concentrated in universities. This clearly indicates that CDSS should be more cost-effective to be more commonly used by an average dental practitioner.

   b. Specialty dentistry practice is less as compared to general dentistry practice, therefore CDSS models should focus on extensive programming to meet the needs of decision making in all wider areas of dentistry.
   c. Every program should be linked to an electronic database to decrease chair time for dentists. Only then would it be cost-friendly and time-saving for the dentists.
4. The main aim of AI technology should be on creating precise and practical models and programs, which would help in covering difficult and specialized areas in dentistry. As the field has great application in facial and maxillofacial surgery departments in treatment of neoplasms and in orthognathic surgery, hospitals delivering academic knowledge in this specific field should be of primary focus (Hung et al., 2020b).
5. Despite all its applications and popularity, the ethical and societal concerns raised due to AI over a wide range of its algorithms in dentistry cannot be overlooked. The information used by AI technology from the patient's data and medical tests are used for developing algorithms for problem solving, which then can become commercialized and may cost the patient another time when they use it. These ethical dilemmas, though they might appear indirectly linked to patients, are important to address and should be the prime focus of concern in analyzing and solving these issues (Mörch et al., 2021).

11.8 Future and scope

Though the impact of AI technology is not considerable in clinical dentistry currently, it has a very promising future. AI can be successfully implemented in clinics for a definitive diagnosis for the patient, for planning treatment with precise decision making in clinical procedures, along with prediction of failures of dental procedures. Therefore AI can serve as a dependable modality in future in the fields of oral and radiological diagnosis, conservative and endodontic dentistry, prosthodontics, orthodontics, oral and maxillofacial surgery, periodontics, and forensic dentistry. Though AI models show an increasing scope in dentistry, there are many areas that are still under development. Further studies and extensive work are required for further assessing AI techniques in dentistry for establishing its use in clinical procedures.

References

Abdolali, F., Zoroofi, R. A., Otake, Y., & Sato, Y. (2016). Automatic segmentation of maxillofacial cysts in cone beam CT images. *Computers in Biology and Medicine*, *72*, 108–119.

Ahmed, N., Abbasi, M. S., Zuberi, F., Qamar, W., Halim, M. S. B., Maqsood, A., & Alam, M. K. (2021). Artificial intelligence techniques: Analysis, application, and outcome in dentistry-A systematic review. *BioMed Research International*, *22*, 9751564.

Alcalá-Fdez, J., & Alonso, J. M. (2015). A survey of fuzzy systems software: Tax- onomy, current research trends, and prospects. *IEEE Transactions on Fuzzy Systems*, *24*(1), 40–56.

Aliaga, I. J., Vera, V., De Paz, J. F., García, A. E., & Mohamad, M. S. (2015). Modelling the longevity of dental restorations by means of a CBR system. *BioMed Research International*, *2015*, 540306.

Aminoshariae, A., Kulild, J., & Nagendrababu, V. (2021). Artificial intelligence in endodontics: Current applications and future directions. *Journal of Endodontics*, *47*(9), 1352–1357.

Arsiwala-Scheppach, L. T., Chaurasia, A., Müller, A., Krois, J., & Schwendicke, F. (2023). Machine learning in dentistry: A scoping review. *Journal of Clinical Medicine*, *12*(3), 937, 25.

Arısu, H. D., Dalkilic, E. E., Alkan, F., Erol, S., Uctasli, M. B., & Cebi, A. (2018). Use of artificial neural network in determination of shade, light curing unit, and composite parameters' effect on bottom/top vickers hardness ratio of composites. *BioMed Research International*, 4856707.

Barr, A., Feigenbaum, E. A., & Cohen, P. R. (1981). *The handbook of artificial intelligence* (vol. 13). Los Altos, CA: William Kaufmann Inc.

Bouchahma, M., Ben Hammouda, S., Kouki, S., Alshemaili, M., &Samara K. (2019). An automatic dental decay treatment prediction using a deep convolutional neural network on X-ray images. In 2019 IEEE/ACS 16th International Conference on Computer Systems and Applications (AICCSA), Abu Dhabi, United Arab Emirates. pp. 1–4.

Bouletreau, P., Makaremi, M., Ibrahim, B., Louvrier, A., & Sigaux, N. (2019). Artificial intelligence: applications in orthognathic surgery. *Journal of Stomatology, Oral and Maxillofacial Surgery*, *120*(4), 347–354.

Breiman, L. (2001). Random forests. *Machine Learning*, *45*(1), 5–32.

Campbell, P. J., Getz, G., Korbel, J. O., et al. (2020). Pan-cancer analysis of whole genomes. *Nature*, *578*, 82–93.

Casalegno, F., Newton, T., Daher, R., Abdelaziz, M., Lodi-Rizzini, A., Schürmann, F., Krejci, I., & Markram, H. (2019). Caries detection with near-infrared transillumination using deep learning. *Journal of Dental Research*, *98*, 1227–1233.

Chamorro-Martínez, J., Soto-Hidalgo, J. M., Martínez-Jiménez, P. M., et al. (2016). Fuzzy color spaces: a conceptual approach to color vision. *IEEE Transactions on Fuzzy Systems*, *25*(5), 1264–1280.

Chang, S. W., Abdul-Kareem, S., Merican, A. F., & Zain, R. B. (2013). Oral cancer prognosis based on clinicopathologic and genomic markers using a hybrid of feature selection and machine learning methods. *BMC Bioinformatics*, *14*, 170.

Chen, H., Zhang, K., Lyu, P., Li, H., Zhang, L., Wu, J., & Lee, C.-H. (2019). A deep learning approach to automatic teeth detection and numbering based on object detection in dental periapical films. *Scientific Reports*, *9*, 3840.

Chen, Q., Wu, J., Li, S., Lyu, P., Wang, Y., & Li, M. (2016). An ontology-driven, case-based clinical decision support model for removable partial denture design. *Scientific Reports*, *6*, 27855.

Chen, W. P., Chang, S. H., Tang, C. Y., Liou, M. L., Tsai, S. J., & Lin, Y. L. (2018). Composition analysis and feature selection of the oral microbiota associated with periodontal disease. *BioMed Research International*, *15*, 3130607.

Chen, Y. W., Stanley, K., & Att, W. (2020). Artificial intelligence in dentistry: current applications and future perspectives. *Quintessence International (Berlin, Germany: 1985)*, *51*(3), 248–257.

Cheng, E., Chen, J., Yang, J., Deng, H., Wu, Y., Megalooikonomou, V., et al. (2011). Automatic Dent-landmark detection in 3-D CBCT dental volumes. *Conference Proceedings: ... Annual International Conference of the IEEE Engineering in Medicine and Biology Society. IEEE Engineering in Medicine and Biology Society. Conference*, *2011*, 6204–6207.

Choi, H. I., Jung, S. K., Baek, S. H., et al. (2019). Artificial intelligent model with neural network machine learning for the diagnosis of orthognathic surgery. *The Journal of Craniofacial Surgery*, *30*, 1986e9.

Cortés-Ciriano, I., Lee, J. J. K., Xi, R., et al. (2020). Comprehensive analysis of chromothripsis in 2,658 human cancers using whole-genome sequencing. *Nature Genetics*, *52*(3), 331–341.

Cozzani, M., Sadri, D., Nucci, L., Jamilian, P., Pirhadirad, A. P., & Jamilian, A. (2020). The effect of Alexander, Gianelly, Roth, and MBT bracket systems on anterior retraction: a 3- dimensional finite element study. *Clinical Oral Investigations*, *24*(3), 1351–1357.

Cui, Q., Chen, Q., Liu, P., Liu, D., & Wen, Z. (2020). Clinical decision support model for tooth extraction therapy derived from electronic dental records. *Journal of Prosthetic Dentistry Press*, *126*, 83–90.

Dar-Odeh, N. S., Alsmadi, O. M., Bakri, F., et al. (2010). Predicting recurrent aphthous ulceration using genetic algorithms-optimized neural net- works. *Advances and Applications in Bioinformatics and Chemistry*, *3*, 7.

Dot, G., Rafflenbeul, F., Arbotto, M., Gajny, L., Rouch, P., & Schouman, T. (2020). Accuracy and reliability of automatic three-dimensional cephalometric landmarking. *International Journal of Oral and Maxillofacial Surgery*, *49*, 1367–1378.

Du, X., Chen, Y., Zhao, J., & Xi, Y. (2018). A convolutional neural network based auto-positioning method for dental arch in rotational panoramic radiography. *Proceedings of the 40th Annual International Conference of the IEEE Engineering in Medicine and Biology Society (EMBC)*, 2615–2618, Honolulu, HI, USA.

Ekert, T., Krois, J., Meinhold, L., et al. (2019). Deep learning for the radiographic detection of apical lesions. *Journal of Endodontics*, *45*, 917–922.

Flores, A., Rysavy, S., Enciso, R., & Okada, K. (2009). Non-Invasive differential diagnosis of dental periapical lesions in cone-beam CT. *IEEE International Symposium on Biomedical Imaging*, 566–599.

Friedman, J., Hastie, T., & Tibshirani, R. (2001). (1st ed.). *Introduction. The elements of statistical learning*, (Vol 1). Springer Series in Statistics New York.

Gálvez, J. M., Castillo, D., Herrera, L. J., et al. (2020). Towards improving skin cancer diagnosis by integrating microarray and RNA-seq datasets. *IEEE Journal of Biomedical and Health Informatics*, *24*(7), 2119–2130.

Grischke, J., Johannsmeier, L., Eich, L., Griga, L., & Haddadin, S. (2020). Dentronics: towards robotics and artificial intelligence in dentistry. *Dental Materials: Official Publication of the Academy of Dental Materials*, *36*(6), 765–778.

Ha, S.-R., Park, H. S., Kim, E.-H., Kim, H.-K., Yang, J.-Y., Heo, J., & Yeo, I.-S. L. (2018). A pilot study using machine learning methods about factors influencing prognosis of dental implants. *Journal of Advanced Prosthodontics*, *10*, 395–400.

Hatvani, J., Horváth, A., Michetti, J., et al. (2018). Deep learning-based super- resolution applied to dental computed tomography. *IEEE Transactions on Radiation and Plasma Medical Sciences*, *3*(2), 120–128.

Herrera, L. J., Pulgar, R., Santana, J., et al. (2010). Prediction of color change after tooth bleaching using fuzzy logic for vita classical shades identification. *Applied Optics*, *49*(3), 422–429.

Hung, K., Montalvao, C., Tanaka, R., Kawai, T., & Bornstein, M. M. (2020a). The use and performance of artificial intelligence applications in dental and maxillofacial radiology: A systematic review. *Dentomaxillofacial Radiology*, *49*(1), 20190107.

Hung, K., Yeung, A. W., Tanaka, R., & Bornstein, M. M. (2020b). Current applications, opportunities, and limitations of AI for 3D imaging in dental research and practice. *International Journal of Environmental Research and Public Health*, *17*(12), 4424.

Hung, M., Voss, M. W., Rosales, M. N., et al. (2019). Application of machine learning for diagnostic prediction of root caries. *Gerodontology*, *36*(4), 395–404.

Kalra, S., Tizhoosh, H., Shah, S., et al. (2020). Pan-cancer diagnostic consensus through searching archival histopathology images using artificial intelligence. *NPJ Digital Medicine*, *3*(1), 1–15.

Khanagar, S. B., Al-Ehaideb, A., Maganur, P. C., Vishwanathaiah, S., Patil, S., Baeshen, H. A., Sarode, S. C., & Bhandi, S. (2021). Developments, application, and performance of artificial intelligence in dentistry—A systematic review. *Journal of Dental Sciences*, *16*(1), 508–522, 1.

Khanday, N. Y., & Sofi, S. A. (2021). Deep insight: Convolutional neural network and its applications for COVID-19 prognosis. *Biomedical Signal Processing and Control*, *69*, 102814, Aug.

Kuwada, C., Ariji, Y., Fukuda, M., Kise, Y., Fujita, H., Katsumata, A., & Ariji, E. (2020). Deep learning systems for detecting and classifying the presence of impacted supernumerary teeth in the maxillary incisor region on panoramic radiographs. *Oral Surgery, Oral Medicine, Oral Pathology, and Oral Radiology*, *130*(4), 464–469.

Lee, J. H., Han, S. S., Kim, Y. H., Lee, C., & Kim, I. (2020). Application of a fully deep convolutional neural network to the automation of tooth segmentation on panoramic radiographs. *Oral Surgery, Oral Medicine, Oral Pathology, and Oral Radiology*, *129*(6), 635–642.

Lee, J. H., Kim, D. H., Jeong, S. N., & Choi, S. H. (2018a). Detection and diagnosis of dental caries using a deep learning-based convolutional neural net- work algorithm. *Journal of Dentistry*, *77*, 106–111.

Lee, J.-H., Kim, D.-H., Jeong, S.-N., & Choi, S.-H. (2018b). Diagnosis and prediction of periodontally compromised teeth using a deep learning- based convolutional neural network algorithm. *Journal of Periodontal & Implant Science*, *48*, 114–123.

Li, M., Xu, X., Punithakumar, K., Le, L. H., Kaipatur, N., & Shi, B. (2020). Automated integration of facial and intra-oral images of anterior teeth. *Computers in Biology and Medicine*, *122*, 103794.

Machoy, M. E., Szyszka-Sommerfeld, L., Vegh, A., Gedrange, T., & Woźniak, K. (2020). The ways of using machine learning in dentistry. *Advances in Clinical and Experimental Medicine: Official Organ Wroclaw Medical University*, *29*(3), 375–384, 1.

Mörch, C. M., Atsu, S., Cai, W., Li, X., Madathil, S. A., Liu, X., Mai, V., Tamimi, F., Dilhac, M. A., & Ducret, M. (2021). Artificial intelligence and ethics in dentistry: A scoping review. *Journal of Dental Research*, *100*(13), 1452–1460.

Nakano, Y., Takeshita, T., Kamio, N., et al. (2014). Supervised machine learning- based classification of oral mal-odor based on the microbiota in saliva samples. *Artificial Intelligence in Medicine*, *60*(2), 97–101.

Ostaszewski, M., Mazein, A., Gillespie, M. E., et al. (2020). COVID-19 disease map, building a computational repository of SARS-CoV-2 virus-host interaction mechanisms. *Scientific Data*, *7*(1), 1–4.

Otani, T., Raigrodski, A. J., Mancl, L., Kanuma, I., & Rosen, J. (2015). In vitro evaluation of accuracy and precision of automated robotic tooth preparation system for porcelain laminate veneers. *The Journal of Prosthetic Dentistry*, *114*(2), 229–235.

Papantonopoulos, G., Takahashi, K., Bountis, T., & Loos, B. G. (2014). Artificial neural networks for the diagnosis of aggressive periodontitis trained by immunologic parameters. *PLoS One*, *9*(3), e89757.

Park, J. H., Hwang, H. W., Moon, J. H., et al. (2019). Automated identification of cephalometric landmarks: part 1-comparisons between the latest deep-learning methods YOLOV3 and SSD. *The Angle Orthodontist*, *89*, 903e9.

Patcas, R., Bernini, D. A. J., Volokitin, A., Agustsson, E., Rothe, R., & Timofte, R. (2019). Applying artificial intelligence to assess the impact of orthognathic treatment on facial attractiveness and estimated age. *International Journal of Oral and Maxillofacial Surgery*, *48*(1), 77–83.

Rahimy, E., Wilson, J., Tsao, T., Schwartz, S., & Hubschman, J.-P. (2013). Robot- assisted intraocular surgery: development of the IRISS and feasibility studies in an animal model. *Eye (London, England)*, *27*(8), 972–978.

Rana, M., Modrow, D., Keuchel, J., Chui, C., Rana, M., Wagner, M., et al. (2015). Development and evaluation of an automatic tumor segmentation tool: a comparison between automatic, semi-automatic and manual segmentation of mandibular odontogenic cysts and tumors. *Journal of Cranio-maxillo-facial Surgery: Official Publication of the European Association for Cranio-Maxillo-Facial Surgery*, *43*, 355–359.

Saghiri, M. A., Garcia-Godoy, F., Gutmann, J. L., Lotfi, M., & Asgar, K. (2012). The reliability of artificial neural network in locating minor apical foramen: a cadaver study. *Journal of Endodontics*, *38*(8), 1130–1134.

Schmidhuber, J. (2015). Deep learning in neural networks: An overview. *Neural Networks: the Official Journal of the International Neural Network Society*, *61*, 85–117.

Sun, M., Chai, Y., Chai, G., & Zheng, X. (2020a). Fully automatic robot-assisted surgery for mandibular angle split osteotomy. *Journal of Craniofacial Surgery*, *31*(2), 336–339.

Sun, M.-L., Liu, Y., Liu, G.-M., Cui, D., Heidari, A. A., Jia, W.-Y., Ji, X., Chen, H.-L., & Luo, Y.-G. (2020b). Application of machine learning to stomatology: A comprehensive review. *IEEE Access*, *8*, 184360–184374.

Takahashi, J., Nozaki, K., Gonda, T., & Ikebe, K. (2021). A system for designing removable partial dentures using artificial intelligence. Part 1. Classification of partially edentulous arches using a convolutional neural network. *Journal of Prosthodontic Research*, *65*(1), 115–118.

Tian, S., Dai, N., Zhang, B., Yuan, F., Yu, Q., & Cheng, X. (2019). Automatic classification and segmentation of teeth on 3D dental model using hierarchical deep learning networks. *IEEE Access*, *7*, 84817–84828.

Wang, L., Wang, D., Zhang, Y., Ma, L., Sun, Y., & Lv, P. (2014). An automatic robotic system for three-dimensional tooth crown preparation using a picosecond laser. *Lasers in Surgery and Medicine*, *46*(7), 573–581.

Xu, X., Liu, C., & Zheng, Y. (2018). 3D tooth segmentation and labeling using deep convolutional neural networks. *IEEE Transactions on Visualization and Computer Graphics*, *25*(7), 2336–2348.

Yamada, M., Saito, Y., Imaoka, H., et al. (2019). Development of a real-time endoscopic image diagnosis support system using deep learning technology in colonoscopy. *Scientific Reports*, *9*(1), 1–9.

Yamaguchi, S., Lee, C., Karaer, O., Ban, S., Mine, A., & Imazato, S. (2019). Predicting the debonding of CAD/CAM composite resin crowns with AI. *Journal of Dental Research*, *98*(11), 1234–1238.

Zimmermann, H. J. (2011). *Fuzzy set theory and its applications*. Springer Science & Business Media.

CHAPTER

A-scan generation in spectral domain-optical coherence tomography devices: a survey

12

Mohammad Hossein Vafaie and Hossein Rabbani

Medical Image and Signal Processing Research Center, School of Advanced Technologies in Medicine, Isfahan University of Medical Sciences, Isfahan, Iran

12.1 Introduction

Optical coherence tomography (OCT) is a noninvasive imaging device with the ability of generating high resolution 2D and 3D images (Fercher et al., 1995) of different biological tissues like brain cytoarchitecture (Magnain et al., 2014), retinal layers in the eye (Nassif et al., 2004), etc (Ikuno & Tano, 2009; Srinivasan et al., 2009). In OCT, a low-coherence light source is coupled to a fiber-based interferometer, which splits the light beam into two different arms: a reference arm and a sample arm. Optical elements in the sample arm guide the light toward a sample and collect the backscattered light. Similarly, optical elements in the reference arm guide the light toward a reference mirror and collect the reflected light. The lights received back from the sample tissue and reference mirror interference inside an optical coupler where the resulted light beam is guided toward a photodetector (Zhang, Zhang, et al., 2015). In the photodetector, the information of the inner layers of the sample is extracted by converting the light beam into an electrical signal like voltage. Finally, by digitizing the electrical signal and executing various signal processing algorithms, the information of the sample layers is converted to a digital data array, which is referred to A-scan (Wolfgang Drexler et al., 2014). In other words, A-scan is the digitized representation of the inner layers of a specified point on the sample along the z-axis direction.

After generation, A-scans are transmitted to a personal computer (PC) where they are merged together to generate a 2D image of the sample, which is referred to B-scan. By merging B-scans together, a 3D image of the sample is obtained, which is referred to C-scan (Wojtkowski et al., 2005).

Based on the theory of interferometry, to have an interference with adequate amplitude in the optical coupler, length of the sample arm must be exactly equal to the length of the reference arm (Fercher et al., 2003). Therefore, the information extracted from the interfered light belongs to a depth in which the length of the

Artificial Intelligence and Image Processing in Medical Imaging. DOI: https://doi.org/10.1016/B978-0-323-95462-4.00012-1

sample arm is equal to the length of the reference arm. As a result, by increasing the length of the reference arm, information of deeper layers of the sample can be obtained and vice versa. This principle is adopted in development of the first generation of OCT devices referred to time domain-OCT (TD-OCT), in which the reference mirror is placed on a movable stand that moves rapidly to sweep a patch with the length of the preferred path length on the sample (Huang et al., 1991).

Due to mechanical displacement of the reference mirror in TD-OCT devices, A-scan rate is not as high as it is required in high performance applications. To overcome this shortcoming, a new generation of OCT devices is invented, which are referred to frequency domain-OCT (FD-OCT) devices (Choma et al., 2003). In FD-OCT devices, A-scans are generated by analyzing the spectral pattern of the light obtained from interference of the back-scattered lights of the reference arm and sample arm, that is, no physical movement is required to acquire OCT images.

FD-OCT devices can be classified into two main categories of spectral domain-OCT (SD-OCT) and swept source-OCT (SS-OCT). In SD-OCT devices, a broadband superluminescent diode (SLD) is used as the light source while the photodetectors used in TD-OCT devices to extract the A-scans are replaced with a spectrometer (Leitgeb et al., 2003). On the other hand, in SS-OCT devices, a wavelength-tunable light source that can sweep its wavelength range quickly is used as the light source while a simple or differential photodetector array detects all wavelengths over the time (Chinn et al., 1997; Spaide et al., 2012).

SD-OCT devices are commonly used in clinical applications due to the following advantages against the SS-OCT devices: (1) cost of a tunable laser source is extremely higher than that of a broadband SLD light source and (2) SD-OCT devices have superior phase stability and better axial resolution in comparison with the SS-OCT devices.

Despite various applications of SD-OCT devices, the technology behind the design and development of SD-OCT devices is not assessed in detail in the literatures. Therefore, the main topic of this chapter is dedicated to the hardware of the SD-OCT devices as well as the procedure of A-scan generation in these devices.

The organization of this chapter is as follows: the optical and electrical parts of SD-OCT hardware is introduced in detail in Section 12.2 while the signal processing algorithms required to extract the final A-scan are introduced in Section 12.3. Section 12.4 is dedicated to the key features of SD-OCT devices as well as the effects of improper hardware design on these key features. Finally, conclusions are presented in Section 12.5.

12.2 Spectral domain-optical coherence tomography hardware and its contribution in A-scan generation

Hardware of a SD-OCT can be classified into optical, electrical, and mechanical parts. Since participation of the mechanical part in A-scan generation is negligible, only the optical and electrical parts of the hardware are introduced in detail in this chapter.

12.2.1 Optical part of spectral domain-optical coherence tomography hardware

The block diagram of optical part of a typical SD-OCT device is illustrated in Fig. 12.1. Based on this figure, optical part of SD-OCT devices can be divided into four sections: (1) light source and its accessories, (2) sample arm, (3) reference arm, and (4) spectrometer arm.

12.2.1.1 Light source and its accessories

In SD-OCT devices, a light ray with specified bandwidth, central wavelength, and intensity is generated by a broadband light source where the generated light ray is transferred to the other sections of the optical part through an optical coupler.

Light source is an influential component in the optical part of SD-OCT devices since its characteristics affect the key features of SD-OCT devices like axial resolution, maximum imaging depth, and depth of focus (DOF) (Adhi & Duker, 2013).

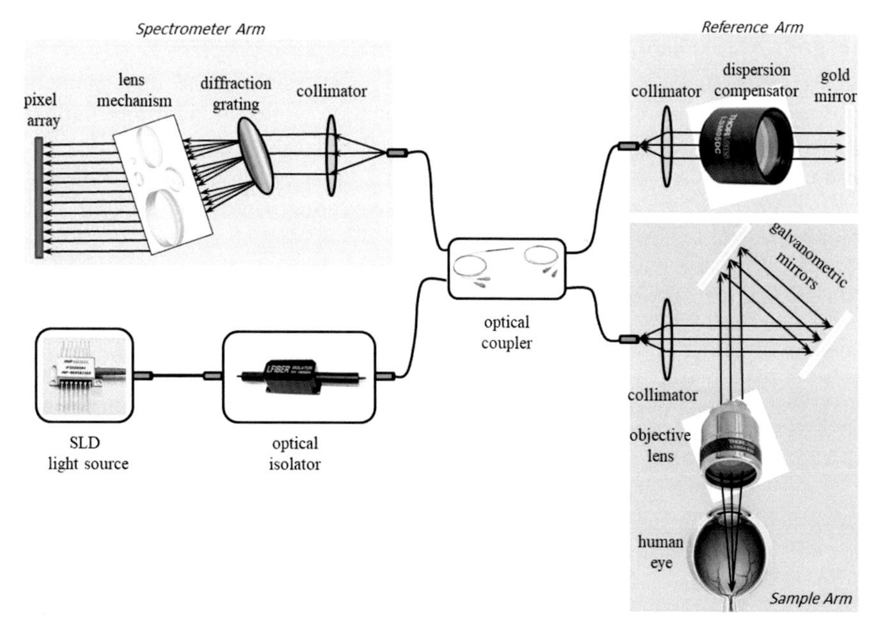

FIGURE 12.1

Block diagram of the optical part of a typical spectral domain-optical coherence tomography.

Moreover, the other optical components of the SD-OCT must be selected by considering the bandwidth and central wavelength of the light source.

In SD-OCT devices, a high bandwidth Gaussian light source must be used to achieve a high axial resolution. SLD is the most suitable choice for the SD-OCT light source due to its higher bandwidth, lower noise level, lower price, and easier configuration compared to the Ti:sapphire and supercontinuum laser light sources (Ishida et al., 2011). Moreover, for ultra-high resolution applications, several SLD light sources can be multiplexed to achieve a higher bandwidth (Braaf et al., 2014).

The light ray produced by the light source is transferred to the optical coupler where the light ray is divided into two parts. One of these parts is transferred to the sample arm while the other is transferred to the reference arm. The schematic of an optical coupler as well as the paths traveled by the light rays in a typical SD-OCT are illustrated in Fig. 12.2.

Optical couplers have different power ratios and different wavelength ranges, for example, in a 75:25 optical coupler, 75% of the light power is transferred to the first output while 25% of the light power is transferred to the second output.

The power ratio of the optical coupler must be selected based on the output power of the light source and the maximum allowable light intensity that can be radiated to the sample tissue.

In practice, a small part of the light ray is returned back to the light source from the optical coupler. This returned ray, regardless of its intensity, leads to damage to the light source (An et al., 2013). To avoid this problem, an optical isolator is placed between the light source and the optical coupler.

By using an optical isolator, the light ray generated by the light source is passed from the isolator with negligible attenuation while the returning light ray is attenuated severely. It must be noted that optical isolators only pass the light rays in a specified wavelength range. Therefore, the wavelength range of the optical isolator must be consistent with the bandwidth and central wavelength of the light source.

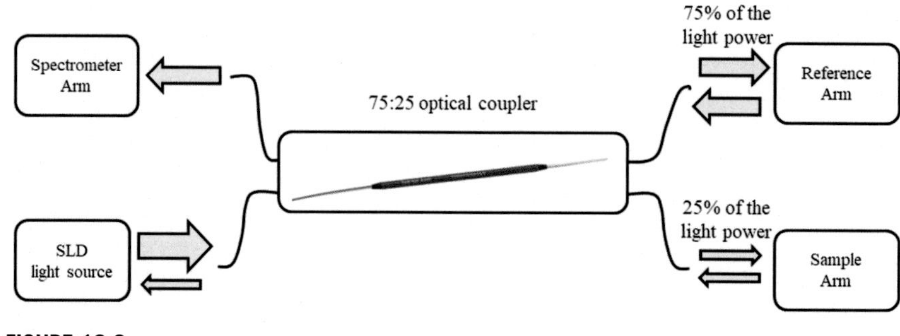

FIGURE 12.2

Schematic of a 75:25 optical coupler used in a spectral domain-optical coherence tomography.

12.2.1.2 Sample arm

As implied by the name, sample arm guides the light ray to depth of the sample tissue. The light ray penetrates into the sample and then travels back to the optical coupler while its characteristics change in accordance with the characteristics of the sample. Therefore by identifying the changes made on the light ray after penetration into the sample, characteristics of the sample can be identified.

Based on Fig. 12.1, sample arm of a SD-OCT device consists of the following optical elements: (1) a collimator, (2) two galvanometric mirrors, and (3) an objective lens (see Fig. 12.3).

12.2.1.2.1 Collimator

Collimator is a package that consists of a lens or a set of lenses that is placed in the entrance of an optical arm to receive a light ray from an optical fiber and produce a collimated light beam. The radius of the collimated light beam depends on the effective focal length (EFL) of the collimating lenses and the mode radius of the fiber at the central wavelength of the light ray (Yang et al., 2003). By increasing (decreasing) the EFL of the collimator, a beam with a larger (smaller) waist is produced, which leads to a smaller (larger) spot size on the sample tissue.

In the sample arm, the collimated light beam penetrates into the sample tissue and travels back to the collimator where the returning light beam is focused to the

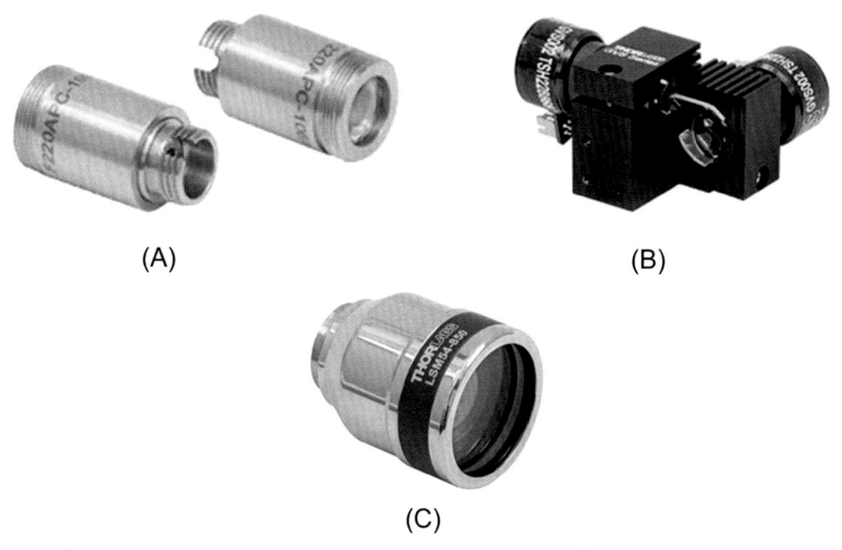

FIGURE 12.3

Optical elements of a sample arm in a typical spectral domain-optical coherence tomography: (A) collimator, (B) two galvanometric mirrors attached to two precise servo motors, and (C) objective lens.

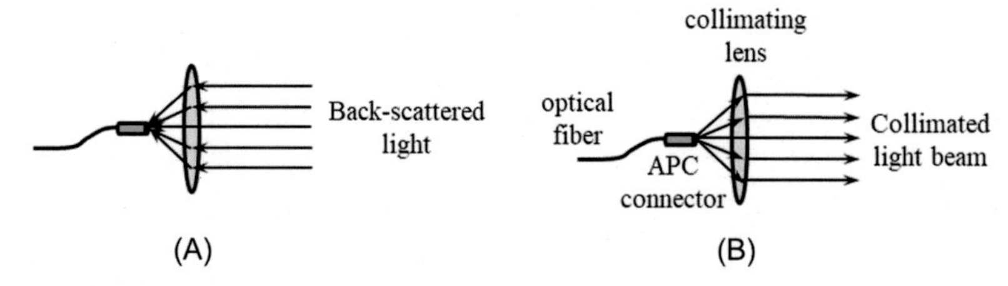

FIGURE 12.4 Different roles of a collimator

(A) collimating an input light ray and (B) focusing the back-scattered light beam.

entrance of the optical fiber to return to the optical coupler. In other words, each collimator has two roles in an optical arm: (1) receiving a light ray from a fiber and collimating the light into a beam with a specified waist and (2) receiving the back-scattered light beam and focusing it into the entrance of the fiber. For more convenience, the operation of a collimator is illustrated in Fig. 12.4. Often, aspherical lenses are used in the collimators to achieve better collimation and higher imaging resolution (Atry & Pashaie, 2016).

It must be noted that position of the collimating lenses must be adjusted precisely with respect to the position of the fiber entrance, since a small misplacement can cause significant aberrations and affects the OCT resolution. Moreover, the fiber and lenses must be aligned carefully since a small tilt can affect the quality of the collimated beam seriously. Since fine tuning of the position of the lenses requires special equipment, manufacturers produce ready to use collimating packages, which deliver a high quality collimated beam without any alignment processes. Each package receives a light ray with a specified wavelength and produces a collimated light beam with a specified beam waist.

12.2.1.2.2 Galvanometric mirrors

To achieve a two-dimensional (2D) image of a sample, the cross section of the sample must be scanned by the beam spot. To do this, the location of the beam spot is moved throughout the preferred cross section of the sample. Often, two galvanometric mirrors with adjustable angles are used to move the beam spot in x- and y-axes of the sample (Kim et al., 2015). By precise control of the mirrors angle, location of the beam spot on the sample can be controlled precisely.

The light beam is guided to a (x, y) position through the mirrors while the beam is penetrated to depth of the sample through an objective lens. The penetrated light beam returns back to the optical coupler to extract the depth profile (or A-scan) of the sample.

By moving the location of the beam spot on the sample, all the A-scans of the preferred cross section are achieved. By merging the obtained A-scans together, a 2D image of the sample (or a B-scan) is achieved.

In OCT devices, a galvanometric system consists of a servo motor, and a high performance driver is adopted to control the angle of each mirror. The servo motor driver is designed such that the mirror angle can be controlled with about one hundredth of a radian precision and very fast dynamic response (Srinivasan et al., 2008). Moreover, to achieve a vibration free movement, each mirror and its servo motor are joined together in a package.

To scan the preferred cross section of a sample, adequate commanding waveforms must be applied to the servo motor drivers. By doing so, the angle of each mirror is changed sequentially in accordance with the applied waveform (Moustapha Hafez et al., 2003); hence, the beam spot is guided in the preferred cross section.

It must be noted that, to penetrate the light beam into an adequate depth, galvanometric mirrors must be placed such that the middle point between them is placed in the back-focal plane of the objective lens (Zhang, Zhang, et al., 2015). Surface quality of the mirrors is another important issue in OCT imaging since any curvature in the surface of the mirrors can reduce the beam quality and the imaging resolution (Kennedy et al., 2014).

12.2.1.2.3 Objective lens

Objective lens is one of the most important optical elements in an OCT device, since the key features like lateral resolution, maximum field of view, and DOF depend on the characteristics of the objective lens.

In OCT devices, lateral resolution (δx) is defined as the radius of the beam spot at the focal point of the sample (Dorrer et al., 2000), while DOF is defined as the range in which the beam waist radius is smaller than $\sqrt{2}\partial x$; that is, DOF is a range in which the imaging resolution is equal to the desired value, while outside this range the resolution drops quickly (Bille, 2019). For more convenience, graphical representation of δx and DOF is illustrated in Fig. 12.5 (Bille, 2019).

Based on mathematical representation of δx and DOF, which are presented in Eqs. (12.1) and (12.2), it is observed that an intrinsic trade-off exists between the lateral resolution and DOF of an OCT.

$$\delta x = 2\sqrt{2\ln 2}\frac{f\lambda}{n\pi d} \tag{12.1}$$

$$b = \mathrm{DOF} = \frac{\pi(\delta x)^2}{\lambda_0} \tag{12.2}$$

where, f is focal length of the objective lens, λ_0 is the central wavelength of the light source, n is the refractive coefficient of retina, and d is diameter of the light beam radiated to the objective lens.

Based on Eqs. (12.1) and (12.2), by using a light source with shorter central wavelength, a better balance can be achieved between the lateral resolution and DOF. However, a shorter central wavelength causes a shorter penetration depth. Moreover, shorter wavelengths are more affected by the biological tissues (Kou et al., 1993). Consequently, the signal-to-noise ratio (SNR) of the OCT as well as the imaging quality is reduced effectively.

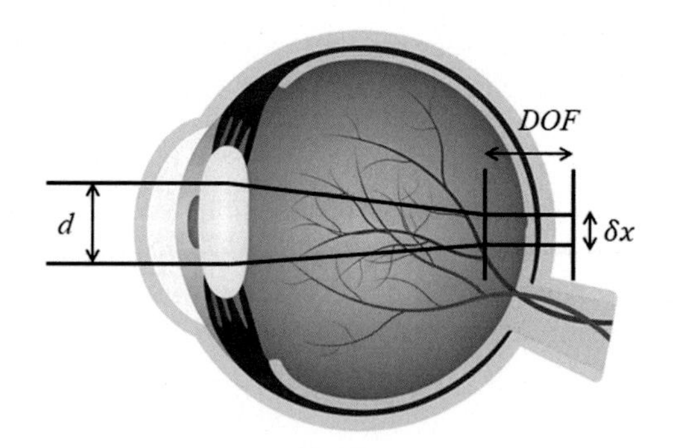

FIGURE 12.5

Graphical representation of lateral resolution and depth of focus.

Often objective lenses are made from several layers of different types of glasses with a complex structure. Therefore cost of an objective lens is considerably higher than that of the other lenses used in an OCT device (Chong et al., 2015).

12.2.1.3 Reference arm

In OCT devices, the light beam traveling back from the sample arm interferes with the light beam traveling back from the reference arm. To achieve the maximum interference, optical length of the two arms as well as the optical behavior of the two arms must be identical.

Based on Fig. 12.1, reference arm of a SD-OCT consists of the following optical elements: (1) a collimator, (2) a dispersion compensator, and (3) a reference mirror (see Fig. 12.6).

A collimator package similar to what is used in the sample arm must be placed in the entrance of the reference arm to collimate the light beam, while a gold mirror must be placed at the end of the reference arm to reflect the light beam without any changes.

Dispersion compensator is an optical element with dispersion properties similar to that of the objective lens but with a significantly lower cost. A dispersion compensator must be placed between the collimator and the reference mirror to mimic the behavior of the objective lens (Fernández et al., 2008). Therefore by using a dispersion compensator, optical behavior of the reference arm becomes similar to the optical behavior of the sample arm.

Often in OCT devices, reference mirror is placed on a movable stand. By using this mechanism, length of the reference arm can be adjusted precisely based on the length of the sample arm, before starting the imaging procedure (Hattori et al., 2015).

It must be noted that the reflection coefficient of the reference mirror is considerably larger than the reflection coefficient of the sample tissue; hence, the

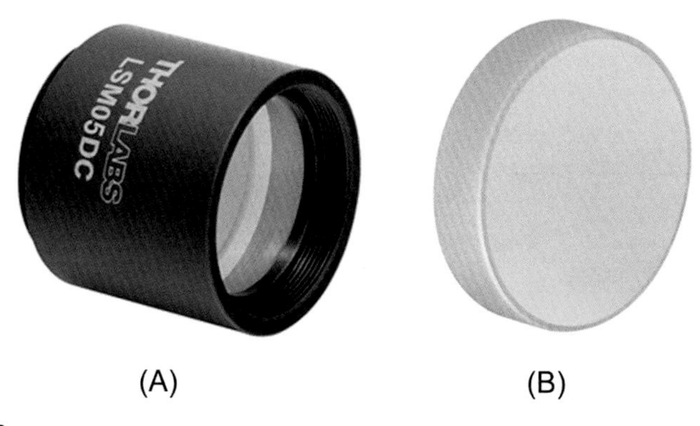

FIGURE 12.6

Optical elements of a typical reference arm: (A) dispersion compensator and (B) reference mirror.

light traveled back from the reference arm is much stronger than the back-scattered light from the sample arm. A strong reference beam causes a high SNR and a poor image quality (de Boer et al., 2003). On the other hand, in imaging from deep layers of the sample, a longer exposure time is necessary to detect the back-scattered light. In this condition, using a strong reference beam will saturate the detector and destroy the imaging quality. To overcome these drawbacks, in some OCT devices, a variable attenuator is placed in the reference arm to adjust the intensity of the reference beam.

12.2.1.4 Spectrometer arm

The interference of the back-scattered light beams is transferred to the spectrometer arm to extract the relevant A-scan of the sample. Based on Fig. 12.1, spectrometer arm of a SD-OCT device consists of the following optical elements: (1) a collimator, (2) a diffraction grating, (3) a set of lenses, and (4) a line-scan CCD camera (see Fig. 12.7).

12.2.1.4.1 Collimator

A collimator package must be placed in the entrance of the spectrometer arm to collimate the light. Often, to achieve a larger spot size on the pixel array of the CCD camera, beam waist of the collimated light in the spectrometer arm must be slightly larger than the beam waist of the collimated light in the sample and reference arms (Zawadzki et al., 2011).

12.2.1.4.2 Diffraction grating

Diffraction gratings have numerous lines on their surfaces. Due to these lines, when a Gaussian light ray is radiated to a diffraction grating, the light ray is diffracted into its constituent wavelengths where each of them leaves the grating by

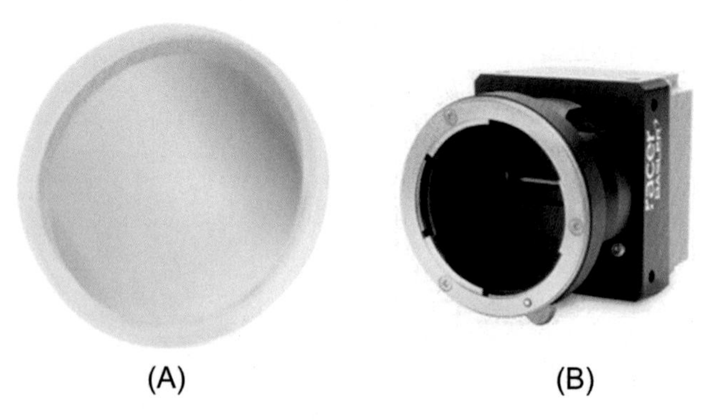

FIGURE 12.7

Optical elements of a typical spectrometer arm: (A) diffraction grating and (B) line-scan CCD camera.

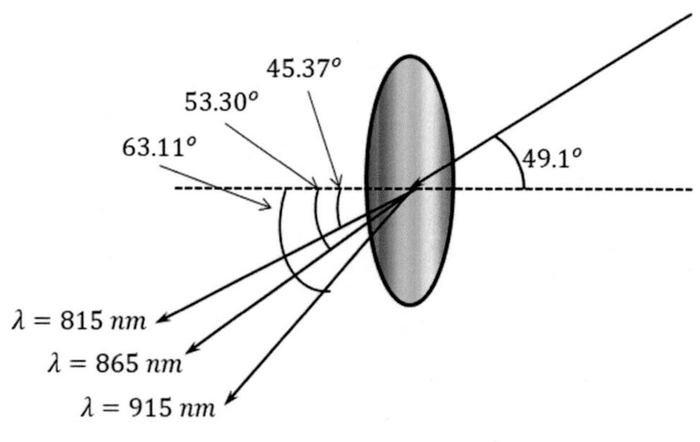

FIGURE 12.8

Diffraction grating operation.

a different angle (Zawadzki et al., 2011); for example, in Fig. 12.8, a Gaussian light ray containing three different wavelengths is radiated to a diffraction grating. As observed, three light rays with different wavelengths are generated where each ray has a different exiting angle.

The exiting angle of a light ray from the surface of a grating depends on the number of lines per millimeter of the grating, central wavelength of the grating, and the wavelength of the exiting light ray (Zawadzki et al., 2005). Since the first two parameters are identical for all of the exiting rays, it is proofed that each wavelength leaves the grating with a different exiting angle. This principle is adopted in the spectrometer arm of the SD-OCT devices to extract the depth profile of the sample.

The penetration depth of a light ray inside a sample tissue depends on the wavelength of the light ray (Unterhuber et al., 2005). Based on this principle, in a spectrometer arm, each of diffracted light rays belongs to a particular depth of the sample; that is, a larger wavelength belongs to a deeper level of the sample and vice versa. Therefore, to achieve the complete information of the depth profile of the sample, all of the diffracted rays must be participated in an A-scan generation (Povazay et al., 2003).

Optical gratings are classified into two main categories of transmission gratings and reflection gratings. Often, transmission gratings are used in SD-OCT devices because the light beam radiated to the grating must be able to travel to the line-scan CCD camera placed at the end of the spectrometer arm (Froehly et al., 2008).

Transmission gratings can be classified into two categories of holographic and blazed gratings. Holographic gratings are better choices for SD-OCT devices due to the following advantages: acceptable efficiency, wide operating wavelength range, and insensitivity to minor misalignments. In contrast, blazed gratings have higher efficiencies around their central wavelengths while the efficiency drops quickly for the wavelengths far from the central wavelength (Ichikawa et al., 2020). Consequently, if a blazed grating is used in the spectrometer arm, the effective bandwidth of the spectrometer arm is reduced, which causes a degradation in the axial resolution of the OCT.

It must be noted that, to achieve the highest diffraction efficiency, light beam must be radiated to the grating surface with a particular angle. Therefore, in practice, the grating is placed on a rotatable stand, which is controlled through a servo motor (Barrick et al., 2016). After development of a SD-OCT device, the angle of the grating is adjusted precisely with respect to the incoming light beam during the factory setting step.

12.2.1.4.3 Lens mechanism

In SD-OCT devices, A-scan is generated by processing the digital output of a line scan CCD camera. Therefore all of the diffracted light rays must be focused on the pixel array of the line CCD camera to participate in A-scan generation. To do this, a lens mechanism consists of various lenses must be placed in the spectrometer arm. The type and dimensions of the lenses as well as their position in the lens mechanism must be designed carefully by considering the following constraints:

1. All the diffracted light rays must be focused on the pixel area of the CCD camera; otherwise, some information of the depth profile of the sample is wasted.
2. Light rays with different wavelengths must be focused on different pixels of the CCD camera; otherwise, information of various depths of the sample is not extractable.
3. Spot size of the focused light rays on the pixel array must be smaller than the pixel size of the CCD camera; otherwise, each light ray is focused on more than one pixel, which leads to reduction in the imaging quality (Ishida & Nishizawa, 2012).

4. The EFL of the lens mechanism must be just equal to the distance between the pixel array of the CCD camera and the position of the lens mechanism; otherwise, light rays are not focused properly on the pixel array of the CCD camera, which leads to significant reduction in the imaging quality.

It must be noted that spot size of a light ray on the pixel array is proportional to the EFL of the lens mechanism as well as the inverse of the collimated light beam radius (Hyle Park et al., 2005). Therefore if a large EFL is required, the radius of the collimated light beam must be increased such that the proper spot size is achieved.

12.2.1.4.4 CCD camera

CCD camera is a key element in the spectrometer arm, since its specifications like line scan rate, spectral response, noise level, and pixel size have significant effects on the A-scan rate, axial resolution, dynamic range, and sensitivity of SD-OCT devices.

A-scan rate of a SD-OCT is mainly limited by the line scan rate of the CCD camera; that is, line scan of the CCD camera is the most time-consuming process in an A-scan generation (Nakamura et al., 2007). Consequently, by selecting a camera with a higher line scan rate, a SD-OCT with a higher A-scan rate can be developed.

CameraLink CCD cameras and Gigabit Ethernet CCD cameras are the two types of cameras that can be used in SD-OCT devices where the line scan rate of Cameralink cameras is significantly higher than that of the Ethernet cameras. However, in contrast with Ethernet cameras that can be connected to the mainboard of a PC directly, the output of CameraLink cameras must be connected to the mainboard through an interface board, which is referred to a frame grabber (Quan et al., 2018). In other words, by using a Cameralink CCD camera, a higher A-scan rate is achieved at the expense of a more complex hardware and a higher development cost. However, CameraLink CCD cameras are widely used in recent SD-OCT devices.

The spectral response of a CCD camera is an influential factor in imaging resolution of a SD-OCT. If bandwidth of the camera is narrower than the bandwidth of the light source, some parts of the information are wasted, which leads to degradation in the imaging resolution. However, thanks to the recent advances in the line-scan cameras, recent CCD cameras have fairly flat responses over a large spectral range, which avoids the aforementioned problem.

Noise level of a CCD camera is another important factor that can affect the OCT resolution. Noise level of a CCD camera specifies the minimum light intensity that can be detected by the camera (Leitgeb et al., 2006). Therefore by using a CCD camera with a lower noise level, a deeper imaging capability can be achieved.

In brief, based on the explanations provided in subsection 2.1.4, the axial resolution of a SD-OCT depends on the overall bandwidth detected by the CCD camera. That is, the actual axial resolution of a SD-OCT depends on the light source bandwidth, radius of the collimated light beam in the spectrometer arm, diffraction grating resolution, EFL of the lens mechanism, and pixel size of the CCD camera. Therefore any mistake in the spectrometer arm design can cause a significant degradation in the axial resolution.

It must be noted that the exiting angle of the diffracted light rays is a nonlinear function of the wavelength. That is, diffracted light rays do not spread uniformly into the pixel array of the CCD camera (Dhalla et al., 2010). Consequently, the output of the CCD camera is not the actual representation of the depth profile of the sample and is referred to "raw A-scan" in the continuation of this chapter.

To achieve the actual A-scans from the raw A-scans, some digital signal processing algorithms are required where the details are presented in Section 12.3 of this chapter.

In some of the recent studies like Gelikonov et al. (2009), instead of adopting signal processing algorithms, a special prism is placed between the grating and the lens mechanism to overcome the nonuniform distribution of the light rays into the pixel array of the CCD camera. However, design and development of this prism is tedious and costly. Therefore this idea is not common in development of SD-OCT devices.

12.2.2 Electrical part of spectral domain-optical coherence tomography hardware

In an OCT device, both the optical and electrical parts of the hardware participate in A-scan generation. The most important tasks performed by the electrical part of SD-OCT hardware to generate an A-scan can be summarized as follows:

- Monitoring the status of the SLD light source as well as its operating point
- Controlling the operating point of the SLD light source
- Controlling the position of the reference mirror
- Controlling the angle of the diffraction grating
- Providing adequate commanding waveforms for controlling the galvanometric mirrors angle
- Providing adequate trigger signal for the CCD camera
- Converting the output of the Cameralink CCD camera into an acceptable format for the mainboard
- Communicating data between the hardware and the user interface software of the SD-OCT.

To perform the duties, mentioned above, various electrical boards are used in a SD-OCT where all of them are introduced in detail in the following subsections.

12.2.2.1 Superluminescent diode monitoring and control hardware

The operating point of a SLD must be controlled continuously because: (1) the intensity of the light beam radiated to the sample tissue must always be below the maximum permissible intensity specified by the ANSI standards (Tomlins &

Wang, 2005) and (2) the operating current as well as the temperature of the SLD must always be below the maximum values specified by its manufacturer (Kowalevicz et al., 2002).

To satisfy the requirements, mentioned above, an electrical board is adopted in SD-OCT devices to control the operating point of the SLD. Often, this board consists of a current sensor, a temperature sensor, a switching power supply, and a microcontroller. The operating current as well as the SLD temperature is measured by the sensors and transmitted to the microcontroller continuously. The microcontroller checks the status of the SLD and changes its operating point if the safety requirements of the device are not satisfied properly.

In SLD control board, a switching power supply is used to control the operating point of the SLD. By doing so, it is guaranteed that in all operating conditions, the sample tissue is protected against the dangerous radiations while the SLD is protected against the overcurrent and overheating.

12.2.2.2 Galvanometric mirrors control hardware

Position of the beam spot on the sample is controlled through two galvanometric mirrors attached to two servo motors. Often, each servo motor has its dedicated electrical driver that receives the commanding signal and controls the mirror angle accordingly. One of the galvanometric mirrors controls the location of the beam spot along the x-axis while the other controls the beam spot along the y-axis. Therefore to scan the desired cross section of the sample, adequate commanding waveforms must be applied to both the servo motor drivers simultaneously.

The servo drivers used in OCT devices receive the commanding signal in the form of an analog or digital signal and rotate the mirror accordingly in a smooth, vibration-free, and high-precision manner (Lee et al., 2009).

In modern SD-OCT devices, cross section of the sample can be scanned according to the various patterns where the most common pattern is illustrated in Fig. 12.9. In this pattern, a high frequency saw-tooth waveform is applied to the servo motor driver belonging to the x-axis while a low frequency saw-tooth (or triangular) waveform is applied to the servo motor driver belonging to the y-axis.

Often, a microcontroller is used in the galvanometric mirror control hardware to generate the adequate commanding waveforms. To do this, at first, the preferred commanding waveforms are stored into various look-up tables (LUTs), which are programmed into the memory of the microcontroller. At the imaging stage, based on the selected scanning pattern and the current position of the mirrors, adequate values are invoked from the LUTs and applied to the servo motor drivers. If digital servo drivers are adopted to control the mirrors, the invoked values are sent to the drivers directly while for analog servo drives, two digital-to-analog converter chips are required to convert the invoked digital values to adequate analog signals. By modifying the shape and characteristics of the commanding waveforms applied to the servo drivers, the scanning pattern of the beam spot on the sample can be set to any desirable pattern (Lee et al., 2020).

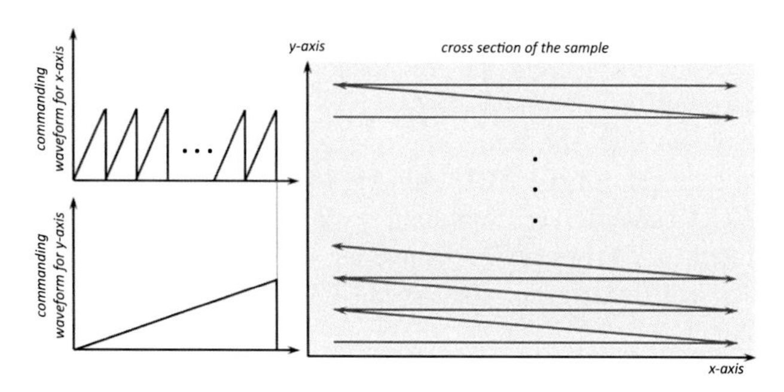

FIGURE 12.9

Scanning pattern in a typical spectral domain-optical coherence tomography device.

12.2.2.3 Reference mirror position control hardware

As mentioned earlier, reference mirror must be placed on a movable stand to adjust the length of the optical path of the reference arm according to the optical path of the sample arm.

In practice, a stepper motor is used to control the position of the reference mirror while a volume potentiometer is provided on the body of the SD-OCT to set the reference position. The value of the potentiometer is sampled by a microcontroller and then is converted to an adequate commanding signal that is applied to the stepper motor driver. In the stepper motor driver, based on the value of the commanding signal, the necessary pulses are generated and applied to the stepper motor to adjust the position of the reference mirror.

12.2.2.4 CCD camera timing control hardware

Each pixel in a CCD camera has its dedicated capacitor. When a light ray is radiated to a pixel, electrical charges are stored in its capacitor. The amount of charges stored in each capacitor depends on the exposure time and the intensity of the light ray radiated to the pixel (Fathipour et al., 2016).

To obtain an A-scan of the sample, the amount of the charges stored in all of the capacitors must be converted to digital values. This process is the most time-consuming part of the A-scan generation. Therefore time scheduling of the other parts of the hardware must be done in accordance with the timing of the CCD camera.

Often, time scheduling of a CCD camera is performed by a square-wave trigger signal (Al-Qazwini et al., 2018). After occurrence of a rising edge on the trigger signal, the amount of the electrical charges stored in each capacitor is converted to a digital value; that is, for a CCD camera with 2048 pixels and a 12 bit analog-to-digital converter, an array consists of 2048 words of 12 bit is generated in each rising edge.

Trigger signal of a CCD camera is generated by the pulse width modulation unit of a microcontroller where the frequency and pulse width of the signal can be adjusted precisely.

It must be noted that, the frequency of the trigger signal must be lower than the maximum scan rate of the CCD camera. Moreover, since the majority of the Cameralink CCD cameras are compliant with low-voltage differential signaling (LVDS) protocol, a TTL-to-LVDS converter chip must be placed between the microcontroller and the trigger input of the CCD camera to balance the voltage level of the signals.

12.2.2.5 Diffraction grating angle control hardware

To achieve the highest diffraction efficiency, the light beam must be radiated to the grating surface with a particular angle. In practice, the grating is placed on a rotatable stand that is controlled through a stepper motor. During the factory settings, the angle of the grating is changed slowly through the stepper motor while the diffraction efficiency is measured through an optical power-meter. The angle in which the maximum diffraction efficiency is obtained is considered as the best grating angle (Hyeon et al., 2015).

It must be noted that, in commercial SD-OCTs, the angle of the grating can only be adjusted by the manufacturer because a small misalignment in the position of the grating can degrade the imaging resolution effectively.

12.2.2.6 Frame grabber

The output of modern high speed CCD cameras is compliant with CameraLink protocol, which is not supported by the common mainboards used in PCs. Therefore an interface board is necessary to convert the Cameralink protocol into an acceptable data format. In these interface boards, which are referred to frame grabbers, a high performance Field Programmable Gate Array (FPGA) is used to receive the data in the Cameralink format, convert the data format, and transmit the converted data through a high speed PCI or PCI Express (PCIe) port. Therefore data can be received in a PC through a PCI/PCIe port of the mainboard.

Often, firmware of frame grabbers is unchangeable and is optimized to convert the Cameralink data format as fast as possible. However, in some of the recent frame grabbers, firmware can be updated by the user such that apart from data conversion, some data processing algorithms can be applied to the received data. In these frame grabbers, expert programming software with a valid license released by the manufacturer is required to upgrade the firmware of the FPGA, which increases the cost. Moreover, a hardware description language like Verilog or VHDL must be used to code the processing algorithms, which is tedious and time-consuming. Therefore using this type of frame grabbers is not common in commercial SD-OCT devices.

In practice, a frame grabber with fixed firmware is used to transfer the data from the CCD camera to a PC while the required signal processing algorithms are coded in a PC and applied to the received data.

12.2.2.7 Main control board

In electrical part of SD-OCT hardware, a main control board is used as a supervisor to manage and coordinate the other control boards of the hardware. This board is connected to the graphical user interface (GUI) software of the SD-OCT through a USB port and receives the operator commands continuously. After receiving each command, the main control board specifies the duties of each part of the hardware and generates the necessary signals for each part accordingly. After that, any required information, including the status of all parts of the hardware, is sent to the GUI for monitoring purposes.

Based on the explanations, above, the duties of the main control board can be summarized as follows:

1. Communicating data with GUI software
2. Receiving the operator commands and specifying the duties of the other hardware parts accordingly
3. Providing time scheduling signals for various parts of the hardware
4. Providing commanding signals for various parts of the hardware
5. Transferring the required information to the GUI (excluding OCT images) for monitoring purposes.

Finally, based on the explanations provided in subsection 2.2, block diagram of the electrical part of SD-OCT hardware can be illustrated as shown in Fig. 12.10.

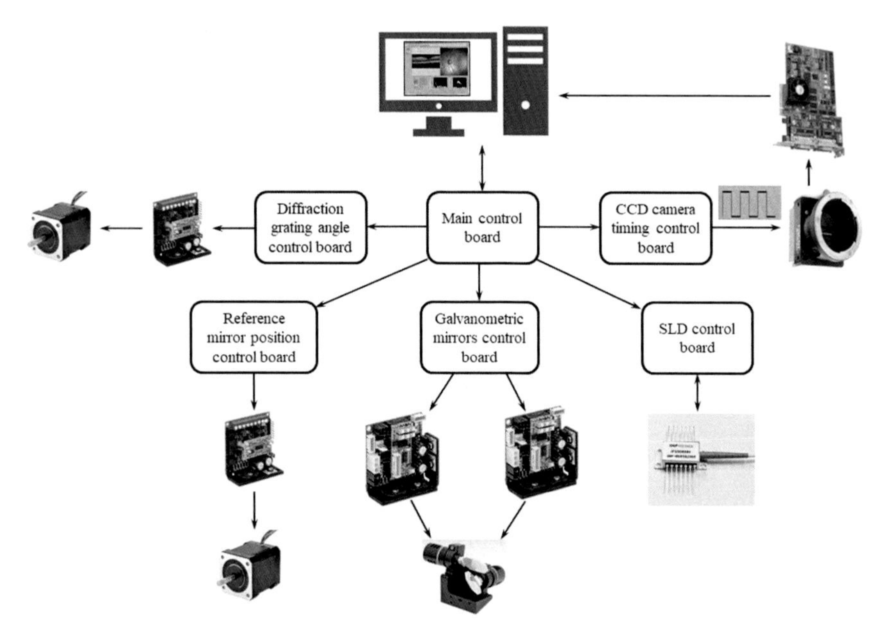

FIGURE 12.10

Block diagram of the electrical part of spectral domain-optical coherence tomography hardware.

12.3 Signal processing algorithms required for A-scan extraction

As mentioned earlier, a nonuniform distribution of A-scans is generated by SD-OCT hardware. After receiving these raw A-scans in a PC, the following signal processing process must be applied to raw A-scans to achieve the final A-scans:

1. The redundant DC offset must be eliminated from the received A-scans. As mentioned in (Chan & Tang, 2010), a redundant DC component is generated when the back-scattered lights from the sample arm and reference arm interfere with each other.
2. Data must be transferred from the wavelength space into the frequency space to achieve the frequency domain representation of A-scans (Davila et al., 2012).
3. Data must be transferred from the frequency space into the z space by conducting the FFT inverse transformation (Leitgeb & Kalkman, 2017).

By doing so, the final A-scan, which is the actual depth profile of the sample along z-axis, can be achieved.

In commercial SD-OCT devices, a GUI software is prepared to show the OCT images on the desktop of a PC where the following tasks must be performed by the GUI sequentially:

(1) Raw A-scans are received from PCI/PCIe port of the mainboard repeatedly, (2) by executing the signal processing process, mentioned above, the final A-scans are extracted from the raw A-scans, (3) the extracted A-scans are merged together to generate a B-scan, and (4) the generated B-scan is shown in the PC desktop.

The GUI software has a duplex communication with the main board of the SD-OCT through a USB port. The desired tasks, configurations, settings, etc., are set by the operator and transferred from the GUI to the main board. On the other hand, the overall status of the SD-OCT as well as the operating parameters is transferred from the main board to the GUI.

Based on the explanations, above, GUI receives data from two different sources: (1) a USB port that provides information about the status of the OCT, operating parameters, etc. and (2) a PCI/PCIe port that provides raw A-scans data. For more convenience, this communication is illustrated in Fig. 12.11.

12.4 Effects of improper hardware design on spectral domain-optical coherence tomography characteristics

The key features of a SD-OCT device like A-scan rate, axial resolution, and maximum imaging depth are highly dependent to the design of the SD-OCT hardware. In other words, although, from theoretical point of view, each of these key features depends on definite parameters, improper hardware or software design can disturb the SD-OCT's performance and cause unpredictable operation. More explanations about this problem are provided in the following subsections.

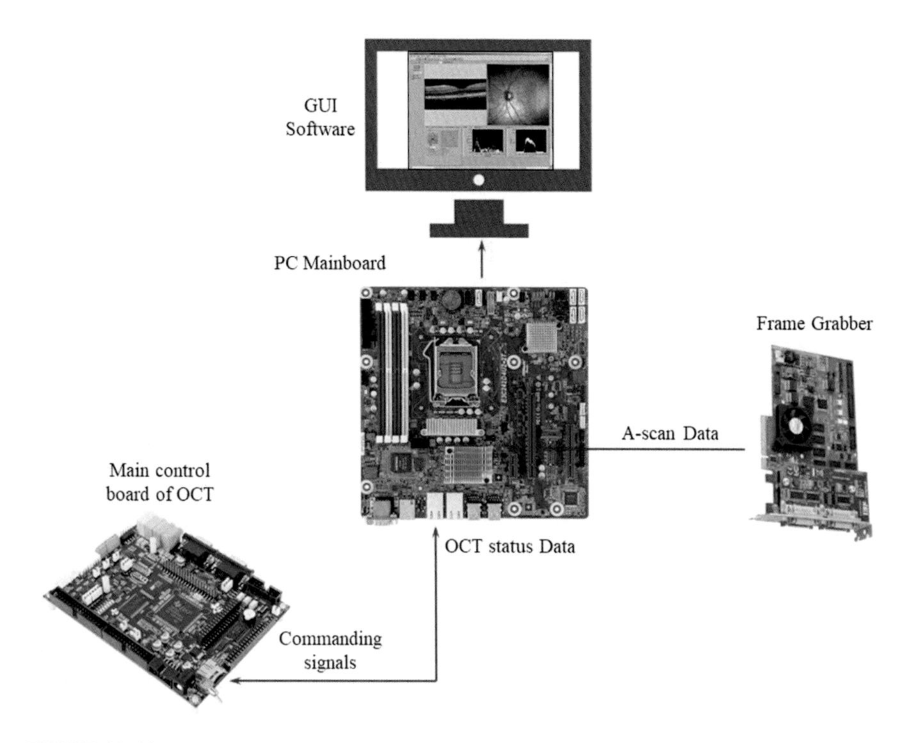

FIGURE 12.11

Communication between graphical user interface software and spectral domain-optical coherence tomography hardware.

12.4.1 A-scan rate

The number of generated A-scans per second depends on the slowest part of OCT hardware participated in A-scan generation; hence, A-scan rate of an SD-OCT is equal to the line-scan rate of the CCD camera.

In CameraLink CCD cameras, A-scan rate is about 80 kHz while in Gigabit Ethernet CCD cameras, A-scan rate is about 40 kHz. However, if the execution time of the signal processing algorithms adopted to extract the final A-scan is higher than the line-scan time of the CCD camera, A-scan rate of the SD-OCT is reduced to a value below the line-scan rate of the CCD camera.

12.4.2 Axial resolution

From theoretical point of view, axial resolution of a SD-OCT is calculated through Eq. (12.3); that is, axial resolution depends only on the bandwidth and central wavelength of the light source (Izatt & Choma, 2008). However, in practice, improper design of the optical arms can reduce the effective bandwidth and degrade the axial resolution, for example, in the following conditions, the effective bandwidth is decayed to a value smaller than the bandwidth of the light source:

1. Bandwidth of the optical elements placed in the SD-OCT arms is smaller than the bandwidth of the light source.
2. Some parts of the diffracted light rays are focused on an area outside the CCD camera pixels array.
3. Spot size of the light beam on the CCD camera array is larger than the pixel size of the CCD camera.

$$\delta z = \frac{2\ln 2(\lambda_0)^2}{n\pi\Delta\lambda} \tag{12.3}$$

where, δz is axial resolution, λ_0 is central wavelength of the light source, n is refractive coefficient of retina, and $\Delta\lambda$ is bandwidth of the light source.

12.4.3 Maximum imaging depth

From theoretical point of view, the maximum imaging depth of a SD-OCT can be calculated through Eq. (12.4); that is, the maximum imaging depth depends on the bandwidth and central wavelength of the light source as well as the number of pixels of the CCD camera (Li et al., 2013). However, in the following conditions the maximum reachable imaging depth is smaller than the value calculated by Eq. (12.4):

1. Bandwidth of the optical elements placed in the SD-OCT arms be smaller than the bandwidth of the light source.
2. Diffracted light rays be focused on a portion of the pixels array of the CCD camera; that is, some part of the CCD camera pixels not participate in A-scan generation.
3. Some parts of the diffracted light rays be focused on an area outside the CCD camera pixels array.

$$z_{max} = \frac{N(\lambda_0)^2}{4n\Delta\lambda} \tag{12.4}$$

where, z_{max} is the maximum imaging depth, N is the pixels number of the CCD camera, λ_0 is central wavelength of the light source, n is refractive coefficient of retina, and $\Delta\lambda$ is bandwidth of the light source.

12.5 Conclusion

In this chapter, the process of A-scan generation is introduced in a comprehensive manner. At first, the optical and electrical parts of SD-OCT hardware that participate in raw A-scan generation are introduced, and then the signal processing

Table 12.1 Summary of the parameters affecting the key features of a spectral domain-optical coherence tomography device.

Feature	Effective parameters						
Axial resolution	Bandwidth of light source	Central wavelength of light source	Light ray collimation in optical arms	Smoothness of galvanometric mirrors	Spot size of radiated ray in pixels array	Angle of radiation to the grating	EFL of lens mechanism
Lateral resolution	Focal length of objective lens	Central wavelength of light source	Collimated light beam waist in sample arm	Smoothness of galvanometric mirrors	Objective lens position		
A-scan rate	Line-scan rate of CCD camera	Transferring rate of Frame Grabber	Execution time of processing algorithms				
Maximum Imaging depth	Bandwidth of light source	Central wavelength of light source	CCD camera pixels number	Angle of radiation to the grating	EFL of lens mechanism	CCD camera position	Spot size of radiated ray in pixels array
Depth of focus	Focal length of objective lens	Central wavelength of light source	Collimated light beam waist in sample arm	Objective lens position			

EFL, effective focal length.

algorithms used to extract the final A-scan are presented. At the next step, effects of improper design of SD-OCT hardware on the key features of a SD-OCT are assessed in detail.

As observed, although from theoretical point of view, the key features depend on the characteristics of the light source and objective lens, improper design of the SD-OCT hardware degrades the performance of the SD-OCT effectively. For more convenience, the summary of the parameters that can affect the key features of the SD-OCT are presented in Table 12.1.

Based on Table 12.1, in comparison with the other optical arms, the spectrometer arm has significant effects on the performance of the SD-OCT; that is, improper design of the spectrometer arm can reduce the imaging quality effectively. Moreover, axial resolution is the most sensitive feature in SD-OCT devices; that is, any mistake in design of the hardware can degrade the axial resolution effectively.

References

Adhi, M., & Duker, J. S. (2013). Optical coherence tomography—current and future applications. *Current Opinion in Ophthalmology*, *3*, 213–221. Available from https://doi.org/10.1097/ICU.0b013e32835f8bf8.

Al-Qazwini, Z., Ko, Z. Y. G., Mehta, K., & Chen, N. (2018). Ultrahigh-speed line-scan SD-OCT for four-dimensional in vivo imaging of small animal models. *Biomedical Optics Express*, *3*, 1216–1228. Available from https://doi.org/10.1364/BOE.9.001216.

An, L., Li, P., Lan, G., Malchow, D., & Wang, R. K. (2013). High-resolution 1050 nm spectral domain retinal optical coherence tomography at 120 kHz A-scan rate with 6.1 mm imaging depth. *Biomedical Optics Express*, *2*, 245–259. Available from https://doi.org/10.1364/BOE.4.000245.

Atry, F., & Pashaie, R. (2016). Analysis of intermediary scan-lens and tube-lens mechanisms for optical coherence tomography. *Applied Optics*, *4*, 646–653. Available from https://doi.org/10.1364/AO.55.000646.

Barrick, J., Doblas, A., Gardner, M. R., Sears, P. R., Ostrowski, L. E., & Oldenburg, A. L. (2016). High-speed and high-sensitivity parallel spectral-domain optical coherence tomography using a supercontinuum light source. *Optics Letters*, *24*, 5620–5623. Available from https://doi.org/10.1364/OL.41.005620.

Bille, J. F. (2019). *High resolution imaging in microscopy and ophthalmology: New frontiers in biomedical opticshigh resolution imaging in microscopy and ophthalmology: New frontiers in biomedical optics* (p. 1) Springer. Available from https://doi.org/10.1007/978-3-030-16638-0.

de Boer, J. F., Cense, B., Park, B. H., Pierce, M. C., Tearney, G. J., & Bouma, B. E. (2003). Improved signal-to-noise ratio in spectral-domain compared with time-domain optical coherence tomography. *Optics Letters*, *21*, 2067–2069.

Braaf, B., Vermeer, K. A., de Groot, M., Vienola, K. V., & de Boer, J. F. (2014). Fiber-based polarization-sensitive OCT of the human retina with correction of system polarization distortions. *Biomedical Optics Express*, *8*, 2736–2758. Available from https://doi.org/10.1364/BOE.5.002736, http://opg.optica.org/boe/abstract.cfm?URI = boe-5-8-2736.

Chan, K. K. H., & Tang, S. (2010). High-speed spectral domain optical coherence tomography using non-uniform fast Fourier transform. *Biomedical Optics Express*, *5*, 1309–1319.

Chinn, S. R., Swanson, E. A., & Fujimoto, J. G. (1997). Optical coherence tomography using a frequency-tunable optical source. *Optics Letters*, *5*, 340–342. Available from https://doi.org/10.1364/OL.22.000340, http://opg.optica.org/ol/abstract.cfm?URI = ol-22-5-340.

Choma, M., Sarunic, M., Yang, C., & Izatt, J. (2003). Sensitivity advantage of swept source and Fourier domain optical coherence tomography. *Optics Express*, *18*, 2183–2189.

Chong, S. P., Merkle, C. W., Cooke, D. F., Zhang, T., Radhakrishnan, H., Krubitzer, L., & Srinivasan, V. J. (2015). Noninvasive, in vivo imaging of subcortical mouse brain regions with 1.7 μm optical coherence tomography. *Optics Letters*, *21*, 4911–4914. Available from https://doi.org/10.1364/OL.40.004911, http://opg.optica.org/ol/abstract.cfm?URI = ol-40-21-4911.

Davila, A., Huntley, J. M., Pallikarakis, C., Ruiz, P. D., & Coupland, J. M. (2012). Wavelength scanning interferometry using a Ti:Sapphire laser with wide tuning range. *Fringe Analysis Methods & Applications*, *8*, 1089–1096. Available from https://doi.org/10.1016/j.optlaseng.2012.02.005, https://www.sciencedirect.com/science/article/pii/S0143816612000498.

Dhalla, A.-H., Migacz, J. V., & Izatt, J. A. (2010). Crosstalk rejection in parallel optical coherence tomography using spatially incoherent illumination with partially coherent sources. *Optics Letters*, *13*, 2305–2307. Available from https://doi.org/10.1364/OL.35.002305.

Dorrer, C., Belabas, N., Likforman, J.-P., & Joffre, M. (2000). Spectral resolution and sampling issues in Fourier-transform spectral interferometry. *Journal of the Optical Society of America. B, Optical Physics*, *10*, 1795–1802. Available from https://doi.org/10.1364/JOSAB.17.001795, http://opg.optica.org/josab/abstract.cfm?URI = josab-17-10-1795.

Drexler, W., Liu, M., Kumar, A., Kamali, T., Unterhuber, A., & Leitgeb, R. A. (2014). Optical coherence tomography today: Speed, contrast, and multimodality. *Journal of Biomedical Optics*, *7*, 1–34. Available from https://doi.org/10.1117/1.JBO.19.7.071412.

Fathipour, V., Bonakdar, A., & Mohseni, H. (2016). Advances on sensitive electron-injection based cameras for low-flux, short-wave infrared applications. *Frontiers in Materials*. Available from https://www.frontiersin.org/article/10.3389/fmats.2016.00033.

Fercher, A. F., Drexler, W., Hitzenberger, C. K., & Lasser, T. (2003). Optical coherence tomography - principles and applications. *Reports on Progress in Physics*, *2*, 239–303. Available from https://doi.org/10.1088/0034-4885/66/2/204.

Fercher, A. F., Hitzenberger, C. K., Kamp, G., & El-Zaiat, S. Y. (1995). Measurement of intraocular distances by backscattering spectral interferometry. *Optics Communications*, *1*, 43–48. Available from https://doi.org/10.1016/0030-4018(95)00119-S, https://www.sciencedirect.com/science/article/pii/003040189500119S.

Fernández, E. J., Hermann, B., Povazay, B., Unterhuber, A., Sattmann, H., Hofer, B., Ahnelt, P. K., & Drexler, W. (2008). Ultrahigh resolution optical coherence tomography and pancorrection for cellular imaging of the living human retina. *Optics Express*, *15*, 11083–11094.

Froehly, L., Ouadour, M., Furfaro, L., Sandoz, P., Leproux, P., Huss, G., & Couderc, V. (2008). Spectroscopic OCT by grating-based temporal correlation coupled to optical spectral analysis. *International Journal of Biomedical Imaging*, 752340. Available from https://doi.org/10.1155/2008/752340.

Gelikonov, V. M., Gelikonov, G. V., & Shilyagin, P. A. (2009). Linear-wavenumber spectrometer for high-speed spectral-domain optical coherence tomography. *Optics and Spectroscopy*, *3*, 459–465. Available from https://doi.org/10.1134/S0030400X09030242.

Hattori, Y., Kawagoe, H., Ando, Y., Yamanaka, M., & Nishizawa, N. (2015). High-speed ultrahigh-resolution spectral domain optical coherence tomography using high-power supercontinuum at 0.8 μm wavelength. *Applied Physics Express*, *8*. Available from https://doi.org/10.7567/apex.8.082501.

Huang, D., Swanson, E. A., Lin, C. P., Schuman, J. S., Stinson, W. G., Chang, W., Hee, M. R., Flotte, T., Gregory, K., & Puliafito, C. A. (1991). Optical coherence tomography. *Science (New York, N.Y.)*, *5035*, 1178–1181.

Hyeon, M. G., Kim, H.-J., Kim, B.-M., & Eom, T. J. (2015). Spectral domain optical coherence tomography with balanced detection using single line-scan camera and optical delay line. *Optics Express*, *18*, 23079–23091. Available from https://doi.org/10.1364/OE.23.023079.

Hyle Park, B., Pierce, M. C., Cense, B., Yun, S.-H., Mujat, M., Tearney, G. J., Bouma, B. E., & Boer, J. F. de (2005). Real-time fiber-based multi-functional spectral-domain optical coherence tomography at 1.3 μm. *Optics Express*, *11*, 3931–3944. Available from https://doi.org/10.1364/OPEX.13.003931, http://opg.optica.org/oe/abstract.cfm?URI = oe-13-11-3931.

Ichikawa, H., Yasuno, Y., & Fujibuchi, H. (2020). *Optical coherence tomography interpreted by diffractive optics: A-scan image formation with wavelength-scale diffraction gratings as samples* (9, pp. 2395–2406). OSA Continuum. Available from http://opg.optica.org/osac/abstract.cfm?URI = osac-3-9-2395, 10.1364/OSAC.393868.

Ikuno, Y., & Tano, Y. (2009). Retinal and choroidal biometry in highly myopic eyes with spectral-domain optical coherence tomography. *Investigative Ophthalmology & Visual Science*, *8*, 3876–3880. Available from https://doi.org/10.1167/iovs.08-3325.

Ishida, S., & Nishizawa, N. (2012). Quantitative comparison of contrast and imaging depth of ultrahigh-resolution optical coherence tomography images in 800–1700 nm wavelength region. *Biomedical Optics Express*, *2*, 282–294. Available from https://doi.org/10.1364/BOE.3.000282.

Ishida, S., Nishizawa, N., Ohta, T., & Itoh, K. (2011). Ultrahigh-resolution optical coherence tomography in 1.7 μm region with fiber laser supercontinuum in low-water-absorption samples. *Applied Physics Express*, *5*, 052501. Available from https://doi.org/10.1143/APEX.4.052501.

Izatt, J. A., & Choma, M. A. (2008). *Theory of optical coherence tomography* (pp. 47–72). Berlin, Heidelberg: Springer Berlin Heidelberg. Available from https://doi.org/10.1007/978-3-540-77550-8_2.

Kennedy, B. F., Kennedy, K. M., & Sampson, D. D. (2014). A review of optical coherence elastography: Fundamentals, techniques and prospects. *IEEE Journal of Selected Topics in Quantum Electronics*, *2*, 272–288. Available from https://doi.org/10.1109/JSTQE.2013.2291445.

Kim, J., Brown, W., Maher, J. R., Levinson, H., & Wax, A. (2015). Functional optical coherence tomography: Principles and progress. *Physics in Medicine and Biology*, *10*, R211–R237. Available from https://doi.org/10.1088/0031-9155/60/10/R211.

Kou, L., Labrie, D., & Chylek, P. (1993). Refractive indices of water and ice in the 0.65- to 2.5-μm spectral range. *Applied Optics*, *19*, 3531–3540. Available from https://doi.org/10.1364/AO.32.003531, http://opg.optica.org/ao/abstract.cfm?URI = ao-32-19-3531.

Kowalevicz, A., Ko, T., Hartl, I., Fujimoto, J., Pollnau, M., & Salathé, R. (2002). Ultrahigh resolution optical coherence tomography using a superluminescent light source. *Optics Express*, *7*, 349–353.

Lee, B. K., Chen, S., Moult, E. M., Yu, Y., Alibhai, A. Y., Mehta, N., Baumal, C. R., Waheed, N. K., & Fujimoto, J. G. (2020). High-speed, ultrahigh-resolution spectral-domain OCT with extended imaging range using reference arm length matching. *Translational Vision Science & Technology*, *7*, 12. Available from https://doi.org/10.1167/tvst.9.7.12.

Lee, S.-W., Jeong, H.-W., Kim, B.-M., Ahn, Y.-C., Jung, W., & Chen, Z. (2009). Optimization for axial resolution, depth range, and sensitivity of spectral domain optical coherence tomography at 1.3 μm. *The journal of the Korean Physical Society*, *6*, 2354–2360.

Leitgeb, R., Hitzenberger, C. K., & Fercher, A. F. (2003). Performance of fourier domain vs. time domain optical coherence tomography. *Optics Express*, *8*, 889–894. Available from https://doi.org/10.1364/OE.11.000889, http://opg.optica.org/oe/abstract.cfm?URI = oe-11-8-889.

Leitgeb, R. A., Villiger, M., Bachmann, A. H., Steinmann, L., & Lasser, T. (2006). Extended focus depth for Fourier domain optical coherence microscopy. *Optics Letters*, *16*, 2450–2452.

Leitgeb, R., & Kalkman, J. (2017). Fourier-domain optical coherence tomography signal analysis and numerical modeling. *International Journal of Optics*, 9586067. Available from https://doi.org/10.1155/2017/9586067.

Li, P., An, L., Lan, G., Johnstone, M., Malchow, D., & Wang, R. K. (2013). Extended imaging depth to 12 mm for 1050-nm spectral domain optical coherence tomography for imaging the whole anterior segment of the human eye at 120-kHz A-scan rate. *Journal of Biomedical Optics*, *1*, 16012. Available from https://doi.org/10.1117/1.JBO.18.1.016012.

Magnain, C., Augustinack, J. C., Reuter, M., Wachinger, C., Frosch, M. P., Ragan, T., Akkin, T., Wedeen, V. J., Boas, D. A., & Fischl, B. (2014). Blockface histology with optical coherence tomography: A comparison with Nissl staining. *Neuroimage*, 524–533. Available from https://doi.org/10.1016/j.neuroimage.2013.08.072.

Moustapha Hafez., Sidler, T. C., & Salathe, R.-P. (2003). Study of the beam path distortion profiles generated by a two-axis tilt single-mirror laser scanner. *Optical Engineering*, *4*, 1048–1057. Available from https://doi.org/10.1117/1.1557694.

Nakamura, Y., Makita, S., Yamanari, M., Itoh, M., Yatagai, T., & Yasuno, Y. (2007). High-speed three-dimensional human retinal imaging by line-field spectral domain optical coherence tomography. *Optics Express*, *12*, 7103–7116.

Nassif, N. A., Cense, B., Park, B. H., Pierce, M. C., Yun, S. H., Bouma, B. E., Tearney, G. J., Chen, T. C., & de. Boer, J. F. (2004). In vivo high-resolution video-rate spectral-domain optical coherence tomography of the human retina and optic nerve. *Optics Express*, *3*, 367–376. Available from https://doi.org/10.1364/OPEX.12.000367, http://opg.optica.org/oe/abstract.cfm?URI = oe-12-3-367.

Povazay, B., Bizheva, K., Hermann, B., Unterhuber, A., Sattmann, H., Fercher, A., Drexler, W., Schubert, C., Ahnelt, P., Mei, M., Holzwarth, R., Wadsworth, W., Knight, J., & Russell, P. St. J. (2003). Enhanced visualization of choroidal vessels using ultra-high resolution ophthalmic OCT at 1050 nm. *Optics Express*, *17*, 1980–1986.

Quan, C., Atry, F., De La Rosa, I. J., Rarick, K. R., & Pashaie, R. (2018). Design and implementation guidelines for a modular spectral-domain optical coherence tomography scanner. *International Journal of Optics*, 3726207. Available from https://doi.org/10.1155/2018/3726207.

Spaide, R. F., Akiba, M., & Ohno-Matsui, K. (2012). Evaluation of peripapillary intrachoroidal cavitation with swept source and enhanced depth imaging optical coherence tomography. *Retina (Philadelphia, Pa.)*, *6*, 1037–1044. Available from https://doi.org/10.1097/IAE.0b013e318242b9c0.

Srinivasan, V. J., Chen, Y., Duker, J. S., & Fujimoto, J. G. (2009). In vivo functional imaging of intrinsic scattering changes in the human retina with high-speed ultrahigh resolution OCT. *Optics Express*, *5*, 3861–3877.

Srinivasan, V. J., Adler, D. C., Chen, Y., Gorczynska, I., Huber, R., Duker, J. S., Schuman, J. S., & Fujimoto, J. G. (2008). Ultrahigh-speed optical coherence tomography for three-dimensional and en face imaging of the retina and optic nerve head. *Investigative Ophthalmology & Visual Science*, *11*, 5103–5110. Available from https://doi.org/10.1167/iovs.08-2127.

Tomlins, P. H., & Wang, R.,K. (2005). Theory, developments and applications of optical coherence tomography. Journal of Physics D, Applied Physics. (15) 2519–2535. Available from: https://doi.org/10.1088/0022-3727/38/15/002, http://inis.iaea.org/search/search.aspx?orig_q = RN:36101439. United Kingdom.

Unterhuber, A., Povazay, B., Hermann, B., Sattmann, H., Chavez-Pirson, A., & Drexler, W. (2005). In vivo retinal optical coherence tomography at 1040 nm - enhanced penetration into the choroid. In *Optics Express* (9, pp. 3252–3258).

Wojtkowski, M., Srinivasan, V., Fujimoto, J. G., Ko, T., Schuman, J. S., Kowalczyk, A., & Duker, J. S. (2005). Three-dimensional retinal imaging with high-speed ultrahigh-resolution optical coherence tomography. *Ophthalmology*, *10*, 1734–1746.

Yang, V. X. D., Gordon, M. L., Qi, B., Pekar, J., Lo, S., Seng-Yue, E., Mok, A., Wilson, B. C., & Vitkin, I. A. (2003). High speed, wide velocity dynamic range Doppler optical coherence tomography (Part I): System design, signal processing, and performance. *Optics Express*, *7*, 794–809. Available from https://doi.org/10.1364/OE.11.000794, http://opg.optica.org/oe/abstract.cfm?URI = oe-11-7-794.

Zawadzki, R. J., Jones, S. M., Olivier, S. S., Zhao, M., Bower, B. A., Izatt, J. A., Choi, S., Laut, S., & Werner, J. S. (2005). Adaptive-optics optical coherence tomography for high-resolution and high-speed 3D retinal in vivo imaging. *Optics Express*, *21*, 8532–8546.

Zawadzki, R. J., Jones, S. M., Pilli, S., Balderas-Mata, S., Kim, D. Y., Olivier, S. S., & Werner, J. S. (2011). Integrated adaptive optics optical coherence tomography and adaptive optics scanning laser ophthalmoscope system for simultaneous cellular resolution in vivo retinal imaging. *Biomedical Optics Express*, *6*, 1674–1686. Available from https://doi.org/10.1364/BOE.2.001674.

Zhang, A., Zhang, Q., Chen, C.-L., & Wang, R. K. (2015). Methods and algorithms for optical coherence tomography-based angiography: A review and comparison. *Journal of Biomedical Optics*, *10*, 100901. Available from https://doi.org/10.1117/1.JBO.20.10.100901.

Zhang, Q., Lu, R., Wang, B., Messinger, J. D., Curcio, C. A., & Yao, X. (2015). Functional optical coherence tomography enables in vivo physiological assessment of retinal rod and cone photoreceptors. *Scientific Reports*, 9595. Available from https://doi.org/10.1038/srep09595.

CHAPTER

13 Medical image super-resolution

Wafaa Abdulhameed Al-Olofi[1] and Muhammad Ali Rushdi[1,2]

[1]*Department of Biomedical Engineering and Systems, Faculty of Engineering, Cairo University, Giza, Egypt*

[2]*School of Information Technology, New Giza University, Giza, Egypt*

13.1 Introduction

Medical imaging has been a powerful tool for inspecting the inner human anatomical structures, evaluating physiological functions, and diagnosing diseases. Based on different underlying biophysical principles, there has been tremendous growth in the available medical imaging modalities including computed tomography (CT), magnetic resonance imaging (MRI), ultrasound, and positron emission tomography (PET). Still, medical image analysis depends in many cases on the availability of high-resolution (HR) images that can provide detailed information on human anatomy and physiology. Obtaining HR images is still a major challenge in medical imaging due to the physical limitations of the imaging systems and also due to noise and blur factors.

Alternatively, computational and algorithmic super-resolution (SR) approaches can be adopted to increase the resolution of medical images. This should significantly enhance and accelerate disease diagnosis and improve treatment outcomes. Furthermore, higher image resolution can considerably boost the results of subsequent stages of medical image analysis such as object detection and segmentation and image classification.

In this chapter, we highlight the need for SR methods in medical imaging, review recent SR methods for key medical imaging modalities, and point out key challenges and potential research directions. For the rest of Section 13.1, we review the concept and types of image resolution, and the need for super-resolution methods in imaging applications. We particularly explore the need of SR methods for different medical imaging modalities and show how research on these methods quickly grew over the past few years. Section 13.2 provides a taxonomy covering different types of super-resolution (SR) algorithms. Section 13.3 reviews representative SR techniques in common medical imaging modalities. Finally, Section 13.4 highlights the limitations, challenges, and open research directions in this field.

Artificial Intelligence and Image Processing in Medical Imaging. DOI: https://doi.org/10.1016/B978-0-323-95462-4.00013-3

13.1.1 Image resolution

The resolution of an imaging system reflects its ability to distinguish between two objects that are spatially, spectrally, or temporally close (Burger & Burge, 2009; González & Woods, 2008).

13.1.1.1 Spatial resolution

Spatial resolution refers to the specific spatial positions where an imaging modality can discriminate between two objects. In other words, the spatial resolution of an imaging system is related to the number of independent pixel values per unit length (lines per image height, lines per mm), and how closely the lines may be spaced while still being resolved. Therefore low-spatial-resolution systems will not be able to differentiate between two objects that are relatively close to each other.

Spatial resolution is typically limited by imaging sensors or acquisition equipment. In fact, common image sensors are charge-coupled devices (CCDs) or complementary metal-oxide semiconductor (CMOS) active pixel sensors. For acquiring two-dimensional images, such sensors are usually stacked in two-dimensional arrays. The spatial resolution of the captured image is defined by the sensor size, that is, the number of sensing elements per unit area (Yue et al., 2016).

13.1.1.2 Spectral resolution

The spectral resolution is a measure of the ability to resolve features of the electromagnetic frequency spectrum. This resolution, usually denoted by $\Delta\lambda$, is the smallest distinguishable spectral difference at a wavelength of λ. Alternatively, spectral resolution can be specified by the difference in velocities Δv associated with the Doppler effect. So, if c is the speed of light, then the resolution can be represented by Δv and the resolving power becomes $R = c / \Delta v$ (Yang et al., 2013).

The spectral resolution is of great relevance in medical imaging as several modalities focus on measuring spectral characteristics. For example, in nuclear medicine, gamma-ray radiations are used to determine the composition of biological tissues by measuring the attenuation of certain X-ray wavelengths using single-photon emission computed tomography (SPECT), positron emission tomography (PET), and spectral photon-counting CT. For instance, spectral CT can identify iodine in bodily tissues (e.g., bone, fat) and can estimate the amount present, when an iodinated contrast is administered to the patient (Heismann et al., 2008). In comparison to traditional X-ray imaging, spectral photon-counting CT significantly provides more information due to the combination of a color spectrum and extremely high spatial resolution.

13.1.1.3 Temporal resolution

Temporal resolution is the discrete resolution with respect to time for a fixed spatial region and viewing angle. This type of resolution is a measure of the temporal sampling frequency. Indeed, a trade-off exists between the temporal and spatial resolutions of a system. In fact, with a finite speed of light, a certain amount of time is needed for information-carrying photons to reach an observer. During this time, the system might have changed. Thus a longer light travel distance is associated with lower temporal resolution (Liu et al., 2022).

Temporal resolution plays a key role in several medical imaging modalities. For example, the evaluation of heart functions in cine MRI highly depends on the reconstruction of HR temporal sequences of MR frames (Lin & Alessio, 2009; Xue et al., 2013). The images showing contractions of the left ventricle can reveal potential anomalies in the wall motion, myocardial strain, and ventricular volumes.

13.1.2 Need for super-resolution in imaging

Low-resolution images may arise from imaging systems due to hardware limitations or acquiring images under conditions of aliasing or degradation.

First of all, many imaging systems use CCDs that can capture and store image intensities using arrays of solid-state detectors. Consequently, the resolution of the captured image depends on the size and number of these detectors. The higher the resolution is, the higher the pixel density is, and the more informative the sensory images are. Actually, HR images are desirable and frequently required in many applications as such images offer sharp details and clear pictorial information for human perception as well as rich features for automatic machine interpretation and representation. The spatial resolution may be increased by minimizing the pixel size. However, as the pixel size shrinks, the quantity of light supplied falls as well, and the image quality deteriorates significantly under the influence of shot noise (Yang & Huang, 2017).

Moreover, image resolution is highly affected by aliasing. According to the Nyquist-Shannon sampling theorem, a perfect recovery of a function from its samples is achievable if the function is sampled at a rate exceeding twice its maximum frequency ($f_s \geq 2f_m$). The aliasing (or distortion) is a type of signal corruption that occurs when two patterns (or spectral segments) overlap, and the function is under sampled. The corruption emerges in the form of additional spurious frequency components called aliasing frequencies. For example, a noisy image may contain spatial frequencies over the Nyquist limit, due to the finite number of detector pixels. In imaging, the aliasing effects can be essentially minimized through decreasing the high-frequency components by blurring the image before sampling (McLeod & Malac, 2013).

Super-resolution (SR) algorithms address these issues through generating HR imagery from systems having lower-resolution imaging detectors. Specifically, such algorithms combine a collection of low-resolution images containing aliasing artifacts in order to reconstruct HR images. Therefore, SR describes the process of increasing the resolution of an image from low resolution (LR) to HR. The key underlying principle of SR algorithms is to combine nonredundant information contained in multiple LR frames to generate an HR image with increased high-frequency components and reduced degradation patterns.

For the last few decades, SR algorithms have been applied in a wide range of real-world applications (Malczewski & Stasiński, 2009; Yue et al., 2016):

- Surveillance and security

  Traffic surveillance and video-based security systems suffer from low-resolution effects due to the fact that huge amounts of video data are collected with limited computational efficiency and complex motion artifacts. Therefore, enhancing the image quality in these systems is a major SR challenge. For example, it is quite important to perform resolution enhancement for face recognition, 3D face reconstruction, and face alignment based on LR face images collected from security monitoring systems (Farooq et al., 2021). Also, SR techniques are important for enhancing the outcomes of biometric identification (e.g., with iris and fingerprint images).
- Medical imaging

  Medical modalities provide anatomical body structures and describe functional information, but resolution limitations undermine medical image quality and associated diagnoses. Thus, SR algorithms have been proposed for medical imaging modalities such as MRI, functional MRI, and PET. For example, SR helps in generating HR MRI from otherwise LR MRI images (Plenge et al., 2012). Also, PET image resolution can be enhanced through combining LR images of different points of view and then averaging to generate HR images (Kennedy et al., 2006).
- Satellite imaging

  The resolution of terrestrial observations obtained from remote-sensing imagery can be enhanced through maximizing the focal length of the optical system or minimizing the pixel size of the CCDs. However, the increase in the focal length will cause a high cost of fabrication while the decrease in the size of the CCD pixel unit will cause serious system noise. Therefore, SR algorithms have been extensively introduced for remote-sensing image analysis tasks. For example, an SR method used a mathematical Bayesian model to estimate the distribution of all unknowns in a remote-sensing system. Then, the statistical correlation between LR and HR images was established via dictionary learning (DL) (Luo et al., 2015). Recently, deep convolutional neural networks (CNNs) have made good results. For instance, local–global combined networks were applied along with single-image SR algorithms to exploit local and global details of remote-sensing images (Lei et al., 2017).
- Astronomical imaging

  The resolution of an astronomical imaging system is typically limited by very low light levels and variable stray background light. The enhancement of astronomical images shall help astronomers with large-scale space and universe exploration. For instance, for the GALEX mission surveys, a

Bayesian approach is applied to HR images as the starting reference for ultraviolet analysis (Llebaria et al., 2008).

13.1.3 Medical image acquisition pipeline

In general, the resolution of each medical imaging system is dictated by the physical limitations of the detectors, the signal-to-noise ratio (SNR), and the acquisition time. Moreover, the noise level and fluctuations in the imaging process are controlled by certain physical constraints specific to each imaging modality. In order to strike a reasonable balance between resolution and SNR, the system design should be guided by signal processing principles as shown in Fig. 13.1.

A common goal in medical imaging systems is to have HR medical images to boost performance on different tasks in medical image analysis (Greenspan, 2009). In practice, the range of the frequencies captured by the image detectors is restricted by the maximal sampling frequency, as defined by the detector pitch, or detector spacing. Higher resolution can be achieved by shrinking the size (width) of the detectors and shortening the interdetector distances. However, this increases the noise, and leads to a significantly lower SNR. In general, medical images suffer from problems of LR, high noise levels, low contrast, geometric deformations, and image artifacts. We discuss next some specific issues for common medical imaging modalities.

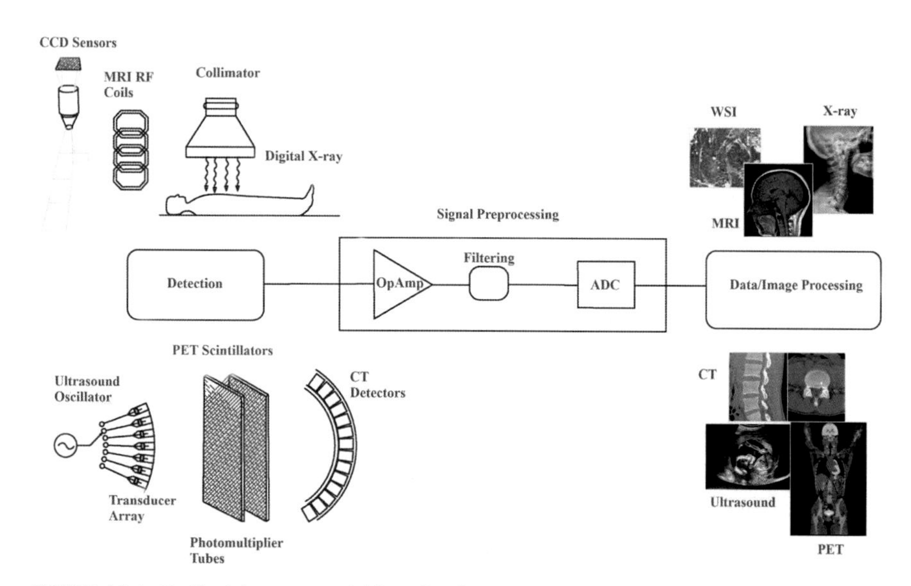

FIGURE 13.1 Medical image acquisition pipeline.

Raw medical images are acquired based on different underlying physical principles, and then basic signal amplification, filtering, and digitization operations are performed to obtain digital medical images.

13.1.3.1 Magnetic resonance imaging

Acquiring HR MRI images is a big challenge due to the complexity and anisotropy of MR data acquisition. For example, when a long signal (with a long repetition time (TR)) is required to enhance image contrast (e.g., T2-weighted imaging in MR angiography (MRA)), it will be much faster to obtain 2D slice stacks than to acquire true 3D MRI data (Mansoor et al., 2018). Also, the MRI pixels have gray-level values proportional to the weighted average of the signals coming from nearby tissues. Thus, the exact location of the boundaries may be shifted, and this introduces a major obstacle for MRI super-resolution and segmentation.

13.1.3.2 Computed tomography

The quality of digital CT images is determined by different factors, such as radiation dosage, slice thickness (Z-axis resolution), low-contrast resolution, and high-contrast resolution (Lambert et al., 2016).

Firstly, the radiation dosage is determined by reducing the milliampere-seconds (mAs), but this is expected to raise the noise by a factor of the standard deviation. Thus, if the mAs is reduced by ½, then the noise will increase by $\sqrt{2} = 1.414$ (40% higher).

Secondly, with the emergence of different CT scanning schemes (axial, helical, and multidetector helical scanning), the artifacts of slice thickness become more complex. This is due to the interaction of different factors including X-ray beam collimation, the detector width, the table speed, and helical interpolation.

Thirdly, the high contrast or spatial resolution of a CT imaging system is dependent on objects that have high SNR. Indeed, the spatial resolution is influenced by geometric resolution parameters such as the focused spot size, the detector width, the radiation sampling rate, the pixel size, the convolution kernel, and the reconstruction filter characteristics.

Fourthly, an object with a very minor variation from its background is frequently used to determine the low-contrast resolution of a system. In this situation, noise is a significant factor in the captured signal. Resolving low-contrast objects of increasingly smaller sizes (increasing spatial frequencies) is a key indicator of the quality of the CT imaging system.

Fifthly, the partial volume effect (PVE) appears when an interface between two separate tissues occurs inside a single voxel. The PVE is a direct result of LR acquisition. The PVE generally blurs the line between tissues and makes tissue characterization more complex. Each voxel in a CT scan indicates the attenuation characteristics of a particular volume. When many tissues are present in a voxel, the result is a (nonlinear) average of the attenuation characteristics of those tissues.

13.1.3.3 Digital mammography

Digital mammography imaging systems effectively obtain digital breast images with the least amount of radiation exposure for the patients. In order to enhance resolution, digital detectors cannot reduce the pixel size without harming the SNR. Mammography imaging artifacts may be caused by several factors (Geiser et al., 2011):

- Detector-related artifacts: Detector artifacts are directly related to single dead pixels, groupings of dead pixels, dead or unread lines, or ghosting. These artifacts can lead to incorrect breast disease diagnosis.
- Machine-related artifacts: The majority of these artifacts are caused by dirt or dust on the compression paddle, and issues with the X-ray tube filtering or grid. Also, machine-based artifacts may appear when the automated exposure control system is not appropriately set. While such control systems may have adequate contrast and grayscale settings, high noise levels may still arise due to processing artifacts associated with the inability to compensate for exposure variations across the detector.
- Patient-related artifacts: Motion, object superimposition, substances on the breast parenchyma, or molecules on the skin can all result in patient-related artifacts. Patient motion is the most frequent patient artifact. Hair, robes, or other foreign items placed over the breast during imaging are examples of additional artifacts. As well, the patient may have placed a hand on the compression paddle or the breast support plate, thus introducing foreign objects in the image field. Many patient-related artifacts can be removed by using proper placement and checking the field of vision before performing the exposure (Chevalier et al., 2012).

To maximize image resolution, LR digital mammograms with small spatial shifts can be merged. These shifts are typically brought on by patient movement, deliberate detector dithering, imaging system vibration, and little gantry movement. In actuality, all of these sources combine to create the motion that is caught in the digital image, implying the need for proper registration of the aliased low-resolution images. Therefore, applying SR processing to X-ray imaging requires overcoming these challenges (Robinson et al., 2017).

Still, there are two limitations that must be addressed while applying SR processing to X-ray and mammography images. First, digital mammography scans contain mainly LR image data (typically in megapixels). Thus, computational efficiency must be considered throughout SR processing. Second, the overall radiation dosage for the set of images must not exceed that of a typical X-ray image. Consequently, the recorded data typically has a very low peak SNR (PSNR), and this makes the SR process more challenging.

13.1.3.4 Ultrasound imaging

In ultrasonic imaging, the Rayleigh diffraction limit is a crucial criterion for the successful application of SR algorithms. This restriction for ultrasonic imaging may be conceived of as roughly being $\lambda/2$ (Contreras Ortiz et al., 2012).

In addition, ultrasound (US) artifacts or noise patterns appear as ultrasound signals pass through media over a short period of time. In particular, speckle noise negatively impacts ultrasound images due to coherent interfacing of constructive and destructive backscattered echoes. Moreover, US waves are distorted by tissue inhomogeneity as well (Kouamé & Ploquin, 2009).

A variety of SR algorithms have been extensively developed to overcome the poor US image quality. The main goals of these methods are decreasing speckle noise, eliminating blurring effects, preserving high-frequency components, and improving resolution (Temiz & Bilge, 2020).

13.1.3.5 Optical coherence tomography

Optical coherence tomography (OCT) is a noninvasive depth-resolved ocular imaging modality used for examining the human retina, and the choroid blood circulation. This modality detects pathological abnormalities through potential HR techniques and three-dimensional imaging. The lateral, axial, and azimuthal resolution levels are frequently related to the characteristics of the illumination source (e.g., bandwidth), the optical path (e.g., the diffraction limit due to the pupil diameter, ocular aberrations, dispersion, etc.), and other physical properties.

However, in OCT imaging, three major artifacts lead to ambiguities of tissue structures. These artifacts include blurring, intensity decay, and speckling. Blurring is the result of poor optical components. Intensity decay is a degradation of light with respect to depth, and speckle is the low coherent light source employed in the OCT configuration (Adabi et al., 2017). Additionally, OCT has a relatively slow imaging speed, and thus obtaining a wide field of view (FOV) would be challenging. Furthermore, the OCT spectrometer cameras set a limitation on the imaging speed. Thus, using fast and efficient CCD detectors is necessary to create aliasing-free 3D images of anatomical structures (Poddar et al., 2017).

13.1.3.6 Endoscopy

A wireless capsule endoscope collects two images every second while moving through the small intestine for two hours. This results in a total number of images of more than 50,000. The quality of the acquired images still faces many degradations, such as undesirable thermal noise in CCD/CMOS chips. Also, the quality of the endoscopic images suffers from inhomogeneous brightness and low contrast due to the convoluting, bending, and waving nature of the stomach organs (Isaac & Kulkarni, 2015).

13.1.4 Growth of publications on image super-resolution

Research efforts have been intensified over the last two decades to develop high-fidelity SR algorithms with reasonable computational costs. This trend has been particularly evident for SR algorithms designed particularly for medical imaging modalities. In order to assess and visualize this research trend, we searched the IEEE Xplore database for papers containing keywords of "medical," "image," and "super-resolution." Fig. 13.2 shows the counts of the conference and journal

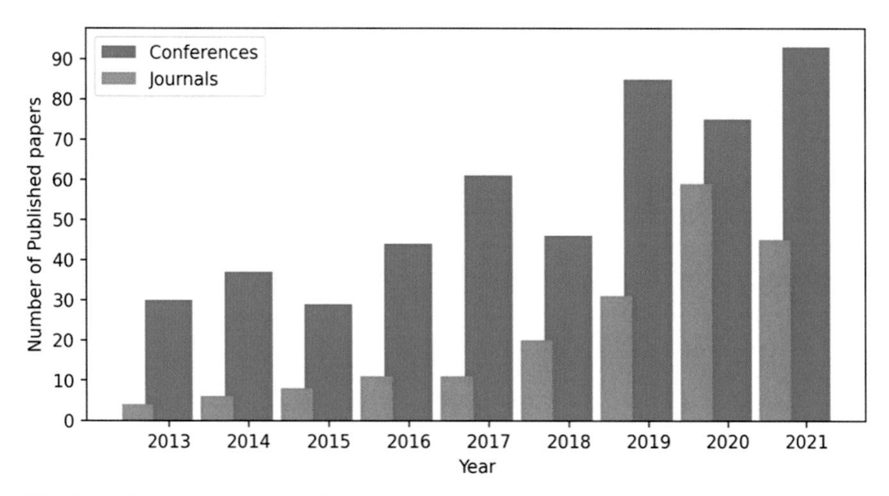

FIGURE 13.2 Growth of medical image super-resolution studies on IEEE Xplore.

Numbers of scientific publications in IEEE Xplore involving keywords of "medical," "image," and "super-resolution" for the years 2013–21.

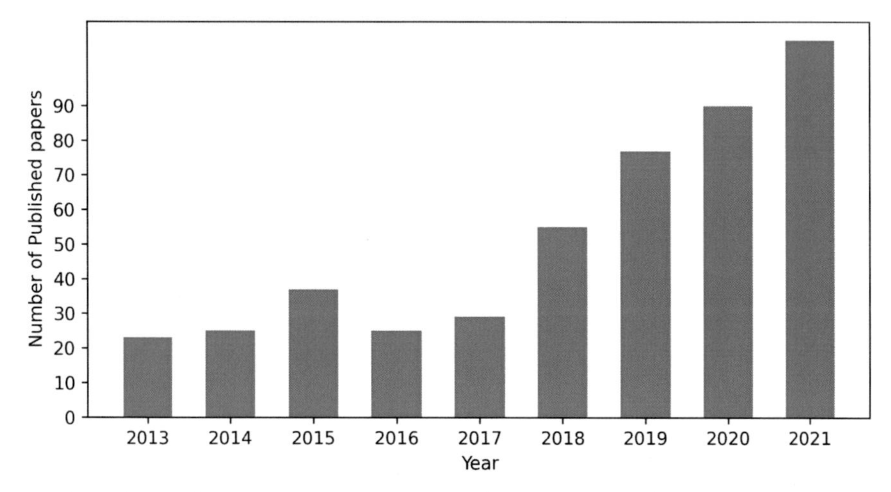

FIGURE 13.3 Growth of medical image super-resolution studies on PubMed.

Numbers of scientific publications in PubMed involving keywords of "medical," "image," and "super-resolution" for the years 2013–21.

papers returned through this search for the years from 2013 to 2021. Although the search results may contain weakly-relevant papers, the growth pattern is evident.

Similar search results were obtained for the PubMed database as shown in Fig. 13.3. In this figure, the increasing trend and the paper counts per year are comparable to those in Fig. 13.2.

Moreover, we explored the publication growth for specific imaging modalities. Again, we searched the IEEE Xplore database for papers published from 2013 to 2021 with "super-resolution" and any of the words "ultrasound," "MRI," "CT," "microscopy," "OCT," or "endoscopy." The search results in Fig. 13.4 show wide variations in the numbers of publications among these modalities, where the US, MRI, and CT modalities have the majority of the publications, as these modalities are widely used. Note that if the "super-resolution" and "microscopy" keywords are used to search for papers on SR algorithms in microscopy, a large number of publications (381) would be returned, but most of these publications pertain to conventional super-resolution microscopy (Erfle, 2017) instead of the computational SR techniques in microscopy (Kaderuppan et al., 2020). To avoid this confusion, we refine the search results in this case by looking for papers that involve "computational" or "algorithm" keywords.

Furthermore, similar search results were obtained from the PubMed database. As shown in Fig. 13.5, there is an overall exponential growth of the papers related to SR in PubMed. A similar search for SR papers in PubMed with specific imaging modalities returns results similar to those obtained for the IEEE Xplore database (See Fig. 13.4).

13.2 A taxonomy of super-resolution algorithms

The SR algorithms can be generally categorized using different criteria or factors, such as the input number of LR images used to reconstruct an HR image

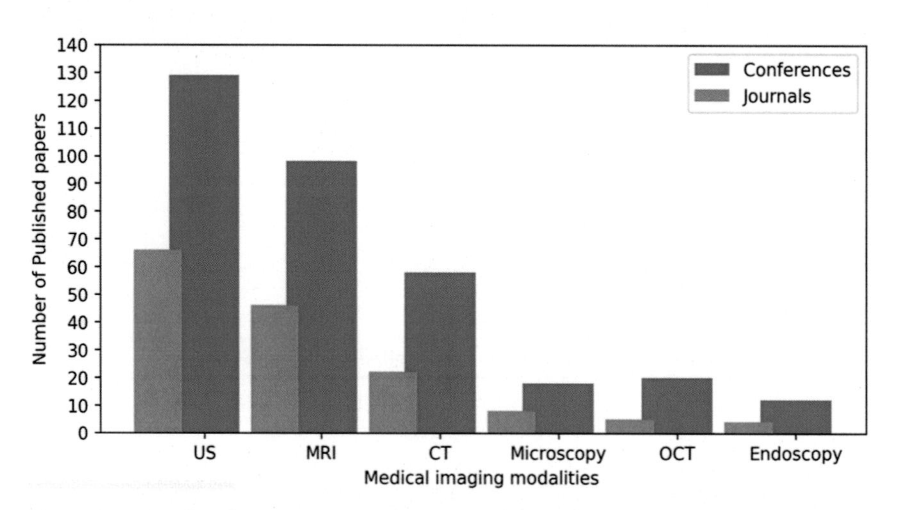

FIGURE 13.4 Medical image super-resolution studies in IEEE Xplore by imaging modality.

Numbers of scientific publications in IEEEXplore involving keywords of "super-resolution" and the names of key imaging modalities. The numbers are for papers published from 2013 to 2021.

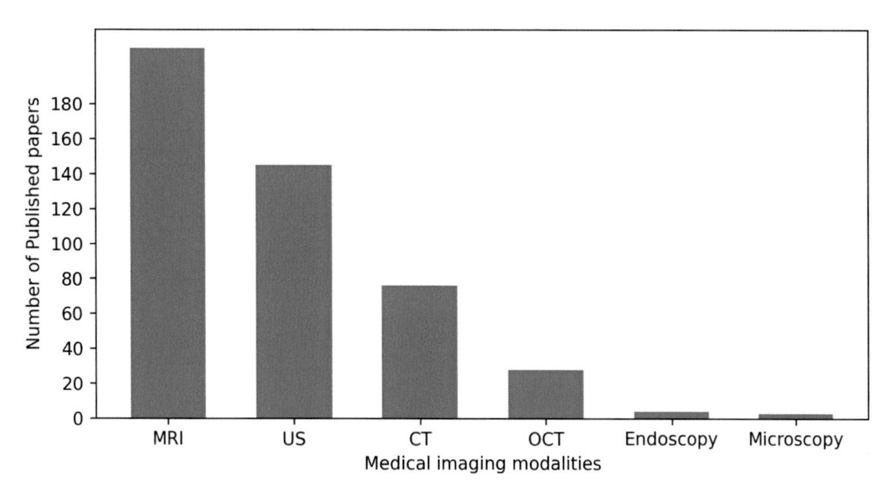

FIGURE 13.5 Medical image super-resolution studies in PubMed by imaging modality.

Numbers of scientific publications in PubMed involving keywords of "super-resolution" and the names of key imaging modalities. The numbers are for papers published from 2013 to 2021.

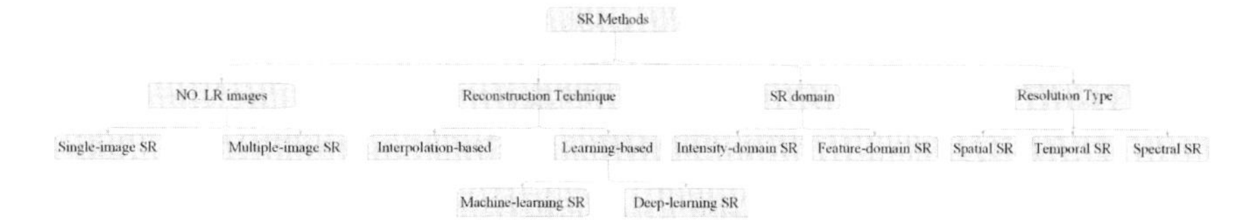

FIGURE 13.6 Taxonomy of super-resolution (SR) methods.

A taxonomy for SR algorithms by the number of LR images, the reconstruction technique, the SR domain, and the resolution type.

(multiple-image SR (MISR) versus single-image SR (SISR)), the reconstruction technique (interpolation-based or reconstruction-based SR (RSR) versus learning-based SR which in turn can be classified into classical machine learning SR (MLSR) and deep learning SR (DLSR) approaches), the SR domain (intensity-domain SR (ISR) versus feature-domain SR (FSR)), and the resolution type (spatial SR, temporal SR, and spectral SR). These different taxonomies of SR methods are summarized in Fig. 13.6. For the rest of this section, we briefly review each of these taxonomies.

13.2.1 Classification by the number of input LR images

As shown in Fig. 13.7, the number of input LR images (i.e., multiple images or a single image) defines two different families of SR techniques. With multiple LR

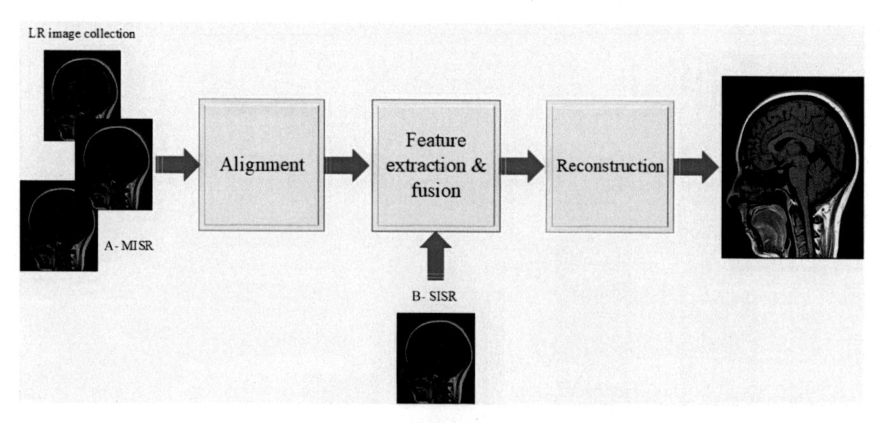

FIGURE 13.7 SR workflows with different numbers of input LR images.

Super-resolution (SR) workflows with different numbers of input LR images. A multiple-image SR (MISR) method aligns multiple LR images and uses them as input for further SR steps. (B) A single-image SR (SISR) method uses an input of a single image.

input images, three main SR tasks should be performed: alignment (or registration) of the multiple LR images, feature extraction and fusion, and finally HR image reconstruction. For a single LR input image, the registration step is not needed. Instead, we just need the feature extraction and fusion module, followed by the HR reconstruction module.

13.2.1.1 Multi-image super-resolution

Given inter-related multiple LR images, a key condition for effective reconstruction of a corresponding HR image is that the input LR images contain complementary information based on subpixel shifts or different viewpoints. The resolution improvement can be justified as follows. Indeed, the learned high-frequency details are available in the LR measurements in an aliased form. The SR process recovers these details by exploiting the complementary information of the LR images, where each image has a different aliasing pattern. In order to uniquely recover the aliased frequencies, a sufficient number of LR images are required (Nguyen et al., 2018). The degradation process can be mathematically expressed as follows in Eq. (13.1):

$$I_i = \phi\left(\hat{I}_i, \{\hat{I}_j\}_{j=1}^{N}; \theta_\alpha\right) \tag{13.1}$$

where ϕ is the degradation function, $\hat{I}_i \in R^{sH \times sW \times 3}$ is the ith HR color image, and $I_i \in R^{H \times W \times 3}$ is the corresponding i^{th} LR color image. Also, $\left\{\hat{I}_i\right\}_{j=1}^{N}$ is a set of NN HR frames.

All frames in this set are assumed to contribute to the generation of the LR image I_i. The parameter vector θ_α denotes the degradation process parameters (e.g., the scaling factor, noise, and motion blur). Generally, the degradation process (i.e., ϕ and θ_α) is unknown and only the LR images are available.

The degradation process can be decomposed as follows in Eq. (13.2):

$$I_j = DBE_{i\rightarrow j}\hat{I}_i + n_j \tag{13.2}$$

where D and B are the downsampling and blur operations, n_j denotes image noise, and $E_{i\rightarrow j}$ is the warping operation based on the motion from $I_i \rightarrow I_j$.

The super-resolution process seeks to reverse the degradation process of Eq. (13.1), and can be formulated as follows in Eq. (13.3):

$$\hat{I}_i = \phi^{-1}\left(I_i, \{I_j\}_{j=1}^{N}; \theta_\beta\right) \tag{13.3}$$

where $\hat{I}_i$ denotes the estimate of the HR i^{th} image, and θ_β denotes the SR model parameters. Clearly, the SR model of Eq. (13.3) uses multiple LR images to reconstruct one HR image. This is why this model is called a multiple-image SR (MISR) scheme.

13.2.1.2 Single-image super-resolution

In contrast to the aforementioned MISR schemes, SR might be carried out on a single LR input image. In such a single-image SR (SISR) scheme, additional complementary information for reconstructing the corresponding HR image may be obtained based on prior knowledge or learning from LR-HR examples. Fig. 13.8 schematically illustrates these two approaches.

The prior-knowledge-based SISR methods typically explore different sources of information within the input image such as pattern redundancy across multiple image scales (Glasner et al., 2009), natural image statistics (Zontak & Irani, 2011), and fractal analysis (Zhang et al., 2018). Similar methods based on prior knowledge have been proposed for medical image SR. For example, Zhao, Carass, Dewey, and Prince (2018) proposed an approach to learn the mapping between LR and HR axial MR slices. The learned mapping was then used to super-resolve LR sagittal and coronal slices. Also, Jog et al. (2016) proposed image SR techniques based on a training-data-free strategy using Fourier burst accumulation (FBA). A single HR image with less noise was recovered using FBA. In particular, FBA and anchored neighborhood regression (ANR) were used for filling in missing Fourier information and recovering an HR image with reduced noise from a single LR image.

The learning-based SISR methods obtain complementary information from external databases of LR-HR image pairs (Timofte et al., 2017). In such methods, the training data is used to learn the mapping between a set of HR samples and their LR counterparts. The key assumption here is that the obtained mapping

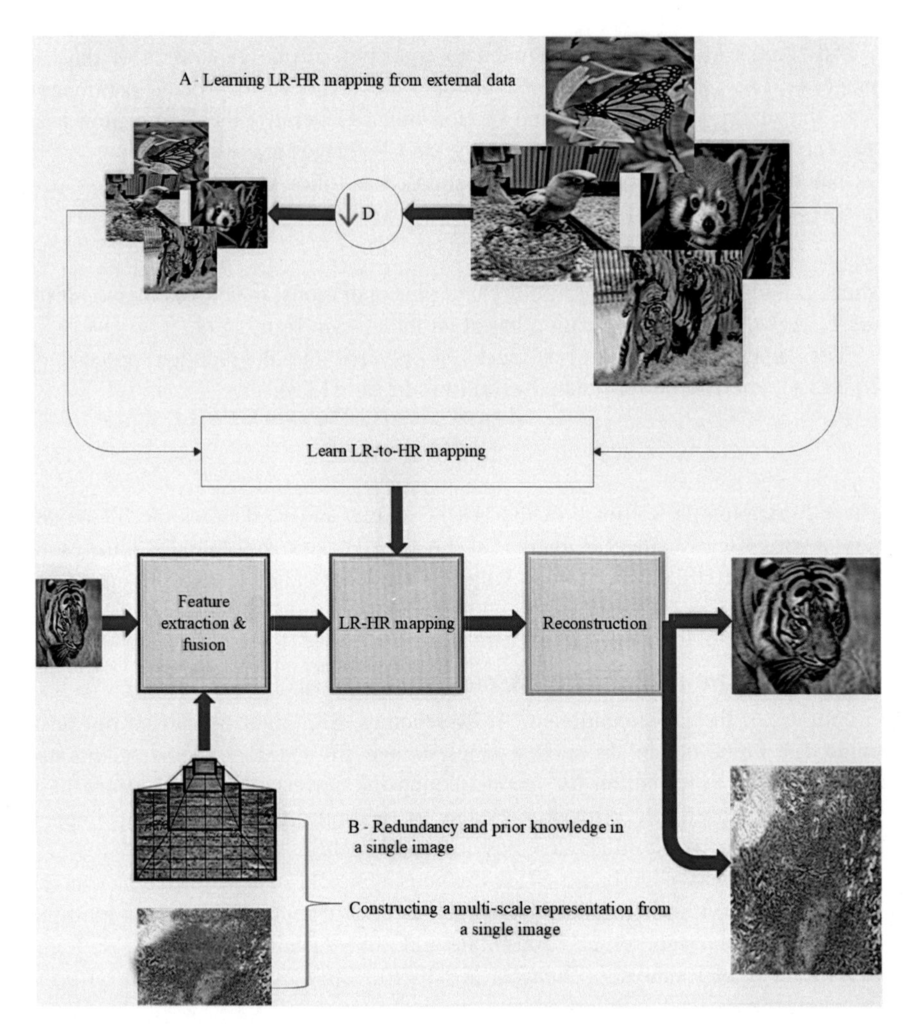

FIGURE 13.8 Single-image super-resolution (SISR) approaches.

Single-image super-resolution (SISR) approaches. (A) A SISR method may depend on learning an LR-HR mapping from an external database or (B) exploiting the redundancies and prior knowledge within a single image.

would still apply to unseen LR images even if they come from different data distributions. Following the creation of such mapping, the prior term for SR reconstruction is updated to reflect the newly acquired knowledge. Different techniques have been investigated to improve the generalizability, sufficiency, and predictability of the learning-based SISR methods (Nasrollahi & Moeslund, 2014). These methods have already been extensively explored in medical image SR (Li, Sixou et al., 2021).

Both types of SISR problems (i.e., the prior-knowledge-based and learning-based problems) can be mathematically formulated as follows in Eq. (13.4):

$$I_x = \phi(I_y; \theta_\alpha), \tag{13.4}$$

where ϕ denotes the image degradation function, I_x and I_y denote a pair of corresponding LR and HR images, and θ_α is a vector of the parameters of the degradation process. Generally, the degradation process function and parameters (i.e., ϕ and θ_α) are unknown and only a single LR image is provided. Although the degradation process is generally unknown and can be affected by various factors (e.g., compression artifacts, anisotropic degradations, sensor noise, and speckle noise), the degradation process may be fairly approximated as Eq. (13.5):

$$I_x = DBI_y + n, \tag{13.5}$$

where D and B denote downsampling and blurring operators, while n denotes image noise.

A SISR method aims to recover a good HR image estimate via reversing the degradation process modeled by Eq. (13.4). The recovery process can be formulated as Eq. (13.6):

$$I_y = R(I_x; \theta_R), \tag{13.6}$$

where R denotes the SR model while θ_R denotes the corresponding parameter set.

In summary, for both multiple-image and single-image SR, complementary information for SR may come from other LR images of the same scene, from prior knowledge of image statistics, or from an LR-to-HR mapping learned from an external set of LR-HR image pairs.

13.2.2 Classification by the super-resolution reconstruction approach

The SR reconstruction methods aim to retrieve the lost high-frequency components either by exploiting multiple complimentary LR images (interpolation-based methods), or learning from pairs of LR-HR images (learning-based methods). When rich and complementary details exist, the interpolation-based approaches are extremely accurate. Otherwise, we should resort to learning-based methods. Fig. 13.9 differentiates between the interpolation-based and learning-based approaches.

13.2.2.1 Interpolation-based super-resolution

Digital image SR is essentially achieved by increasing the discrete image pixels by a predetermined magnification factor. Interpolation-based SR looks for the best estimate of the intensity of a pixel based on the intensity values of nearby pixels. The conventional methods for image interpolation are nearest-neighbor interpolation, bilinear interpolation, and bicubic interpolation (Devaraj, 2019).

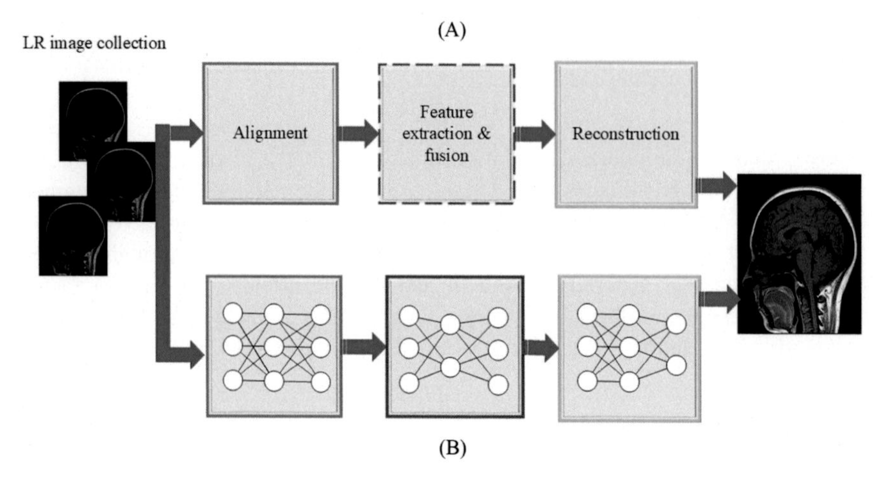

FIGURE 13.9 Super-resolution reconstruction techniques.

Super-resolution reconstruction techniques: (A) interpolation-based techniques and (B) learning-based techniques.

The selection of the most appropriate interpolation method is problem-dependent because each method has advantages and disadvantages of its own.

The nearest-neighbor method is the simplest one where a new pixel is assigned the intensity value of the closest pixel. While this method is very fast, it usually produces low-quality images with blocking artifacts. Bilinear interpolation linearly interpolates the intensity values along two orthogonal image axes. Furthermore, bicubic interpolation fits a second-order polynomial with the existing intensity values in order to estimate the intensity value at the pixel of interest.

More elaborate interpolation methods have been proposed as well. For example, Zhang et al. (2018) proposed a bivariate rational fractal interpolation model for image SR. This model combines rational interpolation and fractal interpolation schemes, and demonstrates competitive results. Also, Chavez-Roman and Ponomaryov performed interpolation in the wavelet domain and employed sparse signal representations for image SR (Chavez-Roman & Ponomaryov, 2014).

Akhtar and Azhar (2010) investigated a new image interpolation technique combining bicubic interpolation with 2D interpolation filtering along image columns and rows to change the interpolated values and create an SR image.

13.2.2.2 Learning-based super-resolution

Learning-based SR methods try to learn high-frequency components lost in LR images from training datasets of pairs of LR-HR images. These methods can use conventional machine learning approaches or more recently deep learning approaches.

13.2.2.2.1 Machine learning super-resolution

The standard ML-based SR techniques extract hand-crafted features (e.g., size, shape, and texture features) of regions of interests or volumes of interests obtained through manual, semi-automatic, or fully-automatic image segmentation stages (Castiglioni et al., 2021). In fact, numerous methods have been recently proposed for ML applications in image SR. For instance, Dharejo et al. (2017) introduced a method for learning LR-HR mapping using coupled DL and the K-SVD algorithm. Then, sparse representations were found with respect to the learned dictionaries, and used to reconstruct HR images.

13.2.2.2.2 Deep learning super-resolution

The DL models automatically extract image features to enhance the SR reconstruction outcomes. Recent deep-learning-based SR methods have helped achieve fully-automatic learning of the LR-HR mapping from training data. With end-to-end training, such mapping can be modeled by a deep neural network. The DL methods typically outperform ML-based SR methods based on hand-crafted features (Wang, Chen et al., 2021).

The most popular deep structures related to SR can be roughly categorized into two groups: convolutional neural networks (CNNs) and generative adversarial networks (GANs). These deep structures have high capabilities to represent abundant information and extract distinctive features through self-learning strategies (Li et al., 2020). Transposed convolutional layers and subpixel layers were introduced into the SR field to overcome the shortcomings of interpolation-based methods and achieve end-to-end SR learning. The transposed convolutional layers had the input expanded to twice its original size and the added pixel values were set to 0. Then, convolution (with a kernel size of 3, a stride of 1, and a padding of 1) was applied to reconstruct high-quality images using a feature extraction network (Tamang & Kim, 2022). Subpixel layers were created by generating a number of channels with various convolution operations and then reshaping and shuffling to obtain HR dynamic MRI images. In particular, a U-Net-based network with perceptual loss was trained on a benchmark dataset and fine-tuned using one subject-specific static HR MRI scan (Sarasaen et al., 2021).

13.2.3 Super-resolution domain

Most of the conventional SR algorithms are applied to slightly-misaligned LR images in the image intensity domain. More recently, modern SR methods sought to carry out SR mapping in some feature domains. The goal of such feature-domain mapping is to either reconstruct HR images or use learned SR mapping to boost performance on other image analysis tasks such as object detection, segmentation, and classification. The SR pipelines for the intensity and feature domains are depicted in Fig. 13.10.

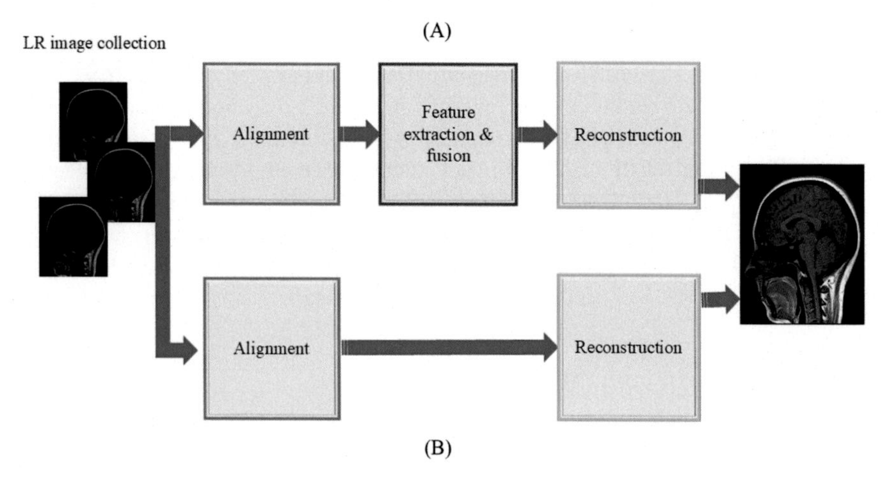

FIGURE 13.10 Super-resolution methods by domain.

Super-resolution (SR) in different domains: (A) intensity-domain SR, and (B) feature-domain SR.

13.2.3.1 Intensity-domain super-resolution

Intensity-domain SR algorithms upscale input images by creating new pixels that resemble the ones in the input images. These algorithms typically detect sharp edges and sudden shifts, and then create additional pixels to emphasize those characteristics. For example, the modified Laplacian filter (MLF) and intensity correction (IC) are two core techniques for image resolution enhancement. A simple 3×3 MLF can restore frequency components that have been attenuated as a result of averaging and downsampling degradation processes. The IC technique iteratively enhances the quality of any resolution-enhanced (enlarged) image (Shen et al., 2006).

Another intensity-based SR technique is intensity estimation using covariance-based $\ell 0$ SR microscopy (COL0RME). This technique assumes a sparse distribution of fluorescence molecules. For this technique, a nonconvex optimization problem is formulated in the covariance domain to ensure temporal and spatial independence between emitters (Stergiopoulou et al., 2021).

13.2.3.2 Feature-domain super-resolution

Feature-domain SR methods are essentially designed for boosting performance on classification tasks by carrying out SR in some feature domains. For example, such methods have been developed to enhance performance in the face and iris recognition tasks (Nguyen et al., 2013). Also, sequences of cascaded multiscale fusion (MSF) blocks have been used to adaptively identify local multiscale features using variable-size convolution kernels. Then, local residual learning (LRL) has been employed to identify effective features from previous MSF blocks and current multiscale features (Wang et al., 2019). Tan et al. (2018) proposed a new feature SR GAN (FSR-GAN) model that super-resolves images in the feature space in order to transform raw poor features of small-sized images into effective features. The FSR-GAN architecture is divided into two subnetworks: a feature generator network G and a feature discriminator network D, which are jointly trained for SR reconstruction.

13.2.4 Resolution type

13.2.4.1 Spatial super-resolution

The spatial resolution of a digital image is defined by the number of independent pixel values per unit length (lines per image height, lines per mm). The spatial-domain SR techniques seek to increase the spatial pixel dimensions of an input image to achieve an enhanced visual perception of fine image details. The majority of the SR methods are designed to enhance spatial resolution (unless another design criterion is stated). Fundamentally, these SR methods account for different motions and degradations, and incorporate prior knowledge to obtain regularized solutions.

13.2.4.2 Temporal super-resolution

The temporal resolution of an image sequence depends on the time duration to capture a single frame of that sequence. In particular, temporal video SR (VSR) tries to reconstruct a video of high temporal resolution from multiple low-resolution frames (Liu et al., 2022; Ren et al., 2021). Specifically, interframe information is typically used. The VSR process can be mathematically expressed as follows. Let $I_i \in R^{H \times W \times 3}$ denote the ith frame in an LR video sequence I, and let $\hat{I}_i \in R^{sH \times sW \times 3}$ denote the corresponding HR frame, where s is the scale factor, $\{\hat{I}_i\}_{j=i-N}^{i+N}$ is a set of $2N + 1$ HR frames for the center frame $\hat{I}_i$ and N is the temporal radius. Then, the degradation process of the HR video sequences can be formulated as follows in Eq. (13.7):

$$I_i = \phi\left(\hat{I}_i, \{\hat{I}_j\}_{j=i-N}^{i+N}; \theta_\alpha\right) \tag{13.7}$$

where ϕ is the degradation function. The parameter θ_α models the degradation process (e.g., the scaling factor, noise, and motion blur). Generally, the degradation process (modeled by ϕ and θ_α) is unknown and only LR images are provided.

The degradation process is expressed as Eq. (13.8):

$$I_j = DBE_{i \rightarrow j} \widehat{I_i} + n \tag{13.8}$$

where D and B are the downsampling and blur operations, respectively. Also, n_j denotes image noise, and $E_{i \rightarrow j}$ is the warping operation based on the motion from $I_i \rightarrow I_j$.

The SR process can be formulated as follows Eq. (13.9):

$$\tilde{I}_i = \phi^{-1}\left(\{I_j\}_{j=i-N}^{i+N}\right); \theta_\beta) \tag{13.9}$$

where $\tilde{I}_j$ denotes the HR image estimate, and θ_β is the model parameter.

13.2.4.3 Spectral super-resolution

Spectral SR methods seek to enhance the frequency resolution of the input image data. Such methods have been investigated in remote-sensing applications and hyperspectral imaging (Mazelanik et al., 2022).

13.3 Applications in medical imaging

The SR algorithms have been steadily introduced in different medical imaging modalities (See Section 13.1). Here, we survey samples of the SR approaches proposed for each of the major medical imaging modalities and identify the key modality-specific issues. Indeed, a recent survey by Li et al. (2021) covered recent deep learning SR approaches for MRI, CT, ultrasound imaging, and electron microscopy. Our survey is different from that of Li et al. (2021) in terms of several aspects. First of all, we make a wider in-depth coverage of different imaging modalities (e.g., ultrasound and fluoroscopic imaging). Secondly, our survey addresses different SR schemes including those based on reconstruction, machine learning, and deep learning methods (See Section 13.2). Thirdly, while most SR approaches focus on SR in the spatial domain, we pay special attention to methods designed for SR in the temporal or spectral domains.

13.3.1 Magnetic resonance imaging

MRI has been a key target modality for the application of SR algorithms. Fig. 13.11 shows the growth of the number of publications related to both "super-resolution" and "MRI" in PubMed from 2013 to 2021. Indeed, the number of publications in 2021 is almost four times that in 2013. Moreover, Figs. 13.4 and 13.5 show that the overall numbers of publications related to MRI and SR in the PubMed and IEEE databases are much higher than the publication numbers for all other modalities. We discuss next some of these approaches (with a summary given in Table 13.1).

This huge growth of SR methods for MRI can be ascribed to several reasons. First of all, MRI acquisition is a time-consuming costly process. This can be alleviated by the acquisition of fewer MRI samples and resorting to SR methods to fill in gaps in the data. Secondly, MRI data contain slices acquired from multiple views, namely, the sagittal, coronal, and axial views. Sometimes, the data collected for one view can be of high resolution while that of the other views are of LR. The SR methods can exploit the HR data from one view to reasonably

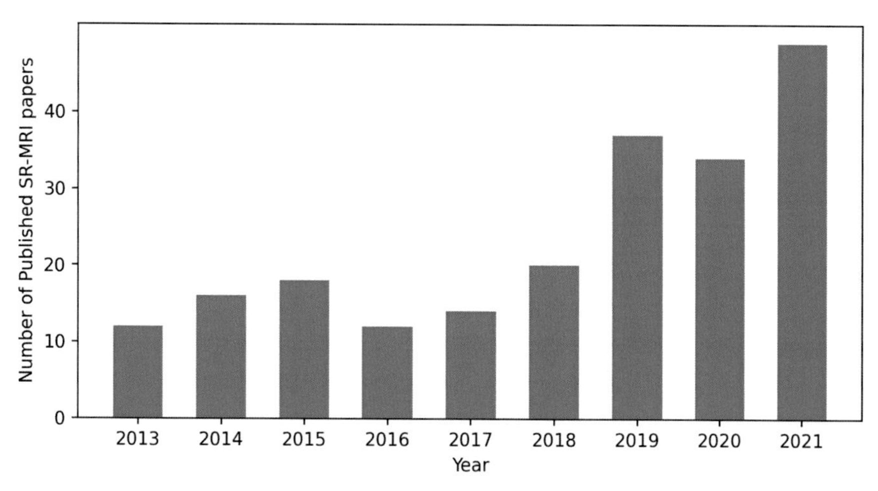

FIGURE 13.11 Growth of MRI super-resolution studies on PubMed.

Numbers of scientific publications in PubMed involving keywords of "super-resolution" and "MRI" for the years 2013–21.

interpolate data in the other LR view (Zhao, Carass, Dewey, & Prince, 2018). Thirdly, MRI data is typically acquired with multiple contrasts (e.g., T1, T2). However, most of these types of MRI are usually collected at low resolutions while few are collected at high resolutions. Thus, SR methods have been proposed to use the HR data of one contrast to help with super-resolving data samples for other lower-resolution MRI contrasts (Zeng et al., 2018).

Most of the recent SR methods for MRI data used the Brainweb dataset (Evans et al., 1997), and focused on enhancing the spatial resolution using deep learning techniques as we explore next.

13.3.1.1 2D magnetic resonance imaging

Mansoor et al. (2018) introduced a method for volumetric MRI SR using GAN architecture with a VGG-like network. This architecture was trained using two-dimensional MRI slices in three orthogonal planes. In particular, features were extracted using a VGG-like network in order to enhance the perceptual quality (rather than focus on pixel-level similarity). The combination of adversarial and perceptual loss functions led to SR image quality exceeding the minimum PSNR threshold of 31.7 dB that can be discerned by the human visual system.

Two different schemes were developed by Zeng et al. (2018) for super-resolving single-contrast and multicontrast MRI images. First, a single-contrast subnetwork was constructed for super-resolving LR T2 images. Second, a multi-contrast subnetwork reconstructed HR estimates of T2 images based on reference HR T1 images and the T2 images obtained by the single-contrast subnetwork.

Table 13.1 Recent approaches for super-resolution in magnetic resonance imaging.

References	Image modality	Organ	Data source	Method	Taxonomy	SR dimension	Results
Mansoor et al. (2018)	MRI	Brain	MRI: For training, 40 pairs of diagnostic-quality LR T1-weighted MR images (with slice thicknesses of 0.3 and 1.2 mm, respectively). Five other pairs were used for testing.	GAN architecture with a VGG-like network and a perceptual loss function.	DLSR	Spatial	Average PSNR: 33.76 ± 0.28 dB
Zeng et al. (2018)	MRI	Brain	1-Brainweb dataset (Mahapatra et al., 2019), (Evans et al., 1997) Simulated MRI data of 1696 training samples and 8 test samples with a resolution of 1 mm^2, a slice thickness of 1 mm, and a size of $181 \times 217 \times 181$. The data samples are of normal and multiple sclerosis categories with two contrasts (T1w and T2w). 2-NAMIC dataset: Ten samples of each of normal and	A CNN-based single-contrast super-resolution (SCSR) subnetwork super-resolves LR T2 images to get SCSR T2 images, which are further enhanced using reference HR T1 images and a CNN-based multi-contrast super-resolution (MCSR) subnetwork.	DLSR	Spatial (2X, 3X, 4X)	Best 2X SR results obtained with the MCSR method: Brainweb: PSNR = 46.58 dB, SSIM = 0.999 NAMIC: PSNR = 38.32 dB, SSIM = 0.945 ECNU_set: PSNR = 41.39 dB, SSIM = 0.986

		schizophrenic subjects with a voxel resolution of $1 \times 1 \times 1\ mm^3$. Also, 2D brain data samples are collected from the 3D data along the transverse, sagittal, and coronal directions. After data augmentation, 1168 and 8 pairs of 2D T1w and T2w images were used for training and testing, respectively. 3. ECNU_set: A real dataset of healthy subjects collected at East China Normal University (ECNU). The dataset has MRI images of 180 people with 3 mm slices and images of 60 people with 4 mm slices (FOV $256 \times 256\ mm^2$). With augmentation, 1280 and 8 pairs of T1w and T2w images were used for training and testing, respectively.				

(Continued)

Table 13.1 Recent approaches for super-resolution in magnetic resonance imaging. *Continued*

References	Image modality	Organ	Data source	Method	Taxonomy	SR dimension	Results
Shi et al. (2018)	2D MRI	Brain	1-Simulated MR images from the Brainweb database (Evans et al., 1997) 2-Real MR image dataset (the ALVIN dataset) (Kempton et al., 2011), (Kempton et al., 2011) 3-Clinical MR samples were obtained from the 2015 Brain Tumor Image Segmentation Benchmark (BRATS) dataset (Menze et al., 2015).	Multiscale global local residual learning (MGLRL) which combines both multiscale global residual learning and block-based shallow networks for local residual learning.	DLSR	Spatial (2X)	Best 2X super-resolution results obtained with the MGLRL method: Brainweb: PSNR = 44.33 dB, SSIM = 0.9958 ALVIN: PSNR = 34.02 dB, SSIM = 0.9527 BRATS: PSNR = 36.53 dB, SSIM = 0.9566
Oktay et al. (2018)	2D MRI	Heart	UK Digital Heart Project Dataset (Bai et al., 2015), (Bernard et al., 2014): 1200 pairs of cine 2D stack short-axis (SAX) and cine 3D high-resolution (HR) cardiac MR images. The voxel sizes for the LR and HR images are $1.25 \times 1.25 \times 10.00$ mm and $1.25 \times 1.25 \times 2.00$ mm, respectively.	Anatomically-constrained convolutional neural networks (ACNN) for image segmentation and super-resolution. Regularization based on TL networks is used to enforce model predictions to follow the distribution of learned low-dimensional representations or priors.	DLSR	Spatial	The ACNN-SR model had an SSIM of 0.796 ± 0.4, better mean opinion scores, and lower running times.

Zhao et al. (2019)	2D MRI	Brain	IXI dataset with 581 T1 volumes, 578 T2 volumes, and 578 PD volumes. The 576 triples of corresponding T1, T2, and proton-density (PD) MR volumes were clipped to the size of $240 \times 240 \times 96$ to fit 3 scaling factors (2, 3, and 4). The volumes were randomly divided into 500, 70, and 6 volumes for training, testing, and validation. With 96 2D MR image samples per volume, the training dataset has $500 \times 96 = 48{,}000$ samples.	A channel-splitting network divides the feature extraction network into a residual branch (for feature reuse) and a dense branch (for exploring new features).	Single image super-resolution (SISR)	Spatial (2X)	The testing performance for 2X: PSNR = 38.70 dB, SSIM = 0.9817.
Liu et al. (2018)	2D MRI	Brain	MRI simulated data obtained from the BrainWeb brain database (Evans et al., 1997), (Mahapatra et al., 2019) 2- Real T1-weighted brain MR images obtained from 30 subjects, with 156 axial	A multiscale fusion convolution network (MFCN) that extracts features of different scales within multiscale fusion units in order to restore the detailed information.	DLSR	Spatial (2X, 3X, 4X)	Best 2X super-resolution results obtained with the MFCN method: BrainWeb: PSNR = 46.57 dB, SSIM = 0.9975 Real MR: Average PSNR $\approx$ 40 dB for 15 real MR images.

(Continued)

Table 13.1 Recent approaches for super-resolution in magnetic resonance imaging. *Continued*

References	Image modality	Organ	Data source	Method	Taxonomy	SR dimension	Results
			slices per anatomical scan (with a size of 256 × 256 pixels for each slice). The low-resolution MR images only included 52 axial slices.				
Giannakidis et al. (2017)	3D MRI	Heart	A dataset of late gadolinium enhancement cardiac magnetic resonance (LGE-CMR) for 28 adult patients with congenital heart disease. Input: LR 2D short-axis (SA) stack volumes of LGE-CMR (resolution: $1.25 \times 1.25 \times 10$mm). Output: MR volumes with a spatial resolution of $1.25 \times 1.25 \times 2$ mm.	A residual CNN for deep learning-based SR reconstruction.	DLSR, SISR	Spatial	The proposed CNN model had better quality compared to the multi-atlas PatchMatch (MAPM) method ($P = .009$, Mann-Whitney), with PSNR = 40.1 dB, SSIM = 0.94.
Oktay et al. (2016)	3D MR	Heart	End-diastolic frames of cine cardiac short-axis (SAX) and long-axis (LAX) MR images were acquired from 1233 healthy adult subjects for different imaging planes. For single-image model training, an image dataset containing 1080 3D SAX cardiac volumes	3D residual CNNs: A single-image de-CNN, a Siamese CNN, and a multichannel (MC) CNN with multiple input images.	DLSR, RSR	Spatial	Best results for the single-image de-CNN: PSNR: 24.45 ± 1.20 dB SSIM: 0.77 ± 0.02 Best results for multiinput MC model: PSNR: 25.26 ± 0.37 dB

			with voxel size $1.25 \times 1.25 \times 2.00$ mm, is randomly split into two subsets and used for single-image model training (930) and testing (150). For multiinput model training, a clinical set of 153 image pairs of LAX cardiac image slices and SAX image stacks are used, with 143 for training and 10 pairs for evaluation.				SSIM: 0.818 ± 0.012
Jog et al. (2016)	3D MRI	Brain	The T1-weighted MPRAGE images from 20 subjects of the Neuromorphometrics dataset. Ground truth: The resolution of HR images is 1 mm^3 isotropic. Input: LR images are obtained by downsampling of the HR images in the z-axis direction by a factor k (where $k = 2$ or 3). Each LR image has a resolution of $1 \times 1 \times 2$ mm or $1 \times 1 \times 3$ mm.	Fourier burst accumulation (FBA) and anchored neighborhood regression (ANR) for filling in missing Fourier information and recovering an HR image with reduced noise from a single LR image.	Frequency-domain SISR	Spatial (2X, 3X)	Best 2X SR results with the FBA method: PSNR = 37.98 dB, Mean S3 = 0.65

(*Continued*)

Table 13.1 Recent approaches for super-resolution in magnetic resonance imaging. *Continued*

References	Image modality	Organ	Data source	Method	Taxonomy	SR dimension	Results
Zhao, Carass, Dewey, and Prince (2018)	3D MRI	Brain	1-T1-weighted magnetization- prepared rapid gradient echo (MPRAGE) images from 20 subjects of the Neuromorphometrics dataset. Ground-truth: Each ground-truth HR image has a resolution of $1 \times 1 \times 1$ mm. Input: LR images are obtained by downsampling of the HR images in the z-axis direction by a factor k (where $k = 2$ or 3). Each LR image has a resolution of $1 \times 1 \times 2$ mm or $1 \times 1 \times 3$ mm. 2-(For evaluation only) Real LR T2-weighted MRI images acquired in 2D at a voxel size of $1.14 \times 1.14 \times 2.20$ mm, and reconstructed with a size of $0.83 \times 0.83 \times 2.20$ mm.	Enhanced deep residual networks (EDSR) (for learning the mapping between LR and HR axial MR slices. The learned mapping is used to super-resolve LR sagittal and coronal slices.	DLSR, SISR	Spatial (2X, 3X)	2X SR results obtained with the EDSSR method: PSNR = 35.14 dB, SSIM = 0.9763 Best 3X SR results obtained with the EDSSR method: PSNR = 34.44 dB, SSIM = 0.9773

Zhao, Carass, Dewey, Woo et al. (2018)	3D MRI	Brain	1-T2-weighted images from 14 multiple sclerosis subjects. Ground-truth data: with an acquired resolution of $1 \times 1 \times 1$ mm. Input: 32×32 patch pairs are extracted from axial MR slices. 2- (For evaluation only) Eight PD-weighted MR images of marmosets, where each image has a resolution of $0.15 \times 0.15 \times 1$ mm (thus $k \approx 6.667$), with HR slices in the coronal plane.	Synthetic multi-orientation resolution enhancement (SMORE) using an EDSR model with two networks: a self-anti-aliasing network (SAA) and a self-super-resolution (SSR) network.	DLSR, SISR	Spatial (2X, 3X, 4X, 5X, 6X)	SR results obtained with the SMORE method: SSIM (at 2X) $\approx$ 0.94 SSIM (at 3X) $\approx$ 0.88 SSIM (at 4X) $\approx$ 0.83 SSIM (at 5X) $\approx$ 0.77 SSIM (at 6X) $\approx$ 0.72
Sánchez and Vilaplana (2018)	3D MRI	Brain	A set of normal T1-weighted MR volumes from Alzheimer's Disease Neuroimaging Initiative (ADNI) database (ADNI). Each volume has voxel dimensions of $224 \times 224 \times 152$. A total number of 589 volumes is divided into 470 training volumes and 119 test volumes.	The architecture, based on the super-resolution generative adversarial network (SRGAN) model, adopts 3D convolutions to exploit volumetric information with a GAN network.	DLSR, SISR	Spatial (2X, 4X)	2X SR with SRGAN: PSNR = 39.28 dB (with nearest-neighbor convolution upsampling) SSIM = 0.9913 (with resize convolution upsampling) 4X SR with SRGAN:

(*Continued*)

Table 13.1 Recent approaches for super-resolution in magnetic resonance imaging. *Continued*

References	Image modality	Organ	Data source	Method	Taxonomy	SR dimension	Results
			For each volume, eight patches are extracted with a patch size of $128 \times 128 \times 92$, and a step of $112 \times 112 \times 76$ (or equivalently an overlap of $16 \times 16 \times 16$).				PSNR = 33.58 dB (with nearest-neighbor convolution upsampling) SSIM = 0.9688 (with resize convolution upsampling)
Chen, Xie et al. (2018)	3D MRI	Brain	A large dataset of T1w structural images for 1113 subjects. Each image has dimensions of $320 \times 320 \times 256$, and an isotropic spatial resolution of 0.7 mm. The whole dataset is randomly split with a ratio of 7:1:1:1 for training (780), validation (111), hyperparameter tuning (111), and testing (111).	3D densely connected super-resolution networks (DCSRN).	DLSR, SISR	Spatial (4X)	Best 4X SR results with the 3D DCSRN method: SSIM = 0.9312 PSNR = 35.05 dB NRMSE = 0.0954
Chen, Shi et al. (2018)	3D MRI	Brain	The same dataset as in (Chen et al., 2018)	A multilevel densely connected super-resolution network (mDCSRN) with a generative adversarial network (GAN).	DLSR, SISR	Spatial (4X)	Best 4X SR results: SSIM = 0.9424 PSNR = 35.88 dB NRMSE = 0.0852 Time = 20.87 *s*

Georgescu et al. (2020)	2D & 3D MRI	Brain	The National Alliance for Medical Image Computing (NAMIC) Brain Multimodality dataset: Twenty 3D MRI images, each composed of 176 slices of 256 × 256 pixels ($1 \times 1 \times 1$ mm^3). The MR images are of the T1-weighted (T1w) and T2-weighted (T2w) contrasts. The dataset is evenly split into training and test subsets with 10 3D images each.	Two CNNs for in-plane and depth super-resolution, with 6 and 4 convolutional layers, respectively.	DLSR, SISR	Spatial (2X, 4X)	T1- weighted: Best 2X 2D SR results with the CNN method: SSIM = 0.9775 PSNR = 39.29 dB IFC = 3.56 Best 2X for 3D SR results with the CNN method: SSIM = 0.9687 PSNR = 37.85 dB IFC = 2.38 T2- weighted: Best 2X 2D SR results with the CNN method: SSIM = 0.9882 PSNR = 42.20 dB IFC = 3.79 Best 2X for 3D SR results with the CNN method: SSIM = 0.9835 PSNR = 40.57 dB IFC = 2.67

(*Continued*)

Table 13.1 Recent approaches for super-resolution in magnetic resonance imaging. *Continued*

References	Image modality	Organ	Data source	Method	Taxonomy	SR dimension	Results
Park and Gahm (2022)	3D MRI	Brain	The Human Connectome Project (HCP) 900 dataset (Van Essen et al., 2013): This dataset has 900 masked T1-weighted brain MR volumes. Of the available 897 volumes, 30 volumes were used for training and the other 867 were used for testing. Each volume has a size of 260 × 311 × 260 and an isotropic spatial resolution of 0.7 mm.	A cost-efficient 3D regression-based SR algorithm where gradient information is used to construct and cluster tensor features of LR images. Pairs of LR tensor features and HR patches are used to train filters for SR reconstruction.	SISR, MLSR	Spatial (2X, 3X, 4X)	Best 2X SR results with the proposed method: SSIM = 0.9827 PSNR = 35.97 dB
Jiang et al. (2022)	2D MRI	Brain	1-A dataset of MRI images of 70 participants (Dataset I), with corresponding pairs of 128 × 128 LR images (0.35 T) and 256 × 256 HR images (3 T). A cross-validation strategy divides the	A hybrid attention residual network (HARN) with dense attention blocks for feature extraction, outer feature fusion, and upsampling.	DLSR	Spatial (2X)	2X SR results with the HARN method Dataset I (Best results) SSIM = 0.9080 PSNR = 25.46 dB MOS = 3.89

			dataset into training, validation, and test subsets with a ratio of 7:2:1. 2-(For evaluation only) IXI dataset: Fifty T1 axial-plane MR images acquired from a 3 T scanner (Dataset II). To generate input LR images, the original 3 T images (HR) were blurred using a Gaussian kernel (with a width of $\alpha = 4$), and then downsampled through averaging every four voxels. The LR images have half the resolution of the HR ones.				Dataset II (3rd best results) SSIM = 0.9731 PSNR = 30.66 dB

Shi et al. (2018) developed a CNN-based method for 2D MRI SR. This method incorporates a multiscale global residual learning strategy along with shallow-network block-based local residual learning (LRL). This method elucidates high-frequency image details from learned local residuals.

Zhao et al. (2019) proposed a channel-splitting network with cascaded channel-splitting blocks. In this network, feature extraction is divided into a residual branch (for feature reuse) and a dense branch (for exploring new features). Moreover, integration between the two branches is promoted through a merge-and-run mapping strategy.

Liu et al. (2018) introduced a multiscale fusion convolution network (MFCN) for MRI image SR. While convolutional networks typically stack several convolution layers, the MFCN architecture stacks multiscale fusion units. Each of these units includes one main path and additional subpaths whose outcomes are combined through a fusion layer.

An anatomically-constrained CNN (ACNN) was proposed by Oktay et al. (2018) for the segmentation and SR of cardiac MRI images. The proposed network has a novel loss function with an anatomical shape prior. This loss function boosts the pixel-level estimation accuracy and overall perceptual quality based on similarity computations among low-dimensional image representations.

13.3.1.2 3D magnetic resonance imaging

Giannakidis et al. (2017) utilized a CNN network to learn the nonlinear mapping and the residuals between pairs of HR and LR images of late gadolinium enhancement cardiac magnetic resonance. Specifically, a deconvolution layer was applied to LR 2D short-axis MR volumes in order to estimate the best initial upscaling filters. Then, the outputs of these filters were fed to an architecture of six concatenated convolutional layers to estimate the nonlinear mapping from LR to HR images.

Similarly, Oktay et al. (2016) employed three types of 3D residual CNNs: a single-image deconvolution CNN (de-CNN), a Siamese CNN, and a multichannel CNN with multiple input images. The deconvolution layer helps obtain the best interpolation outcomes, while image information obtained from different views can be fused through the multiinput CNN architecture.

Jog et al. (2016) proposed a technique for super-resolving MPRAGE images obtained from the Neuromorphometrics dataset. This technique uses no external training data samples of LR-HR image pairs. Instead, training examples are extracted from the input image itself. In particular, patches are extracted from the input image and these patches are then used to create a set of new images, where each image has a higher resolution along a particular spatial direction. After that, these constructed images are manipulated through Fourier burst accumulation (FBA) and anchored neighborhood regression (ANR) in order to fill in missing Fourier information and reconstruct an HR image with reduced noise.

Along the same line, Zhao, Carass, Dewey, & Prince (2018) extended the enhanced deep residual networks (EDSR) (Lim et al., 2017) for the SR reconstruction of MRI brain images. The explored MPRAGE data has HR axial-plane images but LR images in the sagittal and coronal planes. For SR model training, the HR axial-plane images were downsampled and artificially degraded to create corresponding LR axial-plane images. Then, the EDSR architecture was trained using pairs of axially-aligned LR and HR images. The trained EDSR model was thus used to super-resolve the LR images of the sagittal and coronal planes.

Also, Zhao, Carass, Dewey, Woo et al. (2018) suggested an antialiasing self-SR technique for MRI data. This technique essentially estimates 3D SR volumes from 2D slices by learning the SR mapping from HR slices in a particular direction, and applying the learned knowledge to slices in other directions. In particular, two EDSR networks are trained: one for self-SR (SSR), and the other for self-anti-aliasing (SAA). For the training phase, MRI slices of the *xy* plane were considered to be HR images, from which both LR and aliased LR images were obtained. For the reconstruction phase, the trained SSA network was applied to the *xz*-plane slices, while the trained SSR network was applied to the *yz*-plane slices. The combination of the SAA and SSR networks actually led to better performance than the method based on an SSR network only.

An adversarial learning approach was followed by Sánchez and Vilaplana (2018) to generate HR MRI scans from LR ones. Specifically, a GAN architecture was employed with 3D convolutions in order to effectively exploit the volumetric MRI information. On the one hand, the total generator loss is composed of a least-square adversarial loss term and a mean-square-error gradient-based reconstruction loss in order to enhance the generated image quality. On the other hand, the discriminator uses the least-square adversarial loss to stabilize the training outcomes.

A 3D densely connected SR network (DCSRN) was proposed by Chen, Xie et al. (2018) for example-based single-image SR of brain MRI images. This network had several advantages, namely, faster training (due to shorter network paths), small model size (due to weight sharing), and fewer chances of overfitting (due to the reduction of the number of parameters). Furthermore, this work was extended by Chen, Shi et al. (2018) who created a multilevel DCSRN-GAN architecture, in which each single DSCRN block was replaced by a collection of shallow DCSRN blocks, and the multilevel DSCRN is used as a generator within the GAN architecture.

Georgescu et al. (2020) proposed a single-image SR (SISR) approach based on two CNNs for in-plane and depth SR, respectively. Each CNN has 10 convolutional layers with one upscaling layer after the first 6 convolutional layers. An intermediate loss is calculated after the upscaling layer, and a final loss is computed after the last layer. This dual-loss scheme promotes the reconstruction of images of higher quality.

Park and Gahm (2022) have investigated a regression-based 3D SR algorithm for MRI images where gradient information is used to construct and cluster tensor features of LR images. In particular, pairs of LR tensor features and HR patches are used to train filters for SR reconstruction. For testing, a tensor is constructed for each voxel and corresponding features are extracted and used to predict the HR image intensities through the learned filters. Also, the orientation span is reduced and the shape variety is increased through a patch reduction approach.

Jiang et al. (2022) have investigated the construction of a deep learning SR system with a 2X magnification factor. In particular, the authors collected pairs of LR and HR MRI images with magnetic field densities of 0.35 and 3 T, respectively. Then, the authors trained a hybrid attention residual network with dense attention blocks for feature extraction, outer feature fusion, and upsampling. The proposed architecture combines dense and attention blocks in order to effectively extract relevant LR features. In addition to providing quantitative metrics for the superiority of the proposed method, the authors demonstrated the clinical feasibility of their work through collecting the expert opinions of two radiologists and comparing the subjective mean opinion scores for several state-of-the-art methods.

13.3.2 Computed tomography

Computed tomography (CT) is an important modality for examining the anatomical structures and diagnosing the pathologies of the abdomen (to identify disorders of the gastrointestinal system, the liver, and the pancreas), the heart (to diagnose different heart diseases), the head (in order to locate injuries, tumors, or blood clots leading to stroke or hemorrhage conditions), the lungs (to check for tumors, excess fluids, pulmonary embolisms, emphysema or pneumonia), and the musculoskeletal system (imaging complex bone fractures, and severely eroded joints).

In the past few years, there has been increasing interest in creating higher resolution CT data. On the one hand, this goal can be achieved by developing enhanced CT imaging techniques, which generally fall under the umbrella of HR computed tomography (HRCT). In particular, Lynch and Oh (2020) reported that *"HSR CT scanners have 0.25-mm detector elements with a maximum in-plane spatial resolution of 0.15 mm and a through-plane spatial resolution of 0.20 mm, resulting in improved spatial resolution compared with conventional multidetector CT scanners with 0.5-mm detector elements."* On the other hand, algorithmic SR techniques have been proposed in the past few years. Fig. 13.12 shows the growth of scientific publications involving SR and CT from 2013 till 2021. The growth pattern is quite remarkable from about 5 papers in 2013 to more than 20 papers in 2021. Next, we review some representative methods of SR in CT imaging (A summary of the key aspects of these methods is given in Table 13.2).

Umehara et al. (2018) have employed an SRCNN architecture for SR of CT chest images. This architecture includes three stages: feature extraction, nonlinear mapping, and reconstruction. The proposed method outperformed traditional linear interpolation methods on data samples from the Cancer Imaging Archive (TCIA), with statistically significant performance improvements for magnification levels of 2X and 4X.

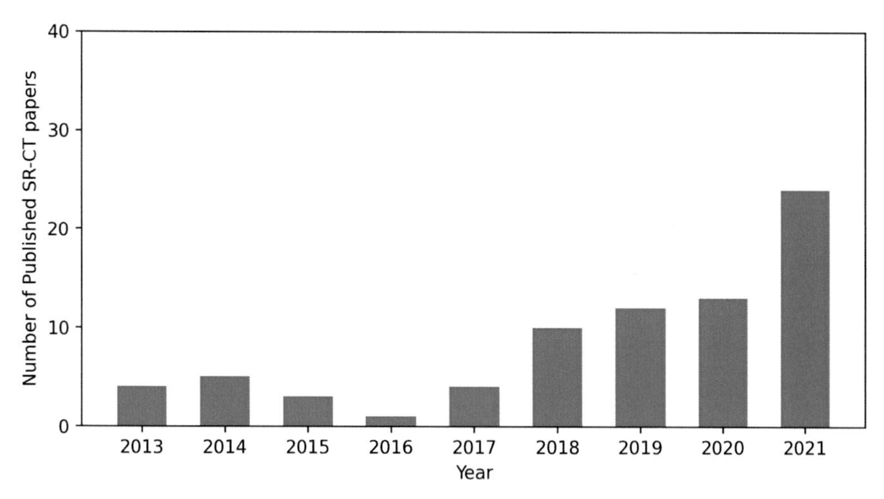

FIGURE 13.12 Growth of CT super-resolution studies on PubMed.

Numbers of scientific publications in PubMed involving keywords of "super-resolution" and "CT" for the years 2013–21.

Park et al. (2018) have used a U-Net variant to construct low-noise HR PET-CT slices from corresponding LR slices. For model training, LR images were collected from ground-truth 3D high-noise HR slices by averaging. The proposed U-Net architecture had two consecutive paths: a contraction path (through which features were extracted, downsized, and pooled), and an expansion path (where features were upsampled and HR images were reconstructed). This method led to a 10% average increase in the PSNR value compared to the thick-slice inputs.

Mansoor et al. (2018) applied a GAN architecture with a VGG-based network in order to super-resolve CT images of lung nodules. The proposed architecture was trained on LR-HR pairs of 2D CT slices in three imaging planes. Training and testing were carried out on forty and five samples of CT scans, respectively. This method resulted in an average PSNR value of 32 dB.

You et al. (2019) addressed the SR problem for CT images using GAN-CIRCLE, which is a semisupervised deep learning architecture composed of two interlinked GAN networks. This proposed architecture ensures matching between the distributions of the input and output images. Good SR results were obtained on the tibia and abdominal CT datasets.

Georgescu et al. (2020) have proposed a SISR architecture for CT images with two CNNs of 6 and 4 intermediate upscaling convolutional layers for in-plane and depth SR, respectively. In association with the intermediate upscaling layer, an intermediate loss term helps in decreasing the residual error between the reconstructed and ground-truth images. This method gives good SR performance in terms of the SSIM, PSNR, and IFC metrics for 2D and 3D CT images.

Table 13.2 Recent approaches for super-resolution (SR) in CT imaging.

References	Image modality	Organ	Data source	Method	Taxonomy	SR dimension	Results
Umehara et al. (2018)	CT	Chest	A total of 89 cases from the Cancer Imaging Archive (TCIA) (Clark et al., 2013). The cases were divided randomly into 45 training cases and 44 external test cases.	Super-Resolution CNN (SRCNN)	DLSR	Spatial (2X, 4X)	The SRCNN scheme yielded significantly higher quality metrics (PSNR and SSIM) than the linear interpolation methods for 2X magnification ($p < 0.05$) and 4X magnification ($p < 0.001$).
Park et al. (2018)	PET CT	Brain	A dataset of 65 clinical PET/CT Parkinson's disease suspected cases (26 males and 39 females) with a total of 7670 slices were used for training. Input: Low-noise low-resolution average of 5 slices (thick slices); Ground truth: High-noise high-resolution middle slice; Output: Low-noise high-resolution (thin slices).	U-Net variant	DLSR, FSR	Spatial	The CNN output yields an approximately 10% higher PSNR and lower NRMSE than the input (thicker slices).

Mansoor et al. (2018)	CT	Lung nodules	For training, forty pairs of HR and LR retrospective scans of diagnostic quality (with slice thicknesses of 1 and 4 mm, respectively) for patients with lung nodules. Five other scans were used for testing.	GAN architecture with a VGG-like network and a perceptual loss function.	DLSR	Spatial	Average PSNR: 32.09 ± 0.24 dB
You et al. (2019)	Contrast-enhanced CT	Ankle, and abdominal structures	Two high-quality sets: 1-Tibia dataset of micro-CT images from 25 cadaveric ankle samples (Chen, Zhang et al., 2018). 2-Abdominal dataset of the 2016 NIH-AAPM-Mayo Clinic Low Dose CT Grand Challenge (Flohr et al., 2005). The dataset includes 5936 full-dose CT scans from 10 patients, with a slice thickness of 1 mm and a reconstruction interval of 0.8 mm.	Two interlinked GAN networks (GAN-CIRCLE) with semisupervised learning	DLSR	Spatial	GAN-CIRCLE had the best overall subjective quality scores. Tibia dataset: 3.79 ± 0.72 Abdominal dataset: 3.62 ± 0.41

(Continued)

Table 13.2 Recent approaches for super-resolution (SR) in CT imaging. *Continued*

References	Image modality	Organ	Data source	Method	Taxonomy	SR dimension	Results
Georgescu et al. (2020)	2D & 3D CT	Brain	The Coltea Hospital (CH) dataset: Ten anonymized 3D images of brain CT, with 6 images (359 slices) and 4 images (238 slices) randomly selected for training and testing, respectively. The slice height and width vary between 192 and 512 pixels, while the 3D image depth varies between 3 and 176 slices. The voxel resolution voxel is $1 \times 1 \times 1$ mm^3.	Two CNNs for in-plane and depth super-resolution, with 6 and 4 convolutional layers, respectively.	DLSR, SISR	Spatial (2X, 4X)	CH dataset: Best 2X 2D SR results with the CNN method: SSIM = 0.9291 PSNR = 36.39 dB IFC = 5.36 Best 2X 3D SR results with the CNN method: SSIM = 0.8926 PSNR = 33.04 dB IFC = 2.83

13.3.3 Ultrasound

The use of ultrasound imaging (US) is crucial in the assessment of anatomical structures, the diagnosis of disorders in breast, abdominal, cardiac, gynecological, urological, and cerebrovascular examinations, as well as in pediatric and operational review (Carovac et al., 2011).

In the past few years, there has been increasing interest in creating higher resolution US data. First of all, this goal can be achieved by developing enhanced US imaging techniques, which generally fall under the umbrella of HR US. A fluorescence microscopy approach known as SR optical fluctuation imaging uses higher-order statistics to analyze the fluctuating optical signals to provide subdiffraction-limit imaging with an enhanced temporal resolution by using contrast-enhanced ultrasound plane-wave scans in order to quickly acquire HR acoustic images (Bar-Zion et al., 2017). A flexible matching layer-based optimum fabrication process of a miniaturized HR 360-degree electronic radial ultrasonic endoscope was created to increase the homogeneity between the elements of the radial array transducer (Peng et al., 2019). By directing and focusing an ultrasonic beam, spatial resolution is gained, and thus the array elements are simultaneously activated by random-sequence-coded signals, resulting in an unfocused ultrasound waveform with random interference. Using the l1-norm minimization and a priori measurements of spatial impulse responses, HR ultrasound images may be reconstructed (Ni & Lee, 2020). Specifically, ultrasound resolution enhancement can be achieved based on controlling key transducer characteristics such as its center frequency, bandwidth, or focus factors (Ploquin et al., 2015). A higher ultrasound frequency or bandwidth leads to better spatial resolution of ultrasound images with different diagnostic purposes. For example, transducers used for imaging fetuses used frequencies between 2 and 18 MHz (Ranganayakulu et al., 2016). For diagnosing regional lymph nodes and soft-tissue tumors, transducers operate in the range of 7.5–15 MHz. Moreover, skin and intravascular assessment typically depend on transducers in the range of 20–50 MHz, while ophthalmic applications use transducers centered in the range of 40–60 MHz. However, increasing the ultrasound frequency decreases the imaging depth.

Alternatively, image resolution can be enhanced without decreasing the imaging depth using algorithmic SR techniques. Such techniques have been proposed in the past few years. Fig. 13.13 shows the growth of the scientific publications involving SR and the US from 2013 till 2021. The growth pattern is quite remarkable from about 5 papers in 2013 to more than 35 papers in 2021. We discuss next some of these approaches (with a summary given in Table 13.3).

For temporal SR of 2D ultrasound videos, Chittajallu et al. (2019) employed a Hilbert transform approach for cardiorespiratory phase estimation. Then, a robust nonparametric regression model was built for respiratory gating. A kernel regression model was thus created for reconstructing temporally-super-resolved images at any cardiac phase. Accurate cardiac phase estimates were obtained with a mean-phase-error range of 3%–6% compared to the ECG-derived cardiac phase. Accurate phase estimation and respiratory gating led to high-quality temporal SR

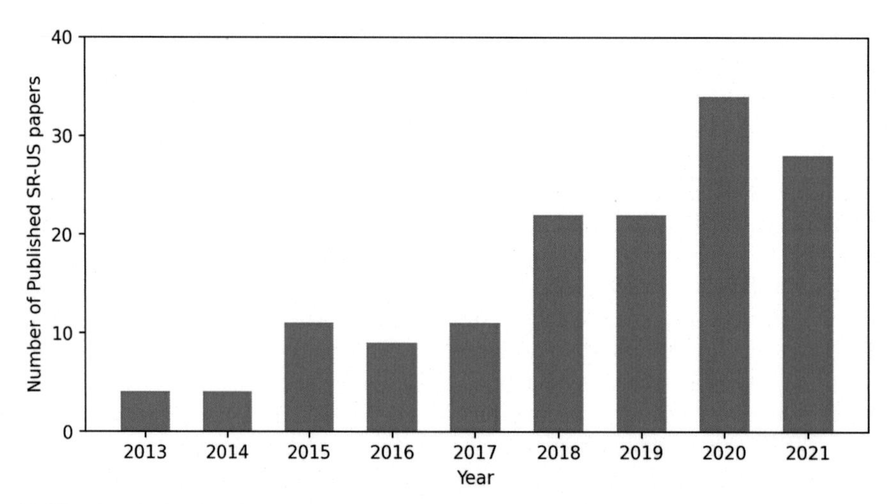

FIGURE 13.13 Growth of ultrasound super-resolution studies on PubMed.

Numbers of scientific publications in PubMed involving keywords of "super-resolution" and "ultrasound" for the years 2013–21.

results for the ultrasound frames. In particular, the average normalized correlation (between reconstructed and ground-truth frames) was experimentally found to be 0.83 and 0.72 on cross-validated non-QRS and QRS frames of 6 ultrasound videos, respectively. Also, SR experiments on videos of different levels of temporal resolution (1X–5X) returned an average normalized correlation of about 0.77.

Temiz and Bilge (2020) used CNN models for 2X, 3X, 4X, and 8X SR of ultrasound B-mode images. These models were trained with a dataset of 27,696 images to learn the nonlinear mappings between LR and HR ultrasound images. Visual and quantitative performance comparisons were made (based on 8 quality metrics) between the proposed method and two other deep learning schemes and four interpolation techniques.

Abdel-Nasser et al. (2017) used SR of LR ultrasound image sequences to boost the performance of CAD systems for differentiating between benign and malignant tumors in breast ultrasound images. A regularized SR reconstruction approach (Farsiu et al., 2004) was employed to create HR ultrasound images from LR sequences of 3, 5, 7, or 9 images.

Lu and Liu (2018) introduced a truly single-image SR algorithm where a CNN is trained on multiscale LR-HR image pairs extracted from the test image. The limited training data is increased through several augmentation operations. Dilated convolution and residual learning are used to improve convergence and accuracy. On the one hand, dilated convolution expands the receptive field without enlarging the network parameters in order to collect important intrinsic image information. On the other hand, residual learning is utilized to directly learn the differences between the HR and LR images, and speed up convergence.

Table 13.3 Recent approaches for super-resolution in ultrasound imaging.

References	Image modality	Organ	Data source	Method	Taxonomy	SR dimension	Results
Chittajallu et al. (2019)	Ultrasound video	Heart	2D cardiac ultrasound videos and ECG recordings of 6 anesthetized mice acquired at 233 frames per second (FPS). Each video has about 300 frames, 11 cardiac cycles, and 2 respiratory cycles. Cross-validation was performed among the frames of each of the 6 videos.	Hilbert transforms for cardiorespiratory phase estimation. Robust nonparametric regression for respiratory gating. A novel kernel regression model for image reconstruction at any cardiac phase and temporal super-resolution.	MLSR	Temporal	Average normalized correlation (between reconstructed and ground-truth frames): 0.83 ± 0.03 (non-QRS frames) 0.72 ± 0.05 (QRS frames)
Temiz and Bilge (2020)	B-mode ultrasound images	Multiple organs	1-A set of 29696 B-mode ultrasound images (Geertsma et al., 2004), with 1000 images for validation and 1000 for testing (TestSet-I), and 27,696 training images. 2-A set of 500 cropped images of a size of 600×450 (TestSet-II).	CNN	DLSR	Spatial (2X, 3X, 4X, 8X)	1-TestSet-I: SSIM = 0.9837 (2X); 0.9470 (3X); 0.8806 (4X); 0.7252 (8X). 2-TestSet-II: SSIM = 0.9404 (2X); 0.8740 (3X); 0.8137 (4X); 0.6963 (8X).

(Continued)

Table 13.3 Recent approaches for super-resolution in ultrasound imaging. *Continued*

References	Image modality	Organ	Data source	Method	Taxonomy	SR dimension	Results
Abdel-Nasser et al. (2017)	Ultrasound images	Breast tumor	A clinical database of 31 malignant and 28 benign B-mode ultrasound (BUS) image sequences (Chen et al., 2010).	A CAD system for ultrasound image classification after ultrasound image SR with a regularized reconstruction approach (Farsiu et al., 2004) from LR image sequences.	MLSR	Spatial	1-SR preprocessing with a 5-image LR sequence: AUC = 0.989 2-No SR preprocessing: AUC = 0.828
Lu and Liu, 2018	Ultrasound	Brachial plexus, heart, and brain.	Three brachial plexus, cardiac, and brain ultrasound images. For each image, HR and LR image pyramids are generated. Each LR-HR pair is vertically and horizontally flipped, and rotated by 45, 90, 135, 180, 225, 270, and 315 degrees, respectively.	Unsupervised super-resolution (USSR) with CNN.	DLSR, SISR	Spatial (2X, 3X, 4X)	Best 2X SR: 1-Brachial Plexus: PSNR: 38.66 SSIM: 0.9809 2-Heart: PSNR: 40.20 SSIM: 0.9718 3-Brain: PSNR: 41.17 SSIM: 0.9575

13.3.4 Whole-slide imaging, cytopathology, and fluoroscopy

In biomedical research, microscopy has been used in a variety of areas, including stem-cell imaging, precancerous lesion detection, receptor imaging, and brain mapping. A better understanding of the prognosis, treatment, and follow-up of many diseases has been enabled by advancements in computational sciences, optics, and chemical engineering. Thus, advanced microscope technologies are becoming more available to biomedical researchers. For example, fluorescence microscopy allowed imaging of molecular and cellular biomarkers, and this significantly aided immunohistopathological research. Also, the annular dark-field imaging mode in electron microscopy may reach resolutions of 50 pm for transmission electron microscopy (TEM) and scanning TEM, and even 20 pm using ptychography (Karhana et al., 2021).

Better resolution levels have been recently sought through computational approaches for images produced in whole-slide imaging, cytopathology, and fluoroscopy. Fig. 13.14 shows the growth of related publications. We discuss next some of these approaches (with a summary given in Table 13.4).

Al-Olofi et al. (2019) investigated a transfer-learning approach to find the scale mappings between WSI levels using partial least-square (PLS) regression. The learned scale mappings can be used to enhance image resolution and hence detect anomalies more effectively in LR images. The authors also explored the effect of different levels of noise on detection performance. In particular, simulations of different scenarios were carried out where WSI images were contaminated with Gaussian noise, and thus several denoising

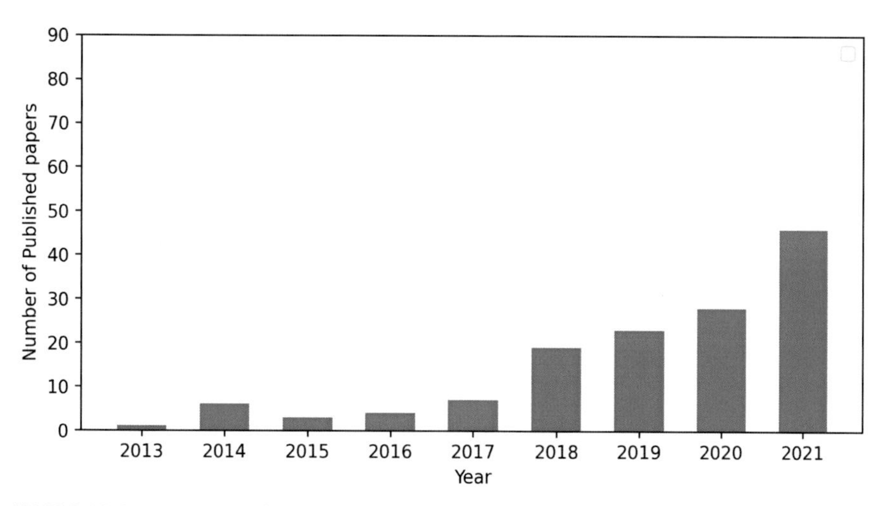

FIGURE 13.14 Growth of microscopy super-resolution studies on PubMed.

Numbers of scientific publications in PubMed involving keywords of "super-resolution," "learning," and "microscopy" for the years 2013–21.

Table 13.4 Recent approaches for super-resolution in microscopy imaging.

References	Image modality	Organ	Data source	Method	Taxonomy	SR dimension	Results
Al-Olofi et al. (2019)	Microscopic pathology whole-slide imaging (WSI).	Cancerous lymph nodes in the breast.	The Camelyon16 dataset (Ehteshami et al., 2016): 40 tumor WSI images (1507 patches) and 40 normal WSI images (1476 patches). These patches were randomly divided into 1990 training patches and 994 testing ones. Each WSI has 5 levels with patch sizes of 32×32, 64×64, 128×128, 256×256, and 512×512, respectively.	Learning scale mappings between WSI levels using partial least-square (PLS) regression and denoising algorithms.	FSR	Spatial	The PLS-based WSI mapping improves the LR WSI classification accuracy by 1%–4%. Under Gaussian noise ($\sigma = 0.8$), the PLS mapping, the BM3D algorithm, and their combination improve the LR detection accuracy from 63.50% to 93.81%, 92.73%, and 97.51%, respectively.
Heinrich et al. (2017)	3D Electron microscopy (EM)	Neural tissues of Drosophila melanogaster	A distortion-free FIB-SEM dataset (Hanslovsky et al., 2017). The volume had a size of $1250 \times 2000 \times 256$ pixels and was divided into a training set (70%), a validation set (15%), and a test set (15%)	Fast super-resolution convolutional neural networks (FSRCNN) and a 3D U-Net architecture for 3D super-resolution.	DLSR	Spatial	3D-FSRCNN model: PSNR = 33.22 and wPSNR = 35.94. 3D U-Net model: PSNR = 36.27 wPSNR = 39.09.

Ma et al. (2020)	Cytopathological slide images	Cervix	A publicly-available dataset of 142 liquid-based cytopathological slides (collected at Tongji Hospital, Huazhong University of Science and Technology). The slides were digitized as whole-slide images (WSIs) with objective lens resolutions of 20X, 10X, and 4X corresponding to physical resolutions of 0.293, 0.586, and 1.465 μm/pixel, respectively. The slides were divided into a training set of 118 slides (145,200 patches) and 24 test slides (29,300 patches).	A pathology super-resolution GAN (PathSR-GAN) with two stages: 1- LR-to-MR reconstruction: The generator in this stage is a densely connected U-Net that achieves 4X to 10X SR. 2- MR-to-HR reconstruction: The generator in this stage is a residual-in-residual DenseBlock with 10X to 20X SR.	DLSR & FSR	Spatial (4X, 10X, 20X)	10X SR results: SSIM: 29.71 PSNR: 0.91 20X SR results: SSIM: 26.92 PSNR: 0.88
Cheng et al. (2022)	Structured illumination microscopy (SIM)	Intracellular structures	1. Images were captured with a Nikon super-resolution microscope (L. Jin et al., 2020). Each sample was imaged under two settings. The first setting is	Fast and Lightweight SIM networks (FLSN) for single-frame structured illumination microscopy.	DLSR	Spatial (2X)	1. RMSE with low exposure (LE) and high exposure (HE): Adhesion: 167.29 (LE), 124.16 (HE) F-actin: 649.27 (LE), 548.23 (HE)

(*Continued*)

Table 13.4 Recent approaches for super-resolution in microscopy imaging. *Continued*

References	Image modality	Organ	Data source	Method	Taxonomy	SR dimension	Results
			standard illumination with a 200 ms exposure time and a laser power of 70 mW. The second setting involves very low-light conditions with only 1% of laser power in use and an ultra-short 20-ms exposure time. Each sample is represented by 15 LR frames and a 2X conventionally-reconstructed HR image. The imaged cellular structures are adhesion (669 training, 99 testing), F-actin (882 training, 126 testing), mitochondria (1007 training, 167 testing), and microtubules (953				Mitochondria: 461.81 (LE), 331.91 (HE) Microtubules: 693.21 (LE), 346.96 (HE) 2- Widefield F-actin image dataset: RMSE: 3008.80 SSIM: 0.732

			training, 135 testing). 2. A dataset of widefield F-actin images (Qiao et al., 2021) with 20,160 images for training and 1920 images for testing.				
Anam et al. (2016)	Fluorescence imaging	Cellular structures	1-The SNPHEp-2 dataset (Wiliem et al., 2013): Forty specimens with five cell patterns: centromere, coarse-speckled, fine-speckled, homogeneous, and nucleolar patterns. Each specimen image was captured using a monochrome high-dynamic-range cooled microscopy camera, which was fitted on a microscope with a Plan-Apochromat 20x/0.8 objective lens and an LED illumination source. A total of 1884 cell	Partial least squares (PLS) to learn the feature-domain LR-to-HR SR mapping	FSR	Spatial	Classification accuracies: MIVIA HEp-2 HR: 65.44% SNPHEp-2 LR: 60.87% LR + PLS: 64.00%

(*Continued*)

Table 13.4 Recent approaches for super-resolution in microscopy imaging. *Continued*

References	Image modality	Organ	Data source	Method	Taxonomy	SR dimension	Results
			images were obtained (905 for training and 979 for testing). 2-The ICPR HEp-2 Cell Classification Contest Dataset (Vento et al., 2016) contains 28 specimens with 1457 cells of six patterns: centromere, coarse-speckled, cytoplasmic, fine-speckled, homogeneous, and nucleolar patterns. Each specimen image was acquired by a 40X fluorescence microscope. For consistency with the SNPHEp-2 dataset, the cytoplasmic images were excluded leaving 1346 cells (663 for training and 683 for testing).				

algorithms were applied, namely, denoising with PLS, Block Matching 3D (BM3D), and the combination of PLS and BM3D.

Heinrich et al. (2017) compared two deep CNN architectures for 3D isotropic SR from nonisotropic electron microscopy (EM). These are a fast SR CNN (FSRCNN) and a 3D U-Net architecture. The proposed architectures were trained and tested on 3D microscopic volumes of the central nervous system for *Drosophila melanogaster*. Both architectures generated high-quality 3D isotropic SR volumes from LR no-isotropic EM data where the U-Net architecture outperformed the FSRCNN one.

Ma et al. (2020) utilized a multisupervised SR approach with a two-stage pathology SR GAN (PathSR-GAN) architecture. First, a densely connected U-Net is used to enlarge LR cytopathological cervical images with a LR of 4X to a medium resolution (MR) of 10X. Second, a residual-in-residual DenseBlock generator was employed to magnify the MR 10X images to HR 20X ones. As expected, the 4X-to-20X registration accuracy and stability were visibly worse than those of 4X-to-10X registration due to the big magnification difference. Still, the proposed system achieved significant improvements in the PSNR and SSIM metrics over other state-of-the-art methods.

Cheng et al. (2022) proposed a single-frame structured-illumination microscopy SR algorithm based on deep learning. This algorithm needs only one shot of a structured illumination frame and generates 2X HR results comparable to the traditional SIM methods that typically use 15 shots per sample. The proposed method was applied to four different intracellular structures: adhesion, F-actin, mitochondria, and microtubules. High-quality results were obtained under both high-exposure and low-exposure settings.

In order to infer HR characteristics from low-resolution HEp-2 images, Anam et al. (2016) suggested a learning-based technique. This method used a partial least-square (PLS) technique to linearly project both LR and HR HEp-2 image data into a shared linear subspace where the features are strongly correlated. This led to better classification performance on LR HEp-2 images.

13.4 Limitations, challenges, and open research directions

13.4.1 Limitations and challenges

Although medical image SR has witnessed remarkable success in the last few years, there are still several limitations that should be addressed in order to advance the state of the art in this field. We briefly discuss here some of these limitations.

13.4.1.1 Data availability

As most of the recent SR methods are based on learning approaches, large datasets of LR-HR pairs are needed. However, the availability of HR data for specific medical imaging modalities still poses a challenge, especially for large-scale magnification goals.

13.4.1.2 Limited studies on some medical imaging modalities

While extensive SR approaches have been proposed for the MRI and CT modalities, few methods and studies dealt with other modalities such as endoscopy, OCT, and microscopy. For example, see the publication growth patterns for different modalities in Fig. 13.4.

13.4.1.3 Real-time performance

A key factor for the feasibility and applicability of SR methods in medical imaging is the ability to reconstruct HR data in real-time to address the critical needs of fast assessment, treatment, and diagnosis. Still, most of the existing methods are experimental in nature with no focus yet on real-time performance.

13.4.1.4 Incorporating anatomical knowledge

The performance of the learning-based SR methods can be well boosted by capturing the constraints of the anatomical objects. However, accounting for such prior knowledge in SR methods is still not straightforward (Almalioglu et al., 2020).

13.4.1.5 Hardware limitations

Several medical image analysis tasks can be accelerated through the design of dedicated hardware systems. However, the availability of such systems for SR methods is still limited (although some attempts have been recently made) (Bae et al., 2020; Lee et al., 2019, 2020).

13.4.1.6 Intrinsic low-resolution characteristics in some imaging modalities

Some imaging modalities typically produce low-resolution images due to the underlying physical principles of image acquisition. For example, due to the natural intrinsic imaging characteristics of ultrasound, it produces poor-quality images of lower resolution compared to other medical imaging modalities. This makes the SR task more challenging for ultrasound and other similar modalities (Temiz & Bilge, 2020).

13.4.2 Open research directions

We point out here several research directions that we believe are worthy of further investigations to create better medical image SR systems.

13.4.2.1 Data augmentation

Data augmentation techniques have received considerable attention in order to compensate for the insufficiency of training data for large-scale machine learning problems. In particular, collecting and/or simulating adequate training data for deep learning-based SR systems is now quite crucial for constructing effective SR systems with large magnification factors (Li et al., 2020).

13.4.2.2 Self-supervised learning

Self-supervised learning schemes have been recently developed to effectively exploit large volumes of unlabeled data through learning supervisory signals and using such signals as automatically generated labels for subsequent supervised learning tasks. Such schemes can potentially help improve the performance of medical SR systems. See for example (Song et al., 2019; Zhao, Dewey et al., 2021).

13.4.2.3 Exploitation of self-similarity and redundancy in medical images

Several existing methods attempted to exploit self-similarity and redundancy in medical images in order to reduce the training data requirements. Examples of this approach include the work of Zhao, Carass, Dewey, and Prince (2018) where HR axial slices were used to reconstruct HR coronal slices from LR ones. Also, Al-Olofi et al. (2019) extracted training examples from WSI pyramids in order to reconstruct and then classify HR WSI patches.

13.4.2.4 Unsupervised super-resolution

Since there is a limited availability of LR-HR pairs of images for training supervised SR systems in specific applications (such as remote sensing), several methods have been recently proposed for SR based on unsupervised learning methodologies (Chen et al., 2020; Lugmayr et al., 2019; Mishra & Hadar, 2023; Prajapati et al., 2021). The majority of these methods are directed towards applications in remote sensing and hyperspectral imaging where HR data can't be easily captured. These methods can be readily extended to medical imaging modalities where the HR data is also scarce.

13.4.2.5 Design of clinically-aware loss functions

The incorporation of prior anatomical and clinical knowledge into loss functions for learning-based SR methods can potentially lead to more clinically plausible results and increase the utility of the reconstructed HR medical images in medical decision support (Li, Yu et al., 2021; Oktay et al., 2018). Also, more localized attention can be made for clinically important image areas.

13.4.2.6 Deep super-resolution network pruning and compression

Most of the state-of-the-art SR algorithms are based on deep learning architectures of high complexity and size. Reducing such complexity of trained SR networks is crucial for the potential deployment of these architectures on systems with limited memory and real-time performance requirements. This reduction can be achieved through modern techniques of network pruning and compression (Chu et al., 2020; Hou & Kung, 2020; Zhan et al., 2021).

13.4.2.7 Interpretability of deep learning models

A key aspect of modern deep learning systems is the ability to interpret and explain network performance (Salimi et al., 2021). Numerous methods and tools have been recently proposed in this direction with a focus on SR tasks for remote sensing and hyperspectral data using primarily deep unfolding networks (Ma et al., 2022; Shi et al., 2022; Wang et al., 2022). Furthermore, an explainable AI (XAI) SR method with deep unfolding networks has been proposed for medical ultrasound image data (Wang, Zhu et al., 2021).

13.4.2.8 Large and arbitrary scale factors

Most of the SR algorithms are still confined to small magnification factors of 2X or 3X. Attempts to achieve higher magnification ratios are typically based on incremental approaches that increase the magnification in small steps in order to avoid significant quality losses at higher magnifications. The quest for large-scale or arbitrary-scale SR essentially depends on the collection of appropriate datasets with wide-scale variations (such as the extreme scale variation textures (ESVaT) dataset) (Liu et al., 2019). In medical imaging, similar datasets are available for specific modalities including, for example, whole-slide imaging (WSI) datasets such as Camelyon (Bandi et al., 2019).

13.4.2.9 Modality-specific degradation processes

Different medical imaging modalities may involve distinct degradation processes whose characterization and incorporation can be of paramount importance for successful HR image reconstruction. For example, MRI data can be subject to Rician noise (Hu et al., 2020) while ultrasound images are typically degraded by speckle noise (Wildeboer et al., 2020).

13.4.2.10 Dealing with different input resolutions and views

The LR data can typically include multiview mixed-resolution images and depth data that could be jointly exploited to reconstruct HR data. Several methods have been introduced to achieve this goal for non-medical data (Garcia et al., 2012; Jin et al., 2016; Lu et al., 2021; Richter et al., 2016). This approach can be extended to medical imaging data (e.g., to reconstruct one MRI view from other views) (Zhao, Carass, Dewey, & Prince, 2018).

13.4.2.11 Quality assessment for super-resolution algorithms

Conventional quality metrics (such as PSNR) might not be suitable for evaluating the quality of the reconstructed HR images produced by SR algorithms. Several alternate methods have been proposed to assess the quality of HR non-medical images (Beron et al., 2020; Jiang et al., 2022; Yan et al., 2019; Zhao et al., 2022). Such methods may be extended to give modality-specific quality assessment metrics.

References

Abdel-Nasser, M., Melendez, J., Moreno, A., Omer, O. A., & Puig, D. (2017). Breast tumor classification in ultrasound images using texture analysis and super-resolution methods. *Engineering Applications of Artificial Intelligence, 59*, 84–92. Available from https://doi.org/10.1016/j.engappai.2016.12.019.

Adabi, S., Turani, Z., Fatemizadeh, E., Clayton, A., & Nasiriavanaki, M. (2017). Optical coherence tomography technology and quality improvement methods for optical coherence tomography images of skin: A short review. *Biomedical Engineering and Computational Biology, 8*, 1179597217713475. Available from https://doi.org/10.1177/1179597217713475.

Akhtar, P., & Azhar, F. (2010). Pakistan a single image interpolation scheme for enhanced super resolution in bio-medical imaging. In *4th International conference on bioinformatics and biomedical engineering, iCBBE 2010*. Available from https://doi.org/10.1109/ICBBE.2010.5518164.

Almalioglu, Y., Bengisu Ozyoruk, K., Gokce, A., Incetan, K., Irem Gokceler, G., Ali Simsek, M., Ararat, K., Chen, R. J., Durr, N. J., Mahmood, F., & Turan, M. (2020). EndoL2H: Deep super-resolution for capsule endoscopy. *IEEE Transactions on Medical Imaging* (12), 4297–4309. Available from https://doi.org/10.1109/TMI.2020.3016744, http://ieeexplore.ieee.org/xpl/RecentIssue.jsp?punumber = 42.

Al-Olofi, W., A., Rushdi, M. A., Islam, M. A., & Badawi, A. M. (2019). Egypt improved anomaly detection in low-resolution and noisy whole-slide images using transfer learning. In *2018 9th Cairo international biomedical engineering conference, CIBEC 2018 – proceedings* (pp. 114–117). Institute of Electrical and Electronics Engineers Inc. Available from https://doi.org/10.1109/CIBEC.2018.8641820, 9781538681541. http://ieeexplore.ieee.org/xpl/mostRecentIssue.jsp?punumber = 8637586.

Anam, A. M., Rushdi, M. A., & Fahmy, A. S. (2016). Enhancement of low-resolution HEp-2 cell image classification using partial least-square regression. In *Proceedings - International conference on image processing, ICIP* (Vol. 2016, pp. 1245–1249). IEEE Computer Society Egypt. Available from https://doi.org/10.1109/ICIP.2016.7532557, 9781467399616.

Bae, S. H., Bae, J. H., Muqeet, A., Monira, M. S., & Kim, L. (2020). Cost-efficient super-resolution hardware using local binary pattern classification and linear mapping for real-time 4K conversion. *IEEE Access*, 224383–224393. Available from https://doi.org/10.1109/ACCESS.2020.3036828, http://ieeexplore.ieee.org/xpl/RecentIssue.jsp?punumber = 6287639.

Bai, W., Shi, W., de Marvao, A., Dawes, T. J. W., O'Regan, D. P., Cook, S. A., & Rueckert, D. (2015). A bi-ventricular cardiac atlas built from 1000 + high resolution MR images of healthy subjects and an analysis of shape and motion. *Medical Image Analysis, 26*(1), 133–145. Available from https://doi.org/10.1016/j.media.2015.08.009, http://www.elsevier.com/inca/publications/store/6/2/0/9/8/3/index.htt.

Bandi, P., Geessink, O., Manson, Q., Van Dijk, M., Balkenhol, M., Hermsen, M., Ehteshami Bejnordi, B., Lee, B., Paeng, K., Zhong, A., Li, Q., Zanjani, F. G., Zinger, S., Fukuta, K., Komura, D., Ovtcharov, V., Cheng, S., Zeng, S., Thagaard, J., … Litjens, G. (2019). From detection of individual metastases to classification of lymph node status at the patient level: The CAMELYON17 challenge. *IEEE Transactions on Medical Imaging, 38*(2), 550–560. Available from https://doi.org/10.1109/tmi.2018.2867350.

Bar-Zion, A., Tremblay-Darveau, C., Solomon, O., Adam, D., & Eldar, Y. C. (2017). Fast vascular ultrasound imaging with enhanced spatial resolution and background rejection. *IEEE*

Transactions on Medical Imaging, *36*(1), 169–180. Available from https://doi.org/10.1109/TMI.2016.2600372, http://ieeexplore.ieee.org/xpl/RecentIssue.jsp?punumber = 42.

Bernard, O., D'hooge, J., & Bosch, J. Challenge on Endocardial Three-dimensional Ultrasound Segmentation. 2014. https://www.creatis.insa-lyon.fr/Challenge/CETUS/organizers.html [Accessed 08 September 2023].

Beron, J., Benitez-Restrepo, H. D., & Bovik, A. C. (2020). Blind image quality assessment for super resolution via optimal feature selection. *IEEE Access*, *8*, 143201–143218. Available from https://doi.org/10.1109/ACCESS.2020.3014497, http://ieeexplore.ieee.org/xpl/RecentIssue.jsp?punumber = 6287639.

Burger, W., & Burge, M. J. (2009). *Principles of digital image processing - Fundamental techniques*. Springer. Available from https://doi.org/10.1007/978-1-84800-191-6, 10.1007/978-1-84800-191-6.

Carovac, A., Smajlovic, F., & Junuzovic, D. (2011). Application of ultrasound in medicine. *Acta Informatica Medica*, *19*(3), 168. Available from https://doi.org/10.5455/aim.2011.19.168-171.

Castiglioni, I., Rundo, L., Codari, M., Di Leo, G., Salvatore, C., Interlenghi, M., Gallivanone, F., Cozzi, A., D'Amico, N. C., & Sardanelli, F. (2021). AI applications to medical images: From machine learning to deep learning. *Physica Medica*, *83*, 9–24. Available from https://doi.org/10.1016/j.ejmp.2021.02.006, http://www.fisicamedica.org.

Chavez-Roman, H., & Ponomaryov, V. (2014). Super resolution image generation using wavelet domain interpolation with edge extraction via a sparse representation. *IEEE Geoscience and Remote Sensing Letters*, *11*(10), 1777–1781. Available from https://doi.org/10.1109/LGRS.2014.2308905.

Chen, C., Zhang, X., Guo, J., Jin, D., Letuchy, E. M., Burns, T. L., Levy, S. M., Hoffman, E. A., & Saha, P. K. (2018). Quantitative imaging of peripheral trabecular bone microarchitecture using MDCT. *Medical Physics*, *45*(1), 236–249. Available from https://doi.org/10.1002/mp.12632, http://aapm.onlinelibrary.wiley.com/hub/journal/10.1002/(ISSN)2473-4209/issues/.

Chen, S, Han, Z., Dai, E., Jia, X., Liu, Z., Liu, X., Zou, X., Xu, C., Liu, J., & Tian, Q. (2020). Unsupervised image super-resolution with an indirect supervised path. In *IEEE computer society conference on computer vision and pattern recognition workshops* (pp. 1924–1933). Hong Kong: IEEE Computer Society. Available from https://doi.org/10.1109/CVPRW50498.2020.00242.

Chen, L., et al. 2010. http://mi.eng.cam.ac.uk/research/projects/elasprj/ [Accessed 08 September 2023].

Chen, Y., Shi, F., Christodoulou, A. G., Xie, Y., Zhou, Z., & Li, D. (2018). *States Efficient and accurate MRI super-resolution using a generative adversarial network and 3D multi-level densely connected network. Lecture notes in computer science (including subseries lecture notes in artificial intelligence and lecture notes in bioinformatics)* (11070, pp. 91–99). United: Springer Verlag. Available from https://doi.org/10.1007/978-3-030-00928-1_11, https://www.springer.com/series/558.

Chen, Y., Xie, Y., Zhou, Z., Shi, F., Christodoulou, A. G., & Li, D. (2018). *Brain MRI super resolution using 3D deep densely connected neural networks. Proceedings - International Symposium on Biomedical Imaging* (2018, pp. 739–742). United States: IEEE Computer Society. Available from http://ieeexplore.ieee.org/xpl/conferences.jsp.

Cheng, X., Li, J., Dai, Q., Fu, Z., & Yang, J. (2022). Fast and lightweight network for single frame structured illumination microscopy super-resolution. *IEEE Transactions on Instrumentation and Measurement*, *71*, 1–11. Available from https://doi.org/10.1109/tim.2022.3161721.

Chevalier, M., Leyton, F., Nogueira, M., Oliveira, M., da Silva, T. A., & Emilio, J. (2012). *Image quality requirements for digital mammography in breast cancer screening*. InTech. Available from https://doi.org/10.5772/30973.

Chittajallu, D. R., McCormick, M., Gerber, S., Czernuszewicz, T. J., Gessner, R., Willis, M. S., Niethammer, M., Kwitt, R., & Aylward, S. R. (2019). Image-based methods for phase estimation, gating, and temporal superresolution of cardiac ultrasound. *IEEE Transactions on Biomedical Engineering*, *66*(1), 72–79. Available from https://doi.org/10.1109/TBME.2018.2823279, http://ieeexplore.ieee.org/xpl/RecentIssue.jsp?reload = true&punumber = 10.

Chu, C., Chen, L., & Gao, Z. (2020). Similarity based filter pruning for efficient super-resolution models. In *IEEE international symposium on broadband multimedia systems and broadcasting, BMSB* (Vol. 2020). China: IEEE Computer Society. Available from https://doi.org/10.1109/BMSB49480.2020.9379712, 21555052. http://ieeexplore.ieee.org/xpl/mostRecentIssue.jsp?punumber = 6255877.

Clark, K., Vendt, B., Smith, K., Freymann, J., Kirby, J., Koppel, P., Moore, S., Phillips, S., Maffitt, D., Pringle, M., Tarbox, L., & Prior, F. (2013). The cancer imaging archive (TCIA): Maintaining and operating a public information repository. *Journal of Digital Imaging*, *26*(6), 1045–1057. Available from https://doi.org/10.1007/s10278-013-9622-7.

Contreras Ortiz, S. H., Chiu, T., & Fox, M. D. (2012). Ultrasound image enhancement: A review. *Biomedical Signal Processing and Control*, *7*(5), 419–428. Available from https://doi.org/10.1016/j.bspc.2012.02.002, http://www.elsevier.com/wps/find/journalbibliographicinfo.cws_home/706718/description#bibliographicinfo.

Devaraj, S. J. (2019). *Emerging paradigms in transform-based medical image compression for telemedicine environment. Telemedicine technologies: Big data, deep learning, robotics, mobile and remote applications for global healthcare* (pp. 15–29). India: Elsevier. Available from http://www.sciencedirect.com/science/book/9780128169483, https://doi.org/10.1016/B978-0-12-816948-3.00002-7.

Dharejo, F. A., Hao, Z., Bhatti, A., Bhatti, M. N., Ahmed, J., & Jatoi, M. A. (2017). China improved dictionary learning algorithm with mappings for single image super-resolution. In *IST 2017 - IEEE international conference on imaging systems and techniques, proceedings* (Vol. 2018, pp. 1–6). Institute of Electrical and Electronics Engineers Inc. Available from https://doi.org/10.1109/IST.2017.8261522, 9781538616208.

Ehteshami, B. B., et al. Camelyon16. 2016. https://camelyon16.grand-challenge.org./ [Accessed 08 September 2023].

Erfle, H. (2017). *Super-resolution microscopy: Methods and protocols*. Humana Press.

Evans, A., et al. (1997). BrainWeb. https://brainweb.bic.mni.mcgill.ca/brainweb/ [Accessed 08 September 2023].

Farooq, M., Dailey, M. N., Mahmood, A., Moonrinta, J., & Ekpanyapong, M. (2021). Human face super-resolution on poor quality surveillance video footage. *Neural Computing and Applications*, *33*(20), 13505–13523. Available from https://doi.org/10.1007/s00521-021-05973-0, http://link.springer.com/journal/521.

Farsiu, S., Robinson, M. D., Elad, M., & Milanfar, P. (2004). Fast and robust multiframe super resolution. *IEEE Transactions on Image Processing*, *13*(10), 1327–1344. Available from https://doi.org/10.1109/TIP.2004.834669.

Flohr, T. G., et al. (2005). https://www.aapm.org/GrandChallenge/LowDoseCT/#trainingData [Accessed 08 September 2023].

Garcia, D. C., Dorea, C., & De Queiroz, R. L. (2012). Super resolution for multiview images using depth information. *IEEE Transactions on Circuits and Systems for Video Technology, 22*(9), 1249–1256. Available from https://doi.org/10.1109/TCSVT.2012.2198134.

Geertsma, T., et al. (2004). https://www.ultrasoundcases.info/ [Accessed 08 September 2023].

Geiser, W. R., Haygood, T. M., Santiago, L., Stephens, T., Thames, D., & Whitman, G. J. (2011). Challenges in mammography: Part 1, artifacts in digital mammography. *American Journal of Roentgenology, 197*(6), W1023–W1030. Available from https://doi.org/10.2214/AJR.10.7246, http://www.ajronline.org/content/197/6/W1023.full.pdf + html, United States.

Georgescu, M. I., Ionescu, R. T., & Verga, N. (2020). Convolutional neural networks with intermediate loss for 3D super-resolution of CT and MRI scans. *IEEE Access, 8*, 49112–49124. Available from https://doi.org/10.1109/ACCESS.2020.2980266, http://ieeexplore.ieee.org/xpl/RecentIssue.jsp?punumber = 6287639.

Giannakidis, A., Oktay, O., Keegan, J., Spadotto, V., Voges, I., Smith, G., & Firmin, D. (2017). Super-resolution reconstruction of late gadolinium enhancement cardiovascular magnetic resonance images using a residual convolutional neural network. In *25th Scientific meeting of the industrial society for magnetic resonance in medicine (ISMRM 2017)*.

Glasner, D., Bagon, S., & Irani, M. (2009). Israel super-resolution from a single image. Proceedings of the IEEE *international conference on computer vision* (pp. 349–356). Available from https://doi.org/10.1109/ICCV.2009.5459271.

González, R. C., & Woods, R. E. (2008). *Digital image processing* (3rd Edition). Pearson Education 1-I-XXII-954. Available from https://www.worldcat.org/oclc/241057034.

Greenspan, H. (2009). Super-resolution in medical imaging. *Computer Journal, 52*(1), 43–63. Available from https://doi.org/10.1093/comjnl/bxm075.

Hanslovsky, P., Bogovic, J. A., & Saalfeld, S. (2017). Image-based correction of continuous and discontinuous non-planar axial distortion in Serial section microscopy. *Bioinformatics (Oxford, England), 33*(9), 1379–1386. Available from https://doi.org/10.1093/bioinformatics/btw794, http://bioinformatics.oxfordjournals.org/.

Heinrich, L., Bogovic, J. A., & Saalfeld, S. (2017). *Deep learning for isotropic super-resolution from non-isotropic 3d electron microscopy. Lecture notes in computer science (including subseries lecture notes in artificial intelligence and lecture notes in bioinformatics)* (10434, pp. 135–143). United States: Springer Verlag. Available from 10.1007/978-3-319-66185-8_16 16113349, http://springerlink.com/content/0302-9743/copyright/2005/.

Heismann, B. J., Henseler, D., Hackenschmied, D. N., Strassburg, M., Janssen, S., & Wirth, S. (2008). Spectral and spatial resolution of semiconductor detectors in medical x-and gamma ray imaging. In *IEEE nuclear science symposium conference record* (pp. 78–83). IEEE.

Hou, Z., & Kung, S. Y. (2020). Efficient image super resolution via channel discriminative deep neural network pruning. In *ICASSP, IEEE international conference on acoustics, speech and signal Processing – Proceedings* (Vol. 2020, pp. 3647–3651). United States: Institute of Electrical and Electronics Engineers Inc. Available from https://doi.org/10.1109/ICASSP40776.2020.9054019, 9781509066315.

Hu, J., Li, X., Wang, X., Li, Y., & Wu, X. (2020). Noise-robust MRI upsampling using adaptive local steering kernel. *IEEE Access, 8*, 158538–158548. Available from https://doi.org/10.1109/ACCESS.2020.3020133, http://ieeexplore.ieee.org/xpl/RecentIssue.jsp?punumber = 6287639.

Isaac, J. S., & Kulkarni, R. (2015). Super resolution techniques for medical image processing. In *Proceedings - International conference on technologies for sustainable development, ICTSD 2015*. India: Institute of Electrical and Electronics Engineers Inc. Available from https://doi.org/10.1109/ICTSD.2015.7095900, 9781479981878.

Japanese ADNI project. (2007). https://adni.loni.usc.edu/data-samples/access-data/. [Accessed 08 September 2023].

Jiang, Q., Liu, Z., Gu, K., Shao, F., Zhang, X., Liu, H., & Lin, W. (2022). Single image super-resolution quality assessment: A real-world dataset, subjective studies, and an objective metric. *IEEE Transactions on Image Processing*, *31*, 2279–2294. Available from https://doi.org/10.1109/TIP.2022.3154588, https://ieeexplore.ieee.org/xpl/mostRecentIssue.jsp?punumber = 83.

Jiang, J., Qi, F., Du, H., Xu, J., Zhou, Y., Gao, D., & Qiu, B. (2022). Super-resolution reconstruction of 3T-like images from 0.35T MRI using a hybrid attention residual network. *Journals & Magazines*, *10*, 32810–32821.

Jin, L., Liu, B., Zhao, F., Hahn, S., Dong, B., Song, R., Elston, T. C., Xu, Y., & Hahn, K. M. (2020). Deep learning enables structured illumination microscopy with low light levels and enhanced speed. *Nature Communications*, *11*(1). Available from https://doi.org/10.1038/s41467-020-15784-x, http://www.nature.com/ncomms/index.html.

Jin, Z., Tillo, T., Yao, C., Xiao, J., & Zhao, Y. (2016). Virtual-view-assisted video super-resolution and enhancement. *IEEE Transactions on Circuits and Systems for Video Technology*, *26*(3), 467–478. Available from https://doi.org/10.1109/TCSVT.2015.2412791.

Jog, A., Carass, A., & Prince, J. L. (2016). *Self super-resolution for magnetic resonance images. Lecture notes in computer science (including subseries lecture notes in artificial intelligence and lecture notes in bioinformatics)* (9902, pp. 553–560). United States: Springer Verlag. Available from https://doi.org/10.1007/978-3-319-46726-9_64 16113349, http://springerlink.com/content/0302-9743/copyright/2005/.

Kaderuppan, S. S., Wong, E. W. L., Sharma, A., & Woo, W. L. (2020). Smart nanoscopy: A review of computational approaches to achieve super-resolved optical microscopy. *IEEE Access*, *8*, 214801–214831. Available from https://doi.org/10.1109/ACCESS.2020.3040319, http://ieeexplore.ieee.org/xpl/RecentIssue.jsp?punumber = 6287639.

Karhana, S., Bhat, M., Ninawe, A., & Dinda, A. K. (2021). *Advances in microscopy and their applications in biomedical research. Biomedical imaging instrumentation: Applications in tissue, cellular and molecular diagnostics* (pp. 185–212). India: Elsevier. Available from https://www.sciencedirect.com/book/9780323856508, https://doi.org/10.1016/B978-0-323-85650-8.00008-5.

Kempton, M., et al. (2011). https://sites.google.com/site/brainseg/. [Accessed 08 September 2023].

Kempton, M. J., Underwood, T. S. A., Brunton, S., Stylios, F., Schmechtig, A., Ettinger, U., Smith, M. S., Lovestone, S., Crum, W. R., Frangou, S., Williams, S. C. R., & Simmons, A. (2011). A comprehensive testing protocol for MRI neuroanatomical segmentation techniques: Evaluation of a novel lateral ventricle segmentation method. *Neuroimage*, *58*(4), 1051–1059. Available from https://doi.org/10.1016/j.neuroimage.2011.06.080.

Kennedy, J. A., Israel, O., Frenkel, A., Bar-Shalom, R., & Azhari, H. (2006). Super-resolution in PET imaging. *IEEE Transactions on Medical Imaging*, *25*(2), 137–147. Available from https://doi.org/10.1109/TMI.2005.861705.

Kouamé, D., & Ploquin, M. (2009). Super-resolution in medical imaging: An illustrative approach through ultrasound. In *Proceedings - 2009 IEEE international symposium on biomedical imaging: From nano to macro, ISBI*, (Vol. 2009, pp. 249–252). France. Available from https://doi.org/10.1109/ISBI.2009.5193030.

Lambert, J. W., Phelps, A. S., Courtier, J. L., Gould, R. G., & MacKenzie, J. D. (2016). Image quality and dose optimisation for infant CT using a paediatric phantom. *European Radiology*, *26*(5), 1387–1395. Available from https://doi.org/10.1007/s00330-015-3951-5, http://www.link.springer.de/link/service/journals/00330/index.htm.

Lee, D., Lee, H., Lee, S., Lee, K., & Lee, H. J. (2019). Institute of Electrical and Electronics Engineers Inc. South Korea hardware design of a context-preserving filter-reorganized CNN for super-resolution. *IEEE Journal on Emerging and Selected Topics in Circuits and Systems*, *9*, 612–622. Available from https://doi.org/10.1109/JETCAS.2019.2950536, 21563365 4, https://ieeexplore.ieee.org/servlet/opac?punumber = 5503868.

Lee, S., Joo, S., Ahn, H. K., & Jung, S. O. (2020). CNN acceleration with hardware-efficient dataflow for super-resolution. *IEEE Access*, *8*, 187754–187765. Available from http://ieeexplore.ieee.org/xpl/RecentIssue.jsp?punumber = 6287639, https://doi.org/10.1109/ACCESS.2020.3031055.

Lei, S., Shi, Z., & Zou, Z. (2017). Super-resolution for remote sensing images via local-global combined network. *IEEE Geoscience and Remote Sensing Letters*, *14*(8), 1243–1247. Available from https://doi.org/10.1109/LGRS.2017.2704122, http://ieeexplore.ieee.org/xpl/RecentIssue.jsp?punumber = 8859.

Li, X., Wu, Y., Zhang, W., Wang, R., & Hou, F. (2020). Deep learning methods in real-time image super-resolution: A survey. *Journal of Real-Time Image Processing*, *17*, 1885–1909. Available from https://doi.org/10.1007/s11554-019-00925-3. 18618219, 6. Springer Science and Business Media Deutschland GmbH China. http://www.springer.com/sgw/cda/frontpage/0.11855.1-40392-70-112907049-0.00.html?changeHeader = true.

Li, Y., Sixou, B., & Peyrin, F. (2021). A review of the deep learning methods for medical images super resolution problems. *IRBM*, *42*(2), 120–133. Available from https://doi.org/10.1016/j.irbm.2020.08.004, http://www.elsevier.com.

Li, Z., Yu, J., Wang, Y., Zhou, H., Yang, H., & Qiao, Z. (2021). DeepVolume: Brain structure and spatial connection-aware network for brain MRI super-resolution. *IEEE Transactions on Cybernetics*, *51*(7), 3441–3454. Available from https://doi.org/10.1109/TCYB.2019.2933633, https://www.ieee.org/membership-catalog/productdetail/.

Lim, B., Son, S., Kim, H., Nah, S., & Lee, K.M. (2017). Enhanced deep residual networks for single image super-resolution. In *IEEE computer society conference on computer vision and pattern recognition workshops* (Vol. 2017, pp. 1132–1140). IEEE Computer Society South Korea. 21607516. Available from https://doi.org/10.1109/CVPRW.2017.151. http://ieeexplore.ieee.org/xpl/conferences.jsp.

Lin, E., & Alessio, A. (2009). What are the basic concepts of temporal, contrast, and spatial resolution in cardiac CT? *Journal of Cardiovascular Computed Tomography*, *3*(6), 403–408. Available from https://doi.org/10.1016/j.jcct.2009.07.003.

Liu, C., Wu, X., Yu, X., Tang, Y. Y., Zhang, J., & Zhou, J. L. (2018). Fusing multi-scale information in convolution network for MR image super-resolution reconstruction. *Biomedical Engineering Online*, *17*(1). Available from https://doi.org/10.1186/s12938-018-0546-9, http://www.biomedical-engineering-online.com/start.asp.

Liu, H., Ruan, Z., Zhao, P., Dong, C., Shang, F., Liu, Y., Yang, L., & Timofte, R. (2022). Video super-resolution based on deep learning: A comprehensive survey. *Artificial Intelligence Review*, *55*(8), 5981–6035. Available from https://doi.org/10.1007/s10462-022-10147-y, https://www.springer.com/journal/10462.

Liu, L., Chen, J., Zhao, G., Fieguth, P., Chen, X., & Pietikainen, M. (2019). Texture classification in extreme scale variations using GANet. *IEEE Transactions on Image Processing*, *28*(8),

3910–3922. Available from https://doi.org/10.1109/TIP.2019.2903300, https://ieeexplore.ieee.org/xpl/mostRecentIssue.jsp?punumber = 83.

Llebaria, A., Magnelli, B., Arnouts, S., Pollo, A., Milliard, B., & Guillaume, M. (2008). France multi-channel 2D photometry with super-resolution in far UV astronomical images using priors in visible bands. *Proceedings of SPIE - The International Society for Optical Engineering, 6812*. Available from https://doi.org/10.1117/12.7703020277786X.

Lu, J., & Liu, W. (2018). Unsupervised super-resolution framework for medical ultrasound images using dilated convolutional neural networks. In *3rd IEEE international conference on image, vision and computing, ICIVC 2018* (pp. 739–744). China: Institute of Electrical and Electronics Engineers Inc. Available from https://doi.org/10.1109/ICIVC.2018.8492821, 9781538649916. http://ieeexplore.ieee.org/xpl/mostRecentIssue.jsp?punumber = 8476690.

Lu, S. P., Li, S. M., Wang, R., Lafruit, G., Cheng, M. M., & Munteanu, A. (2021). Low-rank constrained super-resolution for mixed-resolution multiview video. *IEEE Transactions on Image Processing, 30*, 1072–1085. Available from https://doi.org/10.1109/TIP.2020.3042064, https://ieeexplore.ieee.org/xpl/mostRecentIssue.jsp?punumber = 83.

Lugmayr, A., Danelljan, M., & Timofte, R. (2019). Unsupervised learning for real-world super-resolution. In Proceedings - 2019 *International conference on computer vision workshop*, ICCVW 2019 (pp. 3408–3416). Switzerland: Institute of Electrical and Electronics Engineers Inc. Available from https://doi.org/10.1109/ICCVW.2019.00423, 9781728150239. http://ieeexplore.ieee.org/xpl/mostRecentIssue.jsp?punumber = 8982559.

Luo, Q., Shao, X., Peng, L., Wang, Y., & Wang, L. (2015). SPIE China Super-resolution imaging in remote sensing. *Proceedings of SPIE - The International Society for Optical Engineering, 9501*. Available from https://doi.org/10.1117/12.2176172, http://spie.org/x1848.xml.

Lynch, D. A., & Oh, A. S. (2020). High-spatial-resolution CT offers new opportunities for discovery in the lung. *Radiology, 297*(2), 472–473. Available from https://doi.org/10.1148/RADIOL.2020203473, http://pubs.rsna.org/doi/10.1148/radiol.2020203473.

Ma, J., Yu, J., Liu, S., Chen, L., Li, X., Feng, J., Chen, Z., Zeng, S., Liu, X., & Cheng, S. (2020). PathSRGAN: Multi-supervised super-resolution for cytopathological images using generative adversarial network. *IEEE Transactions on Medical Imaging, 39*(9), 2920–2930. Available from https://doi.org/10.1109/TMI.2020.2980839, http://ieeexplore.ieee.org/xpl/RecentIssue.jsp?punumber = 42.

Ma, Q., Jiang, J., Liu, X., & Ma, J. (2022). Deep unfolding network for spatiospectral image super-resolution. *IEEE Transactions on Computational Imaging, 8*, 28–40. Available from https://doi.org/10.1109/TCI.2021.3136759, https://www.ieee.org/membership-catalog/productdetail/showProductDetailPage.html?product = PER478-ELE.

Mahapatra, D., Bozorgtabar, B., & Garnavi, R. (2019). Image super-resolution using progressive generative adversarial networks for medical image analysis. *Computerized Medical Imaging and Graphics, 71*, 30–39. Available from https://doi.org/10.1016/j.compmedimag.2018.10.005, http://www.elsevier.com/locate/compmedimag.

Malczewski, K., & Stasiński, R. (2009). Super resolution for multimedia, image, and video processing applications. *Studies in Computational Intelligence, 231*, 171–208. Available from https://doi.org/10.1007/978-3-642-02900-4_8.

Mansoor, A., Vongkovit, T., & Linguraru, M. G. (2018). Adversarial approach to diagnostic quality volumetric image enhancement. In *Proceedings - International symposium on biomedical imaging* (pp 353–356). United States: IEEE Computer Society. Available from https://doi.org/10.1109/ISBI.2018.8363591.

Mazelanik, M., Leszczyński, A., & Parniak, M. (2022). Optical-domain spectral super-resolution via a quantum-memory-based time-frequency processor. *Nature Communications*, *13*(1). Available from https://doi.org/10.1038/s41467-022-28066-5, https://www.nature.com/ncomms/.

McLeod, R. A., & Malac, M. (2013). Characterization of detector modulation-transfer function with noise, edge, and holographic methods. *Ultramicroscopy*, 42–52. Available from https://doi.org/10.1016/j.ultramic.2013.02.021.

Menze, B. H., et al. (2015). https://www.smir.ch/BRATS/Start2015. [Accessed 08 September 2023].

Mishra, D., & Hadar, O. (2023). Self-FuseNet: Data free unsupervised remote sensing image super-resolution. *IEEE Journal of Selected Topics in Applied Earth Observations and Remote Sensing*, *16*, 1710–1727. Available from https://doi.org/10.1109/JSTARS.2023.3239758, http://ieeexplore.ieee.org/xpl/RecentIssue.jsp?punumber = 4609443.

Nasrollahi, K., & Moeslund, T. B. (2014). Super-resolution: A comprehensive survey. *Machine Vision and Applications*, *25*(6), 1423–1468. Available from https://doi.org/10.1007/s00138-014-0623-4.

Nguyen, K., Fookes, C., Sridharan, S., Tistarelli, M., & Nixon, M. (2018). Super-resolution for biometrics: A comprehensive survey. *Pattern Recognition*, *78*, 23–42. Available from https://doi.org/10.1016/j.patcog.2018.01.002, http://www.elsevier.com/inca/publications/store/3/2/8/.

Nguyen, K., Fookes, C., Sridharan, S., & Denman, S. (2013). Feature-domain super-resolution for iris recognition. *Computer Vision and Image Understanding*, *117*(10), 1526–1535. Available from https://doi.org/10.1016/j.cviu.2013.06.010, http://www.elsevier.com/inca/publications/store/6/2/2/8/0/9/index.htt.

Ni, P., & Lee, H. N. (2020). High-resolution ultrasound imaging using random interference. *IEEE Transactions on Ultrasonics, Ferroelectrics, and Frequency Control*, *67*(9), 1785–1799. Available from https://doi.org/10.1109/TUFFC.2020.2986588, https://ieeexplore.ieee.org/servlet/opac?punumber = 58.

Oktay, O., Bai, W., Lee, M., Guerrero, R., Kamnitsas, K., Caballero, J., De Marvao, A., Cook, S., O'Regan, D., & Rueckert, D. (2016). *Multi-input cardiac image super-resolution using convolutional neural networks. Lecture notes in computer science (including subseries lecture notes in artificial intelligence and lecture notes in bioinformatics)* (pp. 246–254). United Kingdom: Springer Verlag 16113349. Available from http://springerlink.com/content/0302-9743/copyright/2005/, https://doi.org/10.1007/978-3-319-46726-9_29.

Oktay, O., Ferrante, E., Kamnitsas, K., Heinrich, M., Bai, W., Caballero, J., Cook, S. A., de Marvao, A., Dawes, T., O'Regan, D. P., Kainz, B., Glocker, B., & Rueckert, D. (2018). Anatomically constrained neural networks (ACNNs): Application to cardiac image enhancement and segmentation. *IEEE Transactions on Medical Imaging*, *37*(2), 384–395. Available from https://doi.org/10.1109/tmi.2017.2743464.

Park, J., Hwang, D., Kim, K. Y., Kang, S. K., Kim, Y. K., & Lee, J. S. (2018). Computed tomography super-resolution using deep convolutional neural network. *Physics in Medicine and Biology*, *63*(14). Available from https://doi.org/10.1088/1361-6560/aacdd4, http://iopscience.iop.org/article/10.1088/1361-6560/aacdd4/pdf.

Park, S., & Gahm, J. K. (2022). Super-resolution of 3D brain MRI with filter learning using tensor feature clustering. *IEEE Access*, *10*, 4957–4968. Available from https://doi.org/10.1109/ACCESS.2022.3140810, http://ieeexplore.ieee.org/xpl/RecentIssue.jsp?punumber = 6287639.

Peng, J., Li, X., Tang, H., Ma, L., Zhang, T., Li, Y., & Chen, S. (2019). Miniaturized high resolution integrated 360° electronic radial ultrasound endoscope for digestive tract imaging. *IEEE Transactions on Ultrasonics, Ferroelectrics, and Frequency Control, 66* (5), 975–983. Available from https://doi.org/10.1109/TUFFC.2019.2903308.

Plenge, E., Poot, D. H. J., Bernsen, M., Kotek, G., Houston, G., Wielopolski, P., Van Der Weerd, L., Niessen, W. J., & Meijering, E. (2012). Super-resolution methods in MRI: Can they improve the trade-off between resolution, signal-to-noise ratio, and acquisition time? *Magnetic Resonance in Medicine*, *68*(6), 1983–1993. Available from https://doi.org/10.1002/mrm.24187, http://onlinelibrary.wiley.com/journal/10.1002/(ISSN)1522-2594.

Ploquin, M., Basarab, A., & Kouamé, D. (2015). Resolution enhancement in medical ultrasound imaging. *Journal of Medical Imaging*, *2*(1), 017001. Available from https://doi.org/10.1117/1.jmi.2.1.017001.

Poddar, R., Migacz, J. V., Schwartz, D. M., Werner, J. S., & Gorczynska, I. (2017). Challenges and advantages in wide-field optical coherence tomography angiography imaging of the human retinal and choroidal vasculature at 1.7-MHz A-scan rate. *Journal of Biomedical Optics*, *22*(10). Available from https://doi.org/10.1117/1.JBO.22.10.106018, http://www.spie.org/x866.xml.

Prajapati, K., Chudasama, V., Patel, H., Upla, K., Raja, K., Ramachandra, R., & Busch, C. (2021). Direct unsupervised super-resolution using generative adversarial network (DUS-GAN) for real-world data. *IEEE Transactions on Image Processing*, *30*, 8251–8264. Available from https://doi.org/10.1109/TIP.2021.3113783, https://ieeexplore.ieee.org/xpl/mostRecentIssue.jsp?punumber = 83.

Qiao, C., Li, D., Guo, Y., Liu, C., Jiang, T., Dai, Q., & Li, D. (2021). Evaluation and development of deep neural networks for image super-resolution in optical microscopy. *Nature Methods*, *18*(2), 194–202. Available from https://doi.org/10.1038/s41592-020-01048-5, http://www.nature.com/nmeth/.

Ranganayakulu, S. V., Rao, N. R., & Gahane, L. (2016). *Ultrasound applications in medical sciences*.

Ren, S., Li, J., Guo, K., & Li, F. (2021). Medical video super-resolution based on asymmetric back-projection network with multilevel error feedback. *IEEE Access*, *9*, 17909–17920. Available from https://doi.org/10.1109/ACCESS.2021.3054433, http://ieeexplore.ieee.org/xpl/RecentIssue.jsp?punumber = 6287639.

Richter, T., Seiler, J., Schnurrer, W., & Kaup, A. (2016). Robust super-resolution for mixed-resolution multiview image plus depth data. *IEEE Transactions on Circuits and Systems for Video Technology*, *26*(5), 814–828. Available from https://doi.org/10.1109/TCSVT.2015.2426498.

Robinson, M. D., Chiu, S. J., Toth, C. A., Izatt, J. A., Lo, J. Y., & Farsiu, S. (2017). *New applications of super-resolution in medical imaging super-resolution imaging* (pp. 383–412). United States: CRC Press. Available from http://www.tandfebooks.com/doi/book/10.1201/9781439819319, https://doi.org/10.1201/9781439819319.

Salimi, Y., Akhavanallaf, A., Shiri, I., Sanaat, A., Manesh, A.S., Arabi, H., & Zaidi, H. (2021). Automatic deep learning based calculation of water equivalent diameter from 2D CT localizer image. In *IEEE nuclear science symposium and medical imaging conference record, NSS/MIC 2021 and 28th international symposium on room-temperature semiconductor detectors, RTSD 2022*. Switzerland: Institute of Electrical and Electronics Engineers Inc. Available from https://doi.org/10.1109/NSS/MIC44867.2021.9875506, 9781665421133. http://ieeexplore.ieee.org/xpl/mostRecentIssue.jsp?punumber = 9875398.

Sánchez, I., & Vilaplana V. (2018). *Brain MRI super-resolution using 3D generative adversarial networks*. arXiv. https://arxiv.org.

Sarasaen, C., Chatterjee, S., Breitkopf, M., Rose, G., Nürnberger, A., & Speck, O. (2021). Fine-tuning deep learning model parameters for improved super-resolution of dynamic MRI with prior-knowledge. *Artificial Intelligence in Medicine*, 102196. Available from https://doi.org/10.1016/j.artmed.2021.102196.

Shen, D. F., Chiu, C. W., & Huang, P. J. (2006). Taiwan modified laplacian filter and intensity correction technique for image resolution enhancement. In *IEEE international conference on multimedia and expo, ICME 2006 - Proceedings* (Vol. 2006, pp. 457–460). Available from https://doi.org/10.1109/ICME.2006.262571.

Shi, J., Liu, Q., Wang, C., Zhang, Q., Ying, S., & Xu, H. (2018). Super-resolution reconstruction of MR image with a novel residual learning network algorithm. *Physics in Medicine & Biology*, *63*(8), 085011. Available from https://doi.org/10.1088/1361-6560/aab9e9.

Shi, M., Gao, Y., Chen, L., & Liu, X. (2022). Dual-branch multiscale channel fusion unfolding network for optical remote sensing image super-resolution. *IEEE Geoscience and Remote Sensing Letters*, *19*, 1–5. Available from https://doi.org/10.1109/lgrs.2022.3221614.

Song, T. A., Roy Chowdhury, S., Yang, F., & Dutta, J. (2019). Self supervised super-resolution PET using a generative adversarial network. In *IEEE nuclear science symposium and medical imaging conference, NSS/MIC 2019*. United States: Institute of Electrical and Electronics Engineers Inc. Available from https://doi.org/10.1109/NSS/MIC42101.2019.9059947, 9781728141640. http://ieeexplore.ieee.org/xpl/mostRecentIssue.jsp?punumber = 9039702.

Stergiopoulou, V., De Morais Goulart, J.H., Schaub, S., Calatroni, L., & Blanc-Feraud, L. (2021). COL0RME: Covariance-based l0super-resolution microscopy with intensity estimation. In *Proceedings - International symposium on biomedical imaging* (2021, pp. 349–352). France: IEEE Computer Society. Available from https://doi.org/10.1109/ISBI48211.2021.9433976, 19458452. http://ieeexplore.ieee.org/xpl/conferences.jsp.

Tamang, L. D., & Kim, B. W. (2022). Super-resolution ultrasound imaging scheme based on a symmetric series convolutional neural network. *Sensors.*, *22*(8). Available from https://doi.org/10.3390/s22083076, https://www.mdpi.com/1424-8220/22/8/3076/pdf.

Tan, W., Yan, B., & Bare, B. (2018). Feature super-resolution: Make machine see more clearly. In *Proceedings of the IEEE computer society conference on computer vision and pattern recognition* (pp. 3994–4002). China: IEEE Computer Society. Available from https://doi.org/10.1109/CVPR.2018.00420, 9781538664209.

Temiz, H., & Bilge, H. S. (2020). Super resolution of B-mode ultrasound images withdeep learning. *IEEE Access*, 78808–78820. Available from https://doi.org/10.1109/ACCESS.2020.2990344, http://ieeexplore.ieee.org/xpl/RecentIssue.jsp?punumber = 6287639.

Timofte, R., Agustsson, E., Gool, L. V., Yang, M. H., Zhang, L., Lim, B., Son, S., Kim, H., Nah, S., Lee, K. M., Wang, X., Tian, Y., Yu, K., Zhang, Y., Wu, S., Dong, C., Lin, L., Qiao, Y., Loy, C. C., … Choi, J. S. (2017). Challenge on single image super-resolution: Methods and results, IEEE computer society conference on computer vision and pattern recognition workshops (pp. 1110–1121). *IEEE Computer Society undefined NTIRE 2017*. Available from https://doi.org/10.1109/CVPRW.2017.149. 21607516. http://ieeexplore.ieee.org/xpl/conferences.jsp. 2017.

Umehara, K., Ota, J., & Ishida, T. (2018). Application of super-resolution convolutional neural network for enhancing image resolution in chest CT. *Journal of Digital Imaging*, *31*(4), 441–450. Available from https://doi.org/10.1007/s10278-017-0033-z.

Van Essen, D. C., Smith, S. M., Barch, D. M., Behrens, T. E. J., Yacoub, E., & Ugurbil, K. (2013). The WU-minn human connectome project: An overview. *Neuroimage*, *80*, 62–79. Available from https://doi.org/10.1016/j.neuroimage.2013.05.041.

Vento, M., et al. (2016). https://mivia.unisa.it/datasets/biomedical-image-datasets/hep2-image-dataset/ [Accessed 08 September 2023].

Wang, C., Wang, S., Ma, B., Li, J., Dong, X., & Xia, Z. (2019). Transform domain based medical image super-resolution via deep multi-scale network. In *ICASSP, IEEE international conference on acoustics, speech and signal processing – Proceedings*. (Vol. 2019, pp. 2387–2391). China: Institute of Electrical and Electronics Engineers Inc. Available from https://doi.org/10.1109/ICASSP.2019.8682288, 9781479981311.

Wang, J., Shao, Z., Huang, X., Lu, T., Zhang, R., & Deep, A. (2022). Unfolding method for satellite super resolution. *IEEE Transactions on Computational Imaging*, *8*, 933–944. Available from https://doi.org/10.1109/TCI.2022.3210329, https://www.ieee.org/membership-catalog/productdetail/showProductDetailPage.html?product = PER478-ELE.

Wang, Z., Chen, J., & Hoi, S. C. H. (2021). Deep learning for image super-resolution: A survey. *IEEE Transactions on Pattern Analysis and Machine Intelligence*, *43*(10), 3365–3387. Available from https://doi.org/10.1109/tpami.2020.2982166.

Wang, Z., Zhu, H., Ma, Y., & Basu, A. (2021). XAI feature detector for ultrasound feature matching. In *Proceedings of the annual international conference of the IEEE engineering in medicine and biology society, EMBS* (pp. 2928–2931). Institute of Electrical and Electronics Engineers Inc. undefined. Available from https://doi.org/10.1109/EMBC46164.2021.9629944, 9781728111797.

Wildeboer, R. R., Sammali, F., Van Sloun, R. J. G., Huang, Y., Chen, P., Bruce, M., Rabotti, C., Shulepov, S., Salomon, G., Schoot, B. C., Wijkstra, H., & Mischi, M. (2020). Blind source separation for clutter and noise suppression in ultrasound imaging: Review for different applications. *IEEE Transactions on Ultrasonics, Ferroelectrics, and Frequency Control*, *67*(8), 1497–1512. Available from https://doi.org/10.1109/TUFFC.2020.2975483, https://ieeexplore.ieee.org/servlet/opac?punumber = 58.

Wiliem, A., et al., SNPHEp-2. 2013. https://www.uq.id.au/a.wiliem/datasets/snphep2/index.html. [Accessed 08 September 2023].

Xue, H., Kellman, P., Larocca, G., Arai, A. E., & Hansen, M. S. (2013). High spatial and temporal resolution retrospective cine cardiovascular magnetic resonance from shortened free breathing real-time acquisitions. *Journal of Cardiovascular Magnetic Resonance*, *15*(1). Available from https://doi.org/10.1186/1532-429X-15-102.

Yan, B., Bare, B., Ma, C., Li, K., & Tan, W. (2019). Deep objective quality assessment driven single image super-resolution. *IEEE Transactions on Multimedia*, *21*(11), 2957–2971. Available from https://doi.org/10.1109/TMM.2019.2914883, https://ieeexplore.ieee.org/xpl/mostRecentIssue.jsp?punumber = 6046.

Yang, J., & Huang, T. (2017). *Image super-resolution: Historical overview and future challenges super-resolution imaging* (pp. 1–33). United States: CRC Press. Available from http://www.tandfebooks.com/doi/book/10.1201/9781439819319, https://doi.org/10.1201/9781439819319.

Yang, X., Sun, C., Anderson, T., Moran, C. M., Hadoke, P. W. F., Gray, G. A., & Hoskins, P. R. (2013). Assessment of spectral doppler in preclinical ultrasound using a small-size rotating phantom. *Ultrasound in Medicine and Biology*, *39*(8), 1491–1499. Available from https://doi.org/10.1016/j.ultrasmedbio.2013.03.013, http://www.elsevier.com/locate/ultrasmedbio.

You, C., Li, G., Zhang, Y., Zhang, X., Shan, H., & Wang, G. (2019). *CT super-resolution GAN constrained by the identical, residual, and cycle learning ensemble (GAN-CIRCLE).*

Yue, L., Shen, H., Li, J., Yuan, Q., Zhang, H., & Zhang, L. (2016). Image super-resolution: The techniques, applications, and future. *Signal Processing*, 389–408. Available from https://doi.org/10.1016/j.sigpro.2016.05.002.

Zeng, K., Zheng, H., Cai, C., Yang, Y., Zhang, K., & Chen, Z. (2018). Simultaneous single- and multi-contrast super-resolution for brain MRI images based on a convolutional neural network. *Computers in Biology and Medicine*, *99*, 133–141. Available from https://doi.org/10.1016/j.compbiomed.2018.06.010, http://www.elsevier.com/locate/compbiomed.

Zhan, Z., Gong, Y., Zhao, P., Yuan, G., Niu, W., Wu, Y., Zhang, T., Jayaweera, M., Kaeli, D., Ren, B., Lin, X., & Wang, Y. (2021). Achieving on-mobile real-time super-resolution with neural architecture and pruning search. arXiv. https://arxiv.org.

Zhang, Y., Fan, Q., Bao, F., Liu, Y., & Zhang, C. (2018). Single-image super-resolution based on rational fractal interpolation. *IEEE Transactions on Image Processing*, *27*(8), 3782–3797. Available from https://doi.org/10.1109/TIP.2018.2826139.

Zhao, C., Dewey, B. E., Pham, D. L., Calabresi, P. A., Reich, D. S., & Prince, J. L. (2021). SMORE: A self-supervised anti-aliasing and super-resolution algorithm for MRI using deep learning. *IEEE Transactions on Medical Imaging*, *40*(3), 805–817. Available from https://doi.org/10.1109/TMI.2020.3037187, http://ieeexplore.ieee.org/xpl/RecentIssue.jsp?punumber = 42.

Zhao, C., Carass, A., Dewey, B. E., Prince, J. L. (2018). Self super-resolution for magnetic resonance images using deep networks. In *Proceedings - International symposium on biomedical imaging* (pp. 365–368). United States: IEEE Computer Society. Available from https://doi.org/10.1109/ISBI.2018.8363594, 19458452. http://ieeexplore.ieee.org/xpl/conferences.jsp, 2018.

Zhao, C., Carass, A., Dewey, B. E., Woo, J., Oh, J., Calabresi, P. A., Reich, D. S., Sati, P., Pham, D. L., & Prince, J. L. (2018). *A deep learning based anti-aliasing self super-resolution algorithm for MRI. Lecture notes in computer science (including subseries lecture notes in artificial intelligence and lecture notes in bioinformatics)* (11070, pp. 100–108). United States: Springer Verlag. Available from https://www.springer.com/series/558, https://doi.org/10.1007/978-3-030-00928-1_12.

Zhao, T., Lin, Y., Xu, Y., Chen, W., & Wang, Z. (2022). Learning-based quality assessment for image super-resolution. *IEEE Transactions on Multimedia*, *24*, 3570–3581. Available from https://doi.org/10.1109/TMM.2021.3102401, https://ieeexplore.ieee.org/xpl/mostRecentIssue.jsp?punumber = 6046.

Zhao, X., Zhang, Y., Zhang, T., & Zou, X. (2019). Channel splitting network for single MR image super-resolution. *IEEE Transactions on Image Processing*, *28*(11), 5649–5662. Available from https://doi.org/10.1109/tip.2019.2921882.

Zontak, M., & Irani, M. (2011). Internal statistics of a single natural image. In *Proceedings of the IEEE computer society conference on computer vision and pattern recognition* (pp. 977–984). Israel: IEEE Computer Society. Available from https://doi.org/10.1109/CVPR.2011.5995401, 10636919.

Further reading

Papageorgiou, G., Butler, M., Mobberley, A., Lu, W., Keanie, J., Good, D., Gallagher, K., McNeill, A., & Sboros, V. (2022). A machine learning approach to cancer detection and localization using super resolution ultrasound imaging. In *IEEE international ultrasonics symposium, IUS. 19485727* (vol. 2022). United Kingdom: IEEE Computer Society. Available from https://doi.org/10.1109/IUS54386.2022.9957797, http://ieeexplore.ieee.org/xpl/conferences.jsp.

CHAPTER

14 Class imbalance and its impact on predictive models for binary classification of disease: a comparative analysis

Mubarak Taiwo Mustapha[1,2] **and Dilber Uzun Ozsahin**[1,3,4]

[1]*Operational Research Centre in Healthcare, Near East University, Nicosia/TRNC, Mersin-10, Turkey*

[2]*Department of Biomedical Engineering, Near East University, Nicosia/TRNC, Mersin-10, Turkey*

[3]*Department of Medical Diagnostic Imaging, College of Health Science, University of Sharjah, Sharjah, United Arab Emirates*

[4]*Research Institute for Medical and Health Sciences, University of Sharjah, Sharjah, United Arab Emirates*

14.1 Introduction

In healthcare, predictive models are statistical or machine learning algorithms capable of predicting future occurrences or outcomes based on training data. As a result of their potential to enhance patient outcomes, lower healthcare costs, and optimize resource allocation, they have gained increasing importance in healthcare (Mustapha et al., 2022; Seyer Cagatan et al., 2023; Uzun Ozsahin et al., 2022). Those at high risk of developing specific diseases or health issues can be identified by predictive models, allowing for early screening and preventative intervention (Uzun Ozsahin et al., 2023). By recognizing patterns and relationships that may not be evident to the human eye, they can increase the accuracy of diagnosis and treatment decisions. In addition, predictive models can optimize resource allocation and enhance operational efficiency by forecasting patient care demand and finding process improvement opportunities. They can aid community health management by detecting systemic problems and enhancing treatment delivery to underserved communities. Predictive models can stimulate innovation in the healthcare industry by easing the discovery of new biomarkers, treatments, diagnostic tools, and customized medicine. The value of predictive models is anticipated to expand as the volume and complexity of healthcare data continue to grow, making it imperative for healthcare companies to invest in data science and machine learning capabilities.

Class imbalance is a common problem in machine learning where the distribution of classes in the training data is skewed, meaning that one or more classes are underrepresented compared to others. This can lead to biased models that perform poorly

Artificial Intelligence and Image Processing in Medical Imaging. DOI: https://doi.org/10.1016/B978-0-323-95462-4.00014-5

on the underrepresented class or fail to recognize it. For example, in a binary classification problem where one class represents a rare event, the model may tend to predict the more frequent class as it achieves high accuracy by doing so. In such cases, the model's overall accuracy is high, but its ability to identify the rare event is limited. Class imbalance can be caused by factors such as the event's rarity, biased sampling methods, or data collection processes prioritizing one class over the other. Different techniques can address the class imbalance, including data augmentation, oversampling, under sampling, or combining these methods. Data augmentation involves generating synthetic data to increase the representation of underrepresented classes while oversampling involves duplicating instances of the minority class to create a balanced dataset. On the other hand, under sampling involves reducing the number of instances of the majority class to match the number of instances in the minority class. Other techniques include the use of cost-sensitive learning, which assigns a higher cost to misclassifying instances of the minority class, or the use of ensemble models that combine multiple models to balance the class distribution.

Van Den Goorbergh et al. (2022) evaluated the impact of correcting class imbalance on logistic regression models' performance in terms of discrimination, calibration, and classification. The study used Monte Carlo simulations and a case study on ovarian cancer diagnosis to examine the effect of different imbalance correction methods. The findings showed that correcting class imbalance resulted in poor calibration with a strong overestimation of the probability of belonging to the minority class. However, this correction did not improve discrimination in terms of the area under the receiver operating characteristic curve. The study also found that correcting imbalance improved classification in terms of sensitivity and specificity, but similar results were obtained by shifting the probability threshold. The authors concluded that class imbalance might not necessarily impact model performance, and alternative approaches, such as shifting the probability threshold, may be more effective in improving classification. The study provides insights into the potential drawbacks of correcting the class imbalance in predictive models. Zheng et al. (2022) propose a method to analyze the impact of imbalanced binary data on machine learning models, focusing on the relationship between model performance and imbalance rate (IR) and the stability of performance on imbalanced data. Imbalanced data augmentation algorithms are designed to obtain datasets with gradually varying IR. AFG, an evaluation metric combining AUC, F-measure, and G-mean, is used to evaluate classification performance. A performance stability evaluation method is proposed based on AFG and coefficient of variation. Results show that classification performance decreases with increasing IR, and LR, DT, and SVC are unstable, while GNB, BNB, k-nearest neighbor (KNN), RF, and GBDT are relatively stable. Synthetic minority over-sampling technique (SMOTE) is used for oversampling-based imbalanced data augmentation, and other oversampling methods' consistency requires further research. Future research should use an imbalanced data augmentation algorithm based on under sampling and hybrid sampling to analyze the performance impact of imbalanced binary data.

In another Wisconsin study, Madasamy & Ramaswam (2017) highlight the challenges of dealing with inconsistencies, such as data imbalance and noise in real-time data. The authors conducted experiments that showed that these issues could impact predictors, but each model has its advantages based on its performance metrics. The study concludes that a combination of prediction algorithms is necessary to address these challenges effectively. The authors performed a performance analysis on several standard algorithms and ensemble techniques to identify the best algorithms. Boosting and stacking ensembles were found to be the most promising models. The study suggests combining these techniques can provide more accurate predictions for real-time data with imbalances and noise. Luque et al. (2019) discuss the challenges of classifying imbalanced datasets and the need for appropriate performance metrics. Previous work has shown that imbalance can significantly affect the accuracy and other metrics, and this study provides a systematic analysis of the impact of imbalance on binary classifiers. The study proposes a new way to measure imbalance, which enables the comparison of different performance metrics based on the binary confusion matrix. The results show that Geometric Mean and Bookmaker Informedness are the best null-biased metrics for classification successes, but if classification errors must be considered, then Matthews correlation coefficient is the best choice. Additionally, the study proposes a set of null-biased multiperspective class balance metrics, which extend the concept of class balance accuracy to other performance metrics. The study provides valuable insights into selecting appropriate performance metrics for classifying imbalanced datasets. Hasanin et al. (2019) discussed the challenges of classifying imbalanced datasets and the need for appropriate performance metrics. Previous work has shown that imbalance can significantly affect the accuracy and other metrics, and this study provides a systematic analysis of the impact of imbalance on binary classifiers. The study proposes a new way to measure imbalance, which enables the comparison of different performance metrics based on the binary confusion matrix. The results show that Geometric Mean and Bookmaker Informedness are the best null-biased metrics for classification successes, but if classification errors must be considered, then Matthews correlation coefficient is the best choice. Additionally, the study proposes a set of null-biased multiperspective class balance metrics, which extend the concept of class balance accuracy to other performance metrics. The study provides valuable insights into selecting appropriate performance metrics for classifying imbalanced datasets.

The impact of class imbalance on the performance of predictive models is well established. Although there has been research on class imbalance in various domains, more studies should focus on medical datasets. Also, studies comparing the performance of different predictive models in the context of class imbalance are relatively limited. The few focus on a specific dataset, which limits the generalizability of the results. This study aims to investigate the impact of class imbalance on predictive models for the binary classification of diseases using medical datasets. In the study, we aim to address the research gaps outlined. We aim to use three datasets of medical importance to promote the generalizability of the report. Also, we employ 10 predictive models to provide insight into the strengths and limitations of different models when dealing with imbalanced data.

14.2 Methodology

14.2.1 Dataset and data preprocessing

This study analyzed three medical datasets: the Wisconsin breast cancer dataset (Breast Cancer Wisconsin Diagnostic Data Set. UCI Machine Learning Repository: Breast Cancer wisconsin diagnostic data set, 2023), PIMA Indian diabetic dataset (Pima Indians Diabetes Database, 2017), and the heart dataset (Janosi et al., 1988). The Wisconsin breast cancer dataset contains information about breast cancer tumors and their characteristics. The dataset is grouped into benign or malignant based on these characteristics. The dataset contains 569 samples, with 212 malignant and 357 benign cases, as shown in Table 14.1. The PIMA Indian Diabetes dataset includes information on female patients of PIMA Indian heritage and their medical attributes. The PIMA Indian diabetic dataset comprises 768 samples, with 268 positive and 500 negative cases. The heart dataset contains information about patients and their medical attributes. The heart dataset comprises 303 samples, with 165 positive and 138 negative cases. Before fitting the predictive models to the datasets, some preprocessing steps were performed to ensure the data is suitable for analysis. The datasets were checked for missing values, outliers, and inconsistent data. Missing values were imputed using the mean imputation method. This is done by replacing 0 values in a column with the mean values of the nonzero elements in that column. Outliers were identified and removed using the interquartile range method. Inconsistent data were identified and corrected. The datasets were normalized to ensure that each variable had a similar scale. Normalization was performed using the Z-score normalization method. Feature selection was performed to identify the most important features that have the highest impact on the prediction performance of the models. The feature selection method used was the recursive feature elimination method. Since the datasets are imbalanced and balance class distribution is needed, we first fit

Table 14.1 Data distribution.

Reference	Class		Designation	Count
Breast Cancer Wisconsin Diagnostic (2023)	Wisconsin breast cancer dataset	Malignant	1	212
		Benign	0	357
Learning (2017)	PIMA Indian diabetic dataset	Positive	1	268
		Negative	0	500
Janosi et al. (1988)	Heart dataset	Positive	1	138
		Negative	0	165

the datasets to the predictive models without addressing the class imbalance. Subsequently, we balance the dataset's classes and fit them to the models. SMOTE addressed the class imbalance problem. This technique oversamples the minority class by creating synthetic samples. The synthetic samples are generated by interpolating between the feature vectors of the minority class samples. The SMOTE technique was applied only to the training set to avoid data leakage. The datasets were split into training and testing sets with a ratio of 75:25. The training set was used to train the predictive models, and the testing set was used to evaluate the performance of the models.

14.2.2 Oversampling technique

Oversampling is a technique used in data analysis to address the class imbalance issue in a dataset (Gosain & Sardana, 2017). Oversampling techniques involve creating synthetic samples of the minority class to balance the number of instances in both classes, as shown in Fig. 14.1. The importance of oversampling is that an imbalanced dataset can lead to biased models (Zheng et al., 2022). For instance, if the positive class has very few instances, a machine learning model trained on such data may perform poorly on new data, especially when the positive class is the class of interest. Oversampling the minority class increases the

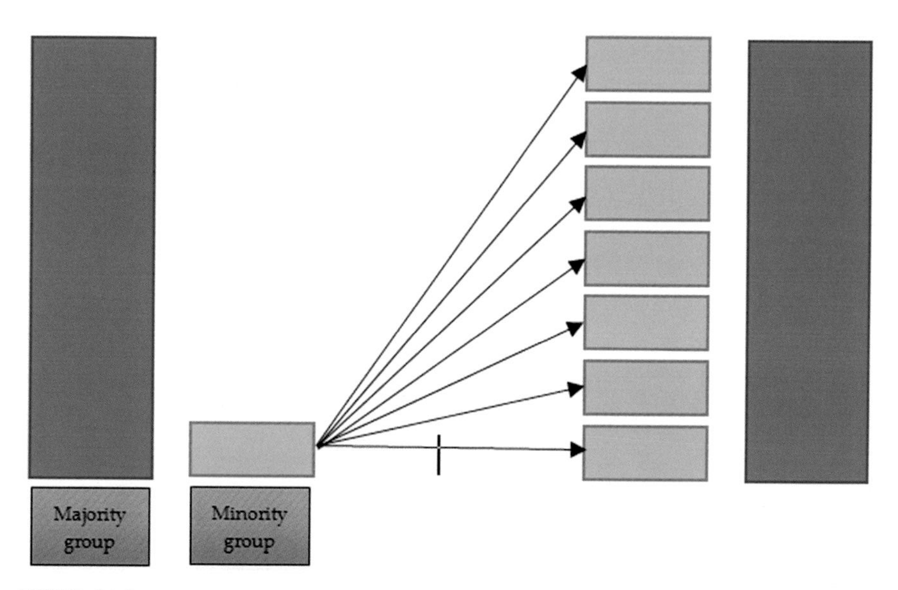

FIGURE 14.1

The basic principle of oversampling technique.

Modified from Agarwal, R. (2020). The 5 most useful techniques to handle imbalanced datasets. https://www.kdnuggets.com/2020/01/5-most-useful-techniques-handle-imbalanced-datasets.html.

number of samples available for training and improves the model's performance. There are several oversampling techniques, including:

1. Random oversampling: This involves randomly replicating samples of the minority class until the number of instances in both classes is balanced.
2. SMOTE: SMOTE is a popular oversampling technique that involves creating synthetic samples of the minority class by interpolating between adjacent instances.
3. Adaptive synthetic sampling (ADASYN): ADASYN is a variant of SMOTE that generates synthetic samples based on the density of minority class instances.
4. Borderline-SMOTE: Borderline-SMOTE is a variant of SMOTE that generates synthetic samples only for the borderline instances of the minority class.

Minority class exaggeration (MCE): MCE is a new oversampling technique that exaggerates the differences between the minority and majority class instances to improve classification accuracy.

14.2.2.1 Synthetic minority over-sampling technique

In this study, the SMOTE oversampling technique was implemented. SMOTE is a widely used oversampling technique that addresses the problem of class imbalance in datasets. SMOTE creates synthetic minority class instances by interpolating between adjacent minority class instances, preserving the distribution of the minority class, as shown in Fig. 14.2. The SMOTE algorithm works as follows:

1. For each minority class instance, SMOTE selects KNNs from the minority class.
2. SMOTE then creates synthetic instances by interpolating between the minority class instance and its KNN.

FIGURE 14.2

Resampling dataset using synthetic minority over-sampling technique.

From Orellana, E. (2020, December 10). Smote. Medium. Retrieved March 14, 2023, from https://emilia-orellana44.medium.com/smote-2acd5dd09948.

One of the main benefits of SMOTE is that it generates synthetic instances instead of replicating existing instances (Saad Hussein et al., 2019). The synthetic instances are added to the minority class, thus balancing the number of instances in both classes. This reduces the risk of overfitting and improves the generalization performance of the machine learning model. SMOTE also preserves the distribution of the minority class, which is important for maintaining the integrity of the dataset. However, SMOTE has some limitations. One limitation is that it may generate noisy instances, particularly when the number of minority class instances is small (Wongvorachan et al., 2023). This is because the interpolation process may create synthetic instances that do not accurately represent the minority class. Another limitation is that SMOTE may not work well when the majority class has a complex distribution (Maclejewski & Stefanowski, 2011). Despite its limitations, SMOTE is one of the most widely used oversampling techniques because of its effectiveness in addressing the class imbalance. SMOTE has been shown to improve the performance of machine learning models on imbalanced datasets. SMOTE is also computationally efficient and works well on high-dimensional data.

14.3 Result and discussion

In this study, we examined the impact of addressing class imbalance on 10 predictive models, including the random forest, artificial neural network (ANN), XGBoost, Naive Bayes, KNN, logistic regression, support vecor machine (SVM), decision tree, AdaBoost, and linear discriminant analysis (LDA). We divided the investigation into two phases. In the first phase, the imbalance Wisconsin breast cancer dataset, PIMA Indian diabetes dataset, and heart disease dataset were all fitted into the models, and their performance was evaluated. In the second phase, the classes in the three datasets were balanced using the SMOTE oversampling technique, and the model performance was evaluated. The balanced dataset generally provided better results compared to the imbalanced dataset. The investigation results demonstrate the impact of class imbalance on the performance of the predictive models.

In the balanced Wisconsin breast cancer dataset, most models achieved high accuracy, precision, recall, and F1 score, indicating that the models performed well in predicting both the minority and majority classes. The logistic regression, SVM, and LDA models achieved the highest accuracy, precision, recall, and F1 score, indicating that these models were most effective in predicting the class labels, as shown in Table 14.2. On the other hand, the decision tree model had a lower performance on this dataset, indicating that it could be more effective in predicting class labels. In contrast, the performance of the models on the imbalanced dataset was generally lower than the balanced dataset. Although most

Table 14.2 Performance evaluation metric of predictive models for the Wisconsin breast cancer dataset.

	Balance dataset				Imbalance dataset			
	Accuracy %	Precision %	Recall %	F1 Score %	Accuracy %	Precision %	Recall %	F1 Score %
Random forest	96	96	97	96	96	96	96	96
ANN	96	94	96	95	95	93	95	94
XGBoost	96	96	96	96	96	96	96	96
Naive Bayes	95	95	95	95	94	94	94	94
KNN	96	96	96	96	96	96	96	96
Logistic regression	98	98	98	98	99	99	99	99
SVM	98	98	98	98	98	98	98	98
Decision tree	89	90	90	89	95	95	95	95
AdaBoost	96	96	96	96	96	96	96	96
LDA	97	97	97	97	95	95	95	95

ANN, *artificial neural network;* KNN, *k-nearest neighbor;* LDA, *linear discriminant analysis;* SVM, *support vector machine.*

models achieved high accuracy, this metric can be misleading in the case of imbalanced datasets as it can be biased toward the majority class. Therefore, examining the precision, recall, and F1 score is essential, which provides a more comprehensive evaluation of the model's performance. In the imbalanced dataset, logistic regression, SVM, and KNN models achieved high performance on most metrics, indicating that they effectively predicted the minority class despite the imbalance. The decision tree model had a lower performance on this dataset, which is consistent with its performance on the balanced dataset.

The results obtained for the balanced and imbalanced PIMA Indian diabetes dataset indicate some differences in performance across the various models. Overall, the models performed slightly better on the balanced dataset, which is unsurprising since imbalanced datasets can pose significant challenges to traditional machine learning algorithms. Looking at the results for each model in detail, it is evident that some models performed consistently well across both datasets, while others showed more variation, as shown in Table 14.3. For instance, the ANN performed well in both datasets, with an accuracy of 81% on the balanced dataset and 77% on the imbalanced dataset. Similarly, the Random Forest and XGBoost models performed consistently well, albeit with slightly lower accuracy scores on the imbalanced dataset. These models are robust enough to handle class imbalance to some extent. Also, some models struggled more with the balanced dataset. For instance, the decision tree model had a noticeable drop in performance on the balanced dataset, with an accuracy of 63% compared to 69% on the imbalanced dataset. Similarly, the Naive Bayes model had a lower accuracy score on the balanced dataset than on the balanced dataset, suggesting that it may be less effective when dealing with class imbalance.

The comparison of the results obtained for the balanced and imbalanced heart disease datasets shows that the performance of most of the models remained consistent across both datasets, except for some variations in the performance metrics. Overall, the Random Forest, XGBoost, and decision tree models performed consistently well on both datasets, with an accuracy of 98% and an F1 score of 98%, indicating that these models are highly accurate in predicting heart disease, as shown in Table 14.4. This suggests that these models are robust to class imbalance and perform well even when the dataset is imbalanced. In contrast, the performance of the other models varied significantly between the two datasets. For instance, the ANN, AdaBoost, and logistic regression models showed a slight improvement in their performance metrics when applied to the imbalanced dataset, while the LDA model showed a decline in their performance metrics. The SVM and Naive Bayes models performed similarly on both datasets. One interesting finding is that the performance of the KNN model remained consistent on both datasets despite the class imbalance. This suggests that KNN is a robust algorithm for imbalanced datasets and can be used for such problems.

Table 14.3 Performance evaluation metric of predictive models for the PIMA Indian diabetes dataset.

	Balance dataset				Imbalance dataset			
	Accuracy %	**Precision %**	**Recall %**	**F1 Score %**	**Accuracy %**	**Precision %**	**Recall %**	**F1 Score %**
Random forest	76	75	77	75	74	73	74	73
ANN	81	78	76	77	77	72	71	72
XGBoost	72	71	72	71	75	73	74	74
Naive Bayes	69	68	70	68	72	70	71	70
KNN	69	70	72	68	71	69	70	69
Logistic regression	68	67	68	67	73	71	70	70
SVM	68	66	67	66	73	71	70	70
Decision tree	63	62	63	62	69	67	68	67
AdaBoost	70	69	70	69	73	71	71	71
LDA	70	68	69	68	74	72	70	71

ANN, *artificial neural network;* KNN, *k-nearest neighbor;* LDA, *linear discriminant analysis;* SVM, *support vector machine.*

Table 14.4 Performance evaluation metric of predictive models for the heart disease dataset.

	Balance dataset				Imbalance dataset			
	Accuracy %	**Precision %**	**Recall %**	**F1 Score %**	**Accuracy %**	**Precision %**	**Recall %**	**F1 Score %**
Random forest	98	98	98	98	98	98	98	98
ANN	87	88	87	87	88	89	88	88
XGBoost	98	98	98	98	98	98	98	98
Naive Bayes	77	77	77	77	77	78	78	77
KNN	81	81	81	81	81	81	81	81
Logistic regression	79	80	79	79	79	80	80	80
SVM	79	79	79	79	78	79	78	78
Decision tree	98	98	98	98	98	98	98	98
AdaBoost	87	87	87	87	88	88	88	88
LDA	79	80	80	79	78	78	78	78

ANN, *artificial neural network;* KNN, *k-nearest neighbor;* LDA, *linear discriminant analysis;* SVM, *support vector machine.*

The confusion matrices for the balanced and imbalanced Wisconsin breast cancer datasets show some differences in the performance of the classification models. In general, the models perform better on the balanced dataset than the imbalanced one. Looking at the balanced dataset (Fig. 14.3), all models achieved high accuracy with true negative (TN) and true positive (TP) values, which is the number of benign and malignant samples in the test set. However, some models achieved slightly better performance than others. For example, the Naive Bayes achieved the highest TP values of 85. Meanwhile, the decision tree and XGBoost had the highest number of false positives (FP) with values of 13 and 3, respectively. This means that these models incorrectly classify benign samples as malignant. In the imbalanced dataset (Fig. 14.4), the performance of the models is slightly lower compared to the balanced dataset. While most models still achieved high TN values, some struggled to correctly identify malignant samples (TP values). For example, the ANN model had a TP value of only 69 in the imbalanced dataset compared to 70 in the balanced dataset. The number of false positives and false negatives also slightly increased in the imbalanced dataset for some models. For example, the ANN model had an increased number of FP in the imbalanced dataset to 5 from 4 in the balanced dataset.

Looking at the confusion matrices for both the balanced and imbalanced PIMA Indian diabetes datasets, we can see that the models performed differently for each dataset. In the balanced dataset, the performance of most models is reasonably good, with random forest, ANN, XGBoost, AdaBoost, and LDA achieving the highest TP rates, as shown in Fig. 14.5. However, some models like Naive Bayes, ANN, and decision tree have relatively high FP rates, indicating that they misclassify some negative cases as positive. Logistic regression, SVM, and LDA show good performance in terms of both TP and TN rates in the imbalanced dataset, as shown in Fig. 14.6. On the other hand, in the balanced dataset, we can see that the models' performance has deteriorated, with most models showing lower TP rates and higher FP rates. The performance of the decision tree model, which performed relatively poorly in the imbalanced dataset, has further degraded in the balanced dataset.

Similar to the previous two datasets, we observe similarities in the model performance for both the balanced and imbalanced heart disease dataset. As shown in Figs. 14.7 and 14.8, the FP, FN, accuracy, precision, and recall measures are all similar. For example, the random forest model achieved perfect precision and recall on balanced and imbalanced datasets. The same can be said about the TP, TN, FP, and FN. Similarly, the XGBoost model achieved perfect precision and recall on both datasets, with similar FP and FN values.

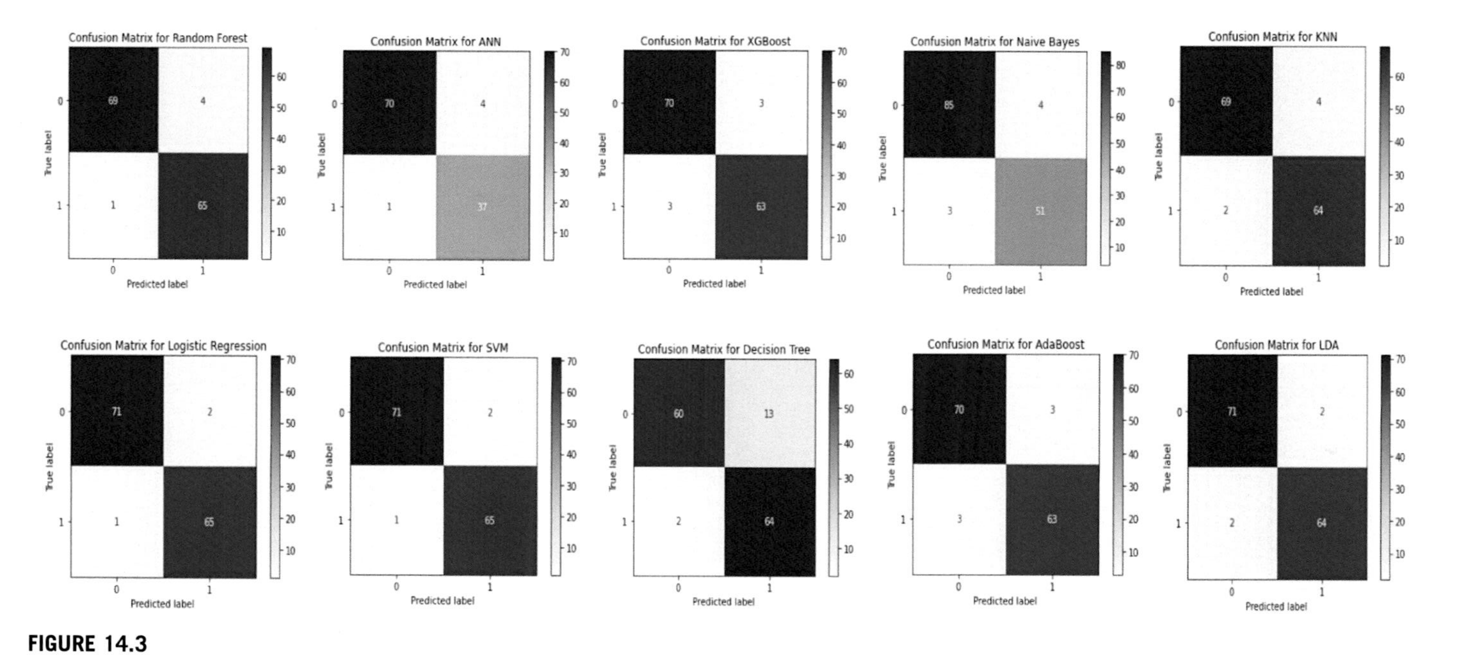

FIGURE 14.3

Confusion metrics of predictive models for balance Wisconsin breast cancer dataset.

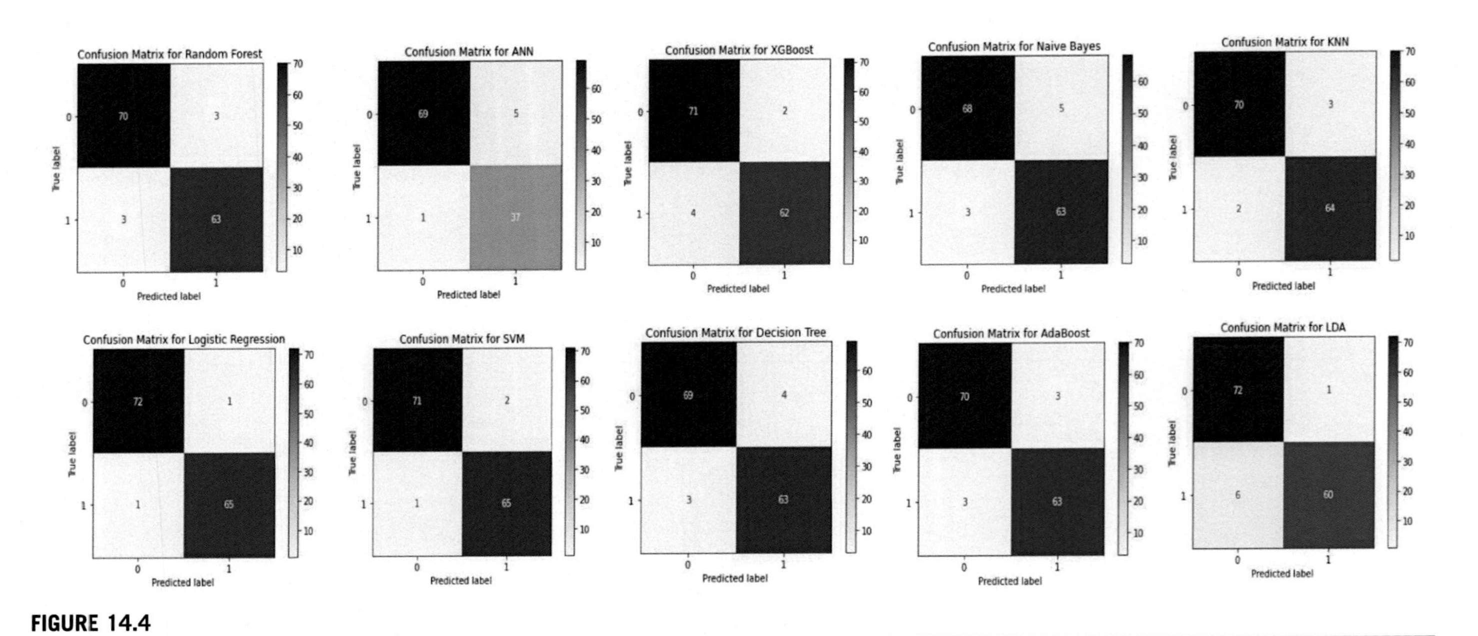

FIGURE 14.4

Confusion metrics of predictive models for imbalance Wisconsin breast cancer dataset.

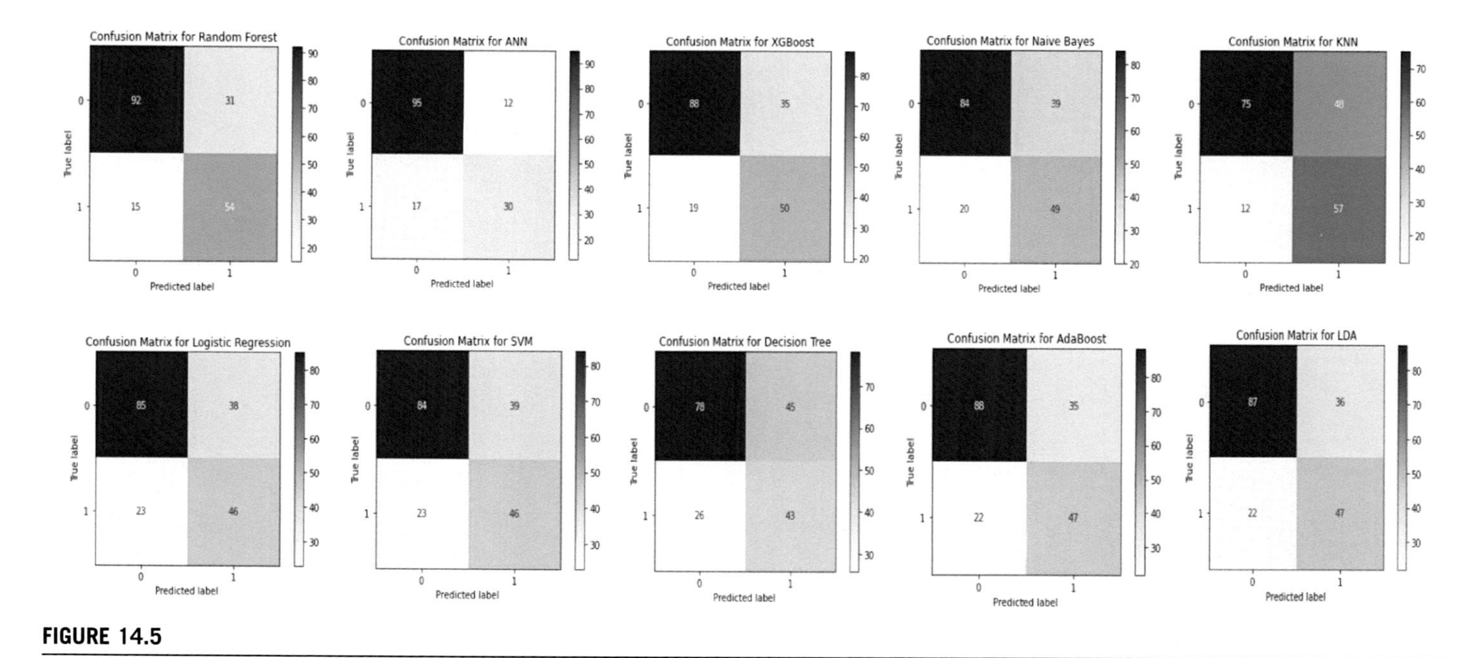

FIGURE 14.5

Confusion metrics of predictive models for balance PIMA Indian diabetes dataset.

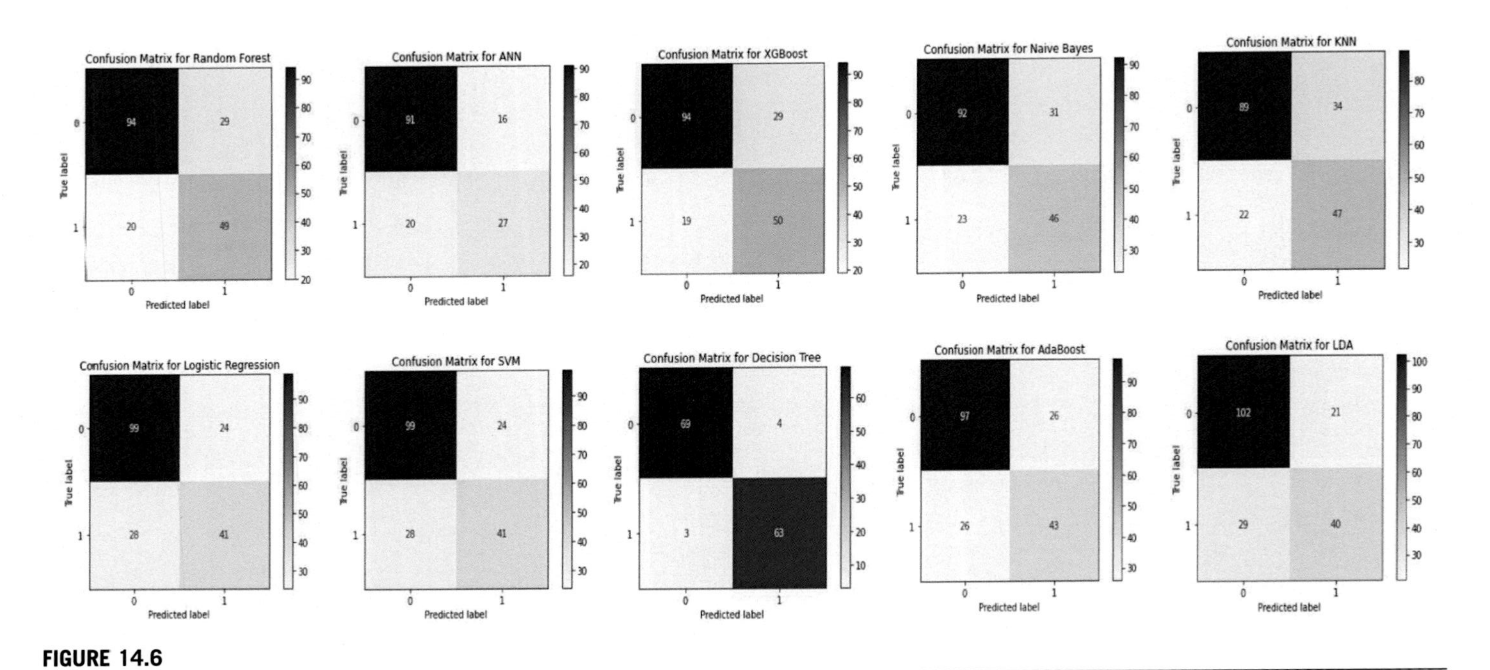

FIGURE 14.6

Confusion metrics of predictive models for imbalance PIMA Indian diabetes dataset.

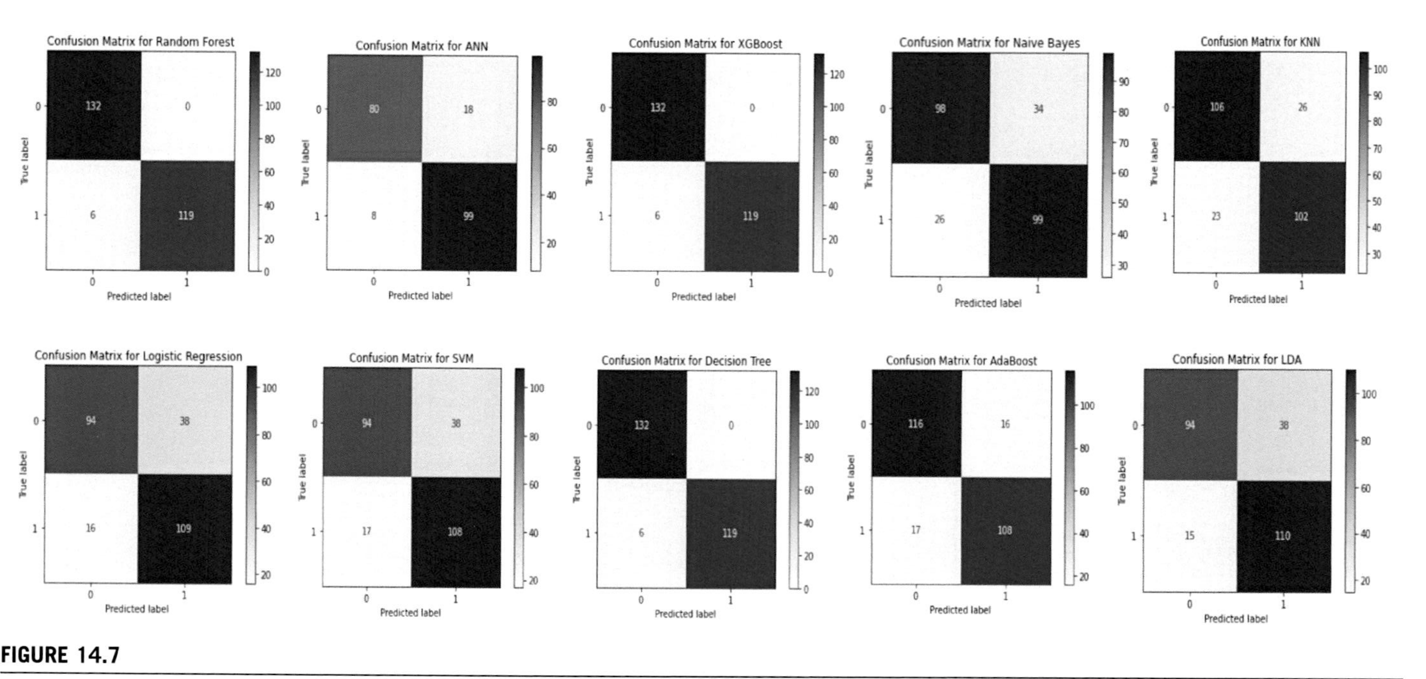

FIGURE 14.7

Confusion metrics of predictive models for balance heart disease dataset.

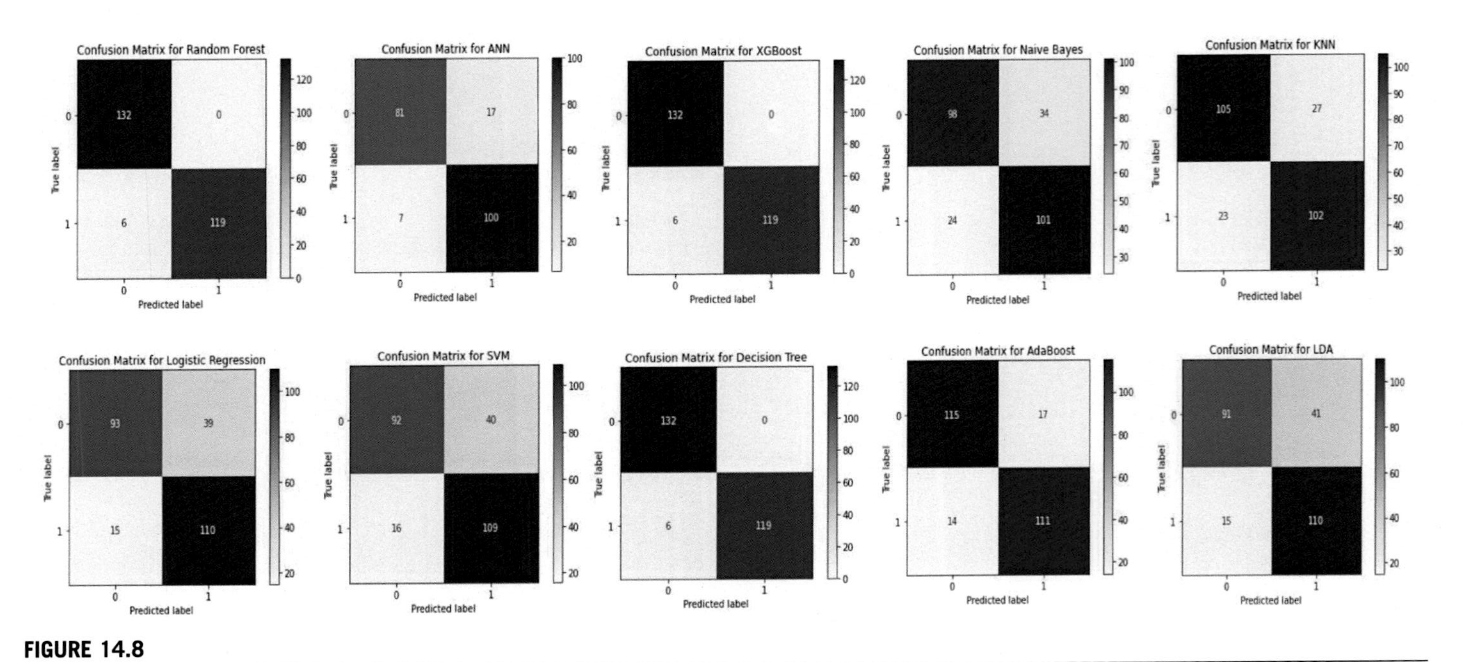

FIGURE 14.8

Confusion metrics of predictive models for imbalance heart disease dataset.

14.4 Conclusion and limitations

This chapter investigated the effect of class imbalance on 10 predictive models, including random forest, ANN, XGBoost, Naive Bayes, KNN, logistic regression, SVM, decision tree, AdaBoost, and LDA, using three datasets: the Wisconsin breast cancer dataset, PIMA Indian diabetes dataset, and heart disease dataset. The investigation was conducted in two phases, first on the imbalanced datasets and then on balanced datasets using SMOTE oversampling. The results indicated that addressing class imbalance through techniques like SMOTE oversampling can improve model performance. The performance of the models on the balanced datasets was generally better than on the imbalanced datasets, indicating that class imbalance can significantly impact the performance of predictive models. Some models showed consistent performance across both datasets, while others showed a significant drop in performance on the imbalanced dataset.

One limitation of this study is that only one class imbalance technique SMOTE was used. Other techniques, such as undersampling and cost-sensitive learning, could also address the class imbalance, and the results may differ. Additionally, the study only focused on binary classification tasks for three specific diseases, and the results may not generalize to other diseases or multiclass classification tasks. Another limitation is that the evaluation metrics used in the study (accuracy, precision, recall, and F1 score) do not account for the cost of misclassification. In real-world scenarios, misclassification can have significant consequences, and the cost of misclassification may vary depending on the disease being diagnosed. Future studies could incorporate cost-sensitive evaluation metrics to reflect the real-world implications of misclassification better.

References

Breast Cancer Wisconsin (Diagnostic) Data Set. UCI Machine Learning Repository: Breast Cancer wisconsin (diagnostic) data set. (2023).

Gosain, A., & Sardana, S. (2017). International Conference on Advances in Computing, Communications and Informatics, ICACCI 2017. https://doi.org/10.1109/ICACCI.2017.8125820. 9781509063673, 79–85. Institute of Electrical and Electronics Engineers Inc. India Handling class imbalance problem using oversampling techniques: A review 2017.

Breast Cancer Wisconsin (Diagnostic). (2023). UCI Machine Learning Repository. Available from https://archive.ics.uci.edu/ml/datasets/breast + cancer + wisconsin + (diagnostic).

Hasanin, T., Khoshgoftaar, T. M., Leevy, J. L., & Bauder, R. A. (2019). Severely imbalanced Big Data challenges: Investigating data sampling approaches. *Journal of Big Data* (1). Available from https://doi.org/10.1186/s40537-019-0274-4, http://journalofbigdata.springeropen.com/.

Janosi, A., Steinbrunn, W., Pfisterer, M., & Detrano, R. (1988). Heart disease data set. https://archive.ics.uci.edu/mL/datasets/heart + disease.

Learning, U.M. Pima Indians Diabetes Database. Kaggle. (2017, October 6). https://www.kaggle.com/datasets/uciml/pima-indians-diabetes-database.

Luque, A., Carrasco, A., & Martín, A. (2019). A. de las Heras, The impact of class imbalance in classification performance metrics based on the binary confusion matrix. *Pattern Recognition*, 216–231. Available from https://doi.org/10.1016/j.patcog.2019.02.023, http://www.elsevier.com/inca/publications/store/3/2/8/.

MacIejewski, T., & Stefanowski, J. (2011). IEEE SSCI 2011: Symposium Series on Computational Intelligence - CIDM 2011: 2011 IEEE Symposium on Computational Intelligence and Data Mining https://doi.org/10.1186/s40537-019-0274-4, 104–111, 8 2011/08. Poland Local neighbourhood extension of SMOTE for mining imbalanced data.

Madasamy, K., & Ramaswam, M. (2017). Data imbalance and classifiers: Impact and solutions from a big data perspective. *International Journal of Computational Intelligence Research*, 2267–2281.

Mustapha, M. T., Ozsahin, D. U., Ozsahin, I., & Uzun, B. (2022). Breast cancer screening based on supervised learning and multi-criteria decision-making. *Diagnostics*. (6). Available from https://doi.org/10.3390/diagnostics12061326, https://www.mdpi.com/2075-4418/12/6/1326/pdf?version = 1653634025.

Pima Indians Diabetes Database. (2017). https://www.kaggle.com/datasets/uciml/pima-indians-diabetes-database, 2023 3 19.

Saad Hussein, A., Li, T., Yohannese, C. W., & Bashir, K. (2019). A-SMOTE: A new preprocessing approach for highly imbalanced datasets by improving SMOTE. *International Journal of Computational Intelligence Systems*, *2*, 1412. Available from https://doi.org/10.2991/ijcis.d.191114.002.

Seyer Cagatan, A., Taiwo Mustapha, M., Bagkur, C., Sanlidag, T., & Ozsahin, D. U. (2023). An alternative diagnostic method for C. neoformans: Preliminary results of deep-learning based detection model. *Diagnostics*, *1*. Available from https://doi.org/10.3390/diagnostics13010081, http://www.mdpi.com/journal/diagnostics/.

Uzun Ozsahin, D., Mustapha, M. T., Bartholomew Duwa, B., & Ozsahin, I. (2022). Evaluating the performance of deep learning frameworks for malaria parasite detection using microscopic images of peripheral blood smears. *Diagnostics*, *11*, 2702. Available from https://doi.org/10.3390/diagnostics12112702.

Uzun Ozsahin, D., Mustapha, M. T., Uzun, B., Duwa, B., & Ozsahin, I. (2023). Computer-aided detection and classification of monkeypox and chickenpox lesion in human subjects using deep learning framework. *Diagnostics*, *2*, 292. Available from https://doi.org/10.3390/diagnostics13020292.

Van Den Goorbergh, R., Van Smeden, M., Timmerman, D., & Calster, Ben Van (2022). The harm of class imbalance corrections for risk prediction models: Illustration and simulation using logistic regression. *Journal of the American Medical Informatics Association*, *9*, 1525–1534. Available from https://doi.org/10.1093/jamia/ocac093, http://jamia.oxfordjournals.org/content/22/e1.

Wongvorachan, T., He, S., & Bulut, O. (2023). A comparison of undersampling, oversampling, and SMOTE methods for dealing with imbalanced classification in educational data mining. *Information*, *1*, 54. Available from https://doi.org/10.3390/info14010054.

Zheng, M., Wang, F., Hu, X., Miao, Y., Cao, H., & Tang, M. (2022). A method for analyzing the performance impact of imbalanced binary data on machine learning models. *Axioms*, *11*, 607. Available from https://doi.org/10.3390/axioms11110607.

Index

Note: Page numbers followed by "*f*" and "*t*" refer to figures and tables, respectively.

C

D

G

H

I

J

K

L

M

T